SURFACE MODIFICATION AND ALLOYING
by Laser, Ion, and Electron Beams

NATO CONFERENCE SERIES

VI MATERIALS SCIENCE

SURFACE MODIFICATION AND ALLOYING
by Laser, Ion, and Electron Beams

Edited by

J. M. Poate
Bell Laboratories
Murray Hill, New Jersey

G. Foti
University of Catania
Catania, Italy

and

D. C. Jacobson
Bell Laboratories
Murray Hill, New Jersey

Published in cooperation with NATO Scientific Affairs Division

PLENUM PRESS · NEW YORK AND LONDON

Library of Congress Cataloging in Publication Data

Main entry under title:

Surface modification and alloying by laser, ion, and electron beams.

(NATO conference series, series VI, Materials science; v. 8)
Proceedings of a NATO Advanced Study Institute on Surface Modification and Alloying, held Aug. 24–28, 1981, in Trevi, Italy.
Includes bibliographical references and index.
1. Metals—Surfaces—Congresses. 2. Metals—Effect of radiation on—Congresses.
3. Alloys—Congresses. 4. Laser beams—Congresses. 5. Ion bombard-
ment—Congresses. 6. Electron beams—Congresses. I. Poate, J. M. North Atlantic
Treaty Organization. Scientific Affairs Division. V. NATO Advanced Study Institute on
Surface Modification and Alloying (1981: Trevi, Italy) VI. Series: NATO conference
series. VI. Materials science; v. 8.
TN689.2.S85 1983   671.7   83-9465
ISBN-13: 978-1-4613-3735-5   e-ISBN:13 978-1-4613-3733-1
DOI:10.1007/ 978-1-4613-3733-1

Proceedings of a NATO Advanced Study Institute on Surface Modification
and Alloying, held August 24–28, 1981, in Trevi, Italy

© 1983 Plenum Press, New York
Softcover reprint of the hardcover 1st edition 1983

A Division of Plenum Publishing Corporation
233 Spring Street, New York, N.Y. 10013

PREFACE

This book is an outcome of the NATO institute on surface modification which was held in Trevi, 1981. Surface modification and alloying by ion, electron or laser beams is proving to be one of the most burgeoning areas of materials science. The field covers such diverse areas as integrated circuit processing to fabricating wear and corrosion resistant surfaces on mechanical components. The common scientific questions of interest are the microstructures and associated physical properties produced by the different energy deposition techniques. The chapters constitute a critical review of the various subjects covered at Trevi. Each chapter author took responsibility for the overall review and used contributions from the many papers presented at the meeting; each participant gave a presentation. The contributors are listed at the start of each chapter. We took this approach to get some order in a large and diverse field.

We are indebted to all the contributors, in particular the chapter authors for working the many papers into coherent packages; to Jim Mayer for hosting a workshop of chapter authors at Cornell and to Ian Bubb who did a sterling job in working over some of the manuscripts. Our special thanks are due to the text processing center at Bell Labs who took on the task of assembling the book. In particular Karen Lieb and Beverly Heravi typed the whole manuscript and had the entire book phototypeset using the Bell Laboratories UNIX™ system.

J. M. Poate

G. Foti

D. C. Jacobson

NATO INSTITUTE ON SURFACE MODIFICATION AND ALLOYING

Trevi, Italy, August 24-28, 1981

The NATO Materials Science Panel recognized early in 1980 that the time was right to hold a NATO Advanced Research Institute devoted to the subject of surface modification and alloying and we were invited by Dr. Alan Chynoweth to direct the Institute. The planning for the meeting was carried out at several committee meetings. Dr. Aquiles Gomes of the NATO Scientific Affairs Division provided considerable advice on all aspects of the planning. We are particularly grateful to our Italian colleagues, Nuccio Foti, Paulo Mazzoldi and Emanuele Rimini for their rigorous scientific input and, moreover, invaluable logistical support before, during and after the meeting. At the US end Jim Mayer provided extensive input on the scientific content as well as organizing a US contingent who could not be directly supported by NATO travel funds. Their participation was made possible by a generous grant from the Defense Advanced Research Projects Agency (DARPA), Washington, D.C. We thank Sven Roosild (DARPA) and Larry Cooper (ONR) for their support and encouragement.

The intent of these Institutes is to assemble a limited number of experts to evaluate and help guide future progress in the field. The format of the meeting consisted of a series of keynote lecturers followed by discussion and short contributions by all participants. The intensive five days of science were balanced by the beautiful Umbrian setting of the Hotel della Torre, Trevi, and the superb hospitality of our local hosts. In particular we wish to thank Carla Carbone of Gruppo Nazionale Struttura della Materia, Rome, for her splendid efforts in helping run the meeting. The efforts of Alfredo Trovato of the Istituto di Struttura della Materia, Catania, helped ensure an efficiently run meeting.

J. M. Poate
J. K. Hirvonen

CONTENTS

CHAPTER 1

AN OVERVIEW OF SURFACE MODIFICATION

J. M. POATE

Bell Laboratories, Murray Hill, New Jersey

G. FOTI

Istituto di Struttura della Materia, Universita di Catania, Italy

CONTRIBUTORS: J. K. Hirvonen, W. Hofer and A. D. Marwick

I. INTRODUCTION

The surfaces of solids influence much of our every day life; the way objects appear and the manner in which they weather. The properties and structures of surfaces and surface layers are of obvious importance in science and technology. Numerous examples come from such diverse worlds as catalysis, semiconductor technology and metallurgy. Catalytic behavior is determined by surface composition and structure. In semiconductors, the electrical properties are usually controlled by the composition and structure of the outermost micron of material. The ability of metals to withstand corrosion or wear is also determined by the composition and structure of the surface layers. It is clear from such examples that

1

technologies which can be used to directly tailor or construct surface layers promise large rewards.

New techniques have been developed within the past decade for modifying surface layers. The techniques fall in two general classes. Firstly, laser and electron beams now have sufficient energy to heat and melt large surface areas in very short times. The heating and cooling rates achievable with pulsed lasers are such that novel metastable surface alloys can be produced and, indeed, a new field of fast crystal growth is emerging. Secondly, atoms can be directly introduced into surface layers by ion implantation or ion-beam mixing techniques. The techniques are sufficiently well developed that high atomic concentrations can be implanted to useful depths. Moreover the atoms can be introduced without any of the usual thermodynamic alloying constraints.

The various surface modification techniques are important weapons in the arsenal of materials science. New materials and alloying phenomena are being investigated. There are, however, important technological and strategic driving forces behind the implementation of such techniques. The composition and structure of the outermost micron of a solid can be tailored in a remarkably controllable and reproducible fashion as witnessed by the fact that implantation is an indispensable part of the processing of Si integrated circuits.

We can illustrate the technological and strategic implications by the following example of Cr implantation. Chromium is a metal of considerable strategic importance with members of NATO having to import over 90% of the Cr they use in alloys. Although Cr is distributed throughout the bulk of the alloys, most of the usage is concerned with the modifications of surface properties. It makes sense therefore to try and fabricate Cr surface alloys directly. Corrosion is one of the most common reasons for the rejection of aircraft bearings during periodic scheduled overhaul in the U.S. Navy. Corrosion also causes premature and catastrophic in-service failures and unacceptably short shelf life for new, unused bearings. In response to these problems the U.S. Naval Research Laboratory, Washington, has carried out research which has demonstrated that ion implantation significantly enhances the corrosion resistance of bearing alloys in both laboratory and simulated field service tests. Ion implantation has also been shown to have no deleterious effects on the fatigue properties of the bearings.

Figure 1 shows aircraft bearings mounted in a vacuum chamber and being implanted with 150 keV Cr ions. The beam is incident from the left and the glow from the far left bearing is due to excited atoms emitting light following their ejection (sputtering) from the surface. Implanted jet aircraft and helicopter engine bearings have passed endurance tests in a test engine at conditions simulating the speeds, loads and temperatures experienced by engines in operational use. Currently several implanted bearings including those shown in Figure 1 are undergoing field service tests in U.S. Navy aircraft. These programs have demonstrated that ion implantation improves the corrosion resistance of the bearings without degrading their mechanical performance. They have also suggested that production line implantation of bearings should be feasible. For the production line, the technique has the important advantage of not affecting the macroscopic dimensions of the precision bearings. The implantation process should be applicable to otherwise finished components and thus can become part of an existing production line. In addition, the cost associated with implanting such bearing components is estimated to be but a small fraction of their initial cost. Further applications of implantation are discussed in Chapter 12.

II. SURFACE HEATING

Laser, electron and even ion beams can be used to heat surface layers so rapidly that the bulk temperature is not affected. The mechanisms by which these beams couple to the solid

Figure 1 Aircraft bearings being implanted with 150 keV Cr ions. From G. K. Hubler, J. K. Hirvonen, C. R. Gossett, I. Singer, C. R. Clayton, Y. F. Wang, H. E. Munson and G. Kuhlman, (1981). "Application of Ion Implantation for the improvement of localized corrosion resistance of M50 field bearings," NRL report 4481.

are different and are discussed in Chapter 2. The main features of laser coupling are illustrated in Figure 2 where the absorption coefficients of Si for photons corresponding to ruby ($\lambda = 0.69\ \mu m$) and Nd ($\lambda = 1.06\ \mu m$) lasers are plotted. In the solid state, there is a wide range of absorption coefficients which depend upon the band structure. However on melting, the differences collapse to a common value of 4×10^5 cm^{-1} (i.e., absorption depths of 250Å) because of the metallic nature of the molten Si. This value of the absorption coefficient is typical of most metals which are in the range 4×10^5 to 10^6 cm^{-1}. The absorption of laser light is very dependent upon the physical state of semiconductors but is relatively independent of the state of metals. However the total deposited energy depends upon the reflectivity of the target which is higher for metals than semiconductors.

Although lasers are most commonly used to heat surfaces, electron and ion beams are also used. The particle beams are more effective in depositing energy because of the basic physical processes of the scattering and energy loss mechanisms. For most particle heating sources with energies up to $\sim$100 keV, the energy is typically deposited over a few microns or so. Unlike lasers, the maximum energy deposition, and associated heating, can occur inside the solid. An important consideration in the use of charged particle beams is that the energy deposited is independent of the state of the target.

Typically energies of a few J/cm^2 are needed to heat and melt surface layers a micron thick. The attraction of beam processing is that the beams can be pulsed with pulse lengths as short as picoseconds for lasers and nanoseconds for particle beams. The deposited power

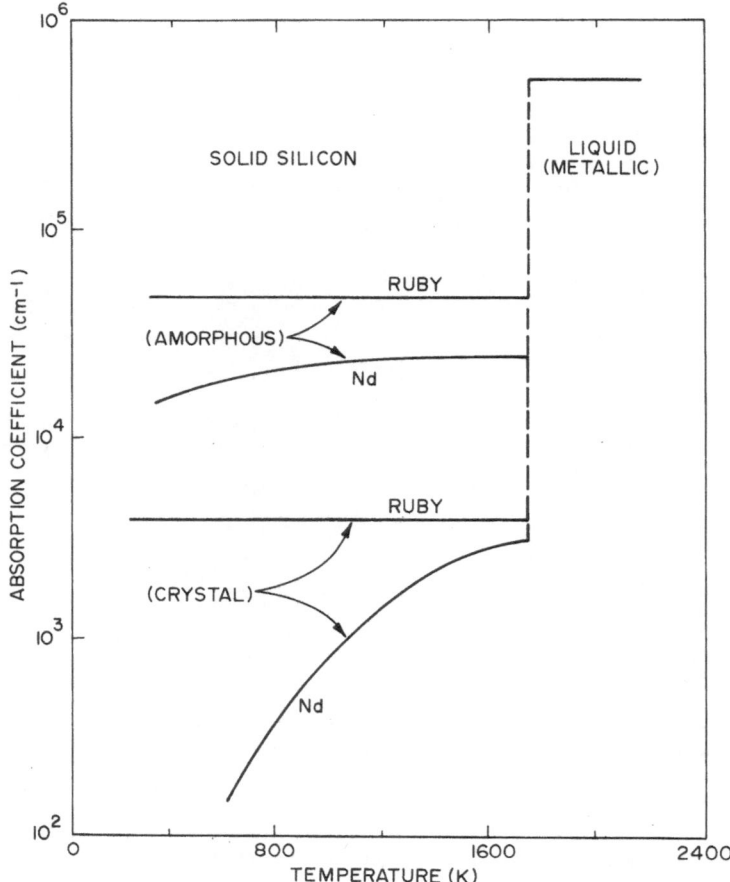

Figure 2   Absorption coefficients of ruby ($\lambda = 0.69\ \mu m$) and Nd ($\lambda = 1.06\ \mu m$) wavelength in amorphous, crystalline and liquid Si as a function of temperature.

can be enormous, for example $1 J/cm^2$ in 10 nsecs corresponds to a power of 100 Megawatts/cm^2. In such cases both the heating and cooling can be extremely rapid as illustrated in Figure 3 for pulsed ruby laser irradiation of Si and Al single crystals (Chapter 2 and 11). The comparison is made for the same melt depths to show the principal differences between semiconductors and metals. The melt front propagates in at approximately the same speed. However the velocity of the resolidification front is an order of magnitude faster for Al than Si because of the thermal conductivity differences. The velocity depends upon the rate at which the latent heat of crystallization can be extracted from the solidifying interface; the rate depends simply on the thermal conductivity and temperature gradients. In these cases the thermal gradients range between 10^6 and 10^7 K/cm with concomitant quenching rates of 10^9–10^{10} K/sec.

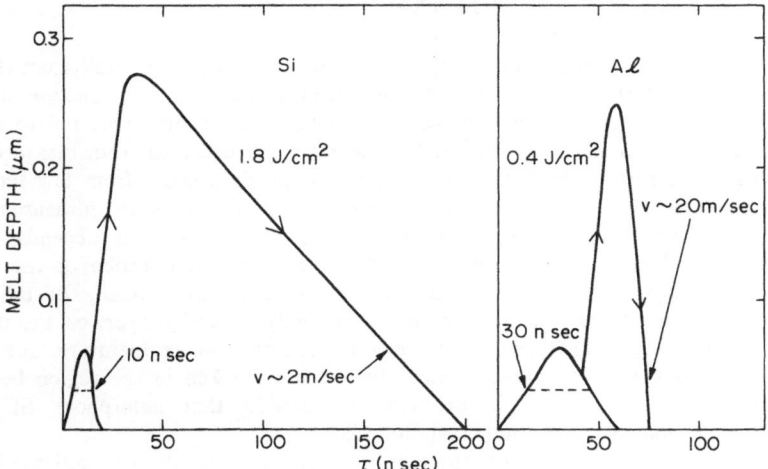

Figure 3   Position of melt-front in Si and Al after irradiation with a pulsed ruby laser.

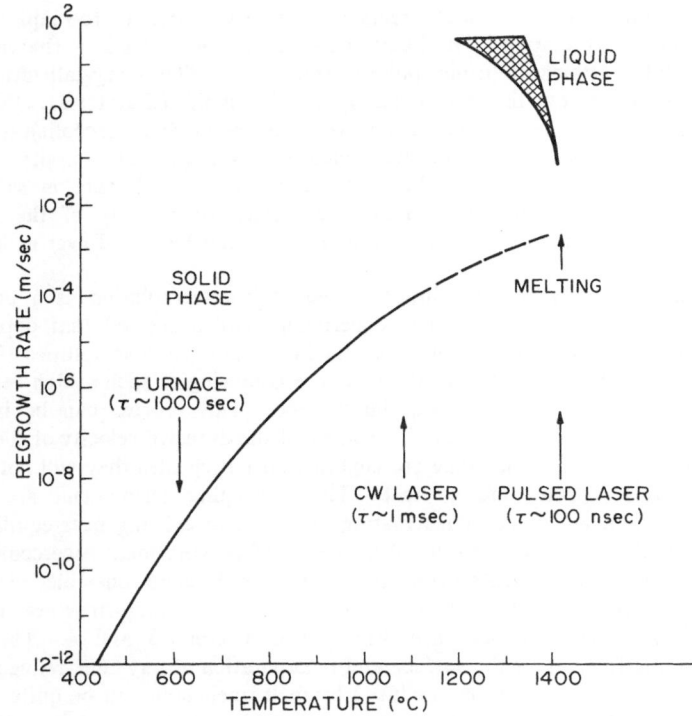

Figure 4   Si crystallization rates as a function of temperature. Various heating schemes are shown with the time required to crystallize 1000Å.

III. CRYSTAL GROWTH: EPITAXY

Some of the more interesting manifestations of surface processing result from the motion of the crystallizing interface. Figure 4 shows Si crystallization rates as a function of interface temperature. The various heating schemes are shown with the time required to crystallize 1000Å indicated in parentheses. The solid phase rates are obtained from measurements of the regrowth of amorphous Si layers on Si; most of the data come from low temperature furnace measurements as discussed in Chapter 5. In the furnace measurements the temperature gradients are essentially zero and the rate of growth depends upon the temperature dependence of bond breaking and the atomic configuration of the interface. Because of its higher free energy, amorphous Si is thermodynamically unstable in the presence of crystalline Si and tends to regrow epitaxially, layer-by-layer, on the underlying crystal. The interface velocity is an exponential function of temperature and the plot represents an extrapolation to higher temperatures. It is broken in the region beneath the melting temperature because it is believed (Chapter 3) that amorphous Si melts at considerably lower temperatures than crystalline Si.

Once the Si is melted, however, there is a sharp jump in the crystallization velocity because of the much greater mobility of the atoms in the liquid and the very high temperature gradients present in laser melting. In the case of conventional liquid phase crystal growth (Czochralski for example) the temperature gradients are, by comparison, low. There the rate of extraction of the seed crystal from the melt is determined by the rate at which latent heat can be removed through the seed crystal. Growth rates, typically 10^{-5} m/sec, are much slower than the recrystallization velocities of the liquid surface layer produced by pulsed irradiation. The hatched region in Figure 4 shows the velocities which are experimentally attainable using pulsed irradiation. The recrystallization velocity is ultimately determined by the undercooling of the melt (Chapter 3); the higher the undercooling the greater the velocity. Definitive estimates of Si undercooling do not yet exist and the hatched region just indicates plausible values. The upper limit to the recrystallization velocity represents the case where the quench rate is so fast that an amorphous layer is produced. It is remarkable that the velocity of the recrystallizing interface, whether from the amorphous or liquid phase, can be varied over at least 14 orders of magnitude.

One of the more striking possibilities of pulsed laser irradiation as a unique tool for surface processing came from the early experiments which showed that dopants could be incorporated on lattice sites at concentrations far in excess of solid solubilities. This subject is discussed fully in Chapter 4. Because the interface is moving at such a high velocity, there is a good probability that impurities will be trapped. This process can be imagined as a competition between the velocity of the interface and the diffusive velocity of the impurities in the liquid at the interface. Once they are engulfed on lattice sites they will not diffuse in the solid because quench rates are so fast. These trapping phenomena are velocity and orientation dependent and make a fascinating extension to existing near-equilibrium crystal growth and solubility data. Due to the phenomenon of constitutional supercooling, at certain velocities and impurity concentrations, the interfaces become unstable resulting in the incorporation of dopants in cellular arrays. Examples of these microstructures that can result from segregation at the interface are discussed in Chapter 3 and 4. The final act of solidification is the freezing of the surface and incorporation of any zone-refined impurities at the surface. The surface structures produced by melt quenching can be quite different from those produced by conventional surface treatments. This rapidly developing field is discussed in Chapter 4.

These studies of pulsed irradiation and the liquid phase stimulated, in fact, interest in the solid phase once more with regard to both furnace and continuous wave laser heating. Epitaxy, segregation and incorporation of impurities in the solid phase are discussed in Chapter 5.

At present there has not been as much interest shown in the rapid crystal growth of metals because, in some ways, Si is a more interesting test case of theory and experiment. For example, the covalently bonded Si can be amorphized, in marked contrast to pure metals. Moreover it has been observed that, in the liquid phase, the epitaxial regrowth of metals is of much poorer quality (Chapter 11) than Si regrowth because of thermal stress effects (Chapter 2).

There is little doubt, however, that the field of rapid crystal growth in semiconductors and metals is proving a challenging test of solidification theory. In particular the segregation and undercooling phenomena must be understood within the framework of any successful theory. The state of the theory is discussed in Chapter 3.

IV. METASTABLE PHASES

One of the central scientific thrusts of surface alloying is the production of metastable phases. Both the pulsed heating and implantation techniques are particularly conducive to forming metastable structures. Alloying, or phase formation, takes place far from equilibrium often in a high density of defects. The potentialities of forming metastable phases have been extensively explored for the case of ion implantation in metals. It turns out that superconductivity is a particularly clean method of studying the electronic and structural aspects of metastability. The ways of producing metastable alloys, from splat-cooling to ion beams, are discussed in Chapter 6 with particular emphasis on superconducting materials.

Ion beam mixing is an intriguing method of producing metastable phases. Deposited films are mixed together using the defects and enhanced atomic mobility produced by an ion beam. Figure 5 shows schematically the mixing of Ag-Cu films by a 300 keV Xe beam with the substrate held at RT or even liquid nitrogen temperatures. The lattice parameter measurements show that a complete series of solid solution have been formed.

The Ag-Cu system occupies a special place in the history of metastability as it was the first system to be rapidly quenched, by conventional techniques from the melt, to form metastable solid solutions (Ag-Cu is an eutectic system with very limited terminal solid solubilities). Subsequently metastable solid solutions have been made by laser quenching, ion implantation and now ion-beam mixing. The subject of ion-beam mixing is discussed in Chapters 8, 9 and 11.

V. ION BEAM EFFECTS

A. Energy Deposition and Cascades

The study of the interaction of ions with solids has been particularly fruitful over the past 40 years and has attracted the attention of such eminent scientists as Bethe, Bohr, Fermi and Wigner. Some fundamental aspects still have to be resolved especially with regard to dense collision cascades. The ions lose energy to the electrons and atoms of the solid by Coulomb interactions. The electronic energy losses, for metals and semiconductors are gentle in nature and ultimately end up as heat. The atomic or elastic interaction can be very violent leading to gross disturbances and damage in the lattice. For example a heavy ion of $\sim$100 keV energy incident on Si can displace up to a thousand atoms. The energy loss process can be

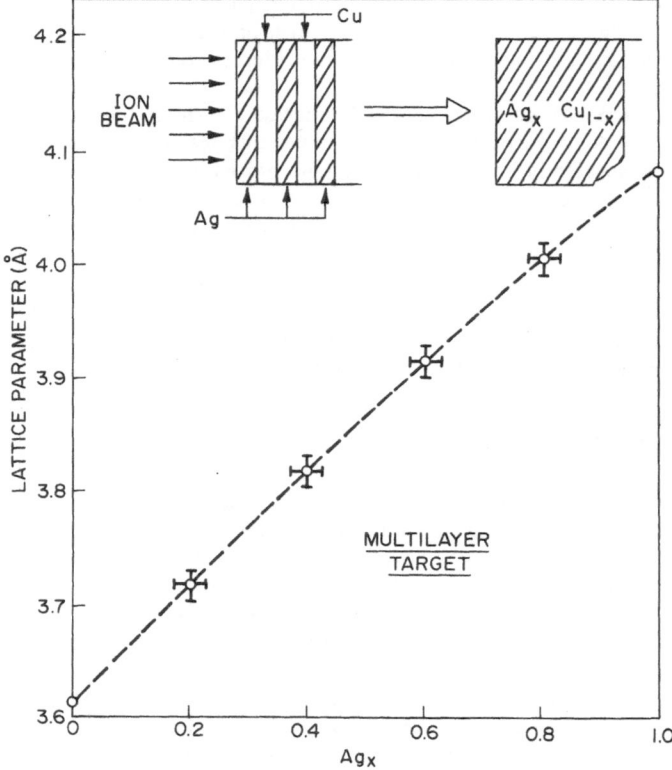

Figure 5   Formation of metastable Ag-Cu solid solution by ion beam mixing as shown by lattice parameter measurements. From B. Y. Tsaur, PhD Thesis, California Institute of Technology (1980).

modelled either in terms of separable electronic energy losses and isolated binary collisions (i.e., linear cascades) or a collective motion of the recoiling atoms (i.e., thermal spike).

Figure 6 shows a calculation of the range and damage profiles of Cr ions implanted into Fe at 50 keV for a linear cascade. The main features of implantation are well illustrated by this calculation. The mean range of the implanted Cr is shallow, ~200Å, with considerable straggle in the range distribution. The damage profile does not mimic the range profile but peaks closer to the surface. This calculation of the damage of energy deposition explains why sputtering yields are so high for this mass combination at these energies; a considerable fraction of the energy is dumped close to the surface and surface atoms can be ejected.

How do we microscopically picture the collision cascade? This subject is fully discussed in Chapter 7. Figure 7 shows a particularly violent form of a collision cascade known as a "displacement spike." This picture was developed some thirty years ago and does illustrate a very important characteristic of ion bombardment in that the deposited energy is localized in a very small volume. Moreover the zone can be considered to be heated (or melted) and quenched in times ~10^{-11} sec giving quench rates of 10^{14} K/sec. This behavior is quite different than the heat flow discussed previously where the energy is transferred to the lattice atoms under equilibrium conditions.

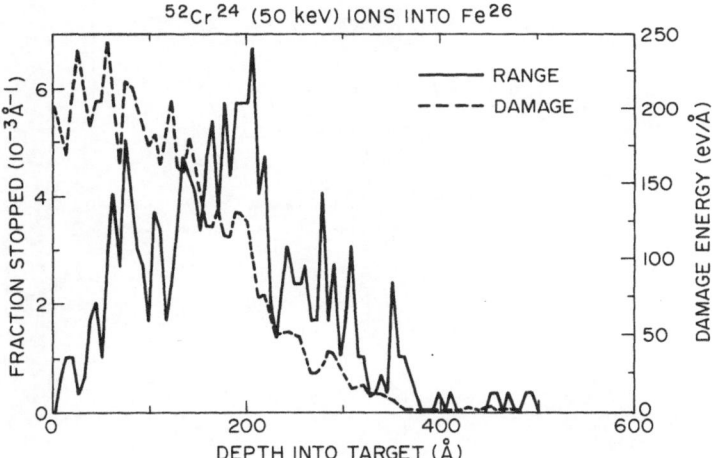

Figure 6   Range and damage profiles of Cr ions implanted into iron at 50 keV, calculated by the TRIM program. Ziegler and Biersack's universal nuclear stopping and LSS electronic stopping powers were used in the calculations, which each generated the trajectories of 500 ions. From J. P. Biersack and L. G. Haggmark, (1980). Nucl. Instr. and Methods *174*, 257 and J. F. Ziegler and J. P. Biersack, unpublished.

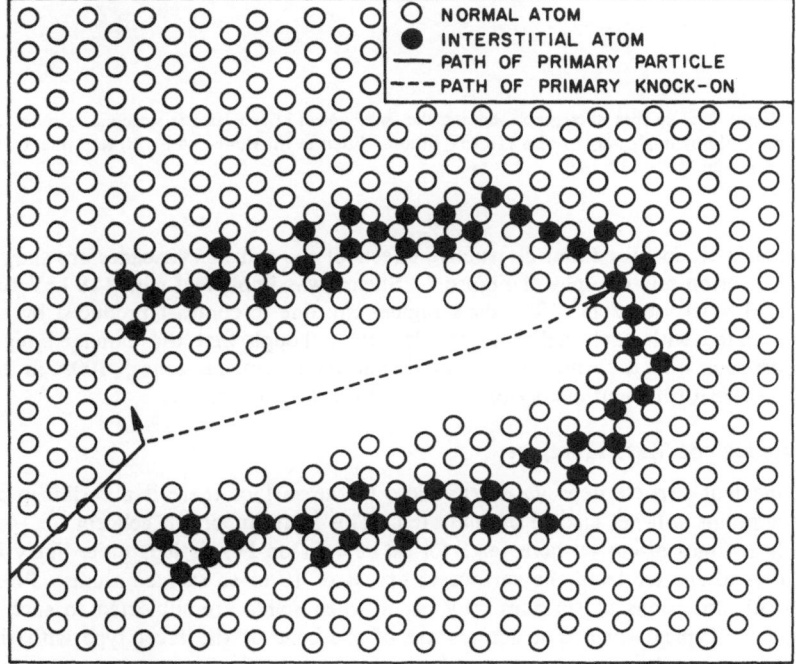

Figure 7   Displacement spike. From J. A. Brinkman, (1954). J. Appl. Phys. *25*, 961.

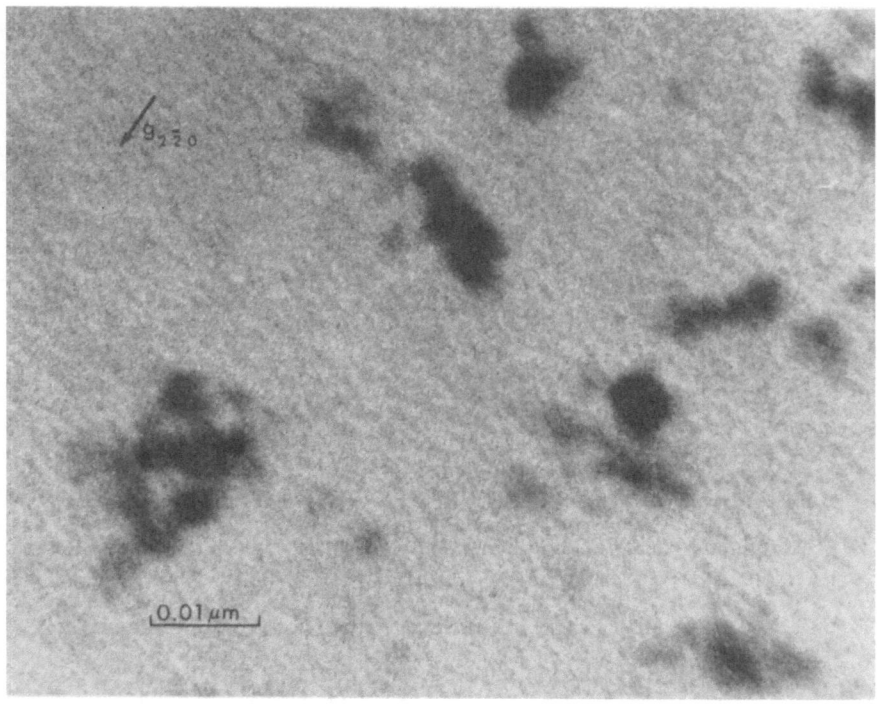

Figure 8   Bright field electron micrograph showing the development of sub-cascade structure in collision cascades produced by implanting Si with 118 keV Bi ions. From L. M. Howe and M. H. Rainville, (1981). Nucl. Instr. and Meth. *182*, 143.

The remarkably localized and destructive effects of the collision cascades are shown in the bright field electron micrographs of Figure 8. Silicon was implanted with 118 keV Bi ions to a fluence of 3×10^{11} ions/cm^2. The dark regions are due to contrast produced by damaged regions associated with each collision cascade. It is though that the damaged regions are indeed amorphous with each incident ion producing an amorphous zone ~100Å in diameter. The Bi ions deposit sufficient energy to develop sub-cascade structures.

B. Sputtering and Mass Transport

When an incident particle deposits energy close to the surface there is a certain probability that surface atoms will get ejected or sputtered. This is an important concept for the energy and particle regimes discussed here (e.g., for 50 keV Cr on Fe) as it is quite likely for each incident atom to sputter from the surface in excess of 10 or so surface atoms. Pronounced morphological alterations of surfaces may occur when solids are subjected to such particle bombardment. Typical surface structures are of the cone or the facet type with their size exceeding the projectile range by orders of magnitude. On single crystals - or on individual grains of polycrystalline samples - these structures show a high degree of regularity, both in lateral separation and in shape. For these reasons surface structures are generally explained in terms of lattice imperfections influencing the sputter-erosion of the surface (intrinsic or radiation-induced).

While surfaces with facets generally appear after high-fluence irradiation ($>10^{17}$ ions/cm^2), surfaces with conical protrusions are often observed when impurities are either present from the beginning, or are simultaneously deposited during irradiation. This latter seeding procedure is a reliable means to produce cone-covered surfaces; in this way the cone density can be controlled to some extent by the intensity of the seeding source and the target temperature. Such surfaces, however, are stable under prolonged irradiation only as long as the seeding source (e.g., co-sputtering, evaporation) is in operation. Under pure conditions when mono-elemental single crystals are bombarded with inert gas or self ions, the situation is different: the structures are strictly related to the beam-lattice orientation and remain stable as long as these conditions are not changed. Conical protrusions show clearly developed crystallographic features for which reason they are often referred to as pyramids. Figure 9 shows the beautiful pyramid-like structures that can be generated on Cu by Ar bombardment. The general topographies on single crystal surfaces, however, are two-dimensional arrays of facets or ridges whose structural regularity has been related to the dislocation network beneath the surface. The strong correlation between surface structure and orientation of the beam with respect to the crystal lattice can often be seen on polycrystals where grain surfaces entirely covered with cones are found in the immediate vicinity of grains with flat or facetted surfaces.

When an alloy is bombarded, the surface can not only undergo morphological changes but also compositional changes. This behavior is illustrated in Figure 10 where the surface composition of a PtSi alloy can be dramatically changed by Ar ion bombardment. The surface becomes Pt rich with the thickness of the altered surface layer depending upon the range of the incident ion. The ways in which surface compositions change are quite complex, depending, for example, on sputtering and defect diffusion mechanisms. This subject is discussed fully in Chapters 8 and 10.

VI. APPLICATIONS AND FUTURE DIRECTIONS

The possibilities of application and future directions for this field of surface processing seem almost limitless when the importance of surface-dominated phenomena are considered. It is worth emphasizing however that the techniques are very useful as basic research tools for surface and materials science. This point is illustrated by the work discussed in Chapter 11 where a single metal, Al, is subject to many forms of surface modification using laser, electron and ion beams.

The modification of surfaces by ion implantation and possible applications are fully discussed in Chapter 12. This field is developing quite vigorously as witnessed by the research into catalysis, corrosion, friction and wear. A not so obvious application is the use of novel surface structures produced by sputtering, as in Figure 9. For instance the distinct change in optical reflectivity could be utilized for selective light absorbers. It is conceivable that the absorption wavelength could be varied by adjusting the appropriate particle bombardment conditions. Another aspect of structured surfaces is their possible application as a sputter-erosion resistant coating on the first wall components of fusion reactors; the sputtering yield of such surfaces is expected to decrease owing to an enhanced recapture probability of ejected particles. Unfortunately however, this reduced escape probability is more than balanced by the increase in yield due to the oblique angle of incidence.

There is little doubt that lasers and other beams which are used for the direct heating of surfaces will occupy an increasingly important role in materials science and technology. At present it appears that much of the research limelight is taken by semiconductors. However, the field of laser alloying of metal surfaces would appear ripe for development as discussed in Chapter 13. An example of the use of lasers in surface alloying is given in Figure 11. Gold

Figure 9   Pyramid structures that develop on Cu single crystal after 40 keV Ar bombardment. From J. L. Whitton, G. Carter and M. J. Nobes, (1977). Radiation Effects, *32*, 129.

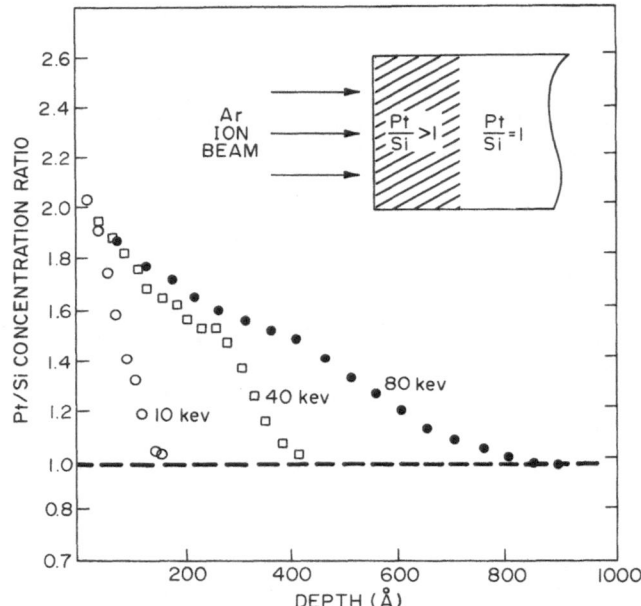

Figure 10   Steady-state Pt/Si profiles of 10-, -40 - and 80 keV Ar-sputtered PtSi samples. From Z. L. Liau, J. W. Mayer, W. L. Brown and J. M. Poate, (1978). J. Appl. Phys. *49*, 5295.

has limited solid solubility in Ni but it is possible to alloy high concentrations in the surface by rapidly melting thin films of Au or Ni with a pulsed Nd:YAG laser; irradiation by a continuous wave CO_2 laser does not produce such high concentrations because of the deeper melt depths. The ability to rapidly melt surfaces of metals and overlaying films should lead to important metallurgical application.

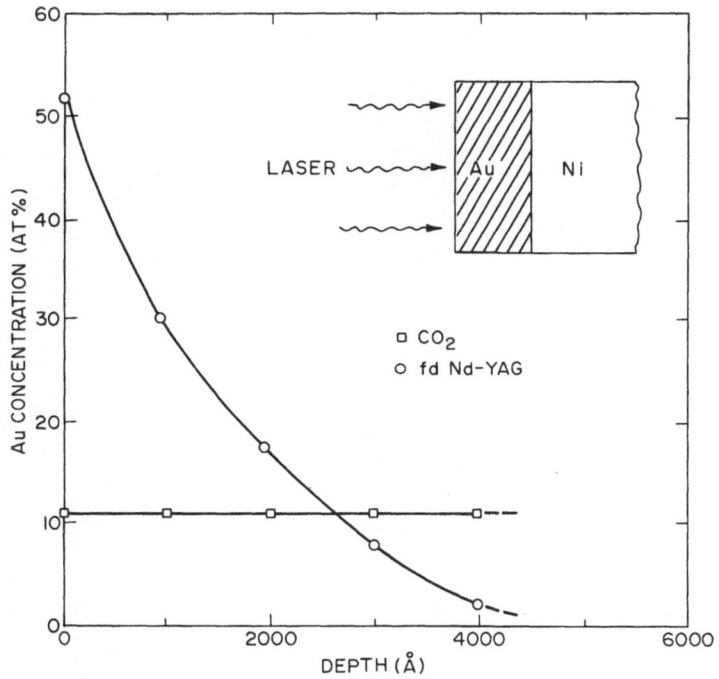

Figure 11 Surface alloys introduced by laser alloying (CO_2 or frequency-doubled Nd:YAG) Au films on Ni. From C. W. Draper, L. S. Meyer, L. Buene, D. C. Jacobson and J. M. Poate, (1981). Applications of Surface Science, 7, 276.

ACKNOWLEDGEMENTS

We are indebted to Larry Howe, Jim Whitton and Jim Ziegler for supplying Figures 8, 9 and 6 respectively.

CHAPTER 2

ENERGY DEPOSITION AND HEAT FLOW FOR PULSED LASER, ELECTRON AND ION BEAM IRRADIATION

E. RIMINI

Istituto di Fisica, Corso Italia 57-I, 95129 Catania, Italy

CONTRIBUTORS: P. Baeri, L. F., Donà dalle Rose, C. W. Draper,
                J. W. Mayer, S. T. Picraux, B. Stritzker, M. Von Allmen,
                C. W. White

I. INTRODUCTION

Energy deposition in times as short as tens of nanoseconds, with densities of joules/cm^2, has emerged in the last few years as a new way of modifying the near surface structures of materials. The initial input came from the use of Q-switched laser pulses to anneal damage in ion implanted semiconductors. Interest in other areas, such as metallurgy, is now growing because of the possibility of forming new phases or new structures. The history of this field is reported in the proceedings of the annual meetings of the Materials Research Society dedicated to this subject since 1978 (Ferris *et al.*, 1979; White and Peercy, 1980; Gibbons *et al.*, 1981).

In these meetings both applied and fundamental aspects were discussed and new possibilities of investigations arose. Fundamentally there is of course the need to understand the mechanisms by which photons, electrons or ions lose their energy inside materials; how this energy is converted into heat and how this heat propagates in the sample. Pulsed laser irradiation causes reordering of damaged semiconductor layers by the formation of a transient liquid layer whose thickness should exceed that of the disordered region. This mechanism implies that the photon energy absorbed by the electrons of the semiconductor is transferred as heat to the lattice atoms in a very short time. The large amount of heat in a thin layer can melt it. At the time of the Trevi meeting there was still some discussion about another possible mechanism of laser annealing: electron-hole plasma annealing in a cold lattice structure. However, the results of a large number of experiments performed both during irradiation and after clearly show that the surface layers melt.

The heating and cooling stages are associated with large thermal gradients $\sim10^6$–10^8K/cm, and quenching rates $\sim10^9$–10^{11}K/sec. During solidification the liquid-solid interface moves with velocities of m/sec. These calculated velocities are required to quantify such subjects as impurity redistribution and determination of segregation coefficients, cell formation, metastable phase formation and the transition from the amorphous phase to an undercooled liquid; these aspects are treated in detail in other chapters of this book. In this chapter, energy deposition and heat flow calculations are presented and discussed bearing in mind the major contributions presented at the Trevi meeting and the relation to the other chapters. The energy deposition of laser pulses into semiconductors is described together with recent *in situ* temperature measurements. In a comparison of laser irradiation effects in semiconductors and metals, the major differences arise because of the different absorption lengths of the light and because of the different thermal parameters responsible for the heat propagation. Thermal gradients during the heating and cooling stage and the time duration of the heating stage are different for semiconductors and metals. The coupling of thermal and concentration gradients (Soret effect) modifies the impurity distribution in metals with respect to the usual mass transport diffusion equation, but this effect is negligible in semiconductors. Moreover the thermal gradients produced by laser irradiations, below the threshold for surface melting, can create stress fields capable of causing plastic deformation in metals. The effect in semiconductors is, however, weaker.

The energy deposition and heat response of pulsed electron and ion beams are considered in Sections IV and V, respectively. Monochromatic electron and ion beams deposit energy at depths determined by the energy loss relationships. The interaction is governed mainly by electron densities and by the mass of the nuclei. Surface structure such as oxides or the differences in bonding between semiconductors or metal, do not play such a crucial effect as they do for laser irradiation. Usually the heated depth in electron or ion beam irradiation is microns thick, while for lasers it can be restricted to 0.1 μm. The calculated thermal gradients both in time and in space are also lower than for laser irradiation. A good introduction to the subjects of this chapter is found in chapters 3 and chapter 4 of the book "Laser Annealing of Semiconductors," (Poate and Mayer, 1982).

II. INTERACTION OF LASER BEAMS WITH SEMICONDUCTORS

A. Absorption of Light and Energy Transfer to the Lattice

The interaction of photons with matter occurs mainly by electronic excitations. In the photon energy range from infrared to ultraviolet wavelengths only valence and conduction electrons participate in the excitation processes. In metals the light is absorbed by free electrons (intraband transition) while in semiconductors, electron transitions from the valence to the conduction band play a crucial role (interband transitions).

In an intrinsic semiconductor crystal the valence band is nearly filled to the top with electrons, while the conduction band is almost empty. Interband transitions by light occur if the photon energy is larger than the energy gap between the valence and conduction band. As a result of light absorption, electron-hole pairs are created. In elemental semiconductors, such as Ge and Si, the maximum of the valence band and the minimum of the conduction band do not occur for the same momentum value in wavevector space and the interband transition is indirect. Momentum conservation requires absorption or emission of phonons for photon energies near the optical forbidden band. Free carriers and band to band transitions are illustrated in Figure 1a.

Disordered semiconductors, such as amorphous Si, have electronic states within the energy gap and the transitions do not require momentum conservation. For instance in Si at the Nd:YAG wavelength $(\lambda = 1.06 \ \mu m, \ \epsilon = 1.17 \ eV)$ the absorption coefficient is

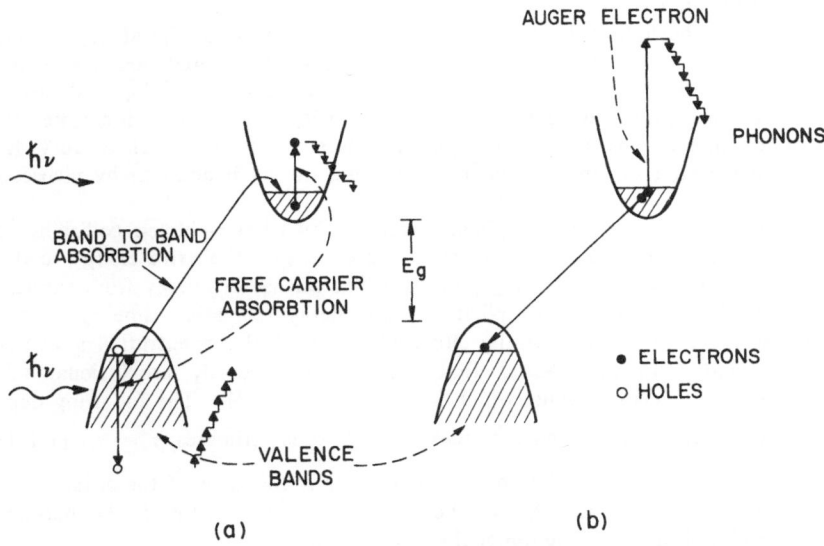

(a)   (b)

Figure 1   Schematic (a) band-to-band and free carriers transitions, (b) Auger recombination between electron-hole pair and phonon emission processes.

(Brodsky *et al.*, 1970) 50 cm^{-1} for crystalline material and 10^3cm^{-1} for amorphous material. The band gap in Si is 1.1 eV.

Impurities present at high concentration in semiconductors modify the electronic properties through band tailing and reduction of the fundamental energy gap in interband transitions. Free carriers generated by the doping also contribute to the optical properties of the semiconductor. In As doped Si samples the optical absorption coefficient, at a photon energy of 1.15 eV, increases from 10 cm^{-1} to 300 cm^{-1} from undoped to a doping level of 4×10^{19}/cm^3. Boron doped Si samples instead show some saturation with doping, indicating the presence of compensation (Schmid, 1981).

The previous considerations describe the main mechanisms of the light absorption in a semiconductor for low intensity. Irradiation of semiconductors with laser pulses of nanosecond or picosecond duration involves a large number of incident photons. For instance a typical annealing pulse from a ruby laser corresponds to 1 J/cm^2 with a duration of 10 nsec. The incident power density is 10^8 W/cm^2. Assuming that half of the incident power density is absorbed within 10^{-5} cm the number of generated electron-hole pairs is 10^{31}/cm^3/sec. Carrier densities exceeding 10^{19}/cm^3 are achieved in times much less than the laser pulse duration. At the melting point of Si the intrinsic carrier density is 3×10^{19}/cm^3. The system created during irradiation is called a free-carrier plasma or an electron-hole plasma; it is the solid state analogue of a gaseous plasma, i.e., a gaseous system of ionized atoms and electrons.

The free carriers generated by the illumination enhance the absorption if their recombination lifetime is not much shorter than the pulse duration. The effect is more pronounced for photon energies near the band gap. The free carrier absorption coefficient is usually calculated according to the Drude model. It is 50 cm^{-1} at n = 5×10^{13}/cm^3. In laser annealing experiments, free-carrier absorption is important during the first stage of the pulse. The effect is pronounced for Nd:YAG wavelengths in Si where the 1.17 eV photon energy is comparable with the optical band gap of 1.1 eV in Si. The free-carrier contribution helps the coupling in increasing the temperature and thus narrowing the forbidden energy gap with a consequent increase of the absorption coefficient.

The energy of the light pulse is transferred to the carriers via the absorption processes. The next step to be considered is then the de-excitation of the excited carrier system and the energy transfer to the lattice atoms. During irradiation both excitation and de-excitation processes occur simultaneously so that the carrier density amounts to a few times 10^{20}/cm^3. Carrier-carrier interactions and carrier-plasmon interaction occur at a very fast rate $\sim 10^{-13}$ sec, the carrier system is then in equilibrium during irradiation by picosecond and nanosecond pulses.

Heat generation occurs through phonon creation by carrier-phonon collisions. Excited carriers, with kinetic energies well above the band edge, give the extra energy to the lattice atoms by a series of collisions creating phonons. The release of energy from carriers to the lattice can be characterized for simplicity by an energy relaxation time τ_e. The rate of energy release is slowed by a screening effect when a critical concentration is exceeded. It has been calculated (Yoffa, 1980a) that the rate is inversely proportional to carrier concentration above a critical value which is 2×10^{21}/cm^3 in Si. The following dependence has been used for calculations and for order of magnitude estimates: $\tau_e = \tau_{e0} [1 + (\dfrac{n}{n_{crit}})^2]$ (Lietoila and Gibbons, 1981). The unscreened time constant τ_{e0} is of the order of 10^{-11} sec. As a result of this energy transfer to the lattice the electrons reach the bottom of the conduction band and the holes the top of the valence band.

Electrons and holes are well above their equilibrium concentration values during irradiation and recombination occurs. The process can be described approximately by a power series as

$$- \frac{dn}{dt} = \gamma_1 \, n + \gamma_2 \, n^2 + \gamma_3 \, n^3 \tag{1}$$

The different terms are expected to dominate over different regimes of carrier concentration. At low concentration values the first term dominates and is associated with phonon-assisted recombination via deep impurity levels. The $1/\gamma_1$ parameter is the minority carrier lifetime. The second term accounts for band-to-band recombination accompanied by photon emission. Radiative processes of this type are highly improbable in indirect gap materials such as silicon, so that this term can be neglected. The third term is present at high injection levels. At high concentrations, electrons and holes recombine directly and the energy is transferred to another electron (or hole). This is the Auger process. The γ_3 value is of order of 10^{31} cm^6/sec (Krausse, 1974). The Auger process can be described by a lifetime τ_A equal to the ratio of the excess carrier concentration to the rate of decay of excess carriers. When the electron and hole concentrations are equal $\tau_A = 1/\gamma_3 n^2$, the measured value of γ_3 is 4×10^{-31} cm^6/sec (Dzievior and Schmid, 1977). At a concentration of 10^{20}/cm^3, τ_A is 2.5×10^{-10} sec.

A sufficiently dense plasma will partially screen the Coulomb interaction between the carriers. The screening becomes important at carrier densities larger than 10^{21}/cm^3 (Yoffa, 1980a; Haug, 1978). The recombination time τ_A decreases with carrier concentration and can reach 6×10^{-12} sec. During laser irradiation the dominant recombination mechanism is then the Auger process where the gap energy is given to an already existing free carrier. These excited carriers deliver the extra energy to the lattice by phonon emission. However, calculation of the energy release to the lattice and Auger recombination do not allow the plasma density to exceed 10^{20}/cm^3 (Combescot, 1981). The de-excitation processes are presented in Figure 1b.

The relevance of the Auger recombination process is illustrated by the following example (Brown, 1980). Suppose that a pulse of 0.2 J/cm^2 of 2 eV photons is suddenly absorbed in a 10^{-5} cm thick layer. The carrier concentration amounts to 6×10^{22}/cm^3. If the carriers were retained in that volume, Auger recombination in 10^{-11} sec would have decreased the concentration to 6×10^{20}/cm^3 and 99% of the energy would have been transformed into heat. The Auger recombination rate becomes predominant at concentrations above 10^{19}/cm^3 and does not allow the carrier concentration to be far out of equilibrium with the lattice temperature. Carrier diffusion may occur in the hot, dense gas of electron holes generated during pulsed irradiation of semiconductors. Electrons and holes diffuse together to avoid space charge effects and their diffusion is governed by the ambipolar diffusion coefficient of the order of 10-100 cm^2/sec. As a consequence the laser energy should be transferred from the carriers to the lattice atoms within a characteristic depth determined primarily by carrier diffusion rather than by the absorption coefficient of the sample. From the point of view of laser annealing the overall effect should be a reduction of the temperature rise and a less efficient coupling (Yoffa, 1980b).

Carrier diffusion in the usual laser annealing conditions has been estimated to be equivalent to an absorption coefficient of 4×10^4 cm^{-1}. The effect of carrier diffusion should then give an upper limit to the absorption coefficient value. So far we have summarized in a simple way some concepts concerning absorption and energy release to the lattice from the excited carriers. It has been suggested that the electron-hole plasma can be decoupled from the lattice with a temperature as high as 10^4K which is, of course, uncorrelated with the crystal temperature. In such conditions, laser annealing would occur in a cold, unbonded lattice (Van Vechten, 1980; Van Vechten and Wautelet, 1981). However it is not clear why the basic electron-phonon collision time of about 10^{-12} sec should be so lengthened as to prevent phonon emission and energy transfer to the lattice during the laser pulse.

B. In Situ Experiments

The interactions of laser beams with semiconductors have been investigated by several *in situ* experiments. The measured quantities determined so far are the optical response, the lattice temperature during irradiation and recently the velocity of the solid-liquid interface during the heating and cooling stage (Galvin *et al.*, 1982).

The reflectivity of semiconductor surfaces, during irradiation with a pulsed ruby laser, was reported in the literature more than ten years ago (Birnbaum and Stocker, 1968). An enhancement of reflectivity was found above a threshold energy density value which for Si was 1.5 J/cm^2. The enhanced reflectivity was due to a molten surface layer. A detailed investigation on the optical response of irradiated Si was performed more recently by Auston *et al.* (1978 and 1979) using time-resolved reflectivity measurements. The measurements have shown that, for sufficient laser energies, the reflectivity of a He-Ne probing beam increases and reaches a flat-top value of 70% lasting up to several hundred nanoseconds, much longer than the pulse duration. Since the enhanced reflectivity is very close to the reflectivity of liquid Si at the probing wavelength, it is assumed that the surface region of the sample melts. The change occurs abruptly, in agreement with a first order phase transition from solid to liquid.

The time dependence of reflectivity has also been checked by changing the wavelength of the probe beam. For Si, 0.633 and 0.480 μm wavelengths have been used (Nathan *et al.*, 1980) and no difference has been found between them in the magnitude and deviation of the high reflectivity stage over a wide range of laser energies. This experimental result rules out the existence of a persistent high density electron-hole plasma.

Lattice temperatures have been reported from time resolved Raman effect measurements (Lo and Compaan, 1980) which indicated that the lattice stayed cold during laser heating. It now appears, however, that these measurements are in question (Compaan *et al.*, 1982).

Recently the time of flight method has been used to measure the temperature of crystalline Si during pulsed laser irradiation (Stritzker *et al.*, 1981). This method allows the determination of the energy of Si atoms which are evaporated from the hot Si surface. The evaporation rate depends on the substrate temperature. The atoms ejected from the Si target were analyzed by a quadrupole mass spectrometer and the time interval between the laser pulse and spectrometer signal was measured. The method requires that the time for emission of atoms should be short compared with their time of flight. Typical times of flight were 50 μsec as compared with the 50 nsec laser pulse duration. Results of these measurements are shown in Figure 2. The lattice temperature measured from the Maxwell-Boltzmann distribution of velocities, ranges between 1200 and 3000K for pulse energy densities between 1.0 and 2.5 J/cm^2. The Si target is then in the molten state during laser irradiation for energy densities higher than 1.4 J/cm^2.

The time of flight measurement provides information on the sample surface temperature during pulsed irradiation. Time resolved x-ray diffraction measurements on silicon, during ruby laser irradiation, provide structural information just below the crystal surface (Larson *et al.*, 1982). The time delay between the Q-switched ruby pulse and the 0.1 nsec synchrotron radiation pulse allows investigation of the dynamics. These measurements provide direct information on the time duration and depth distribution of near surface strains associated with laser annealing. If analyzed in terms of thermal expansion, these data give lattice temperatures just below the surface. Figure 3 shows the measured reflectivities near the (400) reflection at a time of 100 and 195 nsec after laser irradiation as a function of the angular deviation from the Bragg reflection angle for unstrained silicon. The broadening is associated with lattice strains in excess of 0.5% as obtained by a fit of the x-ray data (see the inset) at 100 nsec. The data at 195 nsec are similar to those at 100 nsec, they have a

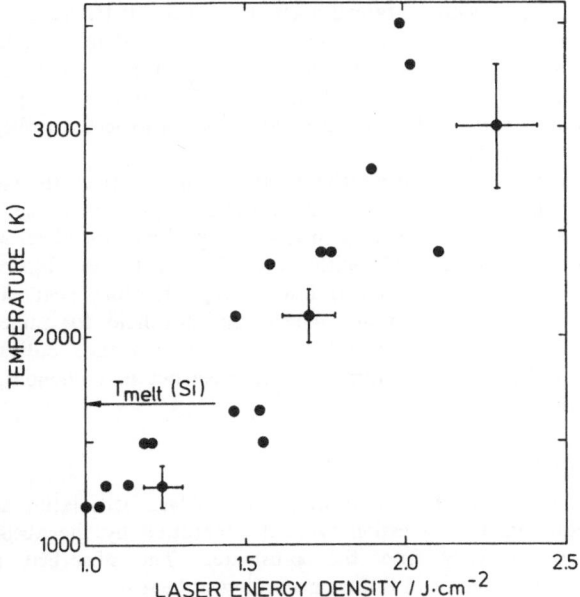

Figure 2   Surface temperature of Si single crystals irradiated with ruby laser pulses of different energy densities (Stritzker *et al.*, 1981).

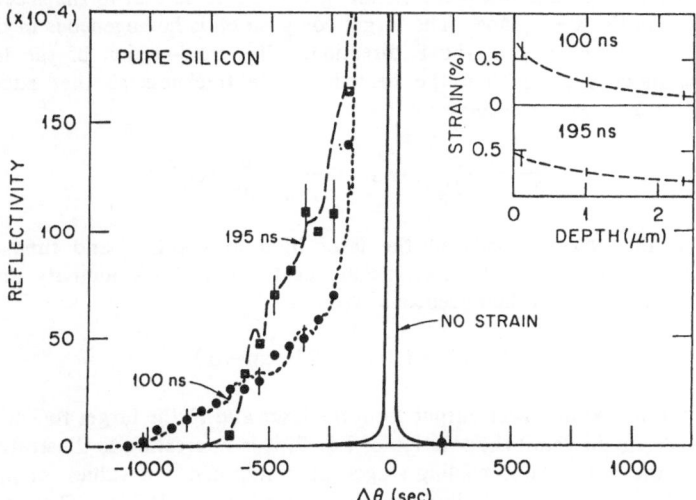

Figure 3   Measured x-ray scattering near the (400) Bragg reflection of pure silicon at 100 and 195 nsec after laser pulses. The dashed lines are fits to the data using the strain profiles shown in the insets (Larson *et al.*, 1982).

smaller angular shift from the Bragg angle. The time resolved temperature distributions are shown in Figure 4. They were obtained from the strain distributions through the use of temperature dependent thermal expansion coefficients. At 100 nsec, surface temperatures of 1150°C are found, while at 195 nsec the surface temperature reaches 750°C. These results while in agreement with the melting model, show several aspects not expected on the basis of calculations. Decrease of the surface temperature was not expected to begin until more than 150 nsec after the laser pulse.

Charged particles (electrons and positive ions) are emitted from the surface of a target at high temperature. The number of emitted particles is related to the temperature of the sample. These particles are measured by means of a positive (for electrons) or of a negative (for positive ions) voltage of few kV with respect to the target. No emission of charged particles was detected in Si irradiated at laser energy densities near the threshold for the amorphous to single crystal transition. Above the threshold for annealing, positive and negative particles were detected in equal amount. These results indicate that the energy relaxation time, establishing equilibrium between electrons and phonons, is of the order of 10^{-11} nsec (Liu *et al.*, 1981).

C. Heating and Cooling

The previous discussion shows that the basic process of laser irradiation and annealing in ion implanted semiconductors is a question of heat generation by the absorption of light and cooling by heat conductivity into the substrate. The absorbed light is converted instantaneously into heat which can diffuse according to heat flow.

The heating and cooling stages can then be determined by numerical solutions of the heat equation, including a source term for the absorption of light at a certain depth, the changes of optical and thermal parameters with temperature and structure of the irradiated layer and finally the latent heat absorbed or generated during the phase transitions (Baeri and Campisano, 1982). The laser beam travels along the z axis normal to the specimen surface, which is uniform in the $x-y$ plane. The target composition is homogeneous in this plane and structural changes occur only in the z direction. The cross-section of the laser beam is assumed to be much greater than the heated sample thickness so that edge effects are negligible. The heat equation becomes

$$\frac{\partial T}{\partial t} = \frac{\alpha}{\rho C_p} I(z,t) + \frac{1}{\rho C_p} \frac{\partial}{\partial z} (\kappa \frac{\partial T}{\partial z}) \qquad (2)$$

where $I(z,t)$ is the power density of the laser light at depth z and time t. T is the temperature, ρ, C_p, κ and α the density, specific heat, thermal conductivity and absorption coefficient of the sample. In a homogeneous medium

$$I(z,t) = I_o(t)(1-R)(\exp-\alpha z) \qquad (3)$$

where $I_o(t)$ is the temporal power output from the laser and R the target reflectivity.

Before considering in detail the solution of Eq. 2, it is interesting to illustrate some simple consequences of the heating and cooling stages, assuming constant values for ρ, α, C, κ and R and a rectangular laser pulse of duration τ_p (Bloembergen, 1979). The heating process involves two characteristic lengths, the absorption length α^{-1} and the heat diffusion length $\sqrt{2D\tau_p} \simeq \sqrt{2\kappa\tau_p/\rho C_p}$ with D the thermal diffusivity. If α^{-1} is smaller than $\sqrt{2D\tau_p}$ the heat source becomes a surface source (Figure 5a) and the average temperature rise ΔT is given approximately by

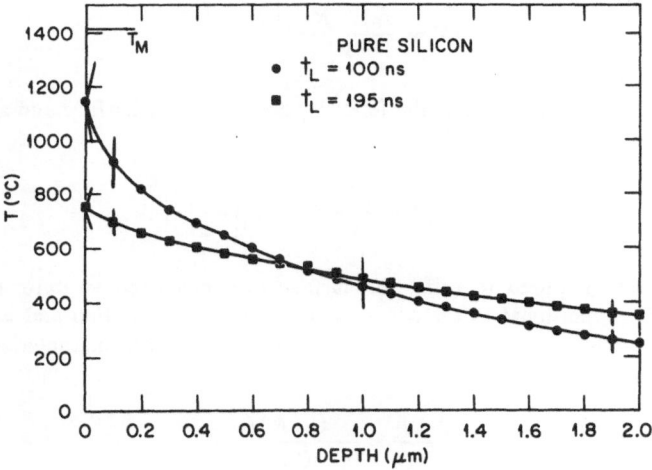

Figure 4   Temperature-depth distributions corresponding to the strain profiles in Figure 3 (Larson *et al.*, 1982).

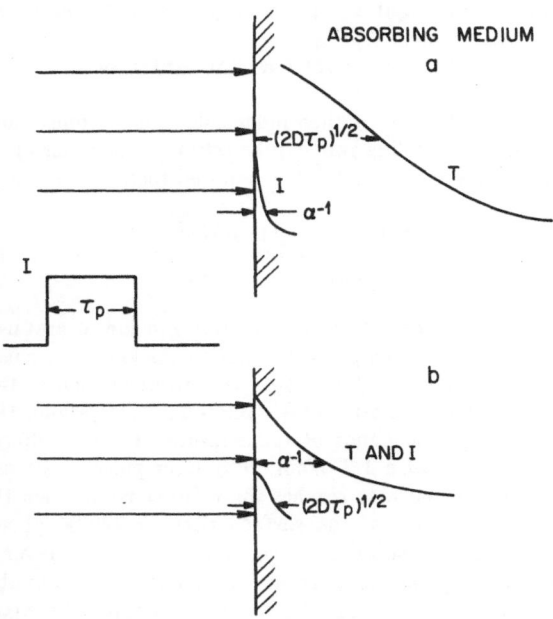

Figure 5   Schematic laser pulse intensity and temperature profiles for a penetration depth, α^{-1}, of the light small (a) and large (b) compared to the thermal diffusion length (Bloembergen, 1979).

$$\Delta T = \frac{I_0(1-R)}{\kappa} \left[\frac{D\tau_p}{2}\right]^{1/2} \tag{4a}$$

The correct analytical solution gives the following for ΔT at the surface and at the end of the pulse:

$$\Delta T(0,\tau_p) = \frac{2I_0}{\kappa}(1-R)\left[\frac{D\tau_p}{\pi}\right]^{1/2} \tag{4b}$$

The energy density required to bring the surface to a given temperature, for instance the melting point, is proportional to the square root of the pulse duration and is independent of the absorption coefficient. Heating and cooling rates are both characterized by τ_p. The heating rate is given by

$$\frac{\Delta T}{\tau_p} = \frac{(1-R)I_0}{C_p\rho(2D\tau_p)^{1/2}} \tag{5}$$

Where the rate is inversely proportional to the square root of the pulse duration.

The other extreme case (Figure 5b) corresponds to $\alpha^{-1} \gg (2D\tau_p)^{1/2}$ and the temperature rise at time τ_p and depth z is given, neglecting heat diffusion, by

$$\Delta T(z) = (1-R)\, I\, \alpha\, \exp(-\alpha z)\, \tau_p/\rho C_p \tag{6}$$

The heating rate, given by $\Delta T(z)/\tau_p$ is then independent of the pulse duration and decreases exponentially with depth. The cooling rate can be estimated considering heat diffusion over a distance α^{-1} so that a cooling time of $\alpha^{-2}/2D$ is required then

$$\left.\frac{\partial T}{\partial t}\right|_{\text{cooling}} \simeq \frac{(1-R)I\alpha^3\tau_p 2D}{\rho C_p} \tag{7}$$

Numerical methods to solve the heat diffusion equation provide several useful quantities, such as time and depth dependence of temperature, molten thickness vs. energy density, thermal gradients and velocity of the solidification front. As an illustration of the method we report the following few examples for irradiation of Si (Baeri and Campisano, 1982).

Figure 6 shows the time dependence of temperature at three different depths (surface, 0.35, and 1.0 μm respectively) for a 1.7 J/cm^2 ruby laser pulse of 10 nsec duration on a Si single crystal. In this case the surface reaches the melting point when the pulse delivers the maximum power; the temperature of the surface rises to 2200K about 18 nsec after the cessation of the pulse. In the subsequent cooling process the layer remains liquid for 180 nsec. The pinning of temperature at the melting point arises from absorption and release of the latent heat for fusion, ΔH_m. At the other two depths the material remains solid. Heating and cooling rates are of the order of 10^{11} and 10^9 K/sec respectively.

For a suitable energy density value, the molten layer propagates inside the sample in a time comparable with the laser pulse duration. The molten layer reaches a maximum thickness which is a function of the pulse energy density. In Si irradiated with ruby laser a typical rate of 0.6 μm/J/cm^2 is obtained. The kinetics of the melt front are shown in Figure 7. The melting proceeds with a planar front at a velocity of about 10 m/sec. The

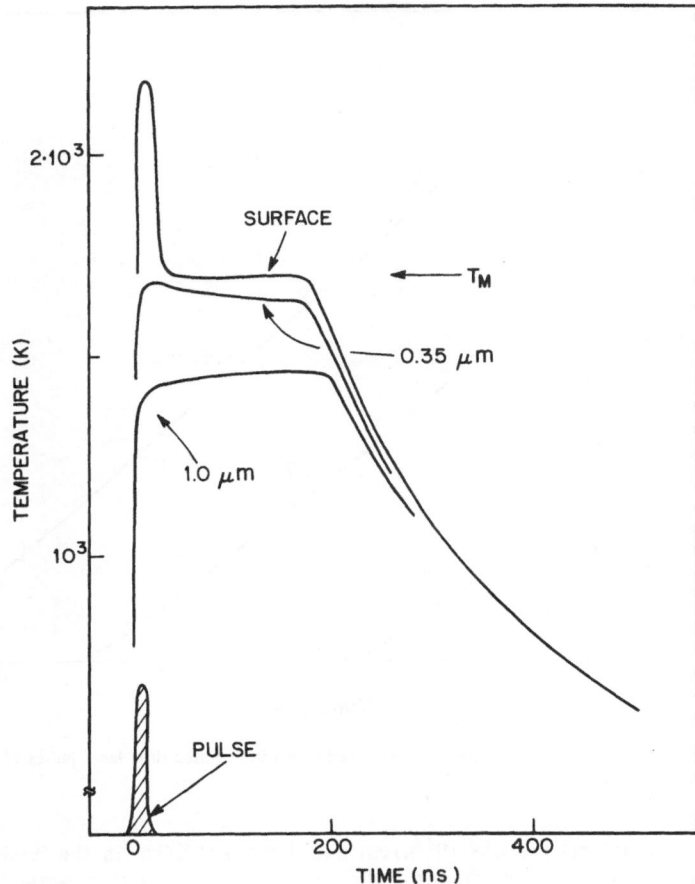

Figure 6   Calculated time dependence of temperature at different depths for Si single crystal irradiated with 10 nsec ruby laser pulse of 1.7 J/cm² energy density (Baeri and Campisano, 1982).

solidification usually occurs in a time interval about ten times the pulse duration. In the solidification process the latent heat of melting liberated at the advancing solid-liquid interface is just balanced by heat conduction into the substrate.

$$\Delta H_m \rho v = \kappa_s \left. \frac{\partial T}{\partial z} \right]_s - \kappa_\ell \left. \frac{\partial T}{\partial z} \right]_\ell \qquad (8)$$

Where v is the liquid-solid interface velocity and the subscripts s and ℓ refer respectively to the solid and liquid phases. The calculated temperature-depth profiles during heating and cooling stages are shown in Figure 8 for Nd and ruby laser irradiation of a 4000Å thick amorphous Si on Si single crystal. The energy density is sufficient to melt all the amorphous layer. During the heating process, when the surface layer is liquid, a large thermal gradient exists because of the small value of the absorption length in liquid silicon (200Å). The

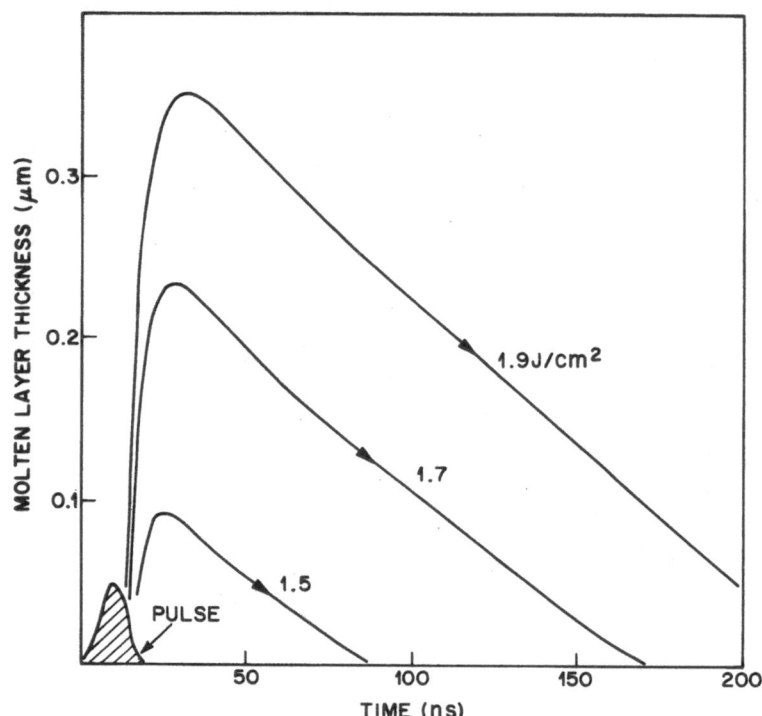

Figure 7   Kinetics of melt front in Si single crystals irradiated with 10 nsec ruby laser pulses of different energy
densities (Baeri and Campisano, 1982).

average thermal gradients are 6×10^7 K/cm and 1.5×10^6 K/cm in the liquid and in the
solid region respectively. During freezing, the temperature is almost constant in the molten
layer, while in the solid the gradient amounts to 5×10^6 K/cm. In Eq. 8 the contribution
due to the term $\kappa_\ell \left. \dfrac{\partial T}{\partial z} \right]_\ell$ can be neglected. The results are similar for both wavelengths.

In the case of $\alpha^{-1} \ll \sqrt{2D\tau_m}$ where τ_m is the melt duration, the temperature gradient
can be estimated by $T_m/\sqrt{2D\tau_m}$. The duration of the liquid phase is nearly equal to the
pulse duration for energy densities just above the annealing threshold. With increasing
energy density the liquid duration increases and the regrowth velocity decreases. In Si, the
solid-liquid interface velocity amounts to few meters per second. In the other case
$\alpha^{-1} > \sqrt{2D\tau_m}$, the regrowth is independent of the pulse duration because the temperature
gradient is determined by the absorption, i.e., $\partial T/\partial z \sim T_m \, \alpha/e$.

Detailed calculations are shown in Figure 9. The average interface liquid-solid velocity
for a 1000Å thick molten layer is plotted vs. pulse duration in the 0.5-100 nsec range and for
three different absorption coefficients which correspond to the absorption of ultraviolet light in
Si single crystal, of ruby light in 1000Å thick amorphous layer on Si and of ruby light in Si
single crystal. The velocity follows the $\tau_p^{-1/2}$ dependence in the case of ruby laser irradiation
of amorphous Si for pulse duration above 10 nsec. At shorter pulses the velocity saturates.
The light absorption length becomes comparable with the heat diffusion length. For

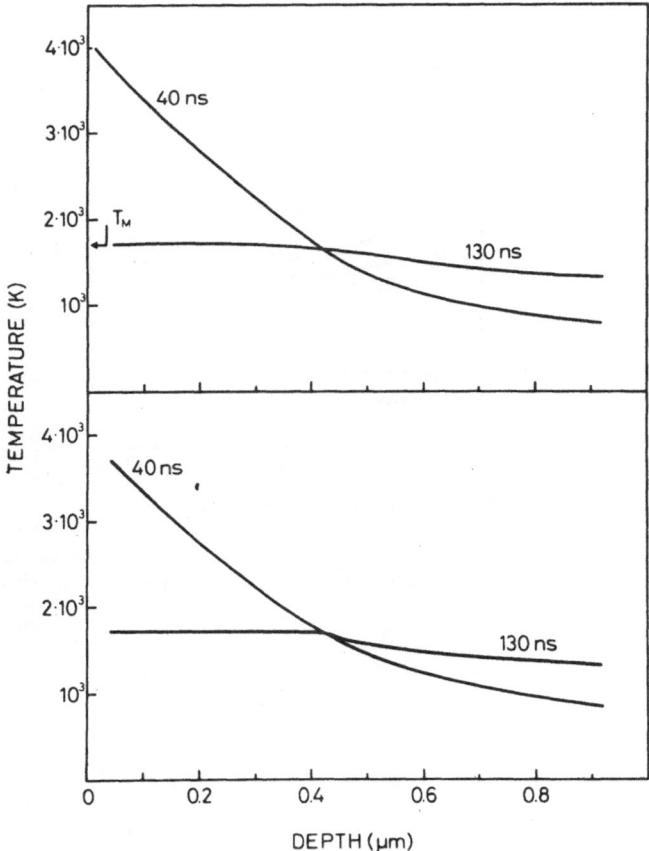

Figure 8   Temperature distribution in a 4000Å thick amorphous Si layer on Si single crystal irradiated with
30 nsec Nd:YAG 2.0 J/cm² (upper part) and with 30 nsec ruby laser 1.9 J/cm² (lower part). The two
distributions are evaluated at 40 and 130 nsec after laser irradiation.

$\alpha = 5 \times 10^4$ cm^{-1} the velocity saturates at 20 m/sec. In the case of a large absorption
coefficient (UV irradiation) the $(\tau_p)^{-1/2}$ dependence is followed up to 1 nsec and the
maximum velocity is above 20 m/sec. The velocity is independent of the pulse duration for
the case of a very low absorption coefficient (lower curve of Figure 9). A saturation value of
2 m/sec is reached for pulses shorter than 100 nsec. These velocities, calculated, on the basis
of Eq. 8, do not include any undercooling at the solid-liquid interface. Deviations from the
melting temperature at the interface should not be relevant for the overall treatment, they
could, however, affect velocity values above 10 m/sec. This point will be detailed in
Chapter 3.

Heat flow calculations for the case of an amorphous layer on single crystal substrates
should also account for the different thermodynamic parameters of amorphous and single
crystal structures. The free energy of the amorphous material is higher than that of the
corresponding crystalline material (see Chapter 3). Calculations for Ge and Si have shown

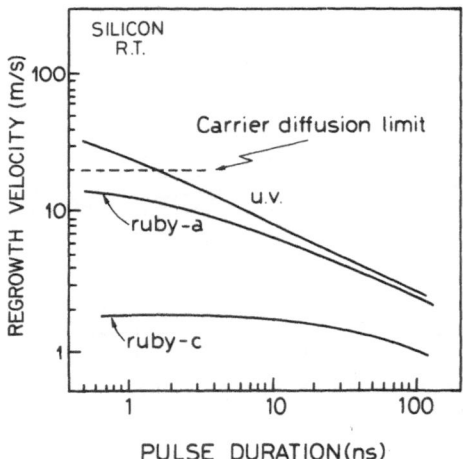

Figure 9   Solid-liquid interface average velocity for a 1000Å molten Si layer vs. pulse duration. The three curves refer to absorption coefficients of ultraviolet, ruby light in amorphous and crystalline silicon respectively (Baeri and Campisano, 1982).

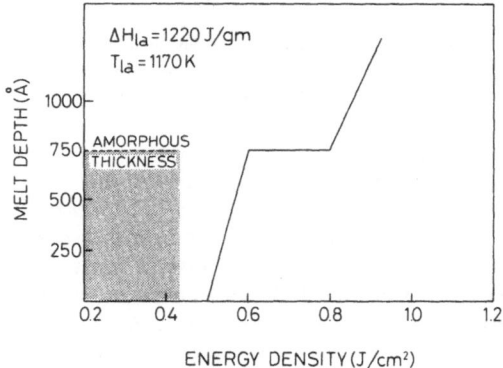

Figure 10   Calculation of the melt front history of 750Å amorphous Si following ruby laser irradiation of 30 nsec pulse length.

that the free energy of the amorphous phase can be 20-30% higher than the crystalline phase (Spaepen and Turnbull, 1979; Bagley and Chen, 1979) and the melting point for the amorphous layer $T_{\ell a}$ can be substantially lower than the crystalline melting temperature T_m. These temperature arguments are based on the premise that the phase transition from the tetrahedrally coordinated amorphous solid to the dense metallic liquid is first order. Experiments using a pulsed electron beam indicated that $T_{\ell a}$ could be as low as 1170K compared to 1685K for T_m (Baeri et al., 1980).

Because of the reduced melting temperatures and latent heat it is possible to melt the amorphous layer but not the underlying crystalline substrate with a suitable choice of laser energy density. The melt thickness vs. energy density is shown in Figure 10. Computations were made assuming for the latent heat $\Delta H_m - 1790$ J/g, $\Delta H_{\ell a} - 1220$ J/g and $T_{\ell a} - 1170$K and an amorphous thickness of 750Å. The threshold for melting the amorphous layer occurs at 0.53 J/cm² ruby laser 30 nsec duration. All the amorphous layer remains molten over a large energy range (0.6-0.8 J/cm²) before the single crystal substrate melts and epitaxial regrowth ensues.

Irradiation with pulsed energy offers then a unique method to investigate the amorphous to liquid transition and the characteristics of an undercooled liquid. The large temperature gradient present in the liquid during the heating stage causes rapid increase in the undercooled liquid temperature in the near surface region. The normal melting temperature is reached very soon and many effects related to an undercooled liquid are washed out (see Figure 11a). Electron beam irradiation, however, produces a flat temperature distribution inside the molten amorphous layer where the undercooling effects were first seen (Baeri et al., 1980). Gradients similar to those produced by e-beam irradiation can be obtained by irradiation of an ion-implanted semiconductor on the back side of the single crystal (Baeri, 1982). The laser should have a low absorptivity in the single crystal semiconductor. The case illustrated in Figure 11b refers to Nd: glass (1.06 μm) irradiation of a few hundred microns thick Si sample. The laser energy is absorbed mainly by the amorphous layer. The heat sink is on the same side as the heat source and the surface represents a barrier to heat diffusion. In this condition, temperature gradients in the liquid layer are practically non-existent as shown in the calculated temperature profiles. For energy densities between 0.9 and 1.4 J/cm² the amorphous layer is molten but in an undercooled regime. The driving force for crystallization is very high and the melt solidifies from both the single crystal interface and random nucleation centers at the free surface (Baeri, 1982). These competitive processes produce layered structures similar to those first seen in electron heating (Baeri et al., 1980).

Mass transport of impurities, if present in the irradiated layer, occurs during melting. The amount of diffusion depends on the heating and cooling stage and is governed by the diffusion equation

$$J_{\text{imp}} - - D_\ell \frac{\partial n}{\partial z} \qquad (9a)$$

with J flux of impurity atoms, D_ℓ diffusion coefficient in the liquid and n concentration of impurities. Diffusion in the solid phase during the time interval involved in laser irradiation is negligible for the low value of the diffusion coefficient with respect to that in the liquid phase. The temperature dependence of the As diffusion coefficient in Si (Brice, 1963; Fair, 1979) is reported in Figure 12, note the several orders of magnitude by which the diffusion coefficient changes at the melting temperature between the solid and the liquid phase. The broadening of the impurity profile is given to first order by $\sqrt{2D_\ell \tau_m}$. As an example, several As profiles are shown in Figure 13, after ruby laser irradiation of different energy densities.

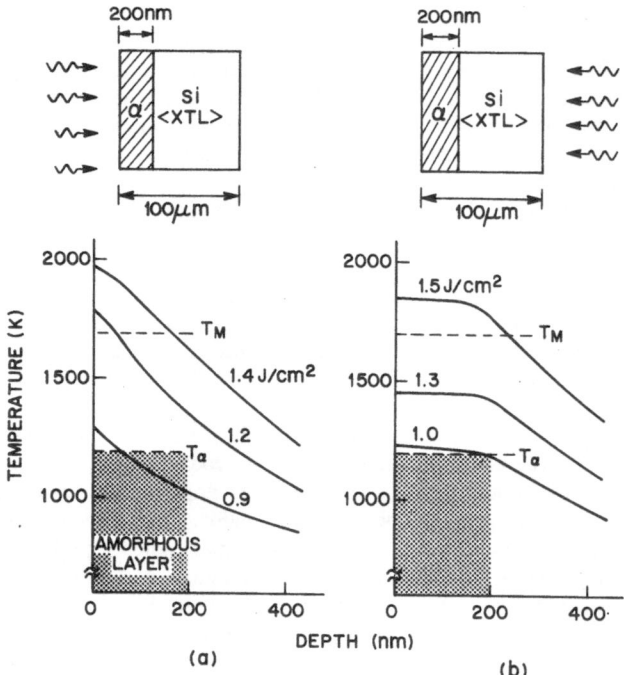

Figure 11 Calculated maximum temperature distributions in Si single crystals covered with 2000Å thick amorphous layer and irradiated with 30 nsec Nd: glass 1.06 μm laser pulses (Baeri, 1982). (a) front irradiation, (b) back irradiation

A diffusion coefficient of 2×10^{-4} cm²/sec fits the experimental data with τ values obtained from the heat flow calculations.

III. INTERACTION OF LASER BEAMS WITH METALS

A. Energy Deposition

In metals the laser is absorbed by the large density of conduction electrons (intraband transition) with a mechanism similar to the free carrier absorption in semiconductors. The excited electrons collide with the lattice atoms and the energy relaxation time is of about 10^{-12} sec. The laser energy is then converted very quickly into atomic motion. The absorption lengths are very small, about 100Å over the whole optical spectrum. The metal reflectivity is very high, of the order of 90% or more, above a certain critical wavelength. Below this value the reflectivity decreases sharply. As an example the frequency dependence of the reflectivity for Au (Wolfe, 1965) and Al is reported in Figure 14. The critical wavelength for Al is below 1 μm, while for Au it is about 0.6 μm. The critical wavelength is connected to the plasma frequency of the electron gas. The reflectivity is high for frequencies lower than the plasma frequency, which represents then a cutoff. From the point of view of laser irradiation, it is more convenient to use laser wavelengths below the critical value so as to increase the coupling efficiency.

Reflection is very sensitive to the status of the surface. Impurities and oxide layers, for example, can change the reflectivity thus producing a considerable effect in the absorbed

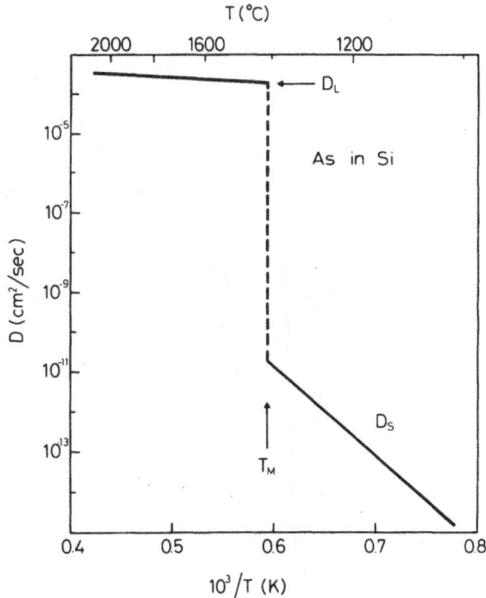

Figure 12   Diffusion coefficient of As in Si as a function of temperature. The discontinuity occurs at the melting point (Brice, 1963; Fair, 1979).

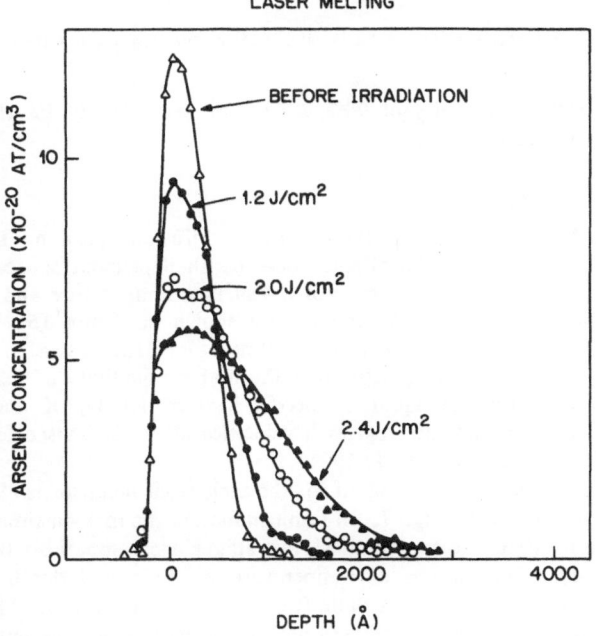

Figure 13   As profile in implanted Si, thermally annealed, and subsequently ruby laser irradiated at different energy density.

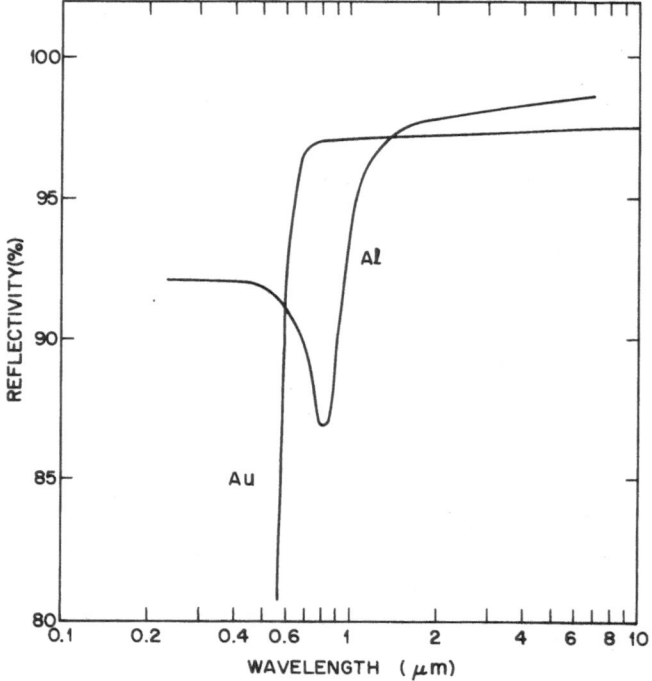

Figure 14   Reflectance of Al and Au (Wolfe, 1965) vs. the wavelength of the incident light.

energy density. If the reflectivity of 90% varies, for example, by 1% the variation in the absorbed energy will amount to 10%.

B. Heating and Cooling

The heat flow can be calculated using the usual heat diffusion equation. The main difference with the semiconductor case is the absence of a depth dependent source term in the heat equation. The calculated time dependence of the surface temperature and the melt depth for various absorbed laser energies in Al crystals are shown in Figure 15. These results were obtained from the heat flow equation to a 30 nsec FWHM Gaussian laser pulse. The calculations assumed temperature-independent thermal conductivity of 2.2 W/cm^{2}K for the solid and 1.1 W/cm^{2}K for the liquid, a specific heat of 1.1 J/g K and a latent heat of 395 J/g. The calculations indicate regrowth velocities above 20 m/sec for absorbed energy lower than 0.6 J/cm^2 (Wampler *et al.*, 1981).

The main difference between Si and Al for the melt front behavior is clearly shown by the comparison of Figure 15b with Fig. 7. In semiconductors the melt-in time is short compared with the solidification time. In metals these two stages are comparable because of the high thermal conductivity. Calculations of temperature vs. time and depth provide a further insight into the thermal history (Donà dalle Rose and Miotello, 1980). The case reported in Figure 16 refers to an Al sample irradiated with a 15 nsec ruby laser pulse for an absorbed energy of 0.65 J/cm^2. The contours of constant value in the plane (z,t) give the overall dynamic behavior of the temperature $T(z,t)$, of the heating-cooling rate $(\partial T/\partial t)$ and of the

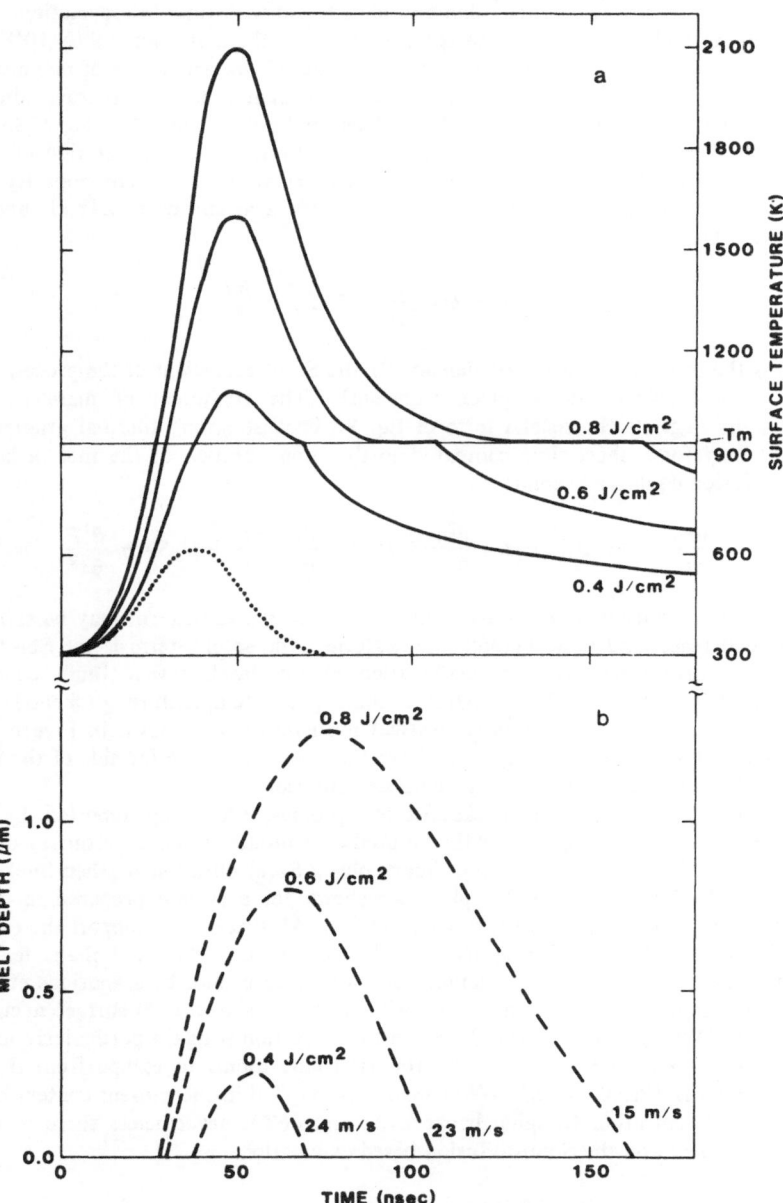

Figure 15   Surface temperature of Al as a function of time for various absorbed energies (J/cm²). The dotted line
shows the shape of the Gaussian laser pulse (a). Melt depth as a function of time with velocity of the
liquid-solid interface (Wampler *et al.*, 1981).

thermal gradient $(\partial T/\partial z)$. The isothermal curve $T - T_M$ is referred as the melt front history and has been redrawn in the other two plots to aid with their interpretation.

High heating and cooling rates and extreme negative thermal gradients (-10^7K/cm) are reached in the heating stage. The values of $\partial T/\partial t$ and $\partial T/\partial z$ are orders of magnitude lower in the cooling stage. These transients can be investigated by the thermal diffusion of implanted impurities during the liquid phase (Donà dalle Rose and Miotello, 1980). In the following the thermal transient will be characterized by the Δt time duration of the liquid stage ($T > T_M$) and by the corresponding average thermal gradient. The impurity flux for a dilute solution expressed in terms of the impurity concentration $n(z,t)$ and sample temperature $T(z,t)$ is

$$J_{\text{imp}} = - D_\ell \, \frac{\partial n}{\partial z} - D_\ell \, S_T \, n \, \frac{\partial T}{\partial z} \tag{9b}$$

where D_ℓ is the ordinary diffusion coefficient, S_T the Soret coefficient of the process (both are supposed to be constant for a given transient). The treatment of mass transport in semiconductors neglects the second term of Eq. 9b because strong thermal gradients in the liquid exist only for a short time compared to the time duration of the molten layer. The impurity diffusion equation becomes

$$\frac{\partial n}{\partial t} = -\text{div} \, J_{\text{imp}} = D_\ell \, \frac{\partial^2 n}{\partial z^2} + D_\ell \, S_T \, \frac{\partial n}{\partial z} \, \frac{\partial T}{\partial z} + D_\ell S_T \, n \, \frac{\partial^2 T}{\partial z^2} \tag{10}$$

In the r.h.s. the first term is an ordinary diffusion term, the third term may be shown not to affect the final result and may therefore be neglected; the second term is the Soret diffusion term and its form shows that a combination of ion implantation (high concentration gradients) and laser pulse melting (extreme liquid phase temperature gradients) allows an enhancement of this effect. The basic physical mechanisms are shown in Figure 17 where the depletion of the near surface region and the thermal drift of the far side of the implanted peak are explained in terms of ordinary and Soret diffusion.

In Figure 18a,b, Rutherford backscattering profiles, previously reported by different groups have been redrawn together with the predicted profiles when an ordinary diffusion of implanted impurities (solid line) or an ordinary plus a Soret diffusion (dashed line) are taken into account. In Figure 18c similar plots are shown for a system prepared in a different manner, i.e., laser mixing a Ni-Au layered structure. All three cases support the existence of Soret diffusion. In the Al(Cu) system (Della Mea *et al.*, 1980) the main feature thus explained is the near surface Cu depletion; such a feature cannot be a spurious effect, since, after the same experimental procedure, the Al(Pb) system shows a Pb surface accumulation. In the Al(Zn) (Wampler *et al.*, 1981) system this depletion is again particularly meaningful since a barrier is known to exist which prevents solute atoms to escape from the surface. Finally the Ni(Au) (Pronko *et al.*, 1981) system is atypical in the present context because of the high Au concentration (ranging between 22 and 66%): nevertheless there is agreement between experiment and the Soret-diffusion based calculation.

C. Heating and Cooling in a Multilayer System

In some cases irradiation experiments are performed on a structure formed by two or more different composition layers to produce alloys and compounds. The fast melting and solidification process obtained by laser irradiation in the Q-switched mode allows the formation of non-conventional alloys or compounds. In the semiconductor field the motivation is the formation of silicides or contacts in GaAs specimens; in metallurgy to improve the

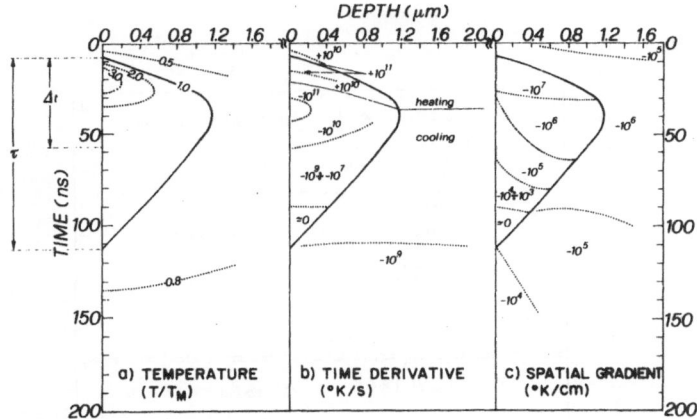

Figure 16   Depth-time plots for the (a) temperature T, (b) time derivative (c) spatial gradient in a laser heated Al sample. Absorbed energy 0.65 J/cm² and 15 nsec pulse duration of ruby pulse. Each isocontour is labelled by the corresponding value of T/T_m in (a), while in (b) and (c) the orders of magnitude are shown (Donà dalle Rose and Miotello, 1980).

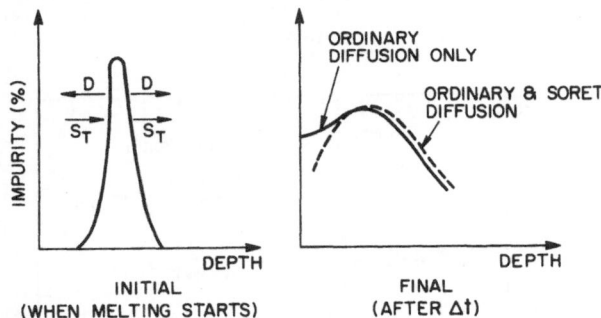

Figure 17   Schematic of the driving processes acting on an implanted impurity profile during laser induced melting of the host metal. D, ordinary diffusion coefficient, S_T, Soret diffusion coefficient.

hardness, the wear resistance, the corrosion resistance of the materials, etc. (Draper, 1981) (see Chapter 13 for a detailed discussion of laser alloying).

From the point of view of the present discussion in a layered structure the reflectance of the outer layer determines the amount of the absorbed energy, the thermal diffusivity of both layers controls the thermal diffusion length and the melt depth. Solidification and interface solid liquid velocity are determined by the release of the latent heat. Controllable melting and solidification can be achieved by selecting materials with melting points that are not too far apart. High vapor pressure elements should be avoided where possible, to reduce the amount of surface vaporization. The use of Q-switched laser pulse in the 10-100 nsec duration implies composition changes over distances less than 1 μm. greater irradiation times are required to increase the depths. The laser wavelength should be chosen in such a way to decrease the reflectance and to increase the amount of absorbed energy.

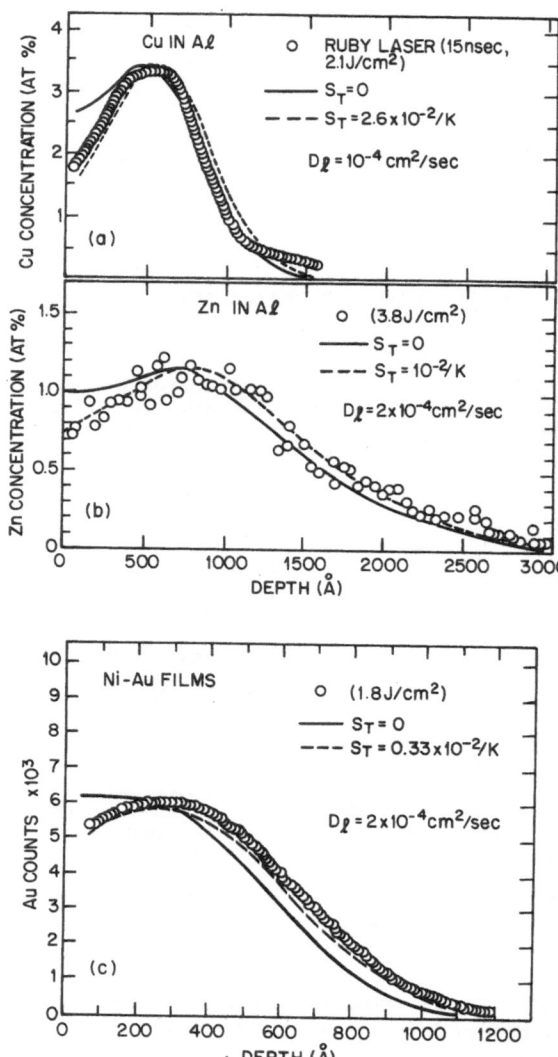

Figure 18   Experimental evidence for Soret diffusion in (a) the Al (Cu) system (Della Mea *et al.*, 1980), (b) the Al (Zn) system (Wampler *et al.*, 1981) and (c) the Ni (Au) system (Pronko *et al.*, 1981).

Interesting effects arise from changes of the thermal diffusivity of the two elements. In an investigation of the effects caused by Nd:YAG laser irradiation on Ni-Au-Ni (110) systems with surface Ni layers 300Å thick and Au layers up to 5000Å thick, it has been found that even though they were irradiated with the same incident power densities, the melt depth increases significantly with the thickness of the Au layer (Draper et $al.$, 1981). The thermal diffusivity of Au is a factor 5 greater than that of Ni. The result is in agreement with the apparent diffusivity change due to the variation of the elemental composition over the thermal diffusion length of the 100 nsec pulse. Another example shown in Figure 19 refers to Au(1600Å)-$Pd_x Ag_{1-x}$(12,000Å)-Ni structures irradiated with a frequency doubled Nd:YAG. The backscattered spectra indicate that the depth of alloying increases with the Ag concentration for the same energy density, or equivalently the melt depth is higher for the higher Ag concentration. The high thermal conductivity of Ag with respect to Pd causes this effect.

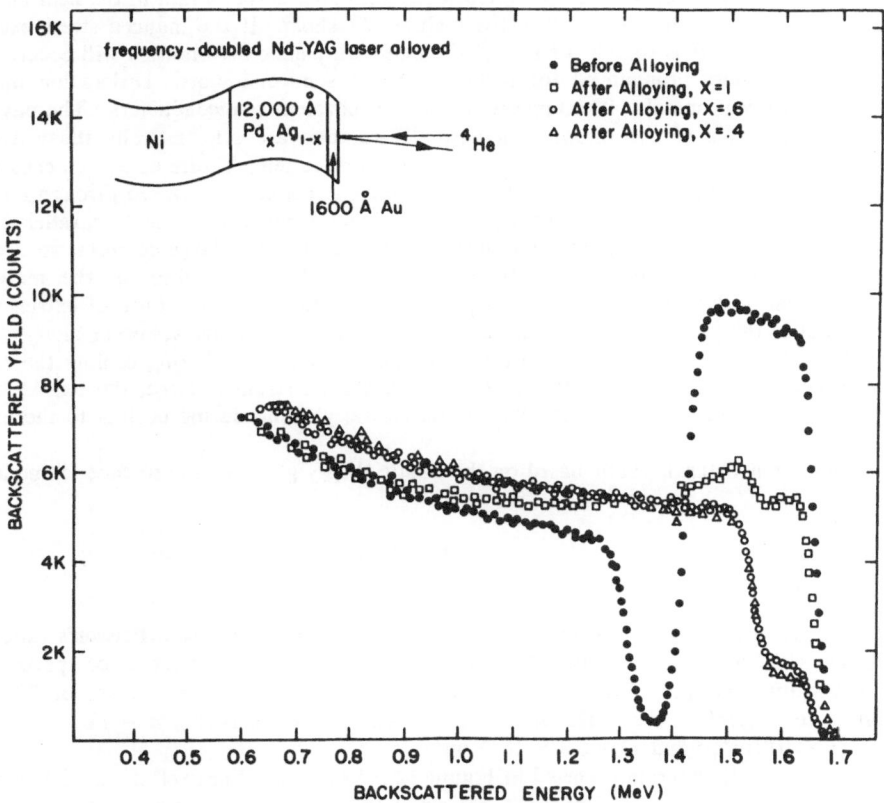

Figure 19   Backscattered yields of Au(1600Å)-$Pd_x Ag_{1-x}$(12,000Å)-Ni multilayer systems alloyed by a frequency-doubled Nd:YAG laser for different concentrations of Ag and Pd (Draper et $al.$, 1981).

The thermal behavior clearly depends also on the change of the thermal parameters such as enthalpy, melting temperature and diffusivity on the composition and structure. As an example Figure 20 shows calculation (von Allmen and Lau, 1982) of the melt front position in a 5000Å thick Si-Pt mixture layer on a sapphire substrate irradiated with a rectangular pulse of 25 nsec duration and 100 MW/cm^2 of absorbed power density with $\alpha = 10^5$ cm^{-1}. Three different cases are considered: i) no reaction occurs between the two elements in the liquid phase; the melting point (1000K) and melting enthalpy (3000 J/g) are equal to the freezing point and freezing enthalpy respectively; ii) a compound is formed with a lower freezing point (800K) and a higher freezing enthalpy (4500 J/g); iii) glass formation with a range of freezing points (800-1000K) and a lower freezing enthalpy (2000 J/g) than the no-reaction phase. In the calculations, the Si-Pt mixture is assumed to have a conductivity of 0.7 W/cm K and the sapphire substrate a conductivity of 0.24 W/cm K. The velocity of the liquid solid interface during solidification is higher in the case of glass formation and is lower in the case of compound formation.

D. Stress Deformation Induced by Laser Irradiation

The large thermal gradients in an irradiated metal introduce severe strain in the near surface region at irradiation values well below the melting threshold. If the induced stress exceeds the elastic limit and thus the yield stress of the material, plastic deformation will occur. The phenomenon is more pronounced for metals than for semiconductors. Dislocation motion occurs very easily in metals. The opposite is true for covalent semiconductors. The physical processes responsible of the thermal, induced deformation are schematically illustrated in Figure 21 (Musal, 1980). The rapid increase of the surface temperature causes an expansion of the material. In the direction normal to the surface, stress can be relaxed through a small outward displacement of the free surface. On the other hand displacements parallel to the surface are inhibited by the surrounding material. During heating, large compressive stresses are induced along these directions. If they overcome the elastic limit of the material, permanent plastic deformation is produced. The deformation takes the form of extrusion of material out of the plane of the free surface. The maximum shear stresses occur along planes inclined at 45° to the free surface, so that slip lines are formed. During cooling the metal will contract elastically at first. If the compressive plastic strain is large, the tensile stress may exceed the yield stress thus causing tensile plastic yielding during cooling to the initial temperature.

A calculation (Musal, 1980) based on the elastic theory gives for the surface temperature rise (ΔT)

$$\Delta T = \frac{(1-\nu)}{E\ a}\ Y \qquad\qquad (11)$$

where E, a, ν, Y are Young's modulus, the coefficient of thermal expansion, Poisson's ratio and the tensile yield stress of the material respectively. (This ΔT value must be compared with the temperature rise given, for example, by Eq. 4a.) For copper, an increase of 20°C is enough to reach the plastic limit, where $\nu = 0.345$, $a = 1.67 \times 10^{-5}$/K, $E = 1.23 \times 10^{11}$N/m^2 and $Y = 6.2 \times 10^7$N/m^2.

The stress-strain diagram reported in Figure 22 is based on a linear-elastic model (Musal, 1980). In the linear elastic regime AOA' the stress increases linearly with the strain, and the behavior is perfectly reversible within A A'. If the applied strain is larger than the yield strain (OAB), the stress increases to the yield stress and then remains constant during plastic contraction from A to B. When the material returns to its original dimensions (zero strain) the material follows the path BC (parallel to AOA') and is left with a residual stress OC.

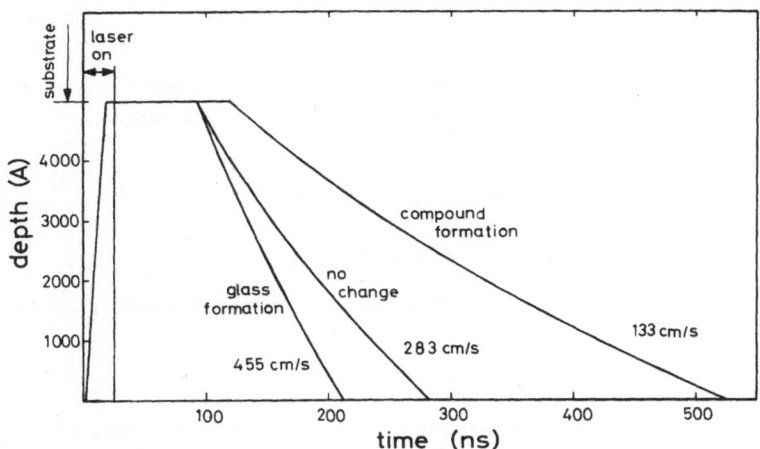

Figure 20   Melt front position vs. time for a 5000Å thick Si-Pt mixture layer on sapphire after laser irradiation. The profiles refer to compound formation, no reaction and glass formation (von Allmen and Lau, 1982).

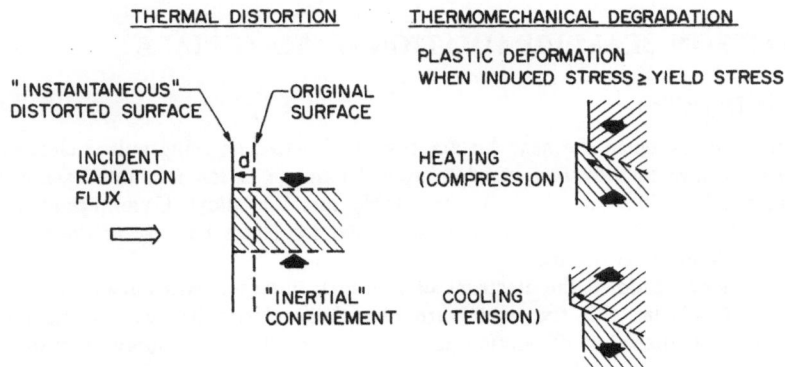

Figure 21   Metal surface thermal distortion (Musal, 1980).

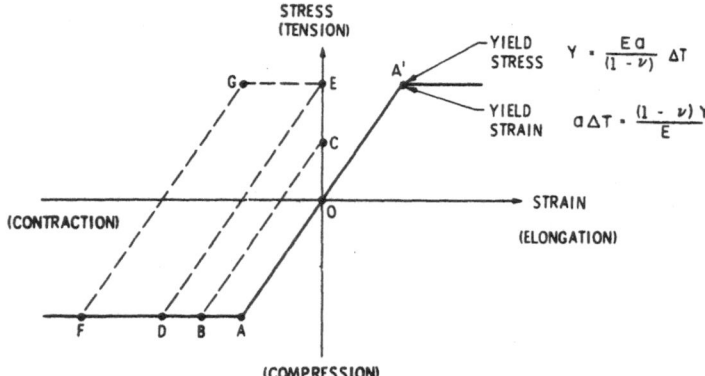

Figure 22 Linear-elastic/perfectly plastic stress-strain diagram for metal surface under pulsed laser irradiation (Musal, 1980).

When a metal is subjected to alternating strains in excess of the yield strain, the plastic deformation tends to occur along closely spaced slip planes that form slip bands. This behavior has been recently observed in laser irradiated Al single crystals at energy densities well below the threshold for surface melting (Follstaedt et al., 1981). The fast propagation of stress inside the sample is accompanied by the propagation of extended defects and is probably responsible of the poor crystallinity obtained after laser irradiation of ion implanted metals.

IV. ELECTRON BEAM IRRADIATION OF MATERIALS

A. Energy Deposition

Energy can be deposited in the near surface region of materials using pulsed electron beams. Pulsed e-beams have already been used to anneal damage in ion implanted semiconductors (Greenwald et al., 1979) and to form metastable metallic alloys (Wampler et al., 1980). The number of experiments is much lower than that for lasers because of the difficulties in obtaining suitable electron beams.

During the slowing down the electrons of the beam interact with nuclei and electrons of the medium. Collision with the nuclei are essentially elastic because of the large mass difference and the direction of motion is changed drastically. Energy is transferred by collisions with the electrons of the medium. As with laser irradiation, the energy from these electrons is transferred nearly instantaneously to the lattice as heat. In the short time of irradiation the heating occurs quasi-adiabatically. The effect of thermal conductivity is negligible and the temperature profile follows the energy loss profile of the electrons in the specimen. The energy dissipation in depth governs the temperature evolution in the heating stage.

The energy dissipation mechanisms were investigated to understand the results obtained with the scanning electron microscope and electron probe microanalyzer for the production of characteristic x-rays (Bishop, 1974). Both Monte Carlo and analytical approaches have been used to calculate the energy deposition. The maximum of the energy occurs at a depth X_p which increases with electron energy and the width of the distribution σ also increases with energy. With laser irradiation, the maximum energy deposition occurs at the surface for an homogeneous medium and absorption and reflection can be very sensitive to the structure of the surface layer. In e-beam irradiation, the energy deposition depends only on incident energy (E) and on the atomic number (Z) of the target. As a first order description the energy deposition profile can be approximated (Spencer, 1959) by a gaussian distribution $\exp - \left[\dfrac{(X-X_p)^2}{2\sigma^2} \right]$ where X_p and σ can be obtained by the following relationships which are valid for the range $Z = 6$ to $Z = 50$ i.e., from C to Sn targets:

$$X_p = (0.143Z + 0.622)r_0 \tag{12}$$

$$\text{and} \quad \sigma = (- 0.0538 \ln Z + 0.374)r_0 \tag{13}$$

The range (r_0) can be obtained from the following expression (Feldman, 1960)

$$r_0 = A \, E^a \tag{14}$$

where $A = 3.92 \times 10^{-6} + 1.562 \times 10^{-7}Z$ and $a = 1.777 - 2.165 \times 10^{-3}Z$. The units of r_0 are g/cm^2 and keV for E. For example, the peak position corresponding to the maximum of energy deposition and the gaussian spread for a 20 keV e-beam in Al target are both 1 μm. The results obtained by means of Monte Carlo procedure are compared with the gaussian approximation in Figure 23 for 20 keV electrons in Al and in Si. The maximum height of the gaussian distribution has been normalized to the Monte Carlo result. The two distributions coincide nicely thus justifying the previous considerations. For a correct application of this relation the amount of energy removed by the back-scattered electrons must be taken into account. The amount of backscattered energy loss depends on the atomic number of the target atoms, on the beam energy and on the impinging angle of the electron beam. The Z dependence is reported in Figure 24 for normal incidence of a 30 keV e-beam (Sternglass, 1954; von Allmen, 1982). For a Si target the fraction of energy loss amounts to 8%. In comparison with laser irradiation, this backscattered energy loss has the same effect as the reflection coefficient. In addition to the Gaussian approximation, polynomial series are also used to describe analytically the energy deposition in Si, Ge and GaAs (Barbier et al., 1981).

So far we have discussed the energy deposition for normal incidence. By varying the electron incidence angle the energy deposition is changed. As an example several energy loss distributions obtained by Monte Carlo calculations are shown in Figure 25 for two values of the electron incidence angle in a Si target (Merli and Rosa, 1981). With decreasing incidence angle, from the 90° value at normal incidence, the peak distribution shifts towards the surface and at 30° the maximum occurs at the sample surface. Not only the energy deposition changes with the incidence angle but also the amount of loss due to backscattered electrons. In a Si target the backscattered energy loss reaches 47% at an angle of 15°, i.e., about half of the energy carried by the electron beam is lost. This value should be compared with the 8% at normal incidence. Both these facts should be considered when, as in most of the irradiation experiments, the beam energy is not monoenergetic and the incidence angle is not normal.

B. Heating and Cooling

The energy deposited by the electron beam is converted instantaneously into heat and the temperature distribution follows that of the energy deposition. For incidence lower than 30°

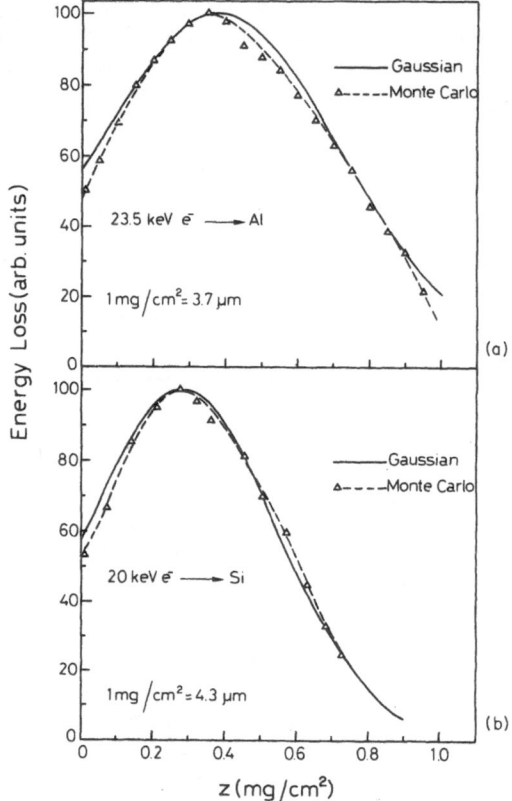

Figure 23   Energy deposition in (a) Al and (b) Si for different e-beam energy irradiations, calculated by
Monte Carlo procedure. Dashed lines represent Gaussian fits (Merli and Rosa, 1981).

in Si, the maximum temperature occurs at the surface, by further decreasing the incidence
angle there is an increase of the thermal gradient on the surface layer. Knowing the energy
deposition profile, it is possible to compute the heating and cooling using the same procedure
discussed for laser irradiation. For example Merli and Rosa (1981) have calculated the
thermal history for Si irradiated with a pulsed electron beam (20 keV, 2.4 J/cm², 120 nsec).
The surface temperature increases during irradiation and melts after 120 nsec and remains
liquid up to 400 nsec. Melting starts beneath the surface at a depth of about 1.2 μm,
coinciding with the maximum energy loss distribution the surface melts after the end of the
pulse and the thickness of the liquid layer reaches 2.5 μm. Cooling down is slow and after
700 nsec a layer 1.7 μm thick is still liquid.

   The main differences with laser irradiation are then related to the depth at which the
maximum temperature is reached and to the minimum molten layer. For e-beam irradiation
the liquid layer is at least a few microns thick. This fact has some consequences in the
cooling stage for the solid-liquid interface velocity. To show these differences, calculations of
temperature, thermal gradient, and time derivative are reported in Figure 26 for an Al target
irradiated with a 12 keV monoenergetic e-beam with triangular shape and 15 nsec duration

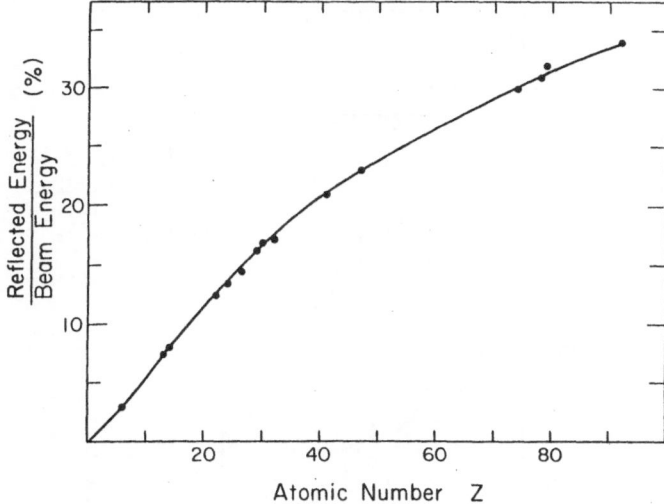

Figure 24   Reflected electron energy (in percent) vs. the target atomic number, at normal incidence of 30 keV e-beam (von Allmen, 1982).

(Donà dalle Rose and Miotello, 1980). These data should be compared with those reported in Figure 16. The absorbed energy density of 0.65 J/cm^2 is chosen in such a way that the maximum surface temperature corresponds closely to the laser irradiated case (Figure 16a). The cooling stage is roughly the same as with a laser source while the heating stage is quite different. The energy is deposited over larger depths than for the laser beams and a thickness about one half the electron range is melted almost instantaneously. The temperature of the liquid is lower than that reached in laser irradiation so that the thermal gradients are smaller. Near the surface the gradients are initially positive and change sign later. Only laser heating allows large temperature gradients for extended periods of time. As a consequence, for example, the Soret effect is quite negligible during e-beam irradiation.

The cooling history depends on the thickness of the melted layer and on the thermal properties of the sample. As an example we consider the irradiation of Al and Si targets. The energy deposition is nearly the same in these two materials, but the melted layer thicknesses are different. For a pulse of 1.5 J/cm^2, 50 nsec and 20 keV, energy calculations predict a melting to a depth of approximately 1.7 μm for Si and 2.6 μm for Al. The average liquid-solid interface velocity is 1.7 m/sec for Si and 8.3 m/sec for Al. This large difference is due to the higher thermal conductivity of Al and the larger heat of fusion for Si. These results are illustrated in Figure 27 (Picraux et al., 1981).

V. ION BEAM IRRADIATION OF MATERIALS

A. Energy Deposition

Monoenergetic pulsed particle beams can also be used to deposit energy into the surface layer of the target in a short time. For pulse durations and energy densities it is possible to melt the surface layer (Hodgson et al., 1981). So far few measurements have been carried out

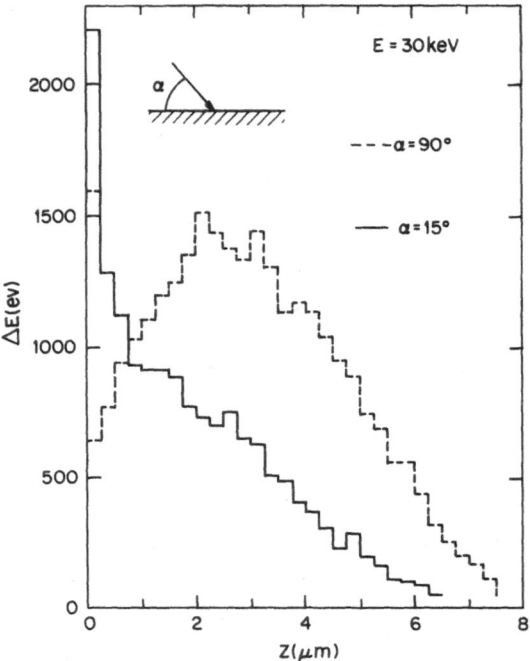

Figure 25   Energy loss distribution in Si for different values of electron incidence angle (Merli and Rosa, 1981).

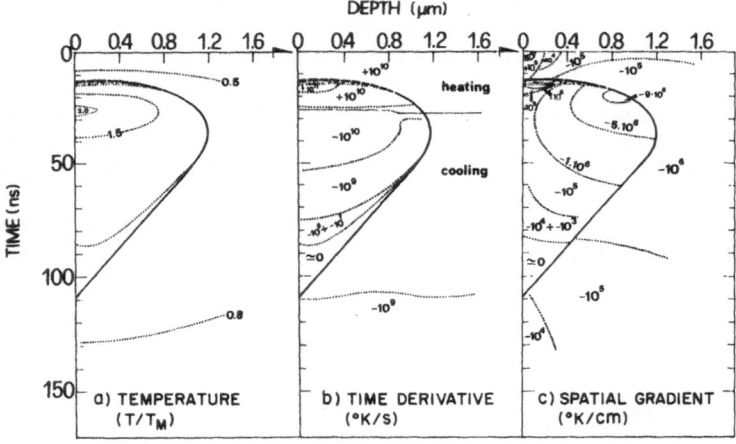

Figure 26   Depth-time plots for the (a) temperature T, (b) time derivative and (c) spatial gradient in Al under e-beam irradiation 0.65 J/cm^2, τ_p = 15 ns, electron energy 12 keV (Donà dalle Rose and Miotello, 1980).

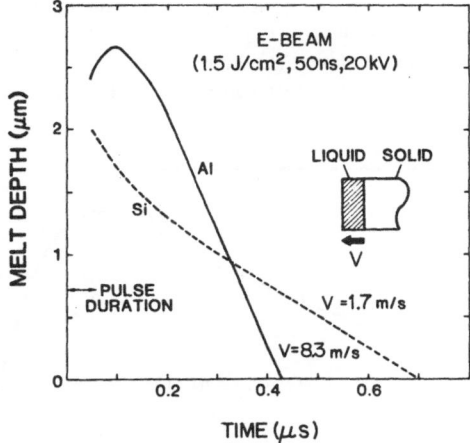

Figure 27   Melt depth history for Si and Al irradiated with e-beam (Picraux *et al.*, 1981).

because of the difficulty in obtaining suitable pulsed ion beams. Ions lose energy to both electrons and nuclei. The relative predominance of two energy loss mechanisms depends on the ion velocity and on the atomic number and mass of the projectile and target atoms. For protons, the two energy loss mechanisms have the same magnitude at an energy of 10 keV. For arsenic projectiles in silicon, the two stopping powers are equal at 750 keV. Stopping powers of any elemental projectile in any atomic target are available so that the depth-dose functions are already known with quite good accuracy. For the case of protons in silicon the following relations are used (Andersen and Ziegler, 1977)

$$-\frac{dE}{dx} = 4.15\ E^{1/2}\ eV/10^{15}\ \text{at/cm}^2, \text{for } E < 10\ \text{keV} \tag{15a}$$

$$-\frac{dE}{dx} = \left[(4.7\ E^{0.45})^{-1} + \{(\frac{3329}{E})\ln \Lambda + \frac{550}{E} + 0.0132\ E\}^{-1} \right]^{-1},$$

$$\text{for } E < 1\ \text{MeV} \tag{15b}$$

The deposited energy for a proton beam of 300 keV in Si is reported in Figure 28. The energy deposition is fairly uniform from the specimen surface down to a depth approaching the particle's range, of about 2.5 μm.

B. Heating and Cooling

Heating and cooling processes are obtained following the usual procedure. It can be anticipated that due to the uniform distribution of deposited energy, thermal gradients during the heating and the cooling stage are lower than or comparable with those of e-beam irradiation. As an example temperature vs. depth curves are shown in Figure 29 for the proton irradiation (300 keV, 300 nsec and 2.1 J/cm²) of crystalline Si. The increase of temperature with time is uniform over a depth of about 2 μm and the layer becomes liquid at about 120 nsec. For greater times, temperatures above the melting point occur in a small region at the maximum stopping power depth. The velocity of the solid-liquid interface during solidification of the molten layer is about 2 m/sec. Because of the flat temperature profile within the sample, the velocity is just controlled by the heat flow in the bulk.

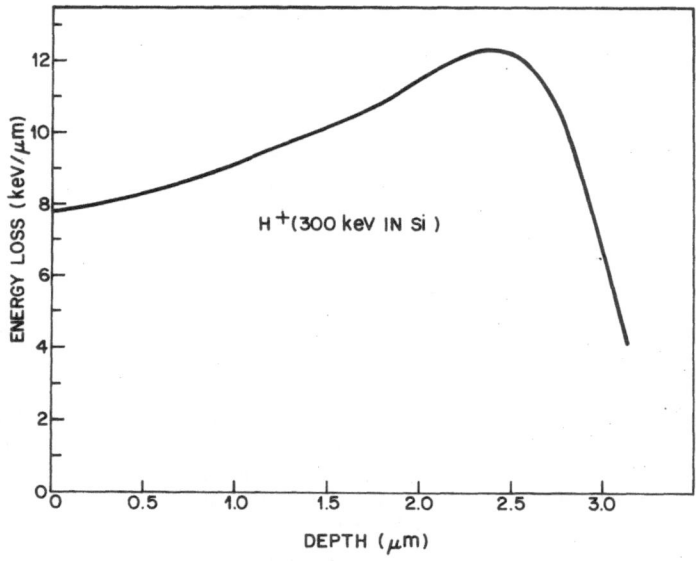

Figure 28   Energy deposition for 300 keV protons in Si.

VI. CONCLUSION

A large body of experiments indicates that during irradiation of semiconductors with nanosecond or picosecond laser pulses the energy in the excited electron-hole pairs is shared with the lattice atoms, so that the two systems are in thermal equilibrium. For suitable laser energy densities, the surface layer of the semiconductor melts; there is, however, still some disagreement between measurements and calculations of the duration of this phase. In this respect the determination of the optical absorption and reflection coefficients of a thin liquid layer will be important. Another quantity of interest is the velocity of the liquid-solid interface during melting and solidification. Recent measurements of the interface velocity in single crystal Si agree with calculations to within 10%. These experiments justify the assumptions used in the theoretical treatments. It will be interesting to extend this method to ion-implanted silicon and to other semiconductors. The velocity of the interface can be changed by varying laser parameters or the target thermal parameters. In this fashion, much work has been carried out in ion implanted semiconductors on the dependence of segregation coefficients and cell formation on the liquid-solid velocity.

In metals, the laser light is absorbed at the surface so that high thermal gradients exist for a long time during the heating stage; moreover this stage lasts for times comparable to the cooling stage, in contrast to the semiconductor case. The coupling of large thermal and concentration gradients during the heating stage causes a mixed term in the diffusion of impurities (Soret effect). Laser irradiation offers a unique way, in combination with ion implantation, of studying these phenomena.

Irradiation of materials with pulsed electron and ion beams has also been used to reorder ion-implanted semiconductors and metals. The energy deposition processes are quite well known and depend mainly on the atomic number and mass of the target atoms. The main difference with laser irradiation is the large depth over which the energy is deposited and the lower values (typically one order of magnitude) of thermal gradients and solid-liquid velocities.

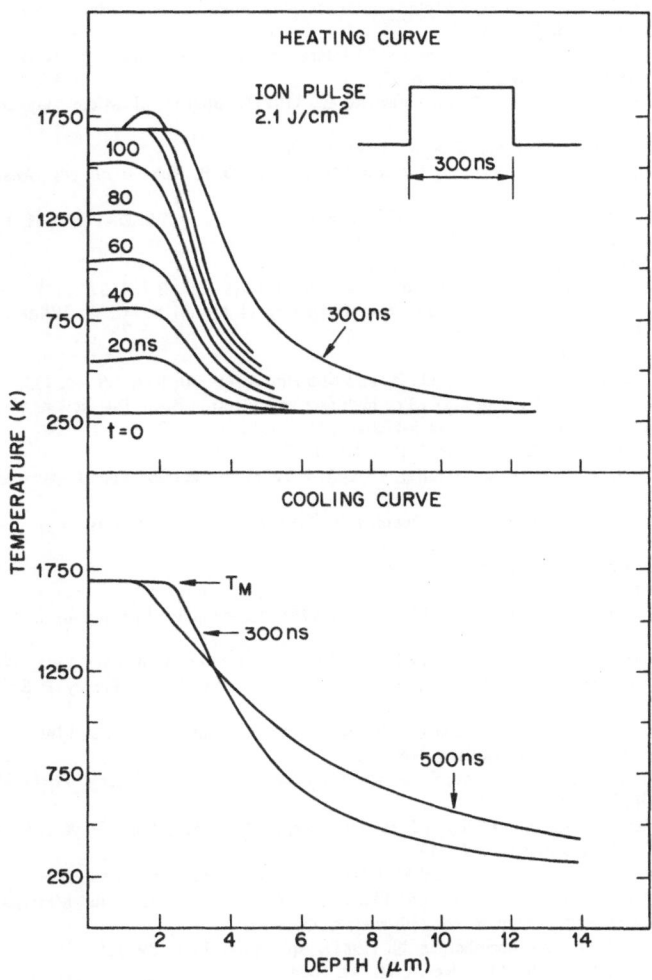

Figure 29   Heating and cooling curves for Si single crystal irradiated with 300 keV proton beam of 2.1 J/cm² and pulse duration 300 nsec.

REFERENCES

von Allmen, M. F. (1982) in "Laser Annealing of Semiconductors," J. M. Poate and J. W. Mayer, eds. Academic Press, N.Y., Chapter 3.

von Allmen, M. F. and S. S. Lau, ibid, Chapter 12.

Andersen, H. H., and Ziegler, J. F., (1977), "Hydrogen Stopping and Ranges in all Elements," Pergamon Press, N.Y.

Auston, D. H., Surko, C. M. Venkatesan, T. N. C., Slusher, R. E. and Golovchenko, J. A., (1978), Appl. Phys. Lett. *33*, 437.

Auston, D. H., Surko, C. M., Venkatesan, T. N. C., Slusher, R. E. and Golovchenko, J. A., (1979), Appl. Phys. Lett. *35*, 635.

Baeri, P. (1982), "Computer Modelling of Laser Annealing," in "Laser and Electron Beam Interaction with Solids," B. R. Appleton and G. K. Celler, eds., Elsevier North-Holland, p. 52.

Baeri, P., Foti, G., Poate, J. M. and Cullis, A. G. (1980), Phys. Rev. Lett. *45*, 2036.

Baeri, P. and Campisano, S. U. (1982), Chapter 4 in "Laser Annealing of Semiconductors," J. M. Poate and J. W. Mayer, eds., Academic Press, N.Y.

Bagley, B. G. and Chen, H. S. (1979), in "Laser-solid Interaction and Laser Processing," S. D. Ferris, H. J. Leamy and J. M. Poate, eds., A.I.P., No. 50, p. 97.

Barbier, D., Baghdadi, M., Langier, A. and Vicario, E., (1981), "Journal of Microscopy and Spectroscopy of Electrons," *6*, 513.

Birnbaum, M. and Stocker, T. L. (1968), J. Appl. Phys. *39*, 6032.

Bishop, H. E. (1974), "Quantitative Scanning Electron Microscopy," D. B. Holts *et al.*, eds., Academic Press N.Y. - Chapter 2.

Bloembergen, N. (1979), in "Laser-solid Interactions and Laser Processing," S. D. Ferris, H. J. Leamy and J. M. Poate, eds., (American Institute of Physics No. 50) p. 1.

Brice, J. C. (1963), Solid State Electronics, *6*, 673.

Brodsky, M. H. Title, R. S., Weiser, K. and Pettit, G. D., (1970), Phys. Rev. B *1*, 2632.

Brown, W. L. (1980), in "Laser and Electron Beam Processing of Materials," C. W. White and P. Peercy, eds. (Academic Press, N.Y.) p. 21.

Combescot, M. (1981), Phys. Lett. *85A*, 308.

Della Mea, G., Donà dalle Rose, L. F. Mazzoldi, P., and Miotello, A. (1980), Rad. Eff. *46*, 133.

Compaan, A., A. Aydiuli, H. W. Lo and M. C. Lee in "Laser and Electron Beam Interactions with Solids," edited by B. R. Appleton and G. K. Celler, North-Holland, N.Y. 1982.

Donà dalle Rose, L. F. and Miotello, A. (1980), Rad. Eff. *53*, 7.

Draper, C. W. (1981), in "Lasers in Metallurgy," edited by K. Mukherjee and J. Mazumder, TMS-AIME, Warrendale.

Draper, C. W., Jacobson, D. C., Buene, L. and Poate, J. M. (1981), unpublished results.

Dzieviour, J. and Schmid, W. (1977), Appl. Phys. *31*, 346.

Fair, R. B. (1979), J. Appl. Phys. *50*, 6552.

Feldman, C. (1960), Phys. Rev. *17*, 455.

Ferris, S. D., Leamy, H. J. and Poate, J. M., eds. (1979) "Laser-Solid Interactions and Laser Processing," (AIP No 50).

Follstaedt, D. M., Picraux, S. T. Peercy, P. S., and Wampler, W. R. (1981), Appl. Phys. Lett. *39*(4), 327.

Galvin, G. F., Thompson, M. O., Mayer, J. W., Hammond, R. B., Paulter, N. and Peercy, P. S. (1982), Phys. Rev. Lett. *48*, 33.

Gibbons, J. F., Hess, L. D. and Sigmon, T., eds. (1981), "Laser and Electron-Beam Solid Interactions and Materials Processing," (Elsevier-North Holland, New York).

Greenwald, A. C., Kirkpatrick, A. R., Little, R. G. and Minnucci, J. A. (1979), J. Appl. Phys. *50*, 783.

Haug, A. (1978), Solid State Electron. *21*, 1281.

Hodgson, R. T., Baglin, J. E. E., Pal., R., Neri, J. H., Hammer, D. A. (1980), Appl. Phys. Lett. *37*, 187.

Krausse, J. (1974), Solid State Electron. *17*, 427.

Larson, B. C., White, C. W., Noggle, T. S. and Mills, D. (1982), Phys. Rev. Lett. *48*, 337.

Lietoila, A. and Gibbons, J. (1981), in "Laser and Electron-Beam Solid Interactions and Materials Processing," J. F. Gibbons, L. Hess and T. Sigmon, eds., Elsevier, p. 23.

Liu, J. M., Yen, R., Kurz, H., and Bloembergen, N. (1981), Appl. Phys. Lett. *39*, 755.

Lo, H. W. and Compaan, A. (1980), Phys. Rev. Lett. *44*, 1604.

Merli, P. G. and Rosa, R. (1981), Optik *58*, 201.

Musal, H. M., Jr., (1980), in "Laser Induced Damage in Optical Materials," NBS-568. Issued July 1980, pp. 159.

Nathan, M. I., Hodgson, R. T. and Yoffa, E. J. (1980), Appl. Phys. Lett. *36*, 512.

Picraux, S. T., Follstaedt, D. M., Knapp, J. A., Wampler, W. R. and Rimini, E. (1981), in "Laser and Electron Beam Solid Interactions and Materials Processing," J. F. Gibbons, L. Hess and T. Sigmon, North-Holland, p. 575.

Poate, J. M. and Mayer, J. W., eds. (1982), "Laser Annealing of Semiconductors," (Academic Press, New York).

Pronko, P. P., Wiedersich, H., Seshan, K., Helling, A. L., Lograsso, T. A. and Baldo, P. M. (1981), in "Laser and Electron Beam Solid Interactions and Materials Processing," J. F. Gibbons, L. Hess and T. Sigmon, eds., North-Holland, pp. 599.

Schmid, P. E. (1981), Phys. Rev. B *23*, 5531.

Spaepen, F. and Turnbull, D. (1979), in "Laser-solid Interaction and Laser Processing," S. D. Ferris, H. J. Leamy and J. M. Poate, eds., A.I.P. n.50, p. 73.

Spencer, L. V. (1958), N.B.S. Energy Dissipation by Fast Electrons-Government Printing office, Washington.

Sternglass, E. J. (1954), Phys. Rev. *95*, 345.

Stritzker, S., Pospieszcyk, P. and Tagle, J. A. (1981), Phys. Rev. Lett. *47*, 356.

Van Vechten, J. A. (1980), in "Laser and Electron Beam Processing of Materials," C. W. White and P. Peercy, eds., (Academic Press, N.Y.) p. 53.

Van Vechten, J. A. and Wautelet, M. (1981), Phys. Rev. B 2923.

Wampler, W. R., Follstaedt, D. M. and Picraux, S. T. (1980), Appl. Phys. Lett. *36*, 366.

Wampler, W. R., Follstaedt, D. M. and Peercy, P. S. (1981), in "Laser and electron beam solid interactions and materials processing," J. F. Gibbons, L. Hess and T. Sigmon, eds., North-Holland, p. 567.

White, C. W. and Peercy, P., eds. (1980), "Laser and Electron Beam Processing of Materials," (Academic Press, N.Y.).

Wolfe, W. L. (1965), "Handbook of Military Infrared Technology," U.S. Government Printing office, Washington.

Yoffa, E. J. (1980a), Phys. Rev. B *21*, 2415.

Yoffa, E. J. (1980b), Appl. Phys. Lett. *36*, 37.

CRYSTAL GROWTH AND PHASE FORMATION

K. A. JACKSON

Bell Laboratories, Murray Hill, New Jersey

CONTRIBUTORS: D. Turnbull, J. M. Poate, P. Baeri, L. Buene and A. G. Cullis

I. INTRODUCTION

Crystal growth usually occurs by a first order phase change: that is, a phase change in which the two phases co-exist and the transformation proceeds by the motion of an interface between them. The interface can traverse the sample substantially at a single temperature and so the finite heat which is released at this temperature results in an infinite specific heat. By contrast, transformations of higher orders are, in general, homogeneous in space. For example, an "ordering" transformation proceeds uniformly throughout a volume, and over a finite temperature interval, so the specific heat remains finite.

The equilibrium between two phases is determined by their bulk properties, but the kinetics of first order phase changes depend on the interface region between the two phases. This region is inhomogeneous by definition, and so the details of first order transformations are more complex than for higher order transformations. For example, crystal growth depends on the nucleation of new layers which is influenced by both the phases present. The cooperative processes involved cannot be modelled adequately using analytical methods. However, they have been investigated in considerable detail using Monte Carlo computer simulation methods.

The computer simulations are necessarily limited in time. As a result, most simulations have been done for conditions which are rather far from equilibrium, giving growth rates which are large compared to the rates at which crystals are grown in the laboratory, in order to obtain a measurable amount of growth during the computer run. Even so, the computer simulations have provided a basic understanding of the processes involved in crystal growth.

The pulsed heating of materials followed by subsequent recrystallization produces growth rates which are several orders of magnitude faster than conventional growth rates and even a few orders of magnitude faster than splat cooling. This is a new regime for the study of crystal growth, one in which the growth rates are much closer to those accessible in computer simulation, so that a direct comparison between the two is possible.

In this new regime, several of the limitations and effects which are present at lower growth rates also occur. Heat flow limitations, impurity diffusion, and interface instability all seem to be extrapolations of effects at slower growth rates. At high growth rates, however, there are also new effects, such as solute trapping. Extensive investigation of this phenomenon, using the powerful analytical capability provided by Rutherford backscattering has resulted in measurements of the velocity and orientation dependence of this phenomenon for several dopants. Theoretical models as well as computer simulation have contributed to understanding this phenomenon.

Glass formation occurs in "good" glass formers at relatively slow cooling rates. Splat cooling and the techniques such as roller quenching and spin quenching for making metallic glasses have extended this to higher speeds, increasing dramatically the number of materials which can be obtained as glasses. In most glass formers, the glass is a frozen form of the liquid structure. However, the elements silicon and germanium exhibit two distinct amorphous structures; one is the liquid and the other is a low temperature amorphous phase. The former is an eleven coordinated metallic liquid, whereas the latter is a four coordinated random network. The transition from one to the other is not understood in detail at present. This transition is responsible for the phenomenon known as explosive crystallization. In this chapter we present an overview of the current understanding of crystal growth and phase formation, and discuss the current work on laser annealing and high speed quenching within this context.

II. CRYSTAL GROWTH

A. Historical Perspective

The growth rate of crystals can be one of their most anisotropic properties. Variations of a factor of 100 or 1000 in different crystal directions under the same growth conditions are not uncommon. These variations result in the "crystalline" external shape of natural crystals. However, not all crystals exhibit large growth rate anisotropies. A qualitative understanding of why this is so has been available for some time, and quantitative predictions of growth rates have resulted from recent computer simulation studies.

Space does not permit a detailed review of this subject, for which the reader is referred to two recent review papers (Leamy et al., 1975; Weeks and Gilmer, 1979), and the extensive literature cited therein. Here we will attempt an overview of the current understanding. Perhaps the best way to do this is through an historical recapitulation, so that some of the early ideas which have contributed to the development of this subject, but which have subsequently proved to be incorrect in detail, can be placed in perspective.

Gibbs (1876-78), in his famous work on the "Equilibrium of Heterogenous Substances" suggested in a footnote that there could be "difficulty in the formation of a new layer" during crystal growth. The earliest theories of crystal growth ignored Gibbs' warning, and assumed that there was no difficulty in forming new layers.

For crystal growth from the vapor phase (Hertz, 1882; Knudsen, 1909), the net growth rate is given by the difference between the arrival rate of atoms at the surface, and the evaporation rate:

$$v = R_{arrival} - R_{evaporation} \qquad (1)$$

At equilibrium, $v = 0$, and if the evaporation rate depends only on the surface temperature, then it can be set equal to the arrival rate at equilibrium, at that temperature:

$$R_{arrival} = \frac{p}{\sqrt{2\pi m k T}} \qquad (2)$$

$$v = \frac{p - p_0}{\sqrt{2\pi m k T}} \qquad (3)$$

Another way of stating this is that the growth rate is proportional to the difference in chemical potential between the two phases. This expression does not take into account the potential difficulty in forming new layers, and so provides an upper limit on the growth rate. A similar expression can be derived for solution growth, with the growth rate depending on the chemical potential difference for the growing species between the parent phase and the crystal.

For melt growth, the corresponding expression was written down by Wilson (1900), in terms of D, the diffusion coefficient in the liquid:

$$v = \frac{Da}{\Lambda^2} f \left[1 - \exp(-\Delta\mu/kT) \right] \qquad (4)$$

where $\Delta\mu$ is the chemical potential difference between the crystal and melt, a is the interatomic spacing, Λ is the diffusional mean free path in the liquid and f is included to account for the fact that not all collisions with the crystal surface result in crystallization. A similar expression was derived later by Frenkel (1932), based on the melt viscosity.

However, there can be a problem in forming new layers. This arises as follows. Suppose we have a perfect crystal surface with a few adatoms on it, in equilibrium with a parent phase. If the adatoms do not interact, the evaporation rate depends on the number of adatoms on the surface, times the evaporation rate per adatom.

$$R_{evap} = N_{adatoms} \times R_0 \qquad (5)$$

Now let us increase the vapor pressure in the gas phase. The arrival rate at the surface increases, the number of adatoms on the surface increases until the evaporation rate is equal to the arrival rate. But the crystal doesn't grow. It will only grow when the adatom density gets large enough so that the adatoms interact to form islands of the new layer.

The nucleation of new layers is the controlling factor in the intrinsic growth rate of crystals. The size of the nucleation barrier depends on the adatom density. As we shall see, in some cases the adatom density is so large that there is no nucleation barrier.

To return to our historical perspective, Frank and his associates (Burton *et al.*, 1951) applied the nucleation theory developed by Volmer and Flood (1934) to the formation of new layers during crystal growth. Using the macroscopic free energy to calculate the free energy associated with a step, the nucleation theory predicted that crystals would not grow at all at small supersaturations, and would only grow at supersaturations or supercoolings of the order of 10 to 20 percent. Applied to ice, for example, this means that water should freeze at a sensible rate only at supercoolings greater than about $\Delta T/T \sim .1$ i.e., at temperatures below about $-27°C$. This is not in accord with experience.

Frank and his co-workers suggested that dislocations in crystals provide continuous steps on crystal surfaces, and that this removes the necessity for the nucleation of new layers. Atoms can interchange between the parent phase and the crystal at the steps on the surface, so that only a fraction "f" of the surface sites are active growth sites. Analysis indicated that f should increase linearly with undercooling for dislocation-controlled growth, resulting in a growth law of the form

$$v = v_0 \, \Delta T^2 \tag{6}$$

These early treatments did not deal with the equilibrium adatom concentration on the crystal surface, and how this changes under dynamic growth conditions. Clearly if the adpopulation is very large, then nucleation will be much easier than if it is very small. Frank and his associates examined this problem by applying two dimensional order-disorder theory to the adatom layer. They concluded that crystals would melt before the surface adatom population got very large, and so this effect should not be important in crystal growth.

Jackson (1958) suggested that the transformation energy, rather than the binding energy should be used to calculate the adpopulation. Thus, for melt growth, where the transformation energy is relatively small, the adpopulation can be very large.

The equilibrium adpopulation coverage on the surface is given approximately by

$$\exp\left[- T_R/T_E\right] \tag{7}$$

where T_E is the temperature of equilibrium between the two phases, and T_R is the surface roughening temperature (Leamy *et al.*, 1975). T_R increases with the transformation energy, but also depends on the geometry of the crystal face, being higher for closer packed faces. Thus, metal crystals or condensed rare gas crystals in contact with their melts have very high adatom populations, (rough surfaces on an atomic scale), whereas the same crystals in equilibrium with their vapors at low temperatures will have low adatom populations, especially on the low index faces.

B. Crystal Grown As A Cooperative Process

In order to treat crystal growth quantitatively, we want to answer the question: what is the largest cluster of atoms amongst the adatoms? This cluster will form the nucleus for the new layer. This question cannot be answered analytically, but has been studied in detail by computer simulation.

The extra free energy to form a cluster on the surface clearly depends on the number and size of clusters which are present spontaneously. As the adpopulation increases, this extra free energy decreases. In order to discuss this situation realistically, we must recognize that adatoms can deposit in the next layer on top of clusters, and that atoms can be missing from nearly complete layers, so that the possibility exists for the interface transition region to be several atom layers thick. When the adpopulation gets large enough, the extra free energy to form a cluster goes to zero. This happens at a singular temperature known as the roughening temperature, T_R. Above this temperature, the interface is not locked into the atom planes, and is free to move like a fluid-fluid interface (Weeks and Gilmer, 1979). Below this temperature, the interface is locked onto the crystal planes, and there is a barrier to the formation of new layers. The further below T_R, the smaller the adpopulation, the larger the barrier to the formation of new layers. The growth rate, predicted by computer simulation depends whether the surface at equilibrium is above or below its roughening temperature. Above the roughening temperature, the growth rate is linear with undercooling. Below the surface roughening temperature the growth rate curves, as illustrated in Figure 1, can be fitted over a limited region by an equation of the form

$$v = v_0 \, \Delta T^n \qquad\qquad (8)$$

with n increasing progressively below T_R. Far below T_R, the growth rates approach those given by classical nucleation theory. Indeed, the nucleation equations work quite well (Gilmer and Jackson, 1977), provided that the proper value for the step free energy is used. Unfortunately, this free energy contains a large entropy contribution, which can only be determined by analysis of the surface configurations. The early equations for crystal growth

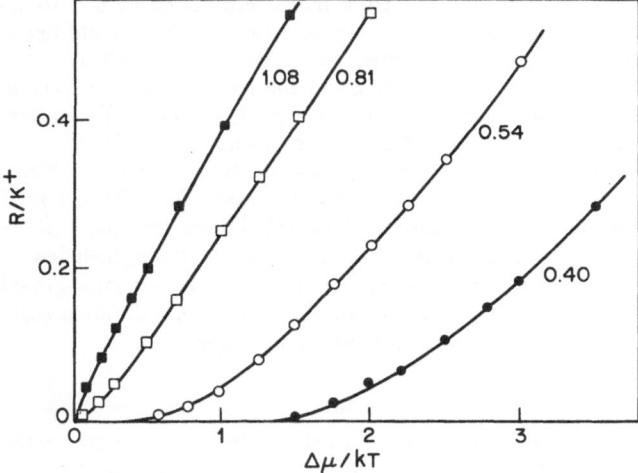

Figure 1   Monte Carlo calculations of the normalized growth rate vs. chemical potential difference for different crystal surfaces. The surfaces are characterized by the ratio T_M/T_R where T_R is the equilibrium roughening temperature for the surface and T_M is the melting point of the crystal.

thus describe the two extremes, one far below T_R where nucleation is important, and the other, above the roughening temperature in the regime where the growth rate is given by Eqn. 3 or 4. In this latter regime, dislocations are unimportant. Dislocations are important only when there is a large nucleation barrier. Many crystals grow in an intermediate regime where new layers form fairly readily, so that dislocations can contribute to, but do not dominate the growth. Large anisotropies in the growth rate occur because various faces of the same crystal have different roughening temperatures, T_R. Small differences in T_R can result in large differences in the rate of nucleation of new layers, thus producing large anisotropies in growth rate.

For growth from the vapor phase, the equilibrium temperature can be above or below T_R, depending on the vapor pressure. For melt growth, however, the melting point is fixed, (unless very large pressures are applied), so that at equilibrium, a particular crystal face is either above or below its roughening temperature (Jackson, 1967).

Facets form during growth only on faces which are below their roughening temperatures. For example, it is well-known that the silicon-melt interface forms a facet only on the (111) plane, and on none of the others.

Detailed computer simulations of crystal growth, and predictions of growth rates have been made (Gilmer and Jackson, 1977) using the Kossel-Stranski model, (Kossel, 1927; Stranski, 1928), which is the crystal growth version of the Ising model. In it, atoms exist only on lattice sites. The actual structures of a melt, say, in contact with a crystal is not considered. The Ising model has proved very successful for describing a great variety of three dimensional critical phenomena, because these are dominated by the cooperative processes involved. It works quite well for the superfluid transition in He, for example, which does not even involve a lattice. The Ising model describes the behavior near T_c, and the critical behavior scales to T/T_c. Similarly for crystal growth, we do not expect the Ising model to predict T_R for a crystal melt interface, for example. But we do expect roughening phenomena to scale with T/T_R. For melt growth, the presence of a close packed liquid might be expected to make it more difficult to form coherent surface clusters, raising the surface roughening temperature above the T_R predicted by the Ising model. For vapor phase, thermal vibrations make it easier to form adatoms thus lowering T_R from the Ising model.

The growth rates predicted by the Ising model are scaled to the arrival rate of atoms (Gilmer and Jackson, 1977). For vapor phase growth, we can estimate this fairly accurately. However, for melt growth systems, the effective arrival rate is not clear.

There are two final points to be made about the computer simulations of crystal growth. The first is that the "simple" Kossel model of a crystal is not so simple after all. It contains many subtle effects which can only be treated using computer simulation.

The second conclusion is based on the fact that the Kossel model is indeed a simplification of the real world. It leaves out the molecular structure and so cannot predict *a priori* the experimental roughening temperature. Chemical effects relating to impurity effects, adsorption, dissociation, molecular rearrangement etc., are not included. In addition the macroscopic transport of matter and heat which accompanies crystal growth, including the phenomena of convection and interface instabilities, occur on a different scale. The interface processes provide a boundary condition for these phenomena.

C. Crystal Growth Rates

The Monte Carlo simulations give us information about the relative rates of arrival and departure of atoms or molecules from the crystal. They cannot give the arrival rate of atoms at the crystal surface. In vapor phase growth, the arrival rate is known from gas kinetics as indicated by Eqn. 2. Similarly, in solution growth, the arrival rate is usually known from diffusivities in the parent phase. For melt growth, however, the arrival rate depends on the structure of the liquid at the interface, and this is not known in detail. A difference of views centering on this point was brought out at Trevi. The arrival rate assumed in Eqn. 4 depends on the mobility in the liquid, through the liquid diffusion coefficient D. The form of this expression has been verified experimentally many times, in a variety of glass forming materials, where the growth rate is slow enough to measure. For example, Vergano and Uhlmann (1969) have made growth rate measurements on GeO_2 over a 300 degree range in undercooling. The viscosity varies by 10^5 over this temperature range. However, the product $\eta \times v$ increases linearly with undercooling over this entire temperature interval. This indicates that the activation energy for atoms joining the crystal is the same as that for liquid viscosity. Or, stated another way, the mobility in the liquid phase determines the rate of arrival at the crystal. The growth rate for GeO_2 can be adequately described by Eqn. 4, using an appropriate value for Λ. Similar results, but usually over a more limited temperature range have been obtained for many materials which have high viscosity melts, including network glass formers, polymers and organic materials (Jackson et al., 1967). For materials with lower viscosity, such as the liquid metals, the growth rate is rapid at very small interface undercooling, so that the growth rate has never been properly measured. Heat flow interferes with such measurements. Growth rates of dendrites into undercooled melts have been measured as a function of the initial bath temperature which is not the same as interface temperature (Chalmers, 1964).

For example, growth rate data for dendrites into supercooled nickel were made by Walker (1964) who observed observed growth at 40 m/sec at 175 degrees undercooling. This is the fastest measured growth rate for dendrites. Most of this undercooling is required to conduct heat away from the dendrite tip. A reasonable estimate is that the actual interface undercooling is at least a factor of five or ten less than this. Eqn. 4 fits this estimate, using a reasonable value of Λ (~0.3 to 0.1 times a). Turnbull suggested an alternative to Eqn. 4. He suggests that the collision frequency per interface site depends on v_s, the velocity of sound, so that the growth velocity is:

$$v = v_s f \, \Delta T/T \qquad (9)$$

This gives agreement with Walker's data for nickel, but not with the data for GeO_2 and other glass forming materials.

Turnbull points out that the experimental observations of Ruhl and Hilsch (1977) indicate that there is a low temperature transformation in deposited metallic films which is not thermally activated. They found that metal films deposited at liquid He temperature undergo a transformation at temperatures as low as 20K. This suggests a very small activation barrier for this transformation. It is not clear that the observed transformation is crystallization.

Eqn. 4 and 9 both give similar results in the small undercooling region ($\Delta T/T < 0.1$), and so we will use Eqn. 4 to convert Monte Carlo data to real growth rates.

Recent molecular dynamics results on the crystallization of a Lennard-Jones fluid (Broughton et al., 1982) can be fitted by an equation of the form.

$$v = \frac{a}{\lambda} \, \frac{\sqrt{3\,kT}}{m} \, f \left\{ 1 - \exp(-\,\Delta\mu/kT) \right\} \qquad (9a)$$

where λ is the distance a liquid atom moves to get to a lattice site during crystallization ($\lambda \approx 0.4a$) and $\dfrac{\sqrt{3\,kT}}{m}$ is the thermal velocity of an atom. This equation is similar to Eqn. 9, except that the velocity of sound is replaced by the thermal velocity, and so the growth rate goes to zero at $0°K$. It is not clear at present whether this result, which applies to a Lennard-Jones system is also applicable to silicon.

Gilmer (1982) performed Monte Carlo simulations for the growth of silicon from the melt. These simulations were made for growth on the (111) (100) and (110) planes of the diamond cubic structure. The reduced equilibrium temperature was adjusted so that (111) was just below its roughening transition, and so could exhibit a "facet" during growth, and all the other faces were above their roughening temperatures and would not exhibit "facets". The growth rates, normalized to an arrival rate of unity, are exhibited in Figure 2. Even though both the (100) and (110) are above their roughening temperatures, a slight difference in growth rates is predicted. Both exhibit a linear dependence on $\Delta\mu$ at small undercoolings, whereas the (111) growth rate does not.

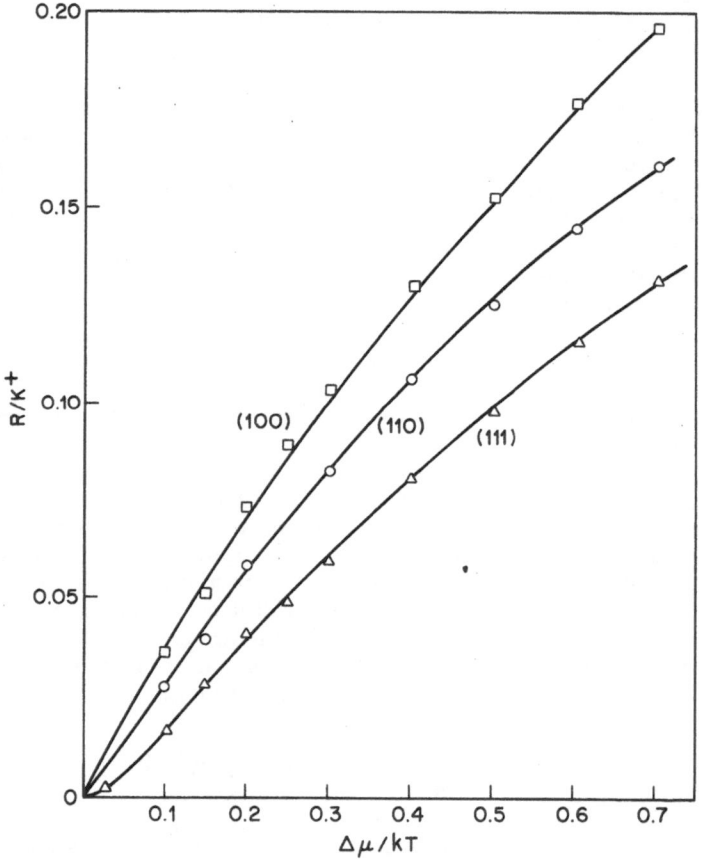

Figure 2   Monte Carlo calculations of the normalized growth rate of silicon from the melt on three different orientations, as a function of the chemical potential difference between the crystal growth.

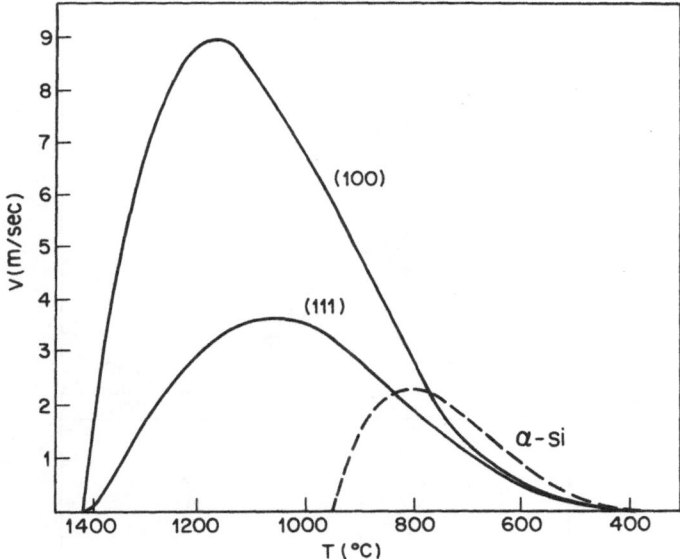

Figure 3   Predicted melt growth rate for silicon on the (100) and (111) faces, based on the data of Figure 2, and
            assuming that the arrival rate of atoms of the crystal depends on the liquid diffusivity, as given by Eq. 4.
            Also shown is a predicted growth rate for amorphous silicon from the melt.

Combining these growth rates with Eqn. 4 to include the temperature dependence of the arrival rate, gives growth rates as shown in Figure 3.

III. SOLUTE TRAPPING

One of the more dramatic observations on laser annealed silicon is the large amount of dopant that can be retained in the lattice (see Chapter 4). This is illustrated in Figure 4 (White *et al.*, 1980). Ions implanted into silicon, in this case, gallium ions accelerated to 100 keV, and implanted to a dose of 6.17×10^{15} ions/cm^2, penetrated into the surface so that the concentration profile below the surface was as shown by the open circles. The composition distribution was determined by Rutherford Backscattering (RBS). The implantation destroyed the crystallinity of the surface layer, and made it amorphous, to a depth of .1 to .15 μm. After laser annealing, the crystallinity of this layer had been restored, and the gallium was distributed as shown by the filled circles. The amount of gallium incorporated into the lattice is significantly above the maximum which can be incorporated under equilibrium conditions. On subsequent furnace annealing, the excess gallium precipitates out of the lattice, verifying that there is indeed gallium significantly in excess of the maximum solid solubility incorporated into the lattice. The phase diagram for Ga-Si is shown in Figure 5. The solubility of Ga in crystalline silicon is retrograde as indicated by the dashed solidus line. That is, with decreasing temperature, the solute concentration in the solid at first increases, and then decreases. This is characteristic of alloy systems in which the solubility of the solute in the solid is very small. The condition for retrograde solid solubility is that the solute has a *higher* enthalpy in the crystal than in the liquid.

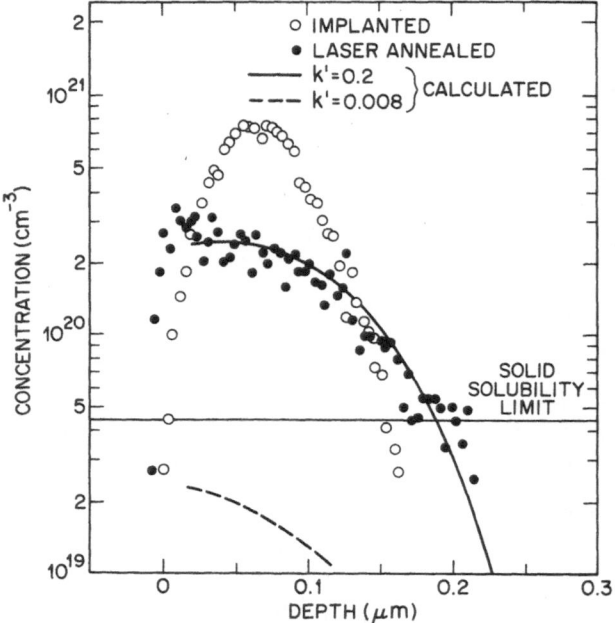

Figure 4   Concentration profile for gallium implanted into silicon as determined by Rutherford backscattering. The open circles are as-implanted, and the filled circles are following laser annealing. Using regrowth rates calculated from heat flow, the laser annealed profile can be fitted assuming $k = 0.2$, but not with the equilibrium k-value, 0.008. The gallium concentration incorporated into the lattice is greatly in excess of the solid solubility limit.

Close to equilibrium, that is, for usual growth conditions, the solute in the solid cannot exceed the maximum solid solubility. The incorporation of solute into the crystal in concentrations far in excess of the equilibrium concentration means that kinetic effects are dominant during regrowth.

It is important to distinguish between what happens at the interface and what happens in the phases adjacent to the interface during regrowth. At the interface, there is a phase transition, where the atoms rearrange, often by small motions, into a lower energy configuration, which also decreases their entropy by limiting their volume in configuration space. This is different from a diffusive jump in the liquid or the crystal, which merely displaces the atoms. The different set of rules which apply in the interface region because of the phase transition provide the boundary conditions for diffusion in the bulk phases.

Transition results in a discontinuity in composition between the two phases, which is conventionally described by the k value, or distribution coefficient:

$$k_e = C_S/C_L \tag{10}$$

where C_S is the composition of impurity given by the solidus line in the phase diagram and C_L by the liquidus line. For Ga in Si, for example (Figure 5) $k_e = .008$, at small concentrations.

For the laser annealing experiments, the duration of the liquid and the regrowth rate can be calculated from heat flow. From this information, the expected distribution of impurities can been calculated as shown in Figure 4. The equilibrium k-value (.008) does not properly predict the observed concentration profile. Rather a much larger k-value, $k = 0.2$ is needed to fit the data.

This effect is not due to the atoms being unable to get out of the way of the rapidly advancing interface. The calculation for $k = .008$ shows that the atoms *can* diffuse fast enough to be pushed to the surface. But they are not. Not because they cannot diffuse fast enough, but rather because the atoms are incorporated into the crystal in greater concentration then predicted by equilibrium thermodynamics.

There have been several models proposed to account for this phenomenon. One of these is a mathematical model, not based on a physical model of the interface. It is capable of fitting many of the data quite well (Wood, 1980). Another is based on diffusion at the interface, which is considered to be distinct from the phase transformation (Aziz, 1982). It does not properly predict either the velocity dependence or the orientation dependence of the k-value. A third model (Baker and Cahn, 1971) is based on a reasonable physical model, but it assumes that the transformation takes place over several atom distances. This model predicts that the k-value should go to unity at a growth rate which is slower than observed. The "solute trapping" model of Jackson *et al.* (1979) (see Chapter 4) successfully describes both the growth rate and orientation dependence of the k-value. It is the only model in which the k-value can saturate with increasing velocity at a value less than unity.

This model starts with the thermodynamically based energy-reaction coordinate diagram shown in Figure 6. The energy differences between the equilibrium configurations (i.e., the ΔH's and E) are known from thermodynamics. The barriers in the reaction paths at the interface (the Q's) are unknown.

In terms of these reaction paths, we can write reaction rates as follows (see Figure 6):

$$R_A^+ = C_A^\ell \left[R_A^+\right]_0 \exp(- Q_A/kT) = C_A^\ell K_A^+ \tag{11}$$

$$R_A^- = C_A^s \left[R_A^-\right]_0 \exp\left[- (Q_A + \Delta H_f^A)/kT\right] = C_A^s K_A^-$$

$$R_B^+ = C_B^\ell \left[R_B^+\right]_0 \exp[- (Q_B + E)/kT] = C_B^\ell K_B^+$$

$$R_B^- = C_B^s \left[R_B^-\right]_0 \exp(- Q_B/kT) = C_B^s K_B^-$$

Here C_i are the compositions at the interface, and the K's are reaction rates. The growth rate is then given by:

$$v = R_A^+ - R_A^- + R_B^+ - R_B^- \tag{12}$$

and the composition of the crystal by:

$$C_B^s/C_A^s = \left[R_B^+ - R_B^-\right]/\left[R_A^+ - R_A^-\right] \tag{13}$$

Algebraic manipulation gives:

$$k = C_B^s/C_B^\ell = K_B^+/\left[v + K_B^-\right] \tag{14}$$

At equilibrium, $k_e = K_B^+/K_B^-$, so that k decreases as the velocity increases. These equations predict that the solid composition cannot exceed the maximum solid solubility, which is not in accord with experiment.

These equations would work if it were not for the discrete nature of the lattice. That is, if the interface were a plane advancing at a rate v, then the composition in the liquid at this mathematical plane could adjust so that the diffusion rate into the liquid satisfied the instantaneous rejection rate from the crystal. However, the impurity atoms at the interface get surrounded by silicon atoms which are part of the crystal. This reduces their mobility, and they get "buried" by the crystallization of other silicon atoms. The rate at which they are incorporated, or trapped, into the crystal depends on the rate of advance of the crystallization front, rather than on the reaction rate R_B^+ (Eq. 11). The simplest way to account for this effect is to assume that a fraction α of the impurity atoms in the liquid at the interface are incorporated into the crystal by this process, so that Eqns. 12, 13 and 14 become:

$$v = R_A^+ - R_A^- + R_B^+ - R_B^- + \alpha \, v \, C_B^\ell \tag{15}$$

$$\frac{C_B^s}{C_B^\ell} = \frac{R_B^+ - R_B^- + \alpha \, v \, C_B^\ell}{R_A^+ - R_A^-} \tag{16}$$

$$k = \frac{K_B^+ + \alpha \, v}{v + K_B^-} \tag{17}$$

Eqn. 17 reduces to k_e for $v = 0$, and k can be greater than k_e at finite velocities.

Using $V \equiv v/K_A^+$, Equation 17 can be rewritten as:

$$k = \frac{\alpha \, V + \exp\left[(Q_A - Q_B - E)/kT\right]}{V + \exp\left[(Q_A - Q_B)/kT + \Delta \, S_B^f/R\right]} \tag{18}$$

Here $\Delta S_B^f = \Delta H_B^f/T_M^B$ is the entropy of fusion that pure B would have if it crystallized in the structure of the silicon matrix. This is unknown, but can be estimated. $E - T\Delta S_B^f$ is known from the phase diagram, since

$$k_e = \exp\left[- E/kT + \Delta S_B^f/R\right] \tag{19}$$

Q_A and Q_B are unknown, but occur only as Q_A-Q_B in the expression for k.

Note that V is a normalized growth velocity, and so it approaches one as V approaches K_A^+, the rate at which the matrix atoms join the crystal.

Equation 18 can be compared with experimental k-values measured at different growth rates. The k-value also depends on orientation (Baeri et al., 1981), as illustrated in Figure 7. Here the same dose of Bi has been implanted into (111) and (100) silicon. After annealing with the same energy laser pulse, the Bi profiles are found to be as shown. The profiles are clearly different and can be fitted with k-values of 0.1 and 0.2 as indicated in Figure 7.

Gilmer (1982) has performed Monte Carlo simulations of impurity trapping. Unfortunately, the range of his calculations is limited by the compositions accessible in this method. He has found results as shown in Figure 8. As a function of undercooling, the (100) incorporates a higher concentration of impurity than (111). However, using the data of Figure 2 to convert $\Delta \mu$ to normalized velocity, results in a prediction that more impurity is incorporated on (111) than on (100) at the same velocity, in accord with experiment.

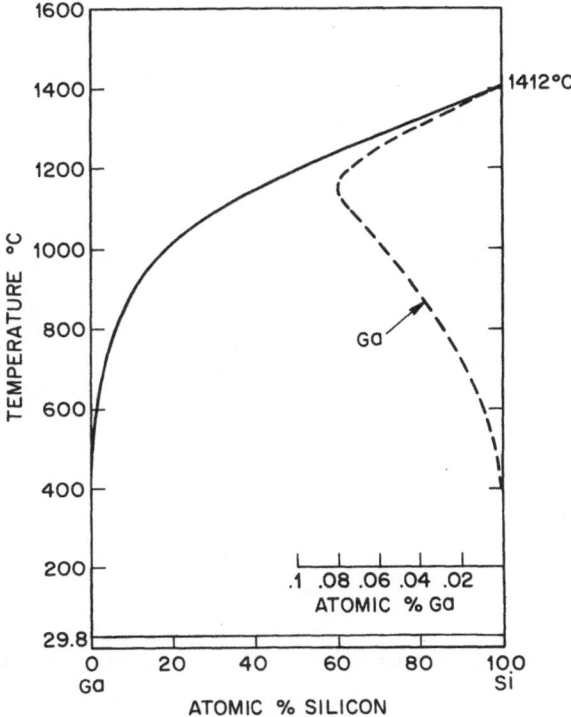

Figure 5   Phase diagram for gallium on silicon. The solidus line is shown dashed, and is referred to the inset scale.

Experimental data for the incorporation of bismuth in silicon (Baeri *et al.*, 1981) are shown in Figure 9 as a function of growth rate, and for the two orientations. The experimental points were derived from data such as Figure 7 with k-values obtained by fitting the observed distributions. The growth rate was varied by changing the substrate temperature, which has a profound effect on the thermal conductivity of silicon. The growth rates are based on heat flow calculations (Baeri *et al.*, 1980). There may be a problem with these calculated rates since the interface was assumed to be at the melting point regardless of the growth rate. However, recent experimental measurements of the growth rate (Galvin *et al.*, 1982) tend to confirm the calculated values.

The solid lines in Figure 9 were calculated using Eqn. 18 in conjunction with the curves shown in Figure 3. Q_B was taken to be 0.7 eV, typical of diffusion activation energies in metallic liquids. $\Delta S_{Bi}^f/R$ was taken to be 3.24, the same value as silicon, which is similar to germanium. The curves were fitted to $k = 0.1$ at $v = 2$ m/sec, resulting in $Q_A = 1.45$ eV. This model accounts well for both the growth rate dependence, including the saturation at $k = 0.3$, as well as the orientation dependence of k.

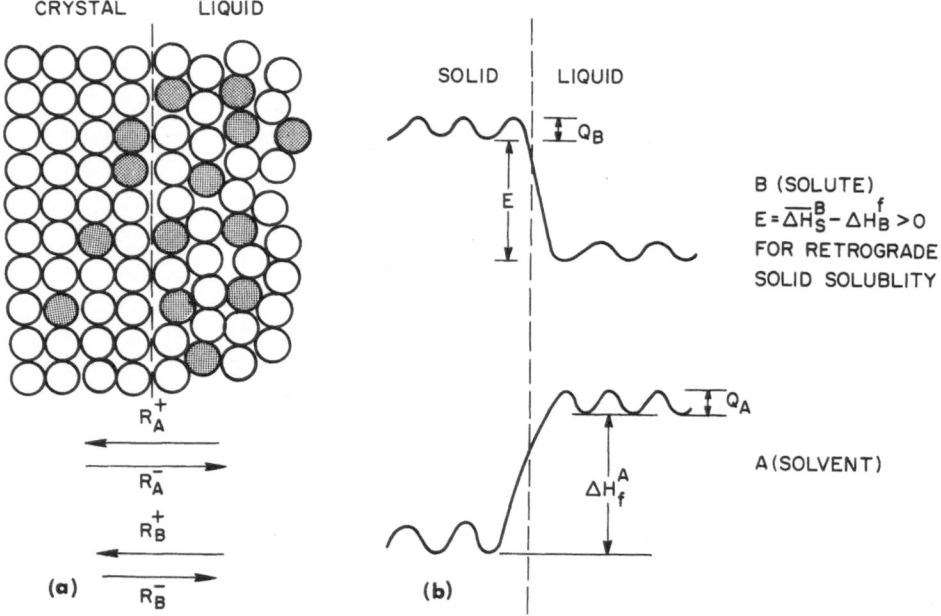

Figure 6a   Schematic drawing of a crystal-melt interface. The rates R_A^+ refers to the rate at which A atoms from the melt join the crystal, and so on.

Figure 6b   Energy versus reaction coordinate for the matrix (A) and solute (B) atoms in an alloy system exhibiting retrograde solid solubility. E is positive as shown for such a system.

This comparison of the theory and experiment gives us, for the first time, some information about the relative mobilities of the silicon atoms and impurity atoms in the interface region.

IV. AMORPHOUS SILICON

Silicon and germanium are unusual materials in that they exhibit two distinct amorphous phases. The low temperature amorphous phase, produced by low temperature ion damage, or low temperature deposition, is a four-coordinated random structure, related to the four-coordinated crystal. The high temperature liquid phase, on the other hand, is a typical metallic liquid, with approximately eleven nearest neighbors.

The relationships between the crystal (x), the liquid (L) and the low temperature amorphous phase (α) are illustrated in Figure 10. The transformations between the liquid and the crystal have been discussed in detail above. The transformation between the amorphous phase and the crystalline phase will be discussed in the next section.

The transformation between the liquid and the amorphous phase is relatively inaccessible from an experimental viewpoint; and the nature of this transition has not been established. The transformation could be either first or second order. In principle, the transformation from the four-coordinated random structure to the eleven-coordinated liquid could either

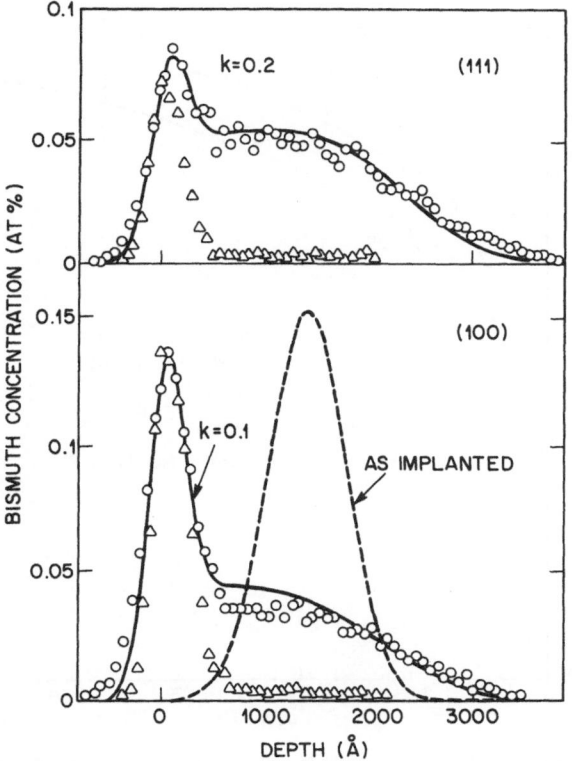

Figure 7   Bi concentration profiles following identical laser irradiation of implanted samples. For the (100) case the profile can be fitted assuming that $k = 0.1$. For the (111) case, $k = 0.2$.   •

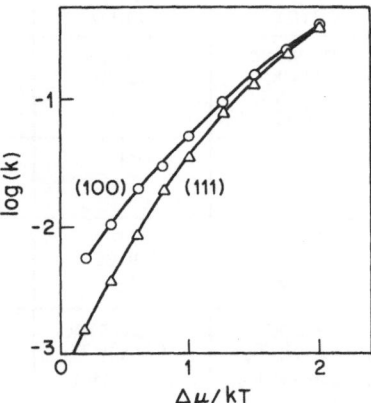

Figure 8   Monte Carlo calculation of the log of the k-value vs. chemical potential difference for the (100) and (111) faces of silicon.

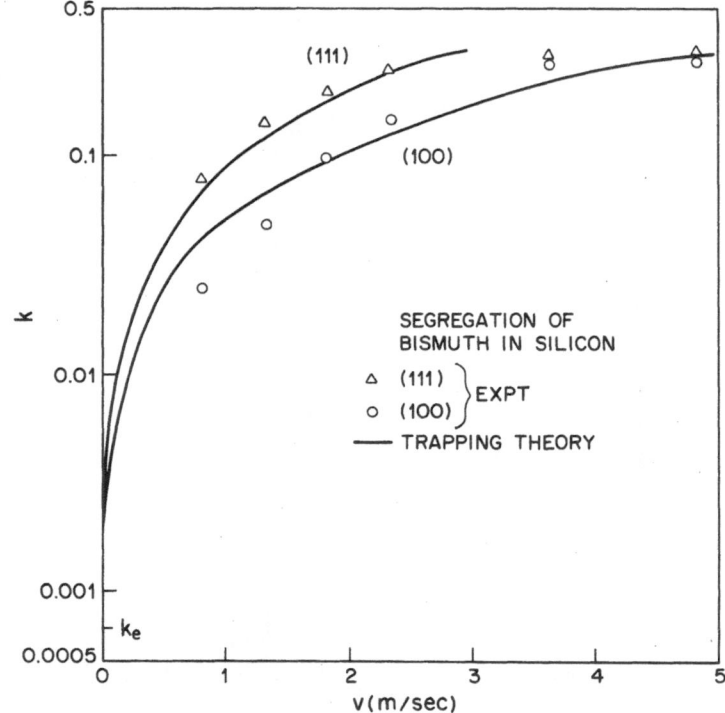

Figure 9   The points are experimental k-values for bismuth implanted into (111) and (100) silicon, plotted against the regrowth velocity calculated from heat flow. The curves were calculated using Eq. 18.

	CRYSTAL	LIQUID	AMORPHOUS
STRUCTURE	DIAMOND CUBIC (REGULAR TETRAHEDRAL)	METALLIC LIQUID	RANDOM TETRAHEDRAL
NUMBER OF NEAREST NEIGHBOURS	4	~11	~4
STABILITY	STABLE BELOW T_M	STABLE ABOVE T_M	METASTABLE
TRANSFORMATIONS	➡LIQUID ABOVE T_M (➡α ON ION IMPLANT)	➡CRYSTAL BELOW T_M (➡α ON VERY RAPID COOLING)	➡CRYSTAL ON HEATING ➡ LIQUID ON VERY RAPID HEATING

Figure 10   The transformations amongst the various forms of silicon.

proceed homogeneously and continuously throughout the volume, or else the two phases could co-exist and the transformation proceed by the motion of an interface between them. Although it has not been established experimentally, the transformation is more likely to be first order than second, since it involves a change of coordination. It could be argued that both structures, being random, have pseudo-spherical symmetry, and so a continuous transformation would be possible. However, the local symmetry around each atom is different, and this suggests a first order transition.

Bagley and Chen (1979) and Spaepen and Turnbull (1979) have suggested that the transition is first order, and estimated some of the thermodynamic properties of the amorphous phase. Their estimates are quite similar and were based on the known thermodynamic properties of the liquid and crystal phases of silicon, and on the properties of germanium. The entropy change associated with the change in disorder in a liquid is similar for many materials, about R (entropy has the same units as the gas constant, R). This value should be the entropy difference between the crystal and the amorphous phase. The entropy difference between the liquid and the crystal is $3.24R$, so the entropy difference between the liquid and the amorphous phase is $2.24R$.

The relative free energies of these phases are illustrated in Figure 11. The liquid and crystal free energies cross at 1420°C, the melting point of the crystal. The liquid and amorphous phase free energies cross at 925°C, the melting point of the amorphous phase. Recent measurements by Donovan et al., (1982) of the transformation heat from the amorphous phase to the solid, obtained calorimetrically using two μm thick amorphous layers was found to be 11.3 kJ/mole, in agreement with the estimates of Bagley and Chen (1979) and Spaepen and Turnbull (1979).

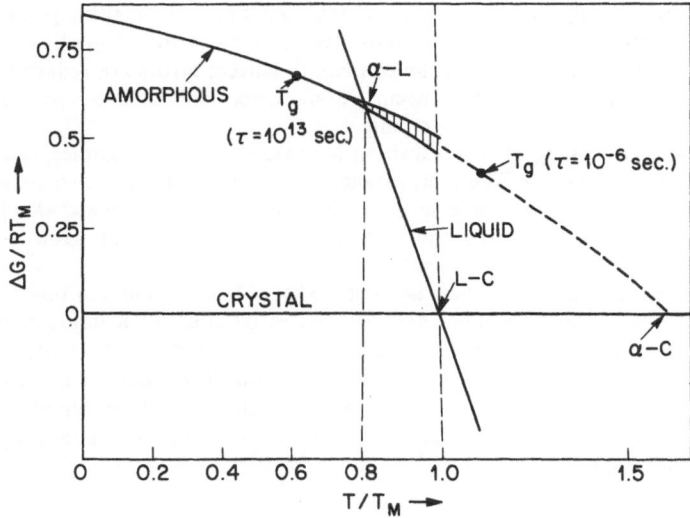

Figure 11   The Gibbs free energies of amorphous and liquid silicon, referred to that of the crystal, as a function of temperature. The crossing point of the curves indicate the temperatures of equilibrium amongst the three phases. The glass temperature of the amorphous phase depends on τ, the time scale during which the measurement is made.

The transformation from the liquid phase to the amorphous phase can occur below 925°C, but it is unlikely to be observed until the undercooling gets large enough so that the liquid to amorphous transition is faster than the liquid to crystal transition. This is illustrated in the growth rate curves in Figure 3, plotted using Eqn. 4. The mobility is assumed to be the same for either transition, but the two curves cross because the entropy change is smaller for the liquid to amorphous transition. The growth rate suggested by Turnbull (Eqn. 9) would not predict that the amorphous phase would grow more rapidly at high rates. However, modification of this expression by the inclusion of an entropy term as in Eqn. 4 would result in a crossing of the growth rate curves at the same temperature as shown in Figure 4.

The growth of the amorphous phase from the liquid phase has recently been observed using very short laser pulses which produce very rapid regrowth rates (Liu *et al.*, 1979; Tsu *et al.*, 1979). Cullis *et al.*, (1982) have measured the energy thresholds for this transition on both the (100) and (111) faces. They also observed a region of twinned growth on the (111) face. These results are illustrated in Figure 12. Using a UV 2.5 nsec pulse, they found no modification of the surface for either orientation for energy levels in the laser pulse below 0.2 J/cm^2. Increasing the energy density heats the sample more, which reduces the regrowth rate. For clarity, let us discuss their results from the other end, in the direction of decreasing energy density, but increasing growth rates. At low growth rates, the silicon regrows as good crystal in both orientations. Above some growth rate, corresponding to an energy density of 0.6 J/cm^2, the (111) orientation no longer regrows as perfect crystal. The regrowth is crystalline, but heavily twinned. This regime is not observed for (100) growth. A similar transition has been observed by Van derSteenhoven and Gilmer (1982) in Monte Carlo computer simulation of the growth of the (111) face of a face centered cubic crystal. In the simulations, the onset of twinning has been observed on both (111) and (100) faces, but it occurs at a much lower growth rate on the (111) face. At faster growth rates, on both the (111) and (100) faces of silicon, a transition to the amorphous phase is observed.

There are two ways of thinking about this transition. In one view this could be a dynamic transition. The growth of a perfect crystal, which occurs near the melting point, requires that any atom that arrives in a "wrong" site must leave, or move to a "right" site before it gets buried. As the growth rate increases the chances to correct errors are reduced. Above some high growth rate, errors will start to occur. Some errors will induce other errors, and the crystal perfection will deteriorate very rapidly.

The other way of looking at this transition is based on thermodynamics. As illustrated in Figure 3, based on estimates of the thermodynamic properties of silicon, we expect that below some temperature, (i.e., above some growth rate) the amorphous phase should grow more rapidly from the liquid than does the crystal. This is because the growth rate (eqn. 4) depends implicitly on the entropy change.

However, these two points of view are not really different. The entropy factor is in the growth rate expression because there are more potential sites for a liquid atom to join the amorphous phase than for a liquid atom to join a perfect crystal. The kinetic view is based on the idea that at high growth rates, it becomes more likely that an atom won't find the "right" site. The thermodynamic argument says that the amorphous phase can grow more rapidly because there are fewer "right" sites in the crystal. These two approaches are different ways of saying the same thing.

V. CRYSTALLIZATION OF AMORPHOUS SILICON

The amorphous phase of silicon can transform to the crystalline phase by a solid state transformation (Csepregi *et al.*, 1977). The amorphous phase has a higher free energy than

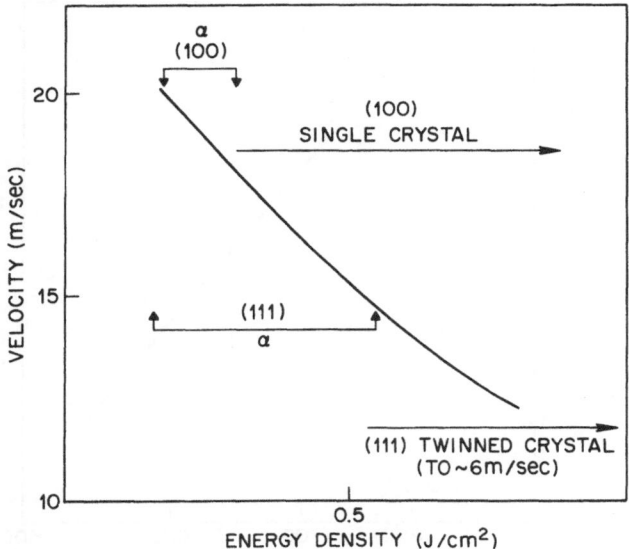

Figure 12   The structure produced by laser annealing depends on the energy in the pulse, as shown. Nothing is observed on either face below a threshold of about 0.2 J/cm². The amorphous phase is observed to form in a thin layer above this threshold. Above about 0.3 J/cm² single crystal silicon regrows on the (100) face. The amorphous phase persists on the (111) face up to about 0.6 J/cm². Above this there is a region in which heavily twinned crystal is formed. At high energy densities, the (111) crystals also regrow as single crystals.

the crystalline phase as shown in Figure 11, and so is unstable with respect to it at all temperatures. However, below about 400°C, the transformation is very sluggish and does not proceed at a measurable rate.

The transformation rate depends strongly on temperature, the temperature dependence being Arrhenius type as shown in Figure 13. The growth rate is very slow compared to rates of crystallization from the melt. The growth rate is orientation and doping level dependent. The dependence on doping level is reminiscent of the doping-level dependence of the velocity of dislocation motion in silicon.

The crystallization of the amorphous phase proceeds very slowly at 500°C. An amorphous layer ~1000Å which has been created by ion implantation, can be regrown by solid state annealing using a scanned laser spot. At 1100-1200°C, a 1000Å layer can regrow in about 1 millisecond. This is the so-called "solid state" regime of laser crystallization.

VI. EXPLOSIVE CRYSTALLIZATION

The amorphous phase can melt to the liquid phase if it is heated above its melting point. The liquid formed will be far below the crystallization temperature, and so will be able to crystallize very rapidly. Below the melting point of the amorphous phase, this cannot happen, and so the growth must proceed by solid state transformation.

The crystallization of the liquid is an exothermic process. The melting of the amorphous phase is endothermic, but the heat absorbed is less than the heat generated by crystallization. Thus, in principle, if the crystallization of the amorphous phase was triggered, the heat of

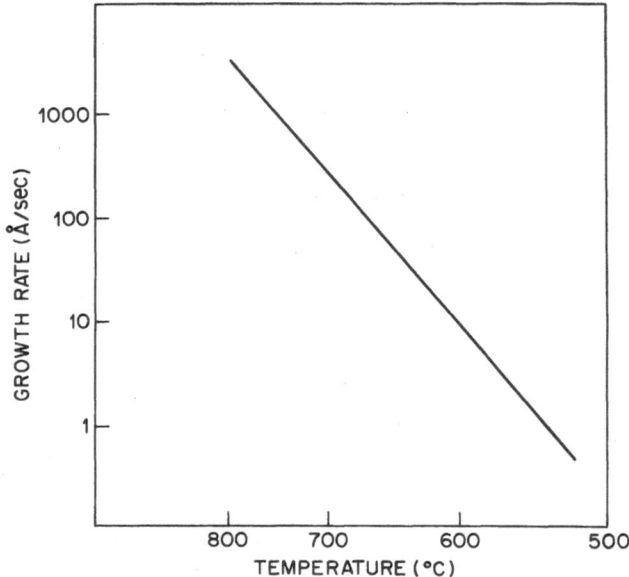

Figure 13   Rate of solid state crystallization of amorphous silicon directly to the crystalline phase.

fusion could heat nearby amorphous material to its melting point and the process could proceed spontaneously. This phenomenon is known as explosive crystallization (Gilmer and Leamy, 1980). It can proceed very rapidly, at rates which have been measured in excess of 1 m/sec.

The process is sensitive to heat loss, and usually a sample must be heated (to a few hundred degrees centigrade) before the crystallization front will propagate spontaneously once it is triggered.

Explosive crystallization proceeds more readily in a thick amorphous layer than in thin one, and proceeds more readily on a substrate of low thermal conductivity than on a high thermal conductivity substrate. These observations are consistent with heat transfer arguments: the heat of fusion must heat the substrate as well as the amorphous phase. The thicker amorphous layer or the poorer thermal conducting substrate both reduce the relative heat dissipation and make it easier to heat the amorphous phase to its melting point.

VII. HEAT FLOW AND CRYSTALLIZATION

The usual crystallization process requires that the latent heat be carried away from the interface region in order for crystallization to proceed. In solution growth, or growth of an alloy, the species which are not incorporated into the crystal increase in concentration in the parent phase, and then can be transported from the vicinity of the crystal by diffusion or other mass transfer processes. The third factor which is important is the intrinsic growth rate, the rate at which atoms or molecules can be added to the crystal; this has been discussed in detail above for silicon. In this section, a brief description of how these three factors interact during crystal growth will be presented. Experimentally, there are two different cases to be considered. The crystal grower can control either the temperature of the crystal growth medium, or he can control the rate at which the crystal grows, for example, by pulling it out of a furnace.

In the first case, illustrated in Figure 14(a), which is usual for growth from solutions, flux growth, LPE, hydrothermal growth, etc., a crystal is placed in a supersaturated (supercooled) medium. The composition and temperature are adjusted (or in many instances, programmed) so that the crystal grows as rapidly as possible while maintaining acceptable crystal quality. The growth rate is not controlled directly, but rather indirectly through the bath composition and temperature. In this case two extremes can be identified, as illustrated in Figure 15. T_M and T_B are the equilibrium and bath temperatures, respectively. The local equilibrium temperature in the vicinity of the crystal can be depressed to T_M' because the growing crystal rejects those species which tend to lower the growth temperature; the actual temperature of the crystal, T, must be above the bath temperature in order to remove the crystallization heat; and the interface temperature must be below the local equilibrium temperature if the crystal is to grow. The heat flow, mass transport and interface undercooling are coupled at the interface by the relation:

$$T_M - T_B = \Delta T_c + \Delta T_I + \Delta T_H \qquad (20)$$

Where $T_M - T'_M = \Delta T_c$, $T'_M - T = \Delta T_I$ and $T - T_B = \Delta T_H$. Any one of these terms can dominate in a particular instance, or any two can be important, or all three can be important. For solution growth, where the growth is usually slow, heat flow is often unimportant. In this case, two limits can be distinguished, one in which $T_M - T_B \approx \Delta T_c \gg \Delta T_I$. In this case mass transport dominates and the growth rate is "diffusion-limited," and $T_M' \sim T_B$. In the other case, $T_M - T_B \approx \Delta T_I \gg \Delta T_c$, so that $T_M' \sim T_M$, and the growth is said to be "interface-controlled."

Experimental control of the supersaturation is used most frequently when the growth is "interface-controlled", so that large supersaturations are necessary to get reasonable growth rates, and the slow intrinsic growth rate of the crystal suppresses interface instabilities.

The other case is where the crystal is placed in a temperature gradient and the growth rate is determined by the motion of the sample with respect to a heat source as illustrated in Figure 14(b). This growth mode is usually preferred when ΔT_I is small, and when ΔT_H is important, so that the growth is controlled by heat flow. The simplest case is Bridgman growth where an ampoule containing the sample is melted in a furnace and then slowly lowered from the furnace. The temperatures of the furnace, and perhaps also the temperatures outside the furnace are adjusted so that the crystal grows as the ampoule is extracted from the furnace. Eqn. 20 still applies, but the interface finds the temperature in the imposed temperature gradient where the interface undercooling is such that the growth rate keeps up with the motion of the isotherms through the sample. In this case, the growth rate is imposed directly on the sample. Of course there are limits within which stable growth of high quality crystal can be maintained.

Czochralski growth, where the crystal is pulled from a melt also falls into this class, and so does MBE, where the growth rate is controlled directly by the flux of atoms to the surface. Laser annealing also falls into this class. The scanned CW spot mode obviously does and so does the pulsed beam mode, at least in the sense that the rate of growth is controlled directly by the heat flow, which is controlled by the experimenter.

VIII. CONSTITUTIONAL SUPERCOOLING

There must always be some undercooling (or supercooling, the two words are used interchangeably) at the interface in order for crystal growth to occur. In solution growth, the

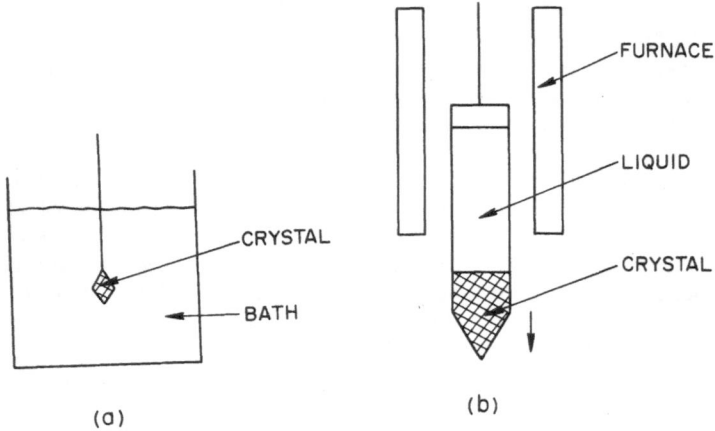

Figure 14a Schematic of crystal growth from an isothermal supersaturated solution.
Figure 14b Schematic of a Bridgman crystal growth apparatus where an ampoule is withdrawn from a furnace at a predetermined rate.

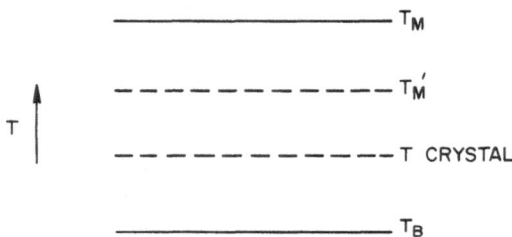

Figure 15 Schematic of the temperatures important for the growth of the crystal shown in Figure 14a. T_M is the saturation temperature of the liquid far from the crystal. T_M' is the saturation temperature of the liquid at the crystal surface. T is the actual temperature of the crystal which must be below T_M' in order for growth to occur. T_B is the bath temperature far from the crystal, which must be below T in order to remove the transformation heat from the crystal.

solution is usually isothermal and the entire solution is undercooled. The composition in the solution is usually enriched in the species rejected by the growing crystal, so that the supercooling is less in the vicinity of the crystal than far away from it. In this situation, the growth tends to be unstable, since the reduced undercooling near the crystal suppresses its growth rate. In this geometry, for example, a metal crystal in its melt, or ice in supercooled water, grows very rapidly and dendritically. The dendrites rapidly fill the container with crystal and remove the undercooling.

Crystals with slow growth kinetics which require large undercooling for growth can be stable in undercooled environments. Good quality crystals can be grown in this configuration only when interface instabilities are suppressed by the slow intrinsic growth rate or by the anisotropy in the growth rate.

For materials which have rapid, isotropic growth rates, i.e, those which are above their surface roughening temperature, a temperature gradient is used to stabilize the growth front. However, in a situation where the temperature increases into the growth medium, there can be a region ahead of the interface which is supercooled. This effect, which is due to the buildup of rejected species at the crystal growth front, is known as "Constitutional Supercooling" (Tiller et al., 1953). As in the isothermal growth case the rejected species have depressed the growth temperature at the crystal so that the supercooling tends to increase away from the crystal. However, in this case, the temperature also increases away from the crystal. This is illustrated in Figure 16. If the temperature increases more rapidly than the compositional effect, then the interface will be stable. However, if the compositional effect wins out then the constitutional supercooling is present, and instabilities can occur. The criterion for this which can be derived from Figure 16, predicts that the growth front should be unstable when

$$\frac{m\, C_0}{D} \left[\frac{1}{k} - 1\right] v > G \tag{21}$$

where m is the slope of the liquidus line on the phase diagram, $k = \dfrac{C_S}{C_L}$ is the distribution coefficient, D is the diffusivity in the liquid, v is the growth rate, C_0 is the composition far from the interface and G is the imposed temperature gradient. A stability analysis of the growth front under these conditions was first performed by Mullins and Sekerka (1964). They analyzed the growth of an infinitesimal perturbation of the interface, and determined the conditions under which such a perturbation would grow or shrink. The condition for growth of the perturbation is given by

$$\frac{m\, C_0}{D} \left[\frac{1}{k} - 1\right] vS > G^* \tag{22}$$

where G^* is a weighted average of the temperature gradients in the solid and liquid, and S is a "stability function." This condition is similar to equation 21 for normal growth conditions. The stability function S implicitly takes into account the effects of surface tension, and effects due to interface undercooling. The first of these is not important at slow growth rates, but the second is quite important for materials which require large undercoolings for growth. At rapid growth rates, the surface tension influences the onset of instability. The instability which results from constitutional supercooling leads to a "cellular" substructure (Chalmers, 1964).

The cellular spacing depends approximately on the diffusion distance D/v. It departs from this more or less for some growth conditions, but is usually within a factor of π or so of this value. At slow growth rates, $D/v = 10^{-4}/10^{-2} = 100\,\mu$ which is typical for cells in crystals grown at usual rates (Chalmers, 1964). For laser annealing, $D/v \sim 10^{-4}/100 \sim 100\,\text{Å}$ (Cullis et al., 1981).

Cellular substructures produced during laser annealing of silicon are discussed in more detail in Chapter 4 of this volume.

IX. DEFECTS PRODUCED BY LASER IRRADIATION

Laser irradiation of silicon using short heat pulses removes radiation damage and even the amorphous layer resulting from ion implantation, provided that the surface layer melting

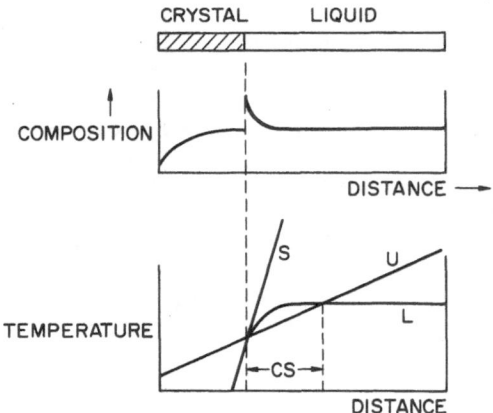

Figure 16   Illustrating constitutional supercooling which exists in the region CS if the actual temperature gradient (u) is shallow compared to the variation in liquidus temperature with distance (L) which is due to the variation in composition with distance. If the actual temperature gradient is steep enough (S) then there is no constitutional supercooling.

penetrates into the good crystal below the damage. The good crystal can then regrow out to the surface. The stresses associated with the temperature gradients do not produce dislocations. Continuous laser heating to melt a column of liquid through a wafer, on the other hand, does result in mechanical damage, which is evidenced by extensive slip surrounding the heated region. These two observations can be explained in terms of the stress generated by the heating in the two cases, and by the mechanical properties of silicon.

Silicon does not deform by dislocation motion at room temperature. It must be heated to several hundred degrees Celsius before the dislocations become mobile. A combination of the short time and small heated volume probably account for the absence of damage after pulsed laser heating. The strong radial temperature gradients associated with melting by a focused CW laser beam produce a high stress level in a volume of crystal which is hot enough to deform.

Metals, on the other hand, can deform readily at room temperature, and so they are also more susceptible to deformation during pulsed laser irradiation. The response of metal surfaces to laser irradiation has been studied in connection with high power laser optics for some time, and single-crystal materials have been investigated (Haessner and Seitz, 1971; Metz and Smith, 1971; Porteus et al., 1976). In general, extensive slip has been observed well below melt thresholds and a crystallographic orientation dependence has been found. The thermomechanical stress degradation of metal mirror surfaces under pulsed laser irradiation and has been studied (Musal, 1979). A surface temperature rise of only 20°K is sufficient under short-pulse large-spot laser irradiation to make pure copper yield plastically due to thermomechanical stress. The damage is manifest as progressive surface degradation in the form of slip bands, intergranular slip and fatigue cracks.

Ni single crystals have been irradiated with laser pulses, and extended defects are observed after irradiation (Buene et al., 1980). Defect generation occurs both at low energy input where the nickel is not melted, but merely rapidly heated and quenched, as well as for high energy inputs where the surface is rapidly melted and recrystallizes epitaxially. Figure 17 shows Rutherford backscattering (RBS) and channeling spectra from Ni (100) and (111) crystal faces irradiated with overlapping spots from a Q-switched ND:YAG laser. The crystals were cut from the same boule. Nomarski interference contrast microscopy

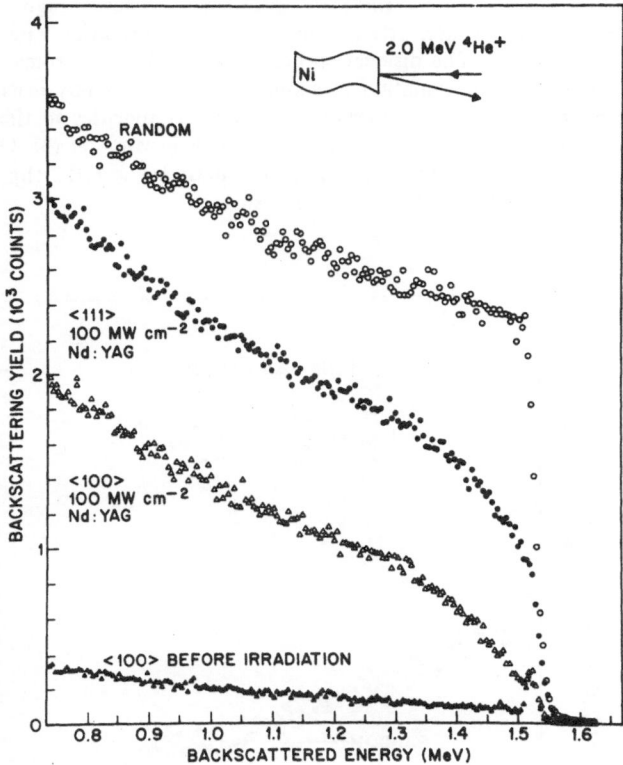

Figure 17   RBS and channeling spectra for Ni <100> and Ni <111> crystal faces irradiated at 100 MW/cm² with a Q-switched Nd:YAG laser. A higher density of defects is indicated for the <111> orientation.

clearly shows that the surface had been melted. The high dechanneling rate in the near-surface region indicates a high density of extended defects. Also, a higher defect density is indicated for the (111) orientation. Similar results were obtained by irradiating with a single pulse using a Q-switched Ruby laser.

Figure 18 illustrates the difference in the channeling behavior in the non-melting (25 MW/cm²) and melting (35 and 65 MW/cm²) regimes, for a (110) oriented surface (Draper et al., 1981). The melting and non-melting regimes were identified by depositing 10Å of Ag on the surface. At 25 MW/cm², the silver remained on the surface. After 35 and 65 MW/cm² irradiation, the silver had penetrated several thousand Angstroms into the nickel: a depth which can only be accounted for by melting.

The nature of the defects observed by channeling have been studied by TEM (Buene et al., 1981). Figure 19 shows TEM micrographs of the laser irradiated (100) Ni crystal from Figure 17. The defects are predominantly in the form of a dislocation cell structure, with two typical cell sizes and an estimated density of $10^{11} - 10^{12}/cm^2$. The TEM observation of the (111) crystal shows a laterally uniform dislocation network with a much higher dislocation density than the (100) crystal. Cross sectional TEM observations of laser irradiated Ni crystals show that slip dislocations are present.

The very high densities of dislocations observed in laser irradiated Ni can be explained in terms of plastic deformation during the rapid temperature cycling. The rapid heating and cooling process introduces high thermal gradients. These high gradients cause thermomechanical stresses which result in plastic deformation by movement and multiplication of dislocations. The dislocation cell structure found in some cases is typical of deformation in metals. The deformation of single crystals is strongly orientation dependent and the dislocation arrangements after deformation bear similarities to those observed here. The damage following laser irradiation in nickel is much greater on the (111) surface than (100) for nickel. For Mo, which has a bcc crystal structure where the slip planes are (110), the (110) surface is damaged more than the (111) surface.

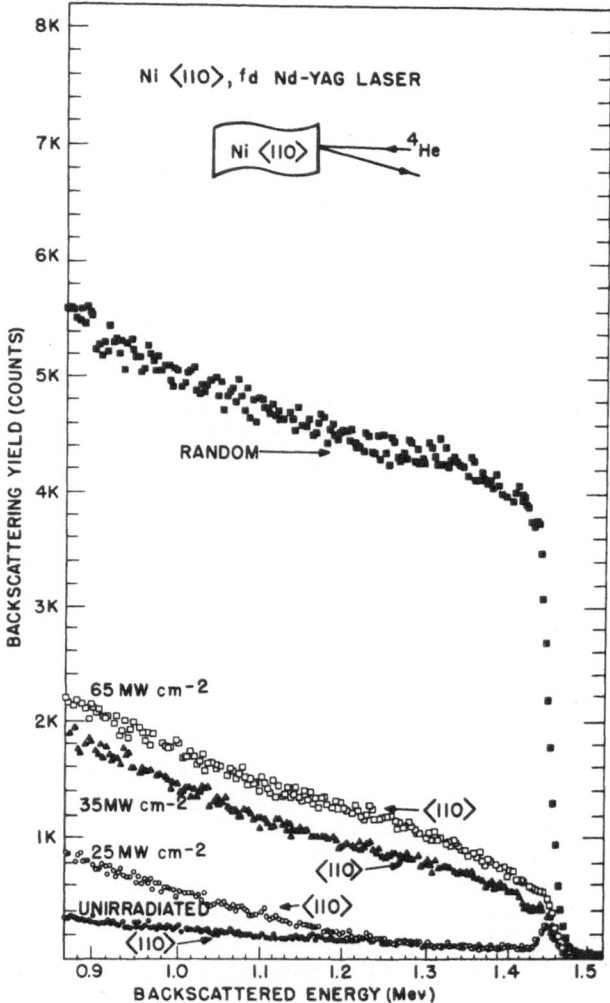

Figure 18   Rutherford backscattering and channeling spectra for Ni <110> with 10Å Ag vacuum deposited. Sample has been irradiated with frequency-doubled Q-switched Nd:YAG laser. The increases in backscattering yield as a function of incident power density indicate an increasing density of defects. At 25 MW/cm^2 the sample has not melted.

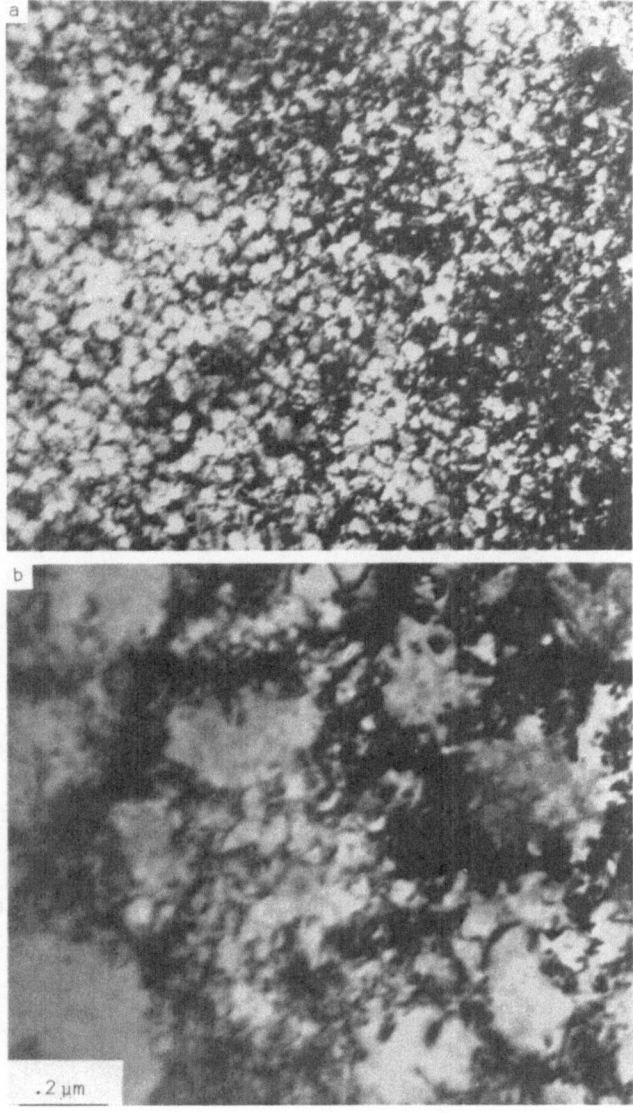

Figure 19   TEM micrograph of laser irradiated <100> Ni crystal. Dislocation free cell structures of strikingly different dimensions are seen.

REFERENCES

Aziz, M. (1982), J. Appl. Phys. *53*, 1158.

Baeri, P., Foti, G., Poate, J. M., Campisano, S. U., Cullis, A. G. (1981), Appl. Phys. Lett. *38*, 800.

Baeri, P., Campisano, S. U., Grimaldi, M. G. and Rimini, E. (1980) "Laser and Electron Beam Processing of Materials," (ed. C. W. White and P. S. Peercy), Academic Press NY, p. 131.

Bagley, B. G. and Chen, H. S. (1979) "Laser-Solid Interactions and Laser Processing," (ed. S. D. Ferris, H. J. Leamy, J. M. Poate), AIP Proc. 50, p. 97.

Baker, J. C. and Cahn, J. W. (1971). Solidification, ASM Metals, Park, Ohio, p. 23.

Broughton, J. Q., Gilmer, G. H. and Jackson, K. A. (1982), to be published.

Buene, L., Poate, J. M., Jacobson, D. C., Draper, C. W. and Hirvonen, J. K. (1980) Appl. Phys. Lett. *37*, 385.

Buene, L., Jacobson, D. C. Nakahara, S., Poate, J. M., Draper, C. W. and Hirvonen, J. K. (1982) "Laser and Electron-Beam Solid Interactions" (ed. J. F. Gibbons, L. D. Hess and T. W. Sigmon) North-Holland, N.Y., p. 583.

Burton, W. K., Cabrerra and Frank, F. C., Phil. Trans. Roy. Soc. (1951) A*243*, 249.

Chalmers, B. (1964) "Principles of Solidification," John Wiley and Sons, p. 103, 154.

Csepregi, E. F., Kennedy, T. J., Gallagher, T. J., Mayer, J. W., Sigmon, T. W. (1977), J. Appl. Phys. *48*, 4234.

Cullis, A. G., Hurle, D. T. J., Webber, H. C., Chew, N. G., Poate, J. M., Baeri, P., Foti, G. (1981) Appl. Phys. Lett. *38*, 642.

Cullis, A. G., Webber, H. C., Chew, N. G., Poate, J. M. and Baeri, P. (1982) Phys. Rev. Lett. *49*, 219.

Donovan, E. P., Spaepen, F., Turnbull, D., Poate, J. M. and Jacobson, D. C. (1982) to be published.

Draper, C. W., Buene, L., Poate, J. M. and Jacobson, D. C. (1981) Applied Optics *20*, 1730.

Frenkel, J., (1932). Physik. Z. Sowjetunion *1*, 498.

Galvin, G. J., Thompson, M. O., Mayer, J. W., Hammond, R. B., Paulter, N., Peercy, P. S. (1982) Phys. Rev. Lett. *47*, 33.

Gibbs, J. W. (1877-78) reprinted in J. W. Gibbs "The Scientific Papers" Vol. 1, p. 325, Dover, N.Y., 1961.

Gilmer, G. H. and Jackson, K. A. (1977) "Crystal Growth and Materials," ed. E. Kaldis and H. J. Scheel, North-Holland p. 79.

Gilmer, G. H. (1982) to be published.

Gilmer, G. H. (in course of publication).

Gilmer, G. H. and Leamy, H. J. (1980) "Laser and Electron Beam Processing of Materials," Academic Press, N.Y., p. 227.

Haessner, F. and Seitz, W. (1971). J. Mater. Sci. *6*.

Hertz, H., Ann. Phys. (1882) *17*, 177.

Jackson, K. A. (1958), "Liquid Metal and Solidification," ASM Cleveland, 1958, p. 174; "Growth and Perfection of Crystals," ed. R. H. Doremus, B. W. Roberts, D. Turnbull, Wiley and Sons, N.Y., p. 319.

Jackson, K. A. (1967) "Progress in Solid State Chemistry," Vol. 4, Pergamon Press, ed. H. Reiss, p. 53.

Jackson, K. A., Uhlmann, D. R. and Hunt, J. D. (1967). J. Cryst. Growth *1*, 1.

Jackson, K. A., Gilmer, G. H., Leamy, H. J. (1980) "Laser and Electron Beam Processing of Materials," (ed. C. W. White and P. S. Peercy), Academic Press, N.Y., p. 104.

Knudsen, M., Ann. Phys. (1909) *29*, 179.

Kossel, W. (1927), Nachr. Ges. Wiss. Göttingen p. 135.

Leamy, H. J., Gilmer, G. H. and Jackson, K. A. (1975) "Surface Physics of Materials I" (ed. J. B. Blakeley) Academic Press, N.Y., p. 121.

Liu, P. L., Yen, R., Bloembergen, N. and Hodgson, R. T. (1979). Appl. Phys. Lett. *34*, 864.

Metz, S. A. and Smith, F. A. (1971). Appl. Phys. Lett. *19*, 207.

Mullins, W. W. and Sekerka, R. F. (1964). J. Appl. Phys. *35*, 444.

Musal, H. M., Jr., (1979) Symp. on Optical Materials for High Power Lasers, Boulder, Colorado.

Porteus, J. O., Soileau, M. J. and Fountain, C. W. (1976). Appl. Phys. Lett. *29*, 156.

Ruhl, W. and Hilsch, P. (1977). Z. Phys. *B26*, 161.

Spaepen, F., Turnbull, D., (1979) "Laser-Solid Interactions and Laser Processing," AIP Proc. 50, p. 73.

Stranski, I. N. (1928). Z. Phys. Chem. *136*, 259.

Tiller, W. A., Jackson, K. A., Rutter, J. W. and Chalmers, B. (1953). Acta. Met. *1*, 428.

Tsu, R., Hodgson, R. T., Tan, T. Y. and Baglin, J. E. (1979). Phys. Rev. Lett. *42*, 1356.

Van DerSteenhoven, G. and Gilmer, G. H. (to be published).

Vergano, P. J. and Uhlmann, D. R. (1969), "Reactivity of Solids," (ed. J. W. Mitchell, R. C. DeVries, R. W. Roberts and P. Cannon) J. Wiley and Sons, N.Y., p. 713.

Volmer, M. and Flood, H., (1934), Z. Physik. Chem. (Leipzig) *A170*, 273.

Walker, J. L. (1964), p. 114 in Chalmers (1964).

Weeks, J. D. and Gilmer, G. H., (1979), Adv. Chem. Phys., *40*, 157.

White, C. W., Wilson, S. R., Appleton, B. R., Young, Jr., F. W. (1980) J. Appl. Phys. *51*, 738.

Wilson, H. A., Phil. Mag. (1900) *50*, 238.

Wood, R. F., (1980), Appl. Phys. Lett. *37*, 302.

CHAPTER 4

SEGREGATION, SUPERSATURATED ALLOYS AND SEMICONDUCTOR SURFACES

C. W. WHITE AND D. M. ZEHNER

Solid State Division, Oak Ridge National Laboratory, Oak Ridge, Tennessee

S. U. CAMPISANO

Istituto di Struttura della Materia, Universita di Catania, Italy

A. G. CULLIS

Royal Signals and Radar Establishment, Malvern, England

CONTRIBUTOR: P. Siffert

I. INTRODUCTION

In pulsed laser annealing of ion implanted semiconductors, the rapid deposition of energy into the near surface region leads to melting of the surface to a depth of several thousand angstroms, followed by liquid phase epitaxial regrowth from the underlying substrate (Ferris *et al.*, 1979; White and Peercy, 1980; Gibbons *et al.*, 1981; Appleton and Celler, 1982). In silicon, the velocity of the liquid-solid interface during solidification is calculated (Wang *et al.*, 1978; Baeri *et al.*, 1978) to be several meters/sec and this is confirmed (Galvin *et al.*, 1982) by recent measurements of time-resolved electrical conductivity of the molten layer. Following solidification, the annealed region is observed to be free of any extended defects (Narayan *et al.*, 1978; White *et al.*, 1979), and Group (III,V) impurities are observed to be highly substitutional in the lattice even when their concentrations greatly exceed equilibrium solubility limits (White *et al.*, 1980).

At the very rapid growth velocities that are achieved during laser annealing of ion implanted silicon, recrystallization of the melted region takes place under conditions that are far from equilibrium at the moving liquid-solid interface. The very high growth velocities which can be achieved during laser annealing and the ability to change the velocity in a predictable manner provide one with a unique opportunity to systematically study nonequilibrium crystal growth phenomena under well defined experimental conditions. These studies show that substitutional impurities can be incorporated into the lattice at concentrations that far exceed equilibrium solubility limits (White *et al.*, 1980; White *et al.*, 1979a). Values for the (nonequilibrium) interfacial distribution coefficient (k') from the liquid can be determined by comparing model calculations for dopant diffusion to experimentally measured dopant concentration profiles. These comparisons show that values for k' are much greater than corresponding equilibrium values k_o for Group (III,V) species in silicon (White *et al.*, 1980). Values for k' are functions of both growth velocity (Cullis *et al.*, 1980; Baeri *et al.*, 1981) and crystal orientation (Baeri *et al.*, 1981a). For each Group (III,V) species there is a maximum concentration (C_s^{max}) which can be incorporated substitutionally into the silicon lattice during laser annealing (White *et al.*, 1980). Values for C_s^{max} are functions of growth velocity, and are approaching predicted thermodynamic limits to solute trapping in silicon (White *et al.*, 1981). Values for C_s^{max} have been found to be limited by lattice strain (for the case of B in Si) and by interfacial instability which develops during regrowth caused by constitutional supercooling at the interface (White *et al.*, 1981). Interfacial instability leads to the formation of a well defined cell structure in the near surface region (Cullis *et al.*, 1980; Narayan *et al.*, 1981; Cullis *et al.*, 1981). Both the solute concentration in the liquid at which instability occurs and the resulting cell size can be predicted with reasonable accuracy using the Mullins and Sekerka perturbation theory treatment of interfacial instability modified to account for the large departures from local equilibrium at the interface during regrowth (Narayan *et al.*, 1981; Cullis *et al.*, 1981).

During liquid phase epitaxial regrowth, perfect planes of atoms grow from the liquid, layer by layer, and the process terminates at the surface. Recent experiments have shown that the laser annealing induced liquid phase epitaxial regrowth process takes place to the very outermost monolayer of the crystal (Zehner *et al.*, 1980, 1980a). These experiments have shown that the high velocities of solidification and extreme quenching rates can lead to metastable surface structures (Zehner *et al.*, 1981) and to surfaces with greatly modified surface electronic properties (Eastman *et al.*, 1981). In addition, it has been found that pulsed laser irradiation in UHV can be used to prepare "atomically clean" surfaces of Si and Ge (Zehner *et al.*, 1980b). The ability to prepare clean, well ordered surfaces in UHV in remarkably short processing times may have far-reaching implications in areas such as molecular beam epitaxy where the surface properties of the substrate determine the quality of the deposited film.

II. SEGREGATION, SUPERSATURATED ALLOYS AND SOLUTE TRAPPING

A. Determination of Distribution Coefficients

Interfacial distribution coefficients (k') can be determined by comparing model calculations for dopant redistribution during laser annealing to measured dopant profiles (White *et al.*, 1980). Interfacial distribution coefficients are defined as

$$k' = \frac{C_s}{C_l}$$

where C_s and C_l are solute concentrations in the solid and liquid phase at the interface. At low growth velocities when solidification occurs under conditions of local equilibrium at the interface, the interfacial distribution coefficient is the equilibrium distribution coefficient k_0 defined as (see Figure 1)

$$k_0 = \frac{C_s}{C_l}\bigg|_{eq}.$$

where C_s and C_l are concentrations in the solid and liquid phase (at fixed temperatures) as determined from the equilibrium phase diagram. During crystal growth, if $k' < 1$, then solute will accumulate in the liquid at the interface, but the ratio of concentrations in the solid and liquid phase at the interface is k'. Under conditions of local equilibrium at the interface (i.e., $k' = k_0 < 1$) if one has a liquid containing a uniform solute concentration n_0 (as illustrated in Figure 1), the first solid to freeze will have a solute composition of $k_0 n_0$. As solidification proceeds, solute accumulates in the liquid at the interface until a steady state concentration n_0/k_0 is reached in the liquid at the interface. Rejected dopant will be transported to the near surface regions and will appear as a terminal spike at the surface. Analytical solutions to the case depicted in Figure 1 (uniform solute concentration in the

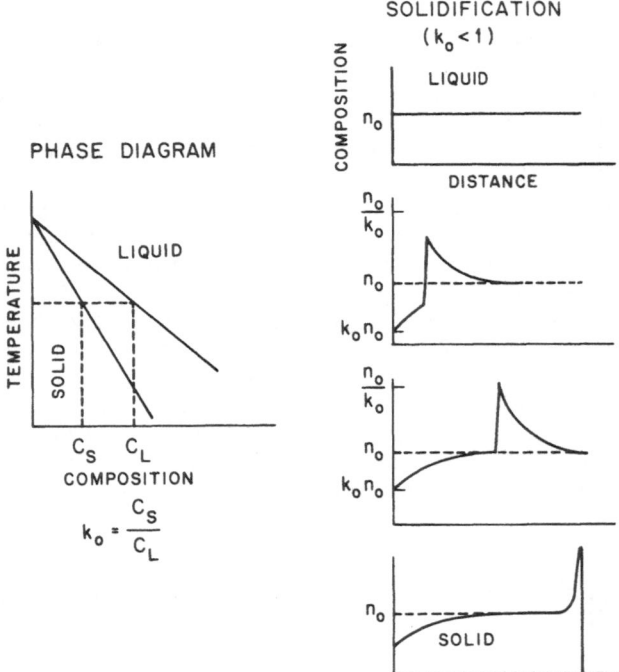

Figure 1   Relationship between phase diagrams, distribution coefficients (k_0), and solidification. The right hand
side shows solute profiles in the liquid and solid at several stages during solidification assuming $k_0 < 1$.

liquid, constant velocity of regrowth under conditions of local equilibrium, mass transport by
liquid phase diffusion only) have been published (Tiller *et al.*, 1953), and one need only
know n_0, k_0, D_l, and v to determine the solute profile in the solid. However, if n_0 is not
initially uniform (as is the case for ion implanted profiles) then numerical solution to the
mass diffusion equation is required in order to determine the solute profile in the solid.

In pulsed laser annealing, the deposited laser energy leads to melting of the near surface
region to a depth dependent on energy density, pulse duration time, and thickness of the
amorphous region. (Details of energy deposition, heat flow, meltfront penetration and
solidification velocity are given in Chapter 2.) As the liquid cools by thermal conduction to
the underlying substrate, the liquid-solid interface recedes toward the surface with a velocity
v which is of the order of meters/sec. During the time the implanted region is molten,
implanted impurities redistribute by diffusion in liquid silicon because liquid phase diffusion
coefficients D_l ($\sim 10^{-4}$ cm^2/sec) are far greater than solid phase diffusivities D_s
($\sim 10^{-12}$cm^2/sec for Group III,V species) even at the melting point.

The interfacial distribution coefficient during regrowth (k') has a profound influence on
the resulting dopant profiles following laser annealing. This is illustrated (Foti, 1981) in
Figure 2 for the case of Te, Cu and Bi in Si. Laser annealing causes all of the implanted Cu
to segregate to the surface. In contrast Te shows some redistribution with a greatly reduced
surface accumulation. Bi is an intermediate example in that it shows symmetric broadening
of the as-implanted profile in addition to a modest amount of surface accumulation. The
differences in these profiles following laser annealing are directly related to the differences in
k' for these three impurities in Si at solidification velocities of several meters/sec.

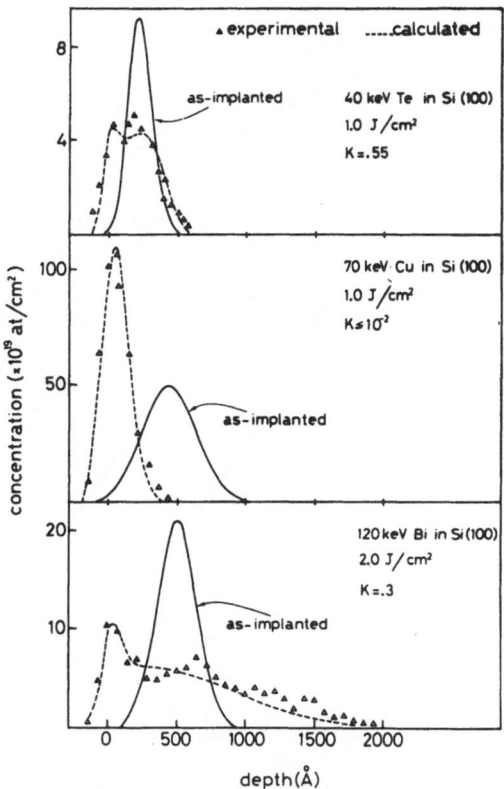

Figure 2   Experimental profiles for Te, Cu and Bi implanted in Si(100) and irradiated with a single ruby laser pulse. Full lines are the as-implanted profiles and dashed lines represent results of numerical calculations of mass transport taking the interfacial distribution coefficient as a fitting parameter. The k_0 values are $<10^{-6}$, 4×10^{-4} and 7×10^{-4} while k' values are 0.55, $<10^{-2}$ and 0.3 for Te, Cu and Bi respectively. From Foti, 1981.

Redistribution of the implanted dopant by liquid phase diffusion can be calculated by numerical solution to the mass diffusion equation expressed in finite differences (Wang *et al.*, 1978; White *et al.*, 1980; Wood *et al.*, 1981). In these model calculations, one includes the distribution coefficient k' as a fitting parameter to describe the partition of dopant between the solid and liquid at each stage during solidification. Values for k' are determined by comparing model calculations to experimental measurements of dopant profiles using least squares analysis.

Although the exact determination of k' requires the fitting of the whole impurity profile, some simple estimates can be obtained by measuring the amount of surface accumulation (Campisano *et al.*, 1980). Figure 3 shows the percent of surface accumulation as a function of the reduced thickness $\xi = vZ/4D_l$, where Z is the maximum liquid layer thickness, v the solidification velocity and D_l the liquid phase diffusion coefficient. Surface accumulation is defined as the amount of dopant remaining in the last 10% of the liquid layer at the conclusion of solidification. The curves are calculated for different k' values and refer to an initial constant distribution of the dopant in the liquid layer. The curves saturate at 90% because only the dopant transported by segregation is taken into account. The k' values

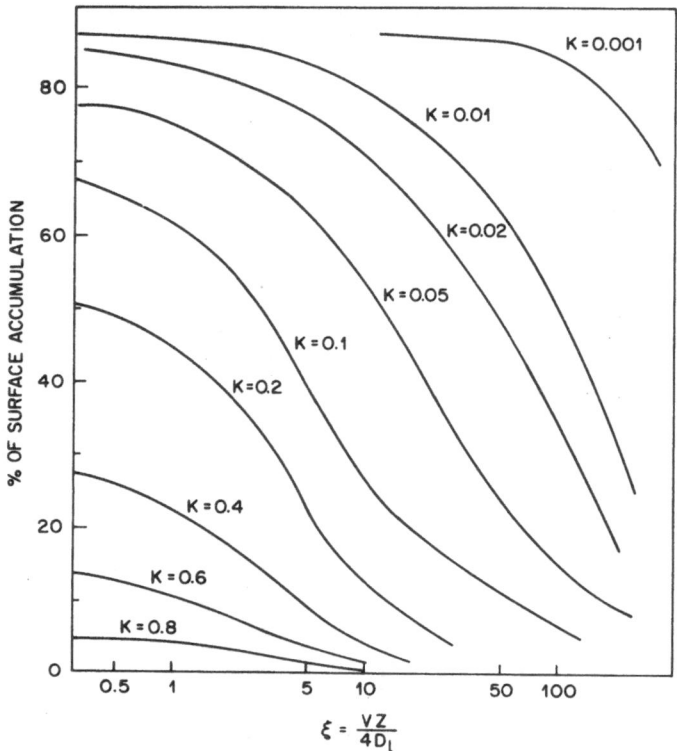

Figure 3   Calculated percentage of dopant transported by segregation to the upper 10% of a liquid layer as a function of the reduced thickness and for different k' values. From Campisano *et al.*, 1980.

determined in this way agree reasonably well with those obtained by fitting the whole profile for an initial gaussian distribution for $0.05 \leqslant k' \leqslant 0.5$ if Z is assumed to be equal to $R_p + \Delta R_p$, where R_p is the projected range and ΔR_p is the range straggling.

Figure 4 shows calculated instantaneous solute profiles in the solid and the liquid at various stages of solidification (Baeri and Campisano, 1982). These profiles are the result of numerical solution to the mass diffusion equation expressed in finite differences, and assuming a fixed value for k'. As depicted in Figure 4, impurities are rejected at the freezing interface and build up to high concentrations in the liquid at the interface. Freezing of the final layer results in the terminal concentration spike at the surface. By performing similar calculations using different values of k', and by comparing the calculated solute profiles in the solid to measured dopant profiles, one can determine the value of k' appropriate to these high speed growth conditions. Figure 5 shows such a comparison (White *et al.*, 1980) for the case of As in Si.

Following laser annealing, Rutherford backscattering-ion channeling measurements show that As is >95% substitutional in the lattice even though the ^{75}As concentration in the near surface region exceeds the equilibrium solubility limit by a factor of 4. This demonstrates the formation of a supersaturated alloy as a consequence of the high speed liquid phase epitaxial regrowth process. The solid line in Figure 5 is a profile calculated assuming a value for the distribution coefficient $k' = 1.0$. The agreement with the experimental profiles measured

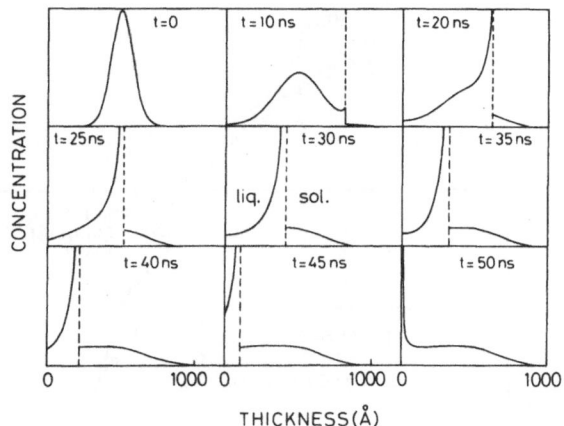

Figure 4   Solute profiles in the liquid and solid at various stages during solidification. For these calculations $k' = 0.1$, $v = 2$ m/sec and $D_l = 10^{-4}$ cm^2/sec. From Baeri and Campisano, 1982.

after laser annealing (solid circles) is excellent. The value determined for k' is considerably higher than the equilibrium value ($k_0 = 0.3$). The increase in the distribution coefficient relative to the equilibrium value is a consequence of the high regrowth velocity which causes a departure from conditions of local equilibrium at the interface during solidification. The fact that As shows no evidence of segregation to the surface is a further indication that k' must be very near to unity.

Figure 6 shows similar results for In in Si. Following laser annealing the terminal concentration spike at the surface shows that the value for k' is less than unity. The In remaining in the bulk of the crystal after laser annealing is >90% substitutional in the lattice (ion-channeling measurements) even though the In concentration in that region exceeds the equilibrium solubility limit by a factor of ~50. The profile in the bulk can be fit with reasonable accuracy using a value for $k' = 0.15$. By contrast, a profile calculated using the equilibrium value for In in Si ($k_0 = 4 \times 10^{-4}$) is shown by the dashed curve in Figure 6. If solidification occurred under conditions of local equilibrium at the interface, almost all of the In would have been zone refined to the surface with very little remaining in the bulk. Clearly, this does not fit the experimental data.

Using similar methods, values for k' have been determined for Group III,V impurities in silicon at the very high growth velocities which can be achieved by pulsed laser annealing (White et al., 1980). The values determined for k' during laser annealing are listed in Table I and compared with corresponding equilibrium values (k_0). The results for k' were determined at a growth velocity of 4.5 m/sec except for the cases of B, P and Sb where a velocity of 2.7 m/sec was used. As shown in Table I, in each case k' is considerably higher than k_0 by factors that extend up to ~600. These large increases in k' relative to k_0 reflect the highly nonequilibrium nature of the laser annealing induced liquid phase epitaxial regrowth process. Results shown in Figures 5 and 6 and in Table I demonstrate the potential for detailed studies of high speed, nonequilibrium crystal growth processes under well controlled experimental conditions.

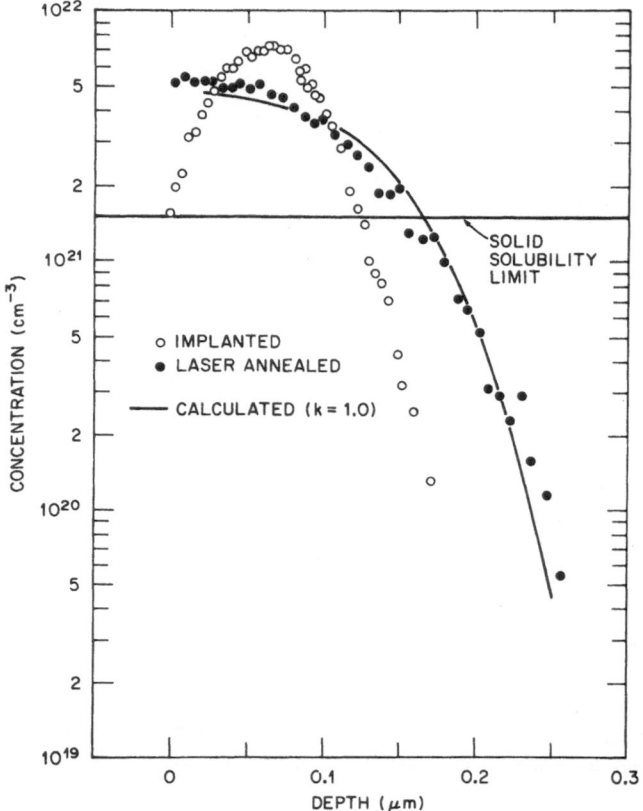

Figure 5   Dopant profiles for ^{75}As (100 keV, 6.4×10^{16}/cm^2) in (100) Si compared to model calculations. The horizontal line indicates the equilibrium solubility limit. From White *et al.*, 1980.

B. Dependence of k' on v

The experimental data presented thus far show that for most of the implanted dopants at large solidification velocities, k' becomes larger than k_0 and eventually approaches unity. However, at a given irradiation condition (i.e., at fixed liquid-solid interface velocity) the resulting k' value is not correlated to the corresponding equilibrium distribution coefficient, k_0, as shown in Figure 2. This suggests that the impurity redistribution process is controlled by kinetics at the interface rather than by thermodynamic equilibrium constraints. In the past, theoretical treatments (Baker and Cahn, 1969; Jindall and Tiller, 1978) have been used to describe nonequilibrium distribution coefficients during rapid crystallization, but no experimental data were available. Laser annealing experiments are providing new insights to such studies and a review is given in Chapter 3 of this volume by Jackson.

From the acquired experimental knowledge and by anticipating some later details one can argue that the interfacial distribution coefficient depends upon velocity as schematically illustrated in Figure 7. A critical velocity value v_{crit} separates the low velocity range, where $k' \simeq k_0$, and the high velocity range where $k' > k_0$ and its value saturates. The determination of v_{crit} and k'_{sat} are crucial points which any theory should deal with. As will

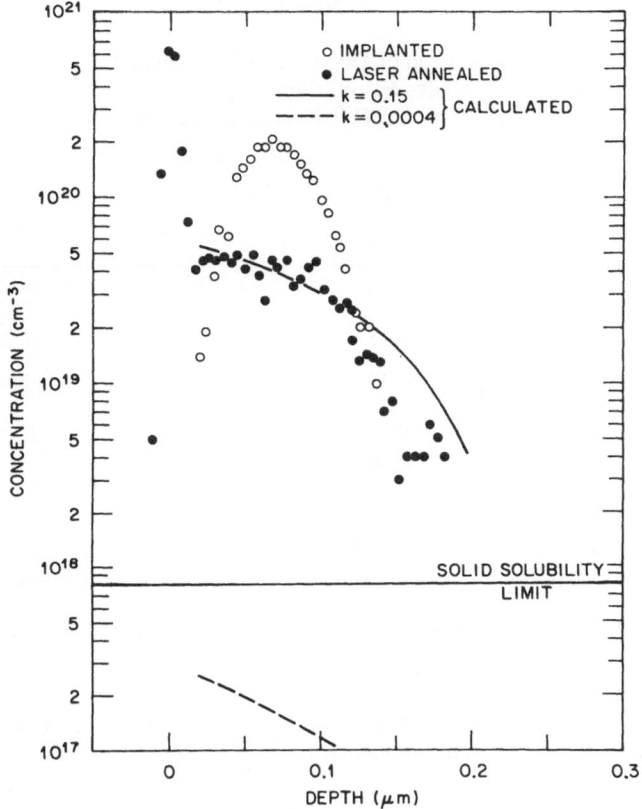

Figure 6   Dopant profiles for ^{115}In (125 keV, 1.2×10^{15}/cm^2) in (100) Si compared to model calculations. The horizontal line indicates the equilibrium solubility limit. The dashed profile is calculated assuming $k' = k_0$. From White et al., 1980.

be discussed later, k' determines the maximum dopant concentration: if $k' < 1$ interfacial instabilities due to constitutional supercooling limit the amount of dopant that can be trapped in lattice sites. If $k'_{sat} = 1$ other mechanisms, such as mechanical strain for the case of B in Si, control the maximum dopant concentration. The determination of v_{crit} and its dependence upon various parameters should clarify the role of thermodynamics and kinetics in determining k'.

To perform such an investigation over a wide velocity range (0.1-20 m/sec) and under well controlled and reproducible conditions it is necessary to clarify the role of the experimental parameters that can be changed during laser irradiation. Most of the functional dependences can be qualitatively derived by simple approximations. However, the exact solidification velocity can be calculated only through numerical solution to the heat conduction equation (see Chapter 2 and Baeri and Campisano, 1982). These calculations have shown that during irradiation the temperature in the near surface region increases and the sample surface melts. After the molten layer has penetrated to its maximum depth and the solidification process begins, the temperature gradient in the liquid is negligible. Under

TABLE I

Comparison of Distribution Coefficients
Under Equilibrium (k_0) and
Laser Annealed (k') Regrowth Conditions

Dopant	(a) k_0	(b) k'
B	0.80	~1.0
P	0.35	~1.0
As	0.30	~1.0
Sb	0.023	0.7
Ga	0.008	0.2
In	0.0004	0.15
Bi	0.0007	0.4

(a)   From Trumbore, 1960.
(b)   Values for k' were determined at a growth velocity of 2.7 m/sec. for B, P and Sb and at 4.5 m/sec for As, Ga, In and Bi.

these conditions, the solidification velocity is given by the rate at which the enthalpy of melting ΔH_m can be extracted from the melted layer:

$$v = \frac{\kappa}{\rho \Delta H_m} \left[\frac{dT}{dx} \right]_s \tag{3}$$

where ρ is the mass density of the material, $\left[\dfrac{dT}{dx} \right]_s$ is the temperature gradient in the solid at the interface, and κ is the thermal conductivity in this region.

The temperature gradient is controlled by energy deposition and diffusion processes. The laser energy is deposited in a depth given by either the absorption length α^{-1} (where α is the absorption coefficient of the laser light) or the thermal diffusion length $\sqrt{2D\tau}$, where D is the thermal diffusivity ($\sim 10^{-1} \text{cm}^2/\text{sec}$ in Si) and τ the pulse duration time. The larger of these two parameters determines the spatial extent of energy deposition. For pulses whose energy density exceeds the threshold for melting the entire amorphous layer, the quantity $\sqrt{2D\tau}$ must be replaced by $\sqrt{2D\tau_s}$, where τ_s is the surface melt duration time. The following parameters which influence the growth velocity have been identified: (a) Pulse Duration: when $\sqrt{2D\tau} > \alpha^{-1}$, the laser energy is strongly absorbed in the near surface region. In this case, for irradiation near the threshold energy density, the temperature gradient in the solid can be approximated by $\left[\dfrac{dT}{dx} \right]_s \sim T_m / \sqrt{2D\tau}$, where T_m is the melting temperature. This approximation is valid for a ruby laser pulse ($\tau > 20$ nsec) incident on a Si crystal with an amorphous layer. (b) Absorption Coefficient: For short pulses, when $\sqrt{2D\tau} < \alpha^{-1}$, the temperature gradient can be approximated as $\left[\dfrac{dT}{dx}\right]_s \sim T_m \dfrac{\alpha}{e}$. For strongly absorbed light this value is an upper limit and cannot be increased appreciably by changing other parameters. This approximation is valid for the case of nanosecond (or less) pulses of ruby or $Nd(\dfrac{\lambda}{2})$ on Si. For strongly absorbed short pulses (i.e., $\sqrt{2D\tau} > \alpha^{-1}$) the free carrier diffusion length determines the thermal deposition profile (Yoffa, 1980). By contrast, the irradiation of an almost transparent solid (Nd on Si single crystal) can be used as a way to reduce the liquid-solid interface velocity (Baeri, 1982) by reducing the thermal gradient in the solid. Finally, if melting occurs on a heated substrate, the solidification process will be rather slow. (c) Energy Density: When the pulse energy density exceeds the threshold energy density for annealing (E_{th}), the near surface region remains melted for a time τ_s larger than τ. Heat flow calculations and *in situ* reflectivity measurements (Auston *et al.*, 1978) have shown that $\tau_s \propto (E - E_{th})^2$ so that $\left[\dfrac{dT}{dx}\right]_s \propto \dfrac{1}{E - E_{th}}$. The variation of velocity with energy density is then very effective just above the threshold for annealing. (d) Substrate Temperature: Historically, changing the substrate temperature was the first method used to change the growth velocity (Cullis *et al.*, 1980). A variation in substrate temperature changes both the thermal diffusivity and the extreme values of the temperature thus affecting the whole temperature profile. In general the velocity increases by decreasing the substrate temperature.

The calculated (Baeri and Campisano, 1982; Baeri, 1982) dependence of velocity on pulse duration and energy density is shown in Figure 8 for the irradiation of a 1000Å thick layer of amorphous Si on a single crystal substrate with a ruby laser. For a pulse duration of 15 nsec, an energy density of 0.7 J/cm^2 is required to melt through the amorphous layer and the corresponding solidification velocity is 6 m/sec. On increasing the energy density to 1.0 J/cm^2, the molten layer thickness increases to 2200Å but the solidification velocity is reduced to 4 m/sec.

Detailed calculations of the liquid-solid interface kinetics show that the solidification velocity is a function of the interface position as shown by the full lines of Figure 9 (Baeri, 1982). The velocity has its maximum value near the beginning of the solidification process and then decreases, due to the decreasing thermal gradients in the solid. The values referred to previously and those reported in Figure 8 refer to the average velocity in the outer 1000Å at the sample surface.

The velocity thus far has been assumed to be determined only by heat flow, with the interface temperature at the melting point. However, some interfacial undercooling must be included to provide the driving force required for a finite solidification rate. No simple estimate of undercooling can be found in the literature at these growth rates. As a first approximation one can assume that the undercooling is proportional to the growth velocity (Baeri, 1982)

$$\Delta T = T_m - T_i = \beta v \qquad (4)$$

This correction has been included in the calculation assuming $\beta = 15$ K/m/sec. Including the effects of undercooling modifies Eq. 3 to

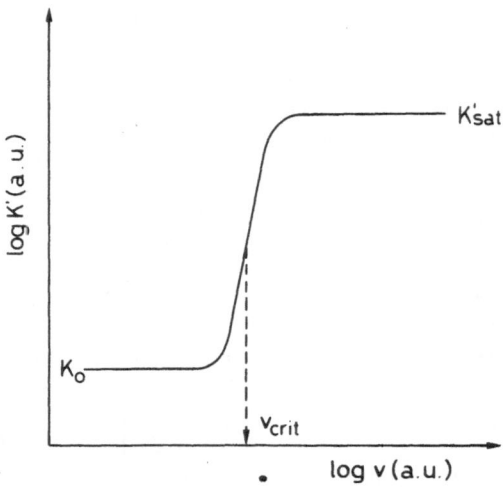

Figure 7   Schematic dependence of the interfacial distribution coefficient k' on the growth velocity v.

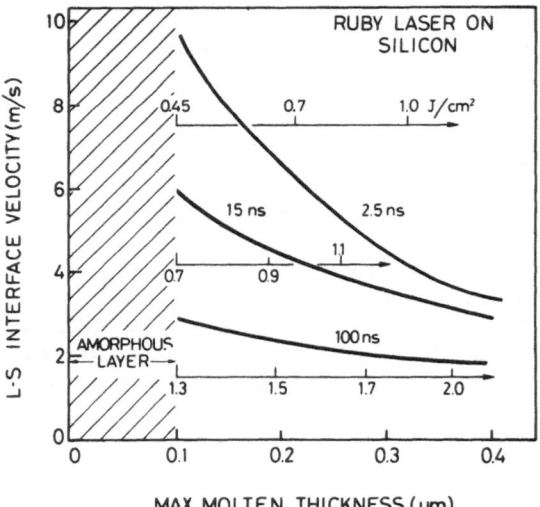

MAX. MOLTEN THICKNESS (μm)

Figure 8   Calculated velocity of the liquid-solid interface as a function of the maximum melted thickness for three different ruby laser pulse duration times. Velocities are calculated in the outer 1000Å thick surface layer, which was originally amorphous. From Baeri and Campisano, 1982; Baeri, 1982.

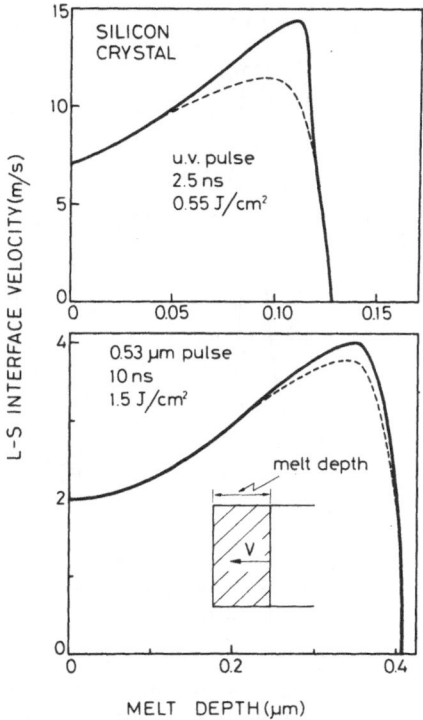

Figure 9   Liquid-solid interface velocity calculated as a function of the interface depth for two different irradiation conditions. Full lines are calculated assuming that the solidification occurs at the melting point. Dashed lines are calculated assuming that solidification occurs with an interfacial undercooling of 15 K/m/sec. From Baeri, 1982.

$$\Delta H_m\,\rho v + \rho C \Delta T v = \kappa \left[\frac{dT}{dx}\right]_s \tag{5}$$

where C is the specific heat. The heat extraction rate must now balance the latent heat and provide the necessary undercooling. The corresponding velocity will then be depressed. The correction will increase with increasing solidification velocity. Dashed lines in Figure 9 are the calculated velocities obtained using a velocity dependent solidification temperature. The assumption of undercooling gives rise to reduced dependence of velocity on depth. Moreover, in the outer 1000Å thick layer velocities calculated with and without undercooling almost coincide, due to the finite thickness of the liquid layer.

The liquid-solid interface velocity can be changed within the range 0.1-20 m/sec through a variation of all these parameters. However, most of the experimental work thus far has been performed in a more restricted range of 1-5 m/sec. As examples, Figures 10, 11 and 12 report impurity distribution profiles after laser irradiation under differing conditions. Figure 10 shows the profiles of Sb (40 keV, $5 \times 10^{15}/cm^2$) in (100) Si and irradiated with 15 nsec ruby laser pulses of 1.0 J/cm² (upper) and 1.8 J/cm² (lower) respectively.

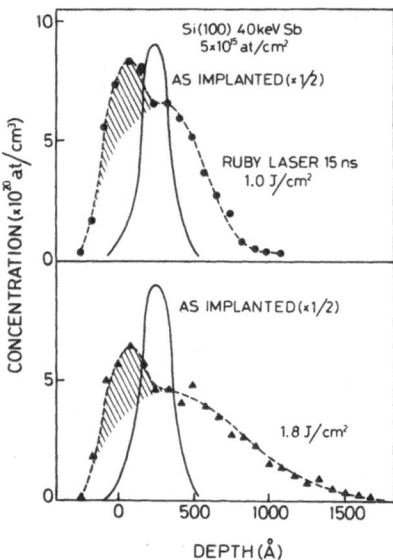

Figure 10   Experimental (dots and triangles) depth profiles for Sb (40 keV, 5×10^{15} at/cm²) Sb implanted in Si(100) and irradiated with 15 nsec ruby laser pulse at 1.0 and 1.8 J/cm². Dashed lines represent calculated profiles assuming $k' = 0.6$ in both cases.

Although the final profiles are different the amount of surface accumulation (dashed area) does not change appreciably. The dashed lines are the numerical fits performed using $k' = 0.6$ in both cases. This value is in good agreement with the estimate one can get from Figure 3. These results show that for Sb in Si, the value for k' does not change appreciably in the velocity range of 1-4 m/sec.

In the same velocity range, however, some impurities exhibit a velocity dependent k' value. Figure 11 shows the effect of changing the liquid-solid interface velocity by changing either the energy density for a fixed pulse duration time (left) or by changing the substrate temperature while keeping pulse duration and energy density fixed (right). These results were obtained for the case of 150 keV In implanted into Si (Baeri, 1982). For the results shown in Figure 11, it is significant to note that the resulting profile and the values for k' are determined only by the growth velocity and not by the parameter used to change the growth velocity.

From these investigations, it is possible to derive experimentally the dependence of k' on solidification velocity (Baeri *et al.*, 1980), as shown for Bi and In in Figure 12. For Bi, the data indicates a saturation in k' at the value of about 0.3. The comparison between Bi and In indicates that different values of v_{crit} (see Figure 7) characterize the two dopants. For a better understanding of the phenomenon however a wider velocity range should be investigated.

C. Picosecond Laser Annealing Results

The upper limit of the velocity range which can be investigated is limited by two physical mechanisms. The liquid-solid interface velocity is controlled by the temperature gradient in the solid and this cannot be arbitrarily large. In the case of picosecond pulses or strongly absorbed radiation the liquid-solid interface velocities which can be achieved are limited by

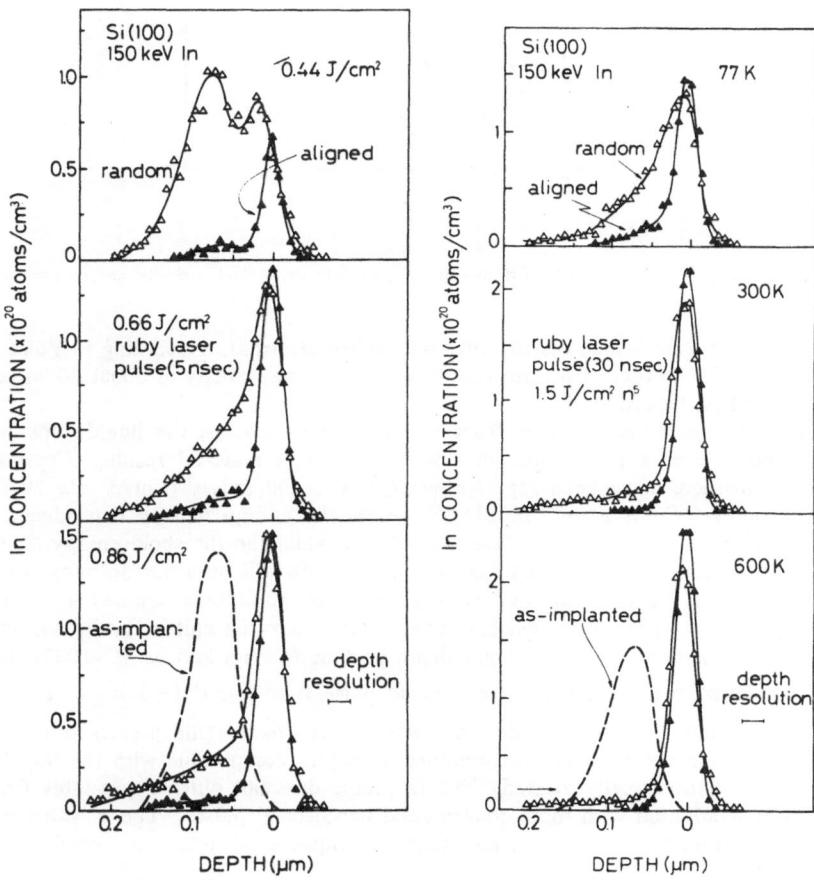

Figure 11 Experimental profiles for In (150 keV, $10^{15}/cm^2$) implanted in (100) Si and irradiated with a ruby laser
pulse of 5 nsec (left) or 30 nsec (right) and changing energy density (left) or substrate temperature
(right). The calculated velocities are (from top to bottom) left: 8.0, 5.5 and 3.5 m/sec; right: 4.0, 3.0
and 2.0 m/sec. The percentage of surface accumulation increases continuously with decreasing velocity.
From Baeri, 1982.

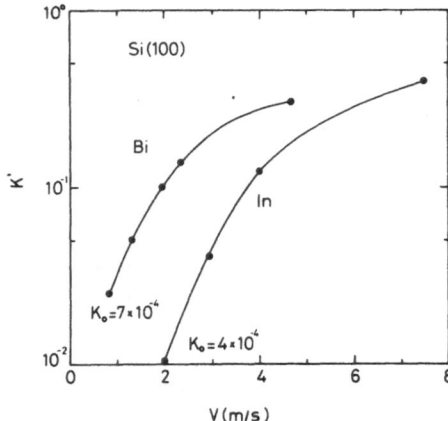

Figure 12   Experimental plot of k' of Bi and In in Si as a function of the liquid solid interface velocity.

the free carrier diffusion which redistributes the excited electrons. According to Yoffa (1980) this diffusion is of the order of 0.1 μm thus giving a limiting velocity of about 20 m/sec even when picosecond pulses are used.

To obtain a crystalline structure from a solidification process, the liquid-solid interface velocity cannot exceed a given limit, otherwise amorphous material results. Crystalline to amorphous transitions have been reported using picosecond pulses (Liu *et al.*, 2979), and 2 nsec UV pulses (Cullis *et al.*, 1982). However, the energy density dependence of the liquid-solid interface velocity will reduce the velocity when the threshold energy density for the amorphous transition is exceeded slightly. Thus there will be a narrow range of energy densities which can be used to produce crystalline to amorphous transitions even when picosecond pulses are used. At somewhat higher energy densities epitaxial regrowth will take place. Figure 13 shows the experimental depth profiles (Campisano *et al.*, 1982) for In in (100) Si irradiated with a single 25 picosecond pulse from a Nd ($\frac{\lambda}{2}$) laser at an energy density $E < 1$ J/cm^2. The crystalline order was restored to the silicon surface layer by the laser irradiation and the In atoms are retained at depths comparable with the implantation range and are substitutionally located. The In profile does not differ appreciably from that obtained after irradiation with more conventional nanosecond pulses. The k' value for In is 0.3 and from Figure 11 we can argue that the liquid-solid interface velocity is about 6 m/sec.

D. Orientation Dependence

The low velocity range ($v < 1$ m/sec) which has hardly been investigated also exhibits some interesting features. The v_{crit} values of many dopants should lie in this range. Moreover, in the proximity of the critical value, the role of all the parameters which determine k' should be emphasized. At high velocities all these dependences should be dominated by rapid kinetics. An example of the importance of these parameters in the low velocity regime is the effect of the substrate orientation. Figure 14 shows the experimental Bi profiles after 2.0 J/cm^2 100 nsec ruby laser pulse on (100) and (111) oriented Si substrates (Baeri *et al.*, 1981a). The velocity in both cases is 1.7 m/sec but the two profiles differ noticeably. The k' fitting values are 0.1 and 0.2 for (100) and (111) respectively.

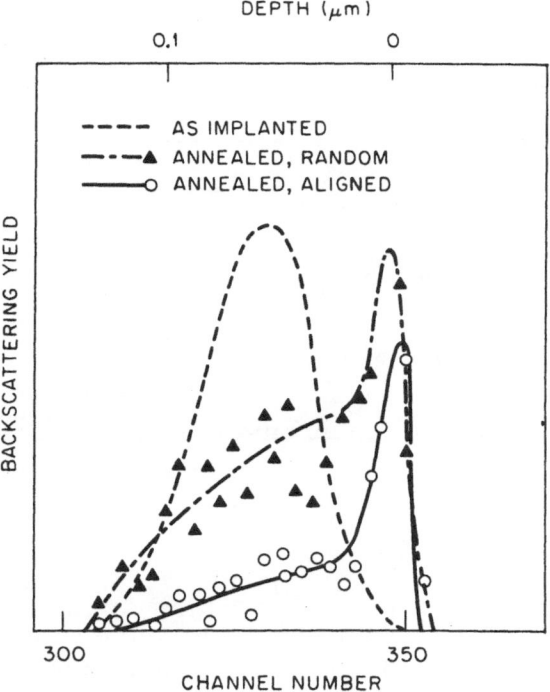

Figure 13   Profiles for In implanted into (100) Si and irradiated with a single 25 picosecond Nd ($\lambda/2$) pulse ($\sim$1 J/cm^2). The dashed line indicates the as-implanted profile and filled and open circles refer to the random and (100) aligned yields respectively after laser annealing. From Campisano *et al.*, 1982.

The effect was investigated in the 0.8-5 m/sec velocity range and results are summarized in Figure 15 (Baeri *et al.*, 1981a). The k' values determined for the (111) substrate are larger than the corresponding values for the (100) orientation. This difference is large at small velocities while above 3 m/sec the values coincide and saturate toward 0.3. The difference in k' with orientation may be related to differences in undercooling on (100) and (111) surfaces and is discussed in Chapter 3.

E. Additional Parameters Affecting k'

The trend schematically depicted in Figure 7 has been demonstrated. However, the v_{crit} value depends on the dopant. For impurities such as Fe, Cu and Ag in Si it has not been possible to reach velocities such that $k' > 0.05$. It is usually believed that for such dopants $k' \simeq k_0$ in the commonly investigated velocity range. The feature that is common to these systems is the high diffusion coefficient in the solid phase, and therefore one may suspect that this last parameter enters into the determination of v_{crit}. To test this hypothesis experiments have been performed using Si and Ge substrates (Baeri and Barbarino, 1982). Dopants that are slow solid phase diffusers in both substrates, such as Te and Sb, are strongly trapped in Si and Ge at concentrations far exceeding the equilibrium solubility. The k' values fitting the impurity profiles are larger than the corresponding k_0 values. A few dopants such as Pb and Zn, have different solid phases diffusivities in Si and Ge. They are slow diffusers in Ge and

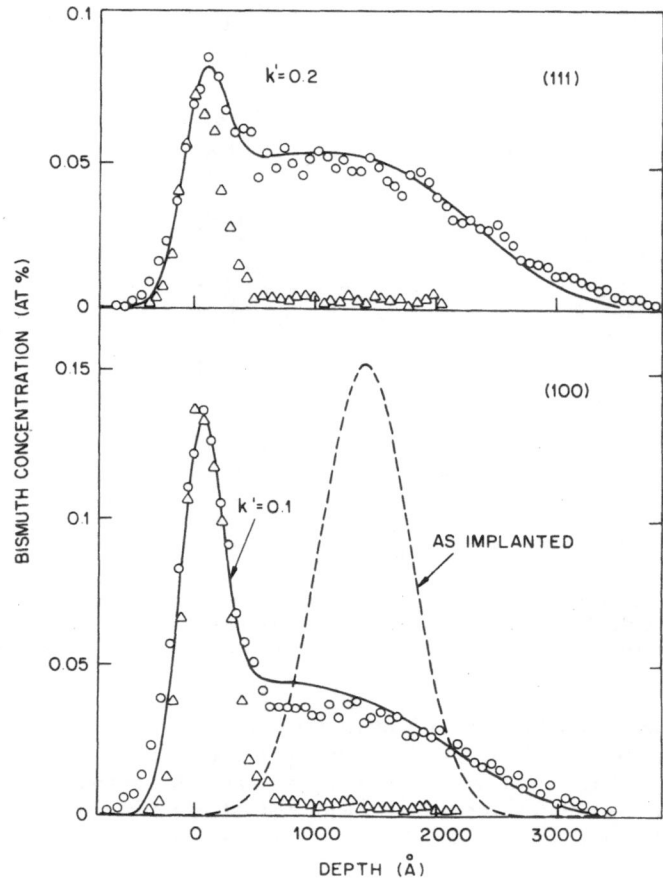

Figure 14   Depth profiles of 250 keV Bi implanted in Si and irradiated with 100 nsec ruby laser pulses of 2.0 J/cm². The full line is the result of the numerical calculation using k' values of 0.2 and 0.1 for the (111) and (100) orientation respectively. Circles (triangles) refer to the random (aligned) backscattering spectrum. From Baeri *et al.*, 1981a.

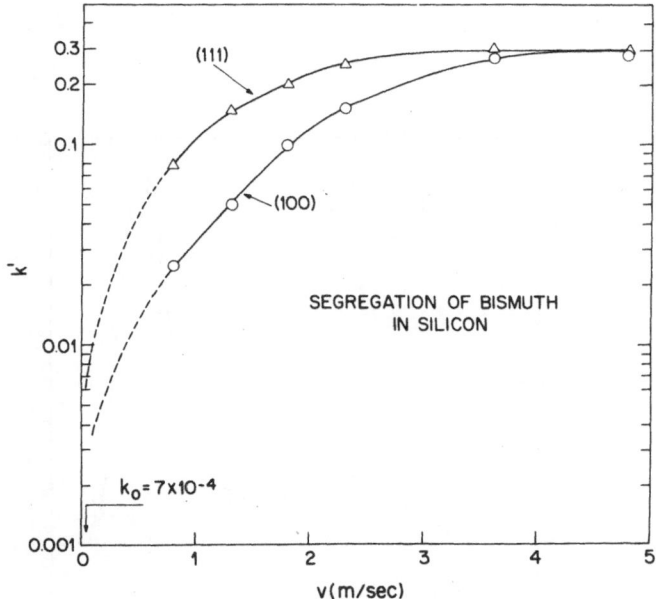

Figure 15   Plot of the k' values of Bi in Si as a function of the liquid-solid interface velocity and substrate orientation. From Baeri *et al.*, 1981a.

fast diffusers in Si. Results are shown in Figure 16 for the case of Pb implanted into Si and Ge and subsequently laser annealed. The impurity profile in Si shows strong surface accumulation after laser irradiation. The small fraction that is not segregated does not show any channeling effect, and therefore does not occupy a specific lattice site. The amount of retained Pb is strongly dependent upon laser energy density i.e., upon liquid-solid interface velocity. This behavior is well understood in terms of interface instability due to constitutional supercooling, as discussed later in this chapter. However, Figure 16 (lower) shows that if Ge substrates are used, the Pb fraction retained inside the crystal is substitutionally located at concentrations well above the solubility limit. The profile can be fit with a value for $k' = 0.2$ which is larger in comparison to $k_0(1.7 \times 10^{-4})$. Very similar results have been obtained using Zn implanted into Si and Ge. Rutherford backscattering analysis indicates complete accumulation of Zn on the Si surface. Ion induced x-ray fluorescence indicates instead a partial attenuation of the Zn yield in channeling conditions for Ge substrates. The substitutional dose was 2.5×10^{15} at/cm^2 and the implanted dose was 10^{16} at/cm^2. The profile should then be very similar to that of Pb in Ge and the corresponding k' much larger than the equilibrium k_0 for Zn in Ge. Therefore, it is clear that trapping can occur in this velocity range for dopants characterized by small diffusivities in the solid phase, while faster diffusers are rejected at the sample surface.

    A description of the trapping process in terms of competition between interface and impurity kinetics has been outlined by Turnbull (1980) and modified by Campisano *et al.*, (1980a). The time required to regrow a layer equal to the interface thickness, $t_i = \dfrac{\lambda}{v}$, is compared with the average time the dopant spent in this region, $t_d = \dfrac{\lambda^2}{D_i}$ where D_i is the

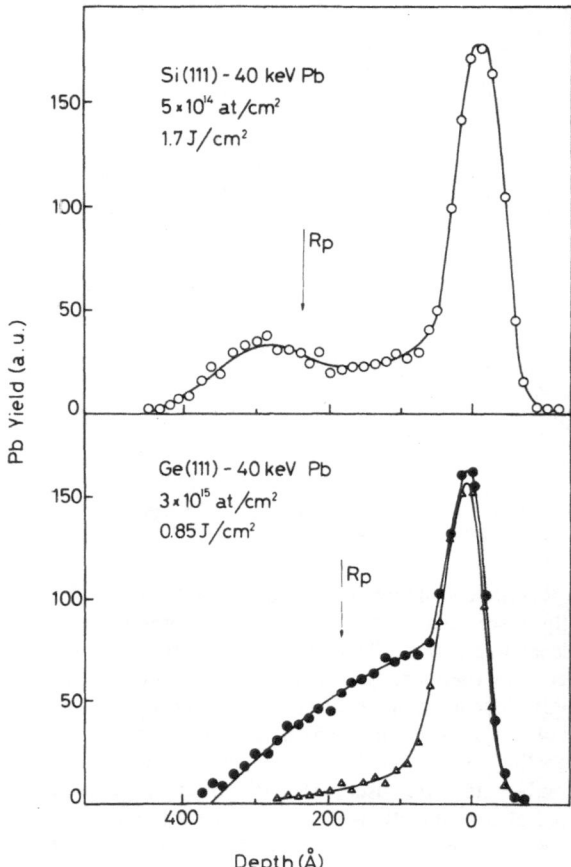

Figure 16   Experimental depth profiles of 40 keV Pb implanted in Si (upper) and Ge (lower) and irradiated with
ruby laser pulses. In the Si substrate the Pb yield does not show any channeling attenuation along the
<111> direction, while in Ge the Pb is substitutional at concentrations which are far in excess of the
equilibrium solubility. For the Ge results, the filled circles represent the random spectrum, and the open
triangles represent the <111> aligned yield.

diffusion coefficient in the interface region. Trapping can occur if the dopant resides in this extended interface region longer than the time required to regrow it, i.e. if $t_i < t_d$. In the opposite case the dopant is rejected into the high solubility phase, i.e. into the liquid. The critical velocity is then

$$v_{\text{crit}} = \frac{D_i}{\lambda} \qquad (6)$$

Assuming that the interface has properties intermediate between those of the two adjacent media, we can put $D_i \sim \sqrt{D_s D_l}$ where the subscript s and l refer to solid and liquid phase respectively. Diffusivities in the solid and liquid phases are calculated at the interface temperature, i.e. close to the melting point. D_l is of the order of 10^{-4} cm²/sec for any dopant in liquid Si; D_s can range from 10^{-11} to 10^{-4} cm²/sec for slow and fast diffusers in solid Si. These values give v_{crit} values of the order of a few cm/sec and a few m/sec for slow and fast diffusers in Si, in good agreement with the experiments.

The simple picture developed thus far is not sufficient to predict the k' values as a function of crystallization velocity. This can be accomplished by means of a more refined model. Examples of these models are:

(a) Jindall and Tiller (1968): This model predicts the correct trend at high velocities but is not adequate at low velocities. The model does appear to predict the correct value for v_{crit} and its dependence on D_s.
(b) Baker and Cahn (1969): This is a thermodynamic approach to the non-equilibrium freezing stating very general conditions for solute trapping to occur.
(c) Wood (1982): This is a model based on rate equations using a very thick interface with properties gradually changing between those of the two adjacent media and a velocity dependent barrier height used as one fitting parameter. Many other parameters can be chosen in order to obtain a good fit to experimental data. The model greatly overestimates the distribution coefficients for fast diffusers such as Cu and Ag.
(d) Jackson (1982): This is a model based on rate equations predicting the k_{sat} value of Bi and the orientation dependence. Full details are given in Chapter 3.

In addition to these models, there are two very recent treatments of solute trapping in silicon (Aziz, 1982; Cline, 1982).

F. Formation of Supersaturated Alloys; Maximum Substitutional Solubilities

As the dose or concentration is increased for each Group (III,V) species, there is a maximum concentration C_s^{max} which can be incorporated substitutionally into the silicon lattice during laser annealing (White *et al.*, 1980). This is shown in Figure 17 for the case of Sb in Si where both the total and substitutional dopant concentration as a function of depth are plotted (obtained from Rutherford backscattering and ion channeling measurements). Up to a concentration of $\sim 2 \times 10^{21}$/cm³ the Sb is almost 100% substitutional in the lattice, but in the near surface region down to a depth of ~ 1300Å, the total concentration is almost a factor of 2 higher than the substitutional concentration which is relatively constant with a value of $\sim 2 \times 10^{21}$/cm³. This value is the maximum substitutional concentration for Sb which can be incorporated substitutionally into Si at this growth velocity (4.5 m/sec).

In the near surface region, where the total concentration exceeds the substitutional concentration, the nonsubstitutional impurity is observed to be located in the walls of a well defined cell structure. This is illustrated in Figure 18 for the case of In in Si where both the

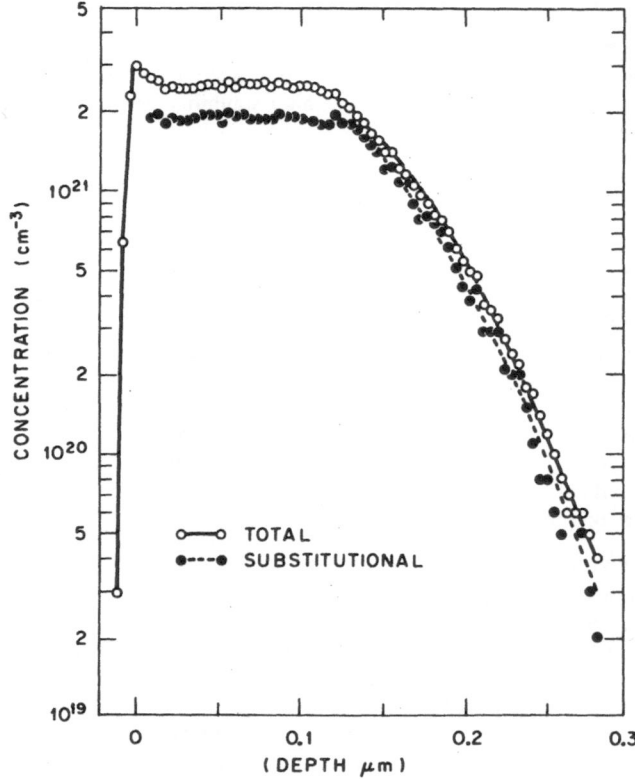

Figure 17   Total and substitutional profiles for ^{121}Sb (200 keV, 4.5×10^{16}/cm^2) in (100) Si after laser annealing. From White *et al.*, 1982.

total and substitutional dopant profiles as well as the microstructure in the near surface region are shown. The concentration profiles demonstrate that following laser annealing, In is substitutional in Si up to a limiting value $C_s^{\max} \sim 1.5 \times 10^{20}$/cm^3 for a growth velocity of 4.5 m/sec.   In the near surface region, TEM results show epitaxial columns of silicon (average diameter $\sim$450Å) extending out to the surface, surrounded by thin cell walls ($\sim$50Å thick) which penetrate to a depth of $\sim$1200Å and contain the pure phase of In. It is the nonsubstitutional In in the near surface region which is located in the cell walls. The substitutional In is trapped in the epitaxial columns of Si which extend to the surface. The formation of a cell structure in the near surface region is due to an interfacial instability which develops during regrowth, caused by constitutional supercooling at the interface (Cullis *et al.*, 1980; Narayan, 1981; Cullis *et al.*, 1981). This is discussed in more detail in Section I.

Using similar techniques, values for $C_s^{\max}$ have been determined (White *et al.*, 1980) for five Group (III,V) species in (100) silicon at a growth velocity of 4.5 meters/sec. These values are listed in Table II and compared with corresponding equilibrium solubility limits (C_s^0). For each dopant, $C_s^{\max}$ is larger than C_s^0 by factors that range from 4 for the case of As to $\sim$500 for Bi. Each of these dopants exhibits retrograde solubility in Si, and as shown by Baker and Cahn (1969), one cannot exceed the retrograde maximum concentration by

TABLE II

Comparison of Equilibrium (C_s^0) and Laser
Annealing Induced (C_s^{max}) Solubility Limits.
Values for C_s^{max} were obtained at a velocity
of 4.5 m/sec.   From White *et al.*, 1980.

Dopant	C_s^0 (cm^{-3})*	C_s^{max} (cm^{-3})
As	1.5×10^{21}	6.0×10^{21}
Sb	7.0×10^{19}	2.0×10^{21}
Bi	8.0×10^{17}	4.0×10^{20}
Ga	4.5×10^{19}	4.5×10^{20}
In	8.0×10^{17}	1.5×10^{20}

* From Trumbore, 1960.

solidification from the liquid unless there is a departure from local equilibrium at the interface. The large values for C_s^{max} relative to C_s^0 therefore convincingly demonstrate the nonequilibrium nature of the laser annealing induced liquid phase epitaxial regrowth process.

It is significant to note also that these solubility limits are independent of the manner in which the dopant is introduced into the silicon crystal. Recent work (Stuck *et al.*, 1980) has shown that solubility limits similar to those in Table II are obtained by depositing a thin layer of the dopant in the pure phase on the surface of the crystal, followed by subsequent laser irradiation. Figure 19 shows the maximum total Sb concentration and the substitutional concentration measured after depositing Sb on the surface followed by ruby laser irradiation. A substitutional concentration of 1.03×10^{21}/cm^3 is measured independent of both the deposited Sb thickness and the energy density used for irradiation. This value is in reasonable agreement with that reported in Table II. These limits are also independent of whether lasers or electron beams are used for annealing. The limits should depend only on growth velocity and possibly on crystal orientation.

As discussed in the previous section, dopant incorporation into the lattice at these high concentrations is a result of "solute trapping" during solidification (Cahn *et al.*, 1980; Jackson *et al.*, 1980; Wood, 1982; Aziz, 1982; Cline, 1982). In the simplest terms this means that impurities can be trapped into the solid if they are unable to exchange rapidly enough between the liquid and solid across the freezing interface to remain in equilibrium. During rapid solidification new layers of atoms are being added to the crystal so rapidly that dopant atoms have a reduced probability of escaping from the solid being formed. It is significant to note that measured values for C_s^{max} are approaching predicted thermodynamic limits to solute trapping (Cahn *et al.*, 1980).

G. Comparison of C_s^{max} in Solid and Liquid Phase Regions

Equilibrium solubility limits can be greatly exceeded also during low temperature solid phase epitaxial regrowth (SPEG) of ion implanted silicon (Campisano *et al.*, 1980a, 1980b;

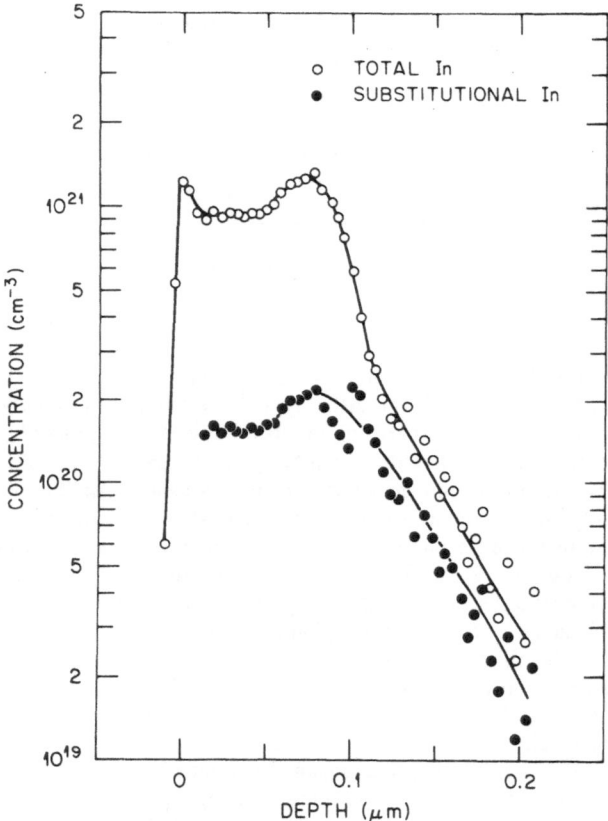

Figure 18a   Profiles for ^{115}In (125 keV, 1.3×10^{16}/cm^2) in (100) Si after laser annealing. From White *et al.*, 1981.

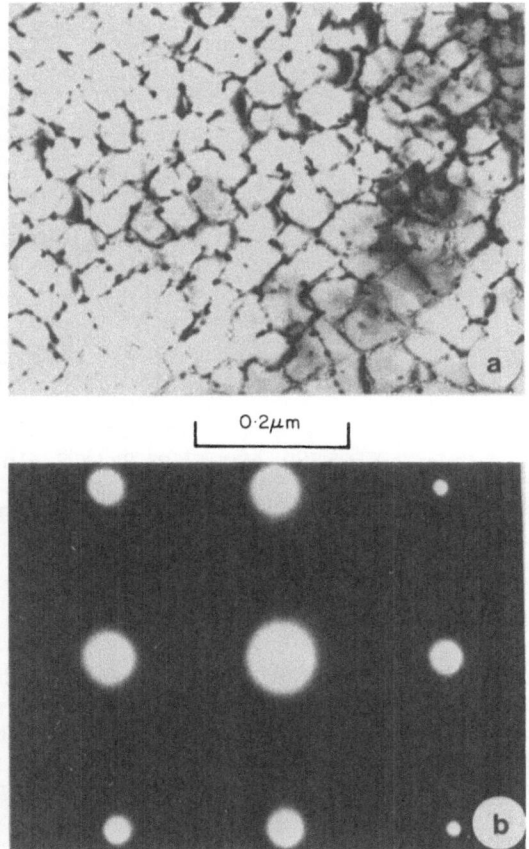

Figure 18b Microstructure for ^{115}In (125 keV, 1.3×10^{16}/cm^2) in (100) Si after laser annealing. From
White *et al.*, 1981.

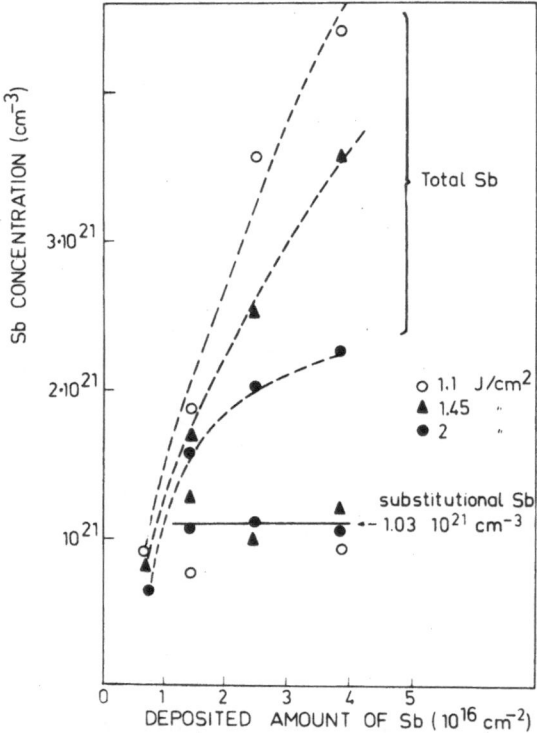

Figure 19   Solubility of Sb as a function of amount of Sb deposited on Si and subsequently laser irradiated. From Stuck *et al.*, 1980.

Williams and Elliman, 1981) (see Chapter 5). Table III compares values for C_s^{max} achieved by solid and liquid phase epitaxy with the corresponding equilibrium value (C_s^0) (Williams and Elliman, 1981). This comparison shows that there is very little difference in the solubility limits which can be achieved using these two recrystallization processes which occur in time scales which differ by many orders of magnitude. Dopant incorporation during SPEG is believed to occur by "solute trapping" at the moving amorphous/crystalline interface (Campisano *et al.*, 1980a). At the temperatures used for annealing ($\sim 550°$C) the velocity of the amorphous/crystalline interface is only $\sim 10^{-10}$ m/sec, but the impurity residence time at the interface is very long because this time is inversely proportional to the solid phase diffusivity which is very low. Consequently, even in the solid phase, a temperature range can be selected such that the impurity residence time at the interface is longer than the monolayer regrowth time and trapping can occur. If higher temperatures are used for annealing, however, precipitation of the dopant concentration in excess of the equilibrium solubility limit at the annealing temperature will occur (Williams and Elliman, 1981). During SPEG when the dopant concentration exceeds C_s^{max}, precipitation of the excess dopant concentration will occur randomly and there will be no well defined cell structures in the near surface region as observed during pulsed laser annealing.

TABLE III

Solubilities in Silicon Obtained by Solid Phase Epitaxy
$[C_s^{max}(SPEG)]$ and Pulsed Laser Annealing
$[C_s^{max}(LPEG)]$ Compared to
Equilibrium Solubility Limits

Dopant	C_s^0 (cm^{-3})	$C_s^{max}(SPEG)$ (cm^{-3})	$C_s^{max}(LPEG)$ (cm^{-3})
As	1.5×10^{21}	9×10^{21}	6.0×10^{21}
Sb	7×10^{19}	1.1×10^{21}	2.0×10^{21}
In	8×10^{17}	5×10^{19}	1.5×10^{20}

H. Mechanisms Limiting Substitutional Solubilities

Maximum substitutional solubilities which can be achieved by pulsed laser annealing of ion implanted Si appear to be limited by three mechanisms (White et al., 1981). The first of these is lattice strain, and this provides a practical limit to the incorporation of B in Si by pulsed laser processing. When B is incorporated substitutionally in Si during pulsed laser annealing, the lattice undergoes a one dimensional contraction in the implanted region in a direction normal to the surface (Larson et al., 1978). Contraction occurs because the covalent bonding radius of B is significantly smaller than that of the Si atom it replaces. The contraction occurs in one dimension only due to the adherence of the near surface region to underlying crystal planes. The magnitude of the contraction is proportional to the local B concentration. Contraction gives rise to strain in the implanted region, and when the strain exceeds the fracture strength of Si, cracks will develop in the implanted region (White et al., 1981). For B in Si this occurs whenever the local B concentration exceed ~4 at.%. If the B concentration is lower than this the implanted region will be strained but cracks will not develop. In order to incorporate more B substitutionally during laser annealing it would be necessary to simultaneously incorporate a dopant which gives rise to lattice expansion. Possible candidates include Ge, Sn, Ga, or In.

A second mechanism which limits the substitutional concentrations achieved by laser annealing is the interfacial instability which develops during regrowth and leads to lateral segregation of the rejected dopant and the formation of a well defined cell structure in the near surface region (Cullis et al., 1980; Narayan 1981; Cullis et al., 1981). This mechanism dominates for the case of Sb, Ga, In and Bi in Si and examples of this are shown in Figure. 17 and 18. Other examples can be found in the following section. Interfacial instability will occur only when the distribution coefficient is less than unity and when the concentration of the rejected impurity is large. At lower concentrations, the interface remains stable during regrowth and the rejected impurity is zone refined to the surface (see Figure 6). The instability leading to cell formation is caused by constitutional supercooling in the liquid at the interface. To delay the onset of instability and to incorporate more dopant substitutionally in the lattice it is necessary to increase the thermal gradient in the liquid. This implies that the regrowth velocity must be increased. Results demonstrating this have been obtained and will be discussed next.

In addition to the limitations on substitutional solubility imposed by lattice strain and by interfacial instability during regrowth (constitutional supercooling), there are predicted thermodynamic limits to solute trapping in Si (Cahn *et al.*, 1980). Basic ideas underlying these predictions are illustrated schematically in Figure 20. On a plot of the Gibbs free energy versus composition at fixed temperature, the solidus and liquidus lines intersect at one point, which is the upper limit for the solid composition which can be formed from the liquid at any composition since nucleation of the silicon phase is not required during laser annealing. Plotting the locus of these points at different temperatures on the equilibrium phase diagram defines the T_0 curve, which is the maximum solid composition which can be formed from the liquid at any temperature even at infinite growth velocity. The T_0 curve thus defines the thermodynamic limit to diffusionless solidification. For retrograde systems, thermodynamic arguments can be used to obtain a simple estimate for the maximum concentration (C_s^l) on the T_0 curve. This maximum concentration on the T_0 curve is the liquidus concentration on the equilibrium phase diagram at the retrograde temperature (Cahn *et al.*, 1980).

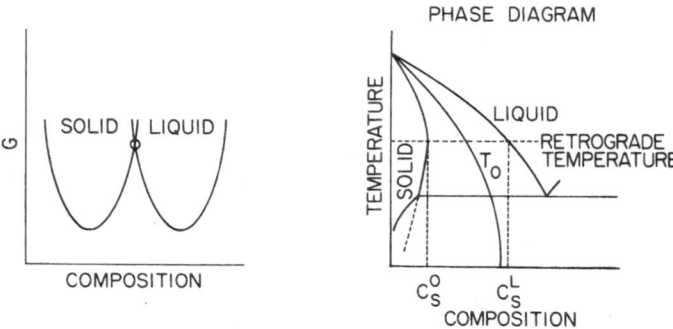

Figure 20   Schematic representation of the method used to determine the thermodynamic limit to solute trapping.

Predictions of (C_s^l) for five dopants in Si (Cahn *et al.*, 1980) are listed in Table IV and compared with measurements of maximum substitutional solubilities (C_s^{max}) obtained at two different growth velocities (White *et al.*, 1981). Laser annealing at temperatures of 300K and 77K results in regrowth velocities of ~4.5 and ~6.0 m/sec. At either growth velocity, values for C_s^{max} are approaching predicted thermodynamic limits for solute trapping. Dopants for which k' is very near to unity (As and Sb from Table I) have measured solubilities which are very close to predicted thermodynamic limits. Measured solubilities for dopants with relatively lower values for k' (Ga, In) are somewhat lower than predictions but are still within an order of magnitude of thermodynamic limits. For Ga, In, and Bi, values for C_s^{max} obtained at a velocity of 6 m/sec are larger by factors of 2-3 than results obtained at a regrowth velocity of 4.5 m/sec. These results are still limited by interfacial instability during regrowth, but they demonstrate that at higher growth velocities the onset of instability can be delayed until higher concentrations accumulate at the interface, as expected from the discussion in the previous section. It is significant to note that at a growth velocity of ~6 m/sec, the Bi concentration in substitutional lattice sites exceeds the equilibrium solubility limit by more than 3 orders of magnitude.

For the case of As in Si, Table IV shows there is no change in the measured maximum substitutional concentration as the regrowth velocity is increased from 4.5 to 6.0 m/sec. This indicates that the thermodynamic limit for As in Si may have been reached. This is reasonable since the measured value for k' is ~1.0 (see Table I). The fact that the

TABLE IV

Maximum Substitutional Dopant Concentrations (C_s^{max}) Obtained at
Growth Velocities of 4.5 and 6.0 m/sec Compared to Equilibrium
Solubility Limits (C_s^0) and Predicted Thermodynamic Limits to Solute
Trapping (C_s^l). Substrate temperatures of 300K and 77K were used
during laser annealing to produce growth velocities of 4.5 and
6.0 m/sec. From White et al., 1981.

Dopant	C_s^0	C_s^{max} ($v = 4.5$ m/sec)	C_s^{max} ($v = 6.0$ m/sec)	C_s^l
		cm^{-3}	cm^{-3}	cm^{-3}
As	1.5×10^{21}	6.0×10^{21}	6.0×10^{21}	5×10^{21}
Sb	7.0×10^{19}	2.0×10^{21}	--	3×10^{21}
Ga	4.5×10^{19}	4.5×10^{20}	8.8×10^{20}	6×10^{21}
In	8.0×10^{17}	1.5×10^{20}	2.8×10^{20}	2×10^{21}
Bi	8.0×10^{17}	4.0×10^{20}	1.1×10^{21}	1×10^{21}

measured value for C_s^{max} exceeds the predicted thermodynamic limit is probably due to uncertainties on the equilibrium phase diagram from which the predictions were made.

In another approach to predicting maximum substitutional solubilities (Stuck et al., 1980; Morehead, 1980), it was postulated that $C_s^{max} = C_s^0 \dfrac{k'}{k_0}$. This expression assumes that one can construct a kinetic phase diagram appropriate to high velocities achieved during laser annealing by a simple displacement of the equilibrium solidus line toward higher concentrations, consistent with the value for k'. This model assumes no displacement of the liquidus line from the equilibrium case. This is reasonable since no nucleation of the Si phase is required. Predictions of maximum substitutional solubilities by this model are in good agreement with measured values obtained at different growth velocities (i.e., different values for k'). In addition, the above expression reduces to the predictions of Cahn et al. (1980) in the limit that k' goes to unity.

I. Constitutional Supercooling and Precipitation

As indicated previously in this chapter, there are quite well defined limits to the amounts of low solubility impurities that can be trapped substitutionally in the Si lattice during laser annealing. At the outset in this discussion it is important to distinguish between those impurities that exhibit reasonable solubility in the transiently molten layer and those that exhibit only a low solubility in liquid Si. This distinctive difference can lead to significantly different precipitation behaviors during the annealing sequence.

The presence of excessively high concentrations of ion implanted impurities that exhibit substantial solubility in molten Si can lead to the formation of a very characteristic cellular pattern of microsegregation upon laser annealing (Cullis et al., 1979; Cullis et al., 1980a). The phenomenon has been ascribed to recrystallization interface instability due to the occurrence of constitutional supercooling in the melt during resolidification (Cullis et al.,

1980a; Cullis *et al.*, 1980). Similar segregation patterns can be produced by the alloying of metal films into semiconductor surfaces (Poate *et al.*, 1978; van Gurp *et al.*, 1979). The type of structure that is produced in an ion implanted layer is well illustrated by work on the laser annealing (Cullis *et al.*, 1980a) of Fe^+ ion implanted (100) Si. In this case, the initial implanted region was amorphous (Figure 21a). However, after ruby laser annealing at 2.2 J/cm^2 the damaged Si was converted back to single crystal although trails of segregated dopant remained within the matrix (Figure 21b). The pattern of microsegregation was not random and this is most clearly shown in the plan-view micrograph of Figure 21c. Here, it is evident that a cell structure had been formed, the boundaries of single crystal Si islands being delineated by segregation trails. The mean cell lateral dimension was ~900Å. The precipitated impurity Fe had formed a silicide phase by combination with matrix Si and this was identified by electron diffraction analysis as ξ_α − FeSi$_2$; see Figure 21d,e (Cullis *et al.*, 1980a). The same phase is formed in the Si lattice under equilibrium conditions achieved by conventional furnace annealing (Cullis and Katz, 1974).

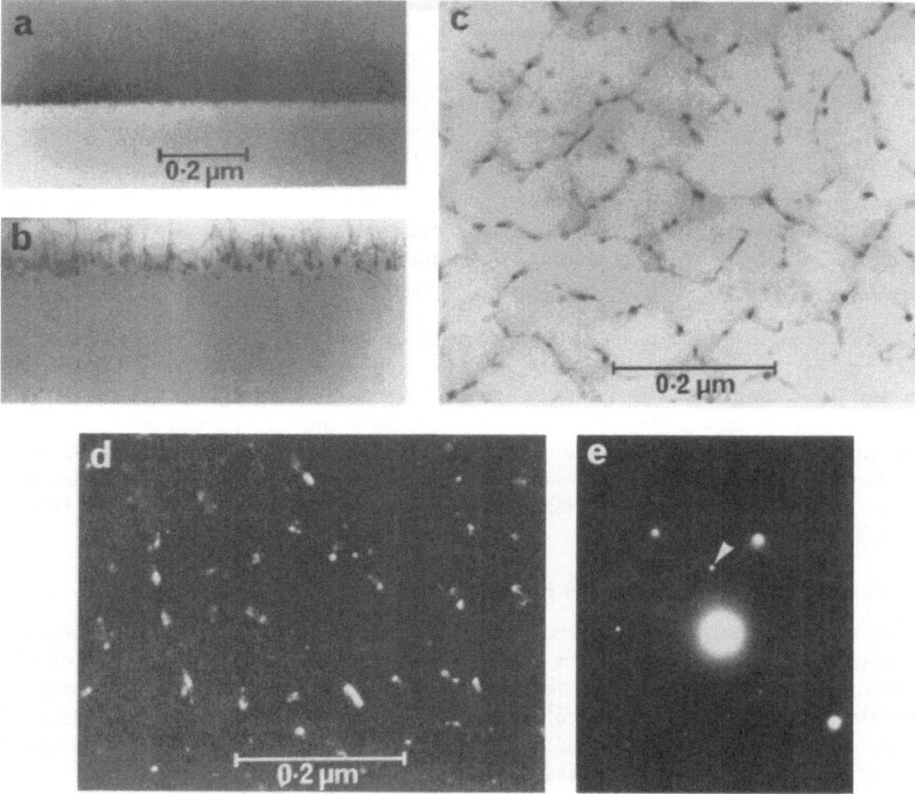

Figure 21 Transmission electron images and diffraction pattern from (100) Si implanted with 5×10^{15} Fe^+ ions/cm^2 at 150 keV. Cross-sectional images (a) as-implanted and (b) annealed at 2.2 J/cm^2. Plan-view images of annealed specimen (c) bright-field showing cell structure and (d) dark-field using 110 ξ_α−FeSi$_2$ reflection arrowed in diffraction pattern (e). From Cullis *et al.*, 1980a.

The formation of cell structures in laser annealed impurity containing Si layers has been observed for a number of other cases, such as Pt, Cu and Co in Si. One system that has been subjected to particularly detailed study is that of implanted In in (100) Si (Narayan, 1981; Cullis et al., 1981). The segregation cells formed by a 1.6 J/cm^2 Q-switched laser pulse are shown in Figure 22a. Once again electron diffraction analysis has led to an identification of the precipitate phase as metallic, elemental In (see Figure 22b). Furthermore, by careful measurements of cell sizes and the monitoring of annealing conditions it has also been possible to demonstrate the applicability of the Mullins and Sekerka (1964) morphological stability theory to cell evolution. The theory has been tested extensively for conventional crystal growth rates ($10^{-5} - 10^{-2}$ cm/sec) where recrystallization interface breakdown occurs on the scale of D/v, D being the diffusion coefficient of the solute in liquid Si and v the growth velocity. Breakdown occurs when the destabilizing effects of solute redistribution outweigh the stabilizing effect of the thermal gradients in the system. However, at very high growth velocities that characterize the laser annealing process, the wavelength scale (D/v) becomes very small and the deformed interface has a high curvature. This provides strong stabilization through the Gibbs-Thomson effect (Sekerka, 1965) so that a larger degree of constitutional supercooling is required for interface breakdown to occur with resulting segregation cell formation. In addition, the strong dependence of the segregation coefficient upon recrystallization velocity has a marked effect upon the observed behavior. Hitherto, these predictions of the theory have not been amenable to test but, with the advent of laser annealing, the new high velocity regrowth regime can be explored.

Correlations between theory and experiment can be made in a number of ways. For example, it is possible to consider the general ranges of conditions which lead to interface stability or instability. This approach (Cullis et al., 1981) is illustrated in Figure 23 where a critical impurity concentration parameter (mC_s/DG) is plotted against the interfacial segregation coefficient (k): m is the slope of the liquidus for the binary alloy, C_s is the concentration of solute in the crystal and G is the weighted temperature gradient. The continuous lines in Figure 23 represent the theoretically computed boundaries between regimes of stability and instability for three different recrystallization velocities in the In/Si system. Complementary experimental data obtained from In$^+$ ion implanted and laser annealed Si are also plotted in Figure 23. It is clear that the experimental stability boundary (running between the crossed and triangular data points) is very close to its theoretically predicted counterpart, especially considering that the calculations were absolute and did not involve the use of fitting parameters. A second illustration of the application of the theory is given in Figure 24 where the computed cell size is shown as a function of recrystallization velocity for the In/Si system (Narayan, 1981). The three curves correspond to values of the effective segregation coefficient taken respectively, as the equilibrium value ($k_0 = 4 \times 10^{-4}$), as an experimental value of 0.15 (corresponding to a fixed recrystallization velocity of 4 m/sec) and as a function $f(v)$ of recrystallization velocity. The latter is, of course, the best approximation and the experimental data point plotted on the figure demonstrates how well the theory accounts for the observations.

In some cases, impurities introduced into Si exhibit a low solubility in both the solid and the liquid and, under such circumstances, different segregation phenomena may occur during laser annealing. A good example is that of implanted C in (111) Si (Cullis et al., 1981a) for which the recrystallization velocity has been found to be a very critical parameter in determining the final state of the system. As shown in Figure 25a, implantation of 10^{16} C$^+$ ions at 40 keV gave an initial amorphous surface layer. Nevertheless, after annealing with a 30 nsec ruby laser pulse at 2 J/cm^2, and with a room temperature substrate, good single crystal Si reformed in the lower part of the layer while the outer regions contained a dense

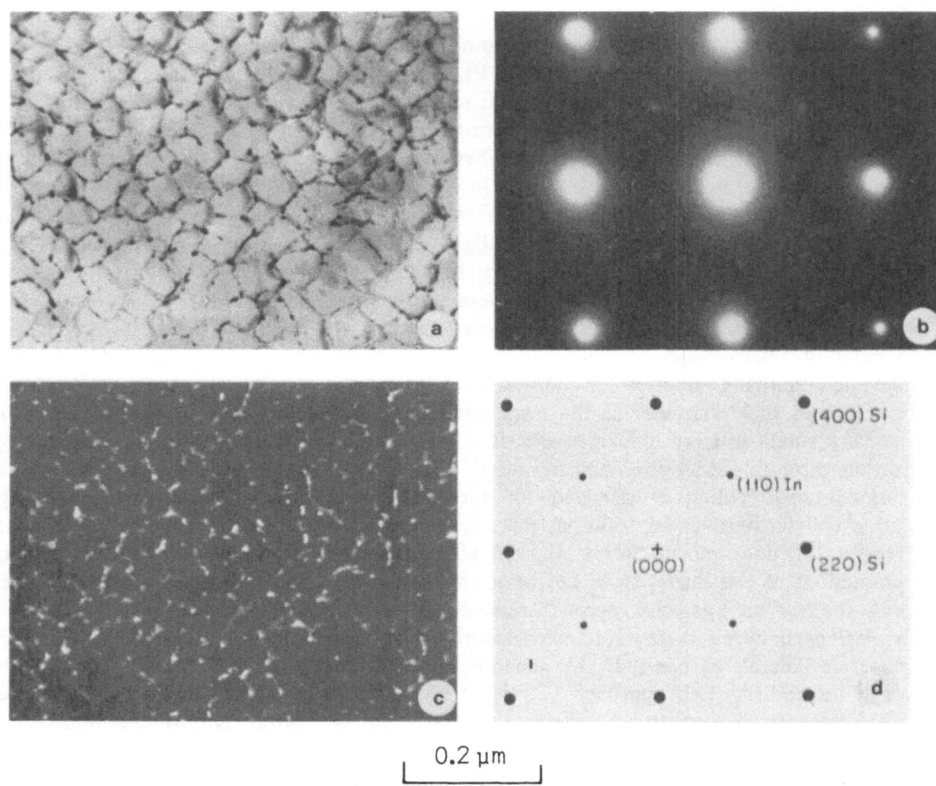

0.2 μm

Figure 22   Transmission electron image (a) and diffraction pattern (b) from (100) Si implanted with 1.25×10^{16} In$^+$ ions/cm^2 and annealed at 1.57 J/cm^2. Image (a) shows cell structure and diffraction pattern (b) shows Si and In spots. Image (c) shows dark-field image using In diffraction spots and (d) is a schematic representation of the diffraction pattern. From Narayan, 1981.

accumulation of $\beta - SiC$ precipitate particles (Figure 25b). The precipitates were randomly distributed with no evidence for the development of a cell structure. However, upon increasing the recrystallization velocity from ~2.5 to ~3.5 m/sec by cooling the sample to 77°K during laser annealing, it was found that C precipitation could be substantially suppressed (see Figure 25). Indeed, a large proportion of the C had been incorporated onto substitutional lattice sites in the Si matrix by solute trapping and this was demonstrated by monitoring the C in Si local mode IR absorption peak at 507 cm^{-1} (Figure 25e). Substantial C was also produced by the 2.5 m/sec anneal. The highest substitutional concentration of C achieved (~2×10^{20} atoms/cm^3) exceeded the equilibrium value by about 2.5 orders of magnitude (Cullis *et al.*, 1981a). The metastability of such solid solutions was demonstrated by use of subsequent furnace anneals at 1000°C which resulted in solid-phase precipitation of the substitutional impurity (Figure 25d). However, the $\beta - SiC$ precipitates produced during the laser annealing sequence were thought to have formed while the Si remained molten, since the solubility of C in liquid Si (Nozaki *et al.*, 1974) is only ~4×10^{18} atoms/cm^3 at the melting point. Precipitation in the liquid was suppressed by an increase in the recrystallization velocity since enhanced trapping denuded the impurity concentration in the melt and simultaneously decreased the overall melt dwell time.

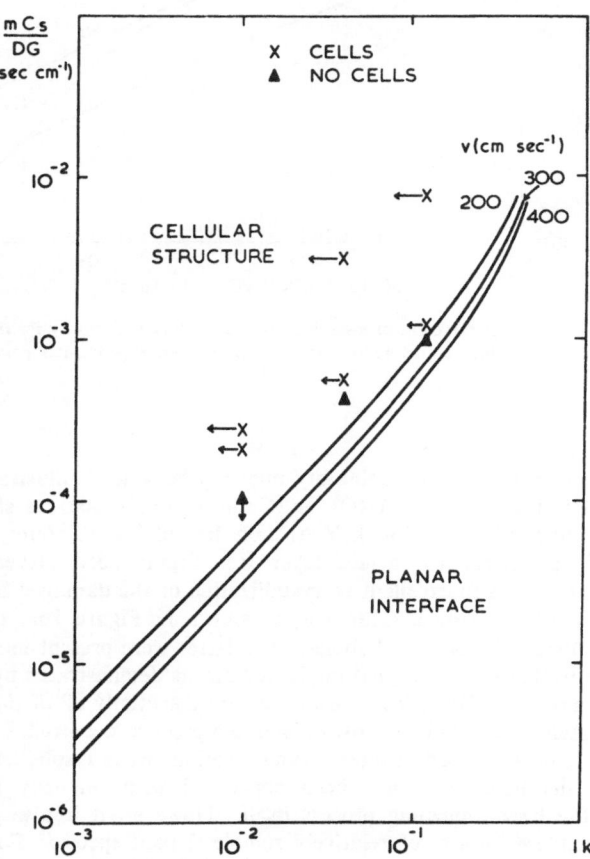

Figure 23   Plot of critical concentration parameter vs. segregation coefficient. Curves are computed boundaries between stable and unstable regimes. Points are experimental values: Δ = planar interface, x = cellular structure. From Cullis *et al.*, 1981.

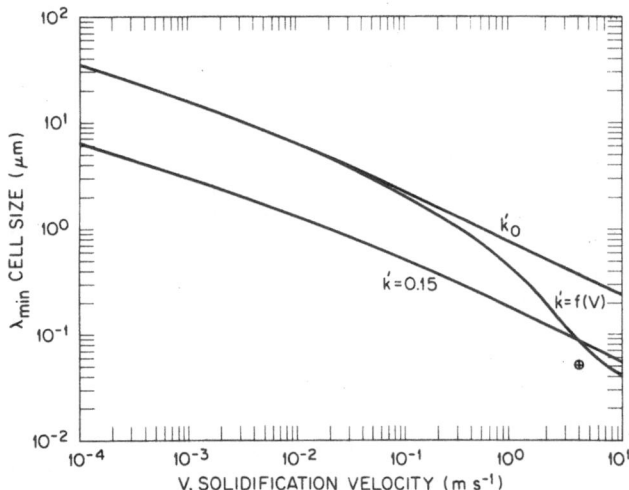

Figure 24   Plot of wavelength at the onset of instability vs. recrystallization velocity for the In-Si system. Curves are for different values of the segregation coefficient. An experimental point is also shown. From Narayan, 1981.

A more extreme example of low solubility impurity behavior is illustrated by work on the laser annealing of implanted Ar in (100) Si (Cullis *et al.*, 1980a) as shown in Figure 26. The initial sample had received a 150 keV Ar^+ ion dose of $2 \times 10^{15}/cm^2$ and this resulted in the formation of an amorphous surface layer (see Figure 26). Nevertheless, ruby laser annealing at 2.2 J/cm^2 led to excellent recrystallization of the damaged Si and few extended crystallographic defects remained. However, as shown in Figure 26a, the annealed region contained a dense array of gas-filled bubbles. The latter were present in a depth distribution which was similar to that of the original implanted Ar, as demonstrated by the cross-sectional micrograph of Figure 26c. Here, it is considered that essentially all of the Ar precipitated in the transient Si melt where bubble growth and coalescence occurred. The bubbles, being relatively immobile, were trapped at their various locations when resolidification took place.

In the above discussion we have been concerned with impurity precipitation which occurred during the laser annealing process itself. However, it is also possible to actually dissolve pre-existing precipitates of relatively soluble dopant species. For example, when P was diffused from a gaseous source into single crystal Si (Narayan, 1979) many dislocation loops and silicide precipitates were introduced at the same time. When such a sample was laser annealed it was found that these defects and precipitates were almost completely removed in the region that was transiently melted. This is illustrated in Figure 27, where different depths of matrix Si were denuded of precipitate particles after annealing with laser pulses of different energy densities. Similar precipitate dissolution behavior has been observed in B-diffused and laser irradiated Si layers (Narayan *et al.*, 1978a). Thus, it is clear that laser melting and high speed recrystallization can be accompanied by impurity precipitation and dissolution phenomena which mark out new regimes of crystal growth.

J. Summary of Crystal Growth

Laser annealing of ion implanted silicon has provided fundamental information on nonequilibrium crystal growth processes. The level of understanding which has been brought

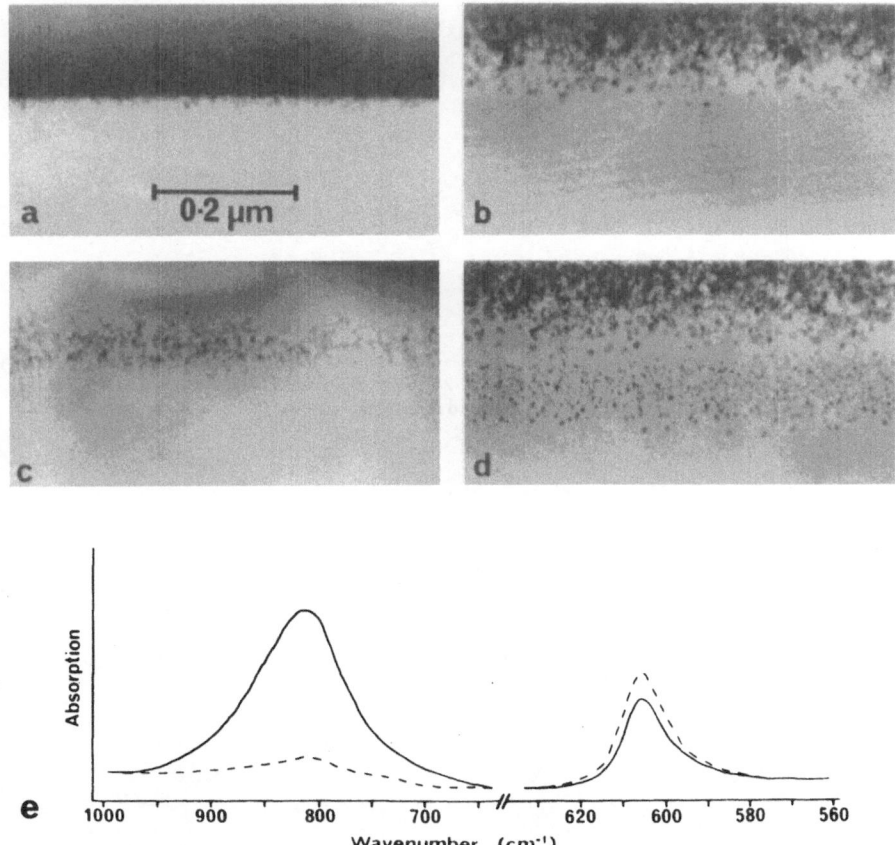

Figure 25   Cross-sectional transmission electron images and IR absorption spectra from (111) Si implanted with
10^{16} C^+ ions/cm^2 at 40 keV. (a) As-implanted; (b) annealed at 2 J/cm^2 with room temperature
substrate; (c) annealed at 2 J/cm^2 with 77K substrate and (d) as for (b) but with additional furnace
anneal at 1050°C for 30 min. Note lower band of newly-formed precipitates. (e) IR absorption spectra
from specimens (b) solid curve and (c) dashed curve. Note local mode peak at 607 cm^{-1} and SiC peak
at 812 cm^{-1}. From Cullis *et al.*, 1981a.

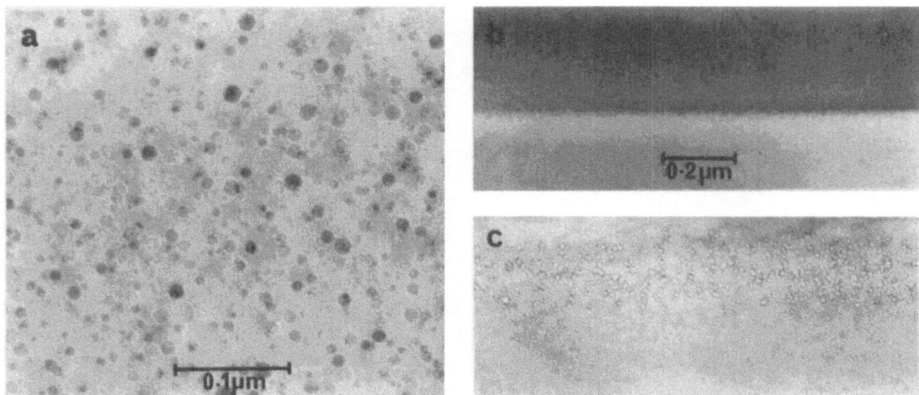

Figure 26   Transmission electron images of (100) Si implanted with 2×10^{15} Ar$^+$ ions/cm^2 at 150 keV. (a) Plan-view, over-focus image of layer annealed at 2.2 J/cm^2 showing bubbles. (b) and (c) Cross-sectional images of as-implanted and 2.2 J/cm^2 annealed layers, respectively. From Cullis *et al.*, 1980a.

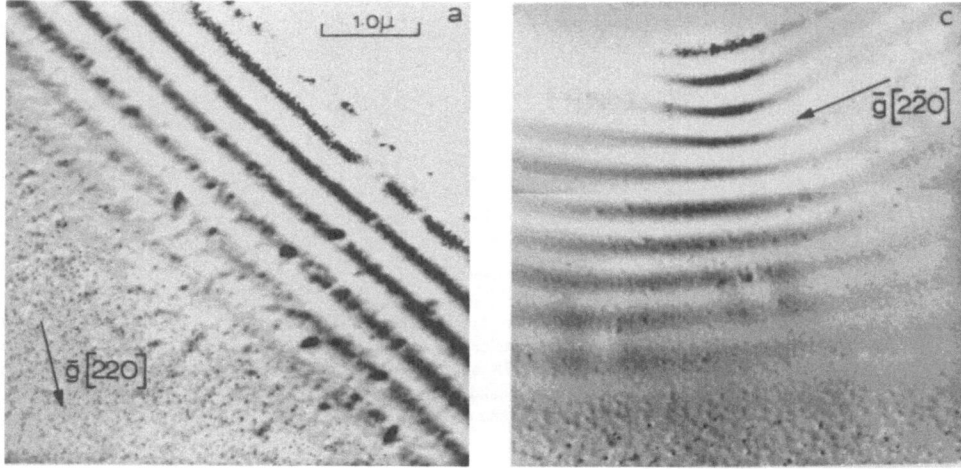

Figure 27   Bright-field transmission electron images showing increase in annealing depth with increase in laser pulse energy density: (a) 2.0 J/cm^2; (c) 3.0 J/cm^2. From Narayan, 1979.

to these unique regimes of crystal growth is impressive considering the fact that the first measurements were made only a few years ago. These advances have been accomplished because one can use two complementary nonequilibrium processing techniques, ion implantation and pulsed laser annealing, in order to carry out experiments under controlled conditions.

During the rapid liquid-phase epitaxial regrowth process, implanted Group (III,V) species can be incorporated into the lattice at concentrations that exceed equilibrium solubility limits by up to the orders of magnitude. Interfacial distribution coefficients, in many cases as a function of velocity, have been determined for a wide variety of impurities in silicon. Theoretical models have been developed which may explain in a quantitative fashion the solute trapping mechanisms. Limits to the substitutional solubility which can be achieved by laser annealing have been established and insight has been gained into the factors that limit substitutional solubility. Measured solubility limits are approaching predicted thermodynamic limits to diffusionless solidification.

The future will see similar experiments carried out at both faster and slower growth velocities. To increase the growth velocity we must use shorter pulse duration times. Slower velocities can be achieved by depositing the energy over a longer time period. Questions that need to be investigated more thoroughly include the saturation value for the distribution coefficients, more conclusive tests of predicted thermodynamic limits to dopant incorporation, orientation effects in solute trapping, incorporation of interstitial species, and the transition to the amorphous state. Results generated in these experiments should provide a sound basis for theoretical understanding of high speed nonequilibrium crystal growth.

III. SURFACE PROPERTIES

The previous section has described the high speed nonequilibrium crystal growth phenomena that can be studied during pulsed laser annealing of semiconductors. During the rapid liquid-phase epitaxial regrowth process induced by pulsed laser annealing new planes of atoms are being added to the crystal on a time scale of 10^{-10} sec, with the process terminating at the surface. Following solidification, the recrystallized region cools toward room temperature with an initial cooling rate of $\sim 10^9$ K/sec. The extreme quenching rates and nonequilibrium nature of the laser annealing induced epitaxial regrowth process suggests that the surface properties (structural and electronic) of laser-annealed semiconductors may be significantly different from those obtained by more conventional treatments. In this section we discuss the surface properties of silicon crystals (intrinsic and highly doped) that have been subjected to pulsed laser annealing.

In order to obtain information about the properties of the surface region (1-20Å) of laser-annealed semiconductors, the irradiation of the sample and subsequent analysis must take place in an ultrahigh vacuum (UHV) environment ($\leqslant 10^{-9}$ Torr). For the investigations to be discussed in this section, the light from a Q-switched ruby laser ($\lambda = 6943$Å) was coupled into a UHV system through a glass window and the samples were irradiated using the single mode (TEM$_{00}$) output of a ruby laser at energy densities that could be varied between ~ 0.2 and ~ 4.0 J/cm^2. The pulse duration time was maintained at a constant value of 15 nsec and the beam diameter was typically between 3.0 and 6.0 mm. The spectroscopic techniques used for examining the surface region employed either electrons or photons as the incident probe. In all cases the detected particle was an electron. Because of the short mean free path of electrons with energies between 20 and 1000 eV, only the outermost surface region, ~ 20Å, is probed. Several different surface sensitive spectroscopic techniques were employed as the various surface analysis facilities used in these investigations. The technique of Auger electron spectroscopy (AES), which is capable of detecting all elements with $Z \geqslant 3$, was used to monitor the levels of both impurities and implanted species in the surface region of the sample. Low-energy electron diffraction (LEED) was employed to investigate geometric order as well as intra- and interatomic spacings in the outermost surface layers. Photoelectron spectroscopy (PES) was used to obtain information about the electronic properties, both valence band and core levels, of the surface region.

A. Surface Cleaning of Si

The production of atomically clean surfaces in UHV is a fundamental requirement in areas of basic research directed toward understanding the physical and chemical properties of surfaces and in device technologies where the presence of contaminants introduced either during fabrication or during application can contribute greatly to the degradation of device performance. Recent work has demonstrated that the fastest and most efficient way to obtain an "atomically clean Si surface" is by pulsed laser irradiation in UHV.

The application of laser irradiation for the purpose of producing atomically clean Si surfaces is demonstrated by the results (Zehner et al., 1980b) shown in Figure 28. The Auger electron spectrum obtained from an air-exposed Si sample after insertion into a UHV system and following bakeout is shown at the top of Figure 28. Oxygen and carbon are readily detected on the surface of the as-inserted sample and the features in the region of the Si $L_{2,3}$ VV transition (70-100 eV) are characteristic of Si in SiO_2. Measurements made by RBS showed O and C concentrations of 8.1×10^{15} and 3.5×10^{15} atoms/cm^2, respectively, in the near surface region. Irradiation with one laser pulse (~ 2.0 J/cm^2) causes a substantial reduction in the levels of O and C present in the surface region. After exposing the same spot to five laser pulses, the Auger electron spectrum shown at the bottom of Figure 28 indicates that for the same detection conditions the O and C signals are within the noise level and correspond to O and C surface concentrations of $\leqslant 0.1\%$ of a monolayer. In addition, the lineshape of the Si $L_{2,3}$ VV transition is that expected from a clean Si surface. Although a H Auger transition cannot be detected with AES, PES results (Zehner et al., 1981a) show the surface region to be free of H. Consequently, by irradiating the crystal with several ($\leqslant 5$) laser pulses, the O and C contaminants have been reduced by factors of at least 500 and 50 respectively. These reduced contaminant levels, which can be obtained in a processing time of ~ 30 nsec are comparable to those obtained by repeated sputtering and conventional thermal annealing over a time period of several days.

The effect of pulsed laser irradiation on samples that had been sputtered with Ar$^+$ ions (1000 eV, 5 μA, 30 min) was also investigated. Irradiation of the sample with laser pulses of ~ 2.0 J/cm^2 produced a similar reduction in the contaminant level Auger signals for O and C as observed for the unsputtered surface. Complete elimination, even under increased sensitivity conditions, of the Ar Auger signal occurred after two or three pulses. In addition, if an atomically clean surface is exposed to O_2 or CO, the surface can be returned to an atomically clean state by irradiation with five laser pulses.

The effect of energy density on the efficiency of removal of impurities was also qualitatively investigated over the range ~ 0.3 to ~ 3.2 J/cm^2. Two general trends were observed as a function of energy density. First, the higher the energy density, the more extensive the removal of the impurities by the first pulse. Second, at any pulse energy density the larger the number of pulses, the more complete the removal of impurities. At energy densities above ~ 2.0 J/cm^2, little additional change could be observed after five pulses. Visible damage, as observed with optical techniques, occurred at energy densities $\geqslant 3.2$ J/cm^2. Energy densities greater than 1.0 J/cm^2 were necessary to produce atomically clean surfaces using ruby laser radiation.

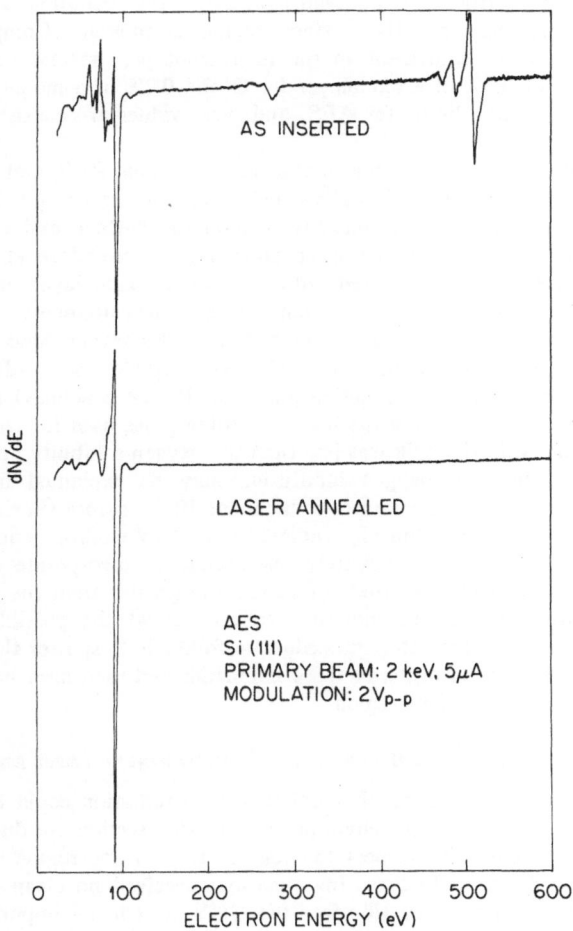

Figure 28 Auger electron spectra obtained from an uncleaned Si surface and after pulsed laser annealing at
~2.0 J/cm². From Zehner *et al.*, 1980b.

There are two possibilities as to the ultimate fate of the original oxygen and carbon surface contaminants. A pronounced pressure rise is observed during the first laser pulse, suggesting that contaminants are desorbed from the surface during irradiation. Alternatively, since pulsed laser annealing results in the formation of a melted region to a depth of several thousand angstroms, impurities can undergo substantial redistribution by means of liquid-phase diffusion during the time the surface region is molten. Complete and uniform redistribution over a depth equivalent to the melt front penetration would give rise to a surface concentration of 0.3% of a monolayer for O and 0.1% of a monolayer for C, both of which are near the detection limits for AES, and these values are consistent with the above measurements.

Recently, an investigation has been performed in which RBS and resonance nuclear reaction techniques utilizing the $^{16}O(\alpha,\alpha)^{16}O$ nuclear reaction in conjunction with AES have been used to determine the oxygen concentration on the surface and in the near surface region both prior to and after pulsed laser annealing (Westendorp et al., 1981). After irradiation of a Si(100) sample, covered with the native oxide layer, in UHV with eight pulses at an energy density of ~1.55 J/cm^2, Auger measurements showed an oxygen concentration at the surface of $\leqslant$0.3% of a monolayer. The oxygen concentration at a mean depth of 1100Å, in the depth interval 500–1700Å, was determined to be $\leqslant$3.1 × 10^{18} atoms/cm^3 by nuclear reaction analysis. It was concluded that there was less than 1% indiffusion of oxygen during the laser annealing pulse used for cleaning and that the concentration determined in the bulk was less than the oxygen solubility limit.

The absence of significant oxygen indiffusion may be explained by noting that the dissolution rate of the oxide in Si is known to be about 10^{-5} cm/min (Saris, 1981). Since the thickness of the native oxide layer on Si, irradiated in a UHV environment, is typically on the order of 20Å, it would take approximately one second to incorporate this layer into the melted region. This is at least six orders of magnitude greater than the time the irradiated region remains above ~600K. To minimize concern about the possible redistribution of impurities such as O and C, the safest procedure to follow is to sputter the surface mildly to reduce substantially the concentration of these impurities and then laser anneal the crystal to produce an atomically clean surface region.

B. Surface Studies of (100), (110) and (111) Si Subsequent to Laser Annealing

In the previous section it was shown that pulsed laser irradiation could be used to produce atomically clean surfaces in a UHV environment. In this section we discuss the annealing capability of this technique with respect to order in the surface region of a single crystal. Since the silicon samples used in these investigations received no cleaning treatment other than a rinse in alcohol prior to insertion into the UHV system, all impurities were removed during the laser irradiation. A comparison of the LEED patterns obtained following laser annealing with those obtained using conventional annealing techniques is shown in Figure. 29 and 30.

1. Si(100) face

After irradiation with one laser pulse of ~2.0 J/cm^2, a well defined (2 × 1) LEED pattern with moderate background intensity was obtained (Zehner et al., 1980). Improvement in the quality of the diffraction pattern occurred with subsequent laser pulses. After five pulses the LEED pattern shown in Figure 29 was observed. The fact that well defined LEED patterns are obtained indicates that the crystalline order extends to the outermost monolayers after the termination of the liquid-phase epitaxial regrowth process. No detectable change in the LEED patterns was observed with additional pulses. A similar (2 × 1) LEED pattern is obtained on Si(100) surfaces by using conventional sputtering followed by thermal annealing treatment (see Figure 29b) and the observation of this pattern

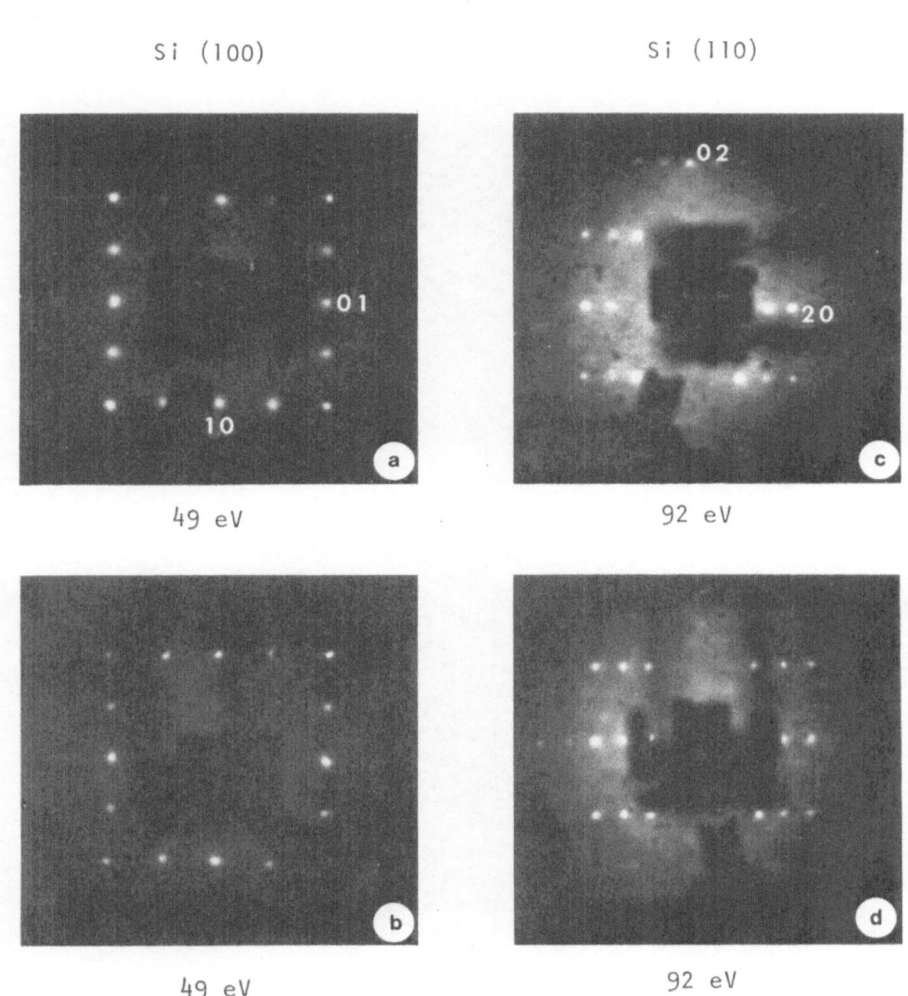

Figure 29   LEED patterns from clean (100) and (110) Si surfaces at primary beam energies of (a,b) 49 eV and (c,d) 92 eV. (a,c) Laser-annealed, (b,d) thermally annealed. From Zehner *et al.*, 1980c.

Si (111)

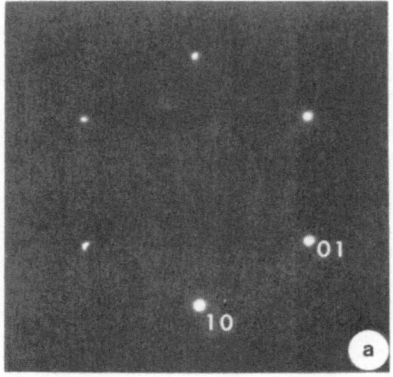

40 eV

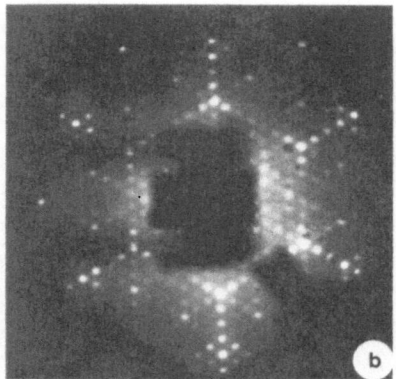

68 eV

Figure 30   LEED patterns from clean (111) Si surfaces at a primary beam energy of (a) 40 eV and (b) 68 eV. (a) laser-annealed, (b) thermally annealed. From Zehner *et al.*, 1980a.

shows that the surface is reconstructed (the atoms in the outermost monolayers do not have the same geometric spacings as do those in the bulk). The above results indicate that the atoms in the outermost layers have enough time under the laser annealing conditions to reorganize into the reordered arrangement from which the (2×1) LEED patterns are obtained. This is consistent with the proposed surface structure models for this surface that involve only small displacements of the atoms in the filled outermost monolayers (Eastman, 1980).

2. Si(110) face

The (1×2) LEED pattern obtained from the Si(110) face following irradiation with five laser pulses of ~ 2.0 J/cm^2 is shown in Figure 29c (Zehner et al., 1980c). A similar (1×2) LEED pattern obtained from a thermally annealed (110) surface is shown in Figure 29d for purposes of comparison. While a detailed description of the atomic arrangements in the surface region of the (1×2) structure does not exist, the similarity in the LEED patterns shows that, as with the (100) surface, the same reordered arrangement can be obtained following either laser or thermal annealing procedures.

3. Si(111) face

A sharp, well-defined (1×1) LEED pattern, shown at the top of Figure 30, was obtained from the Si(111) face subsequent to irradiation with five laser pulses of ~ 2.0 J/cm^2 (Zehner et al., 1980a,c). In contrast to the patterns obtained from the (100) and (110) surfaces, this observation suggests that as a result of laser annealing the normal surface structure (truncation of the bulk) is obtained, and there is no evidence of any ordered lateral reconstruction. By contrast, the (7×7) LEED pattern (shown at the bottom of Figure 30) is always observed on a clean thermally annealed Si(111) crystal surface and indicates that reconstruction is present. Additional investigations showed that the (1×1) LEED pattern could be obtained from the (111) surface when the crystal was held at temperatures in the range 100-700K during laser annealing. The observation of seventh order diffraction spots, indicative of the reconstructed surface, occurred after either annealing the laser-irradiated surface at temperatures greater than ~ 800K or holding the crystal at this temperature during the laser annealing process. By heating for a sufficient time (> 30 min) at temperatures greater than ~ 800K, it was possible to convert the laser-annealed surface to one from which a well-defined (7×7) LEED pattern was observed. This indicates that the structure giving rise to the (1×1) LEED pattern is metastable and by combining laser annealing with conventional thermal annealing it is possible to cycle back and forth between the two structures.

Since the initial observation that the (111) surface of Si is reconstructed subsequent to a sputter-anneal treatment, much effort has been expended in determining geometric models appropriate for this atomic structure and in obtaining a variety of experimental data to support these models. Structural models for this reconstructed surface have been divided into two types, smooth and rough (Eastman, 1980). Smooth models are characterized by a complete outermost monolayer with small atomic displacements from bulklike positions. Rough models incorporate either an outermost monolayer or double layer that contains a large number of vacancies. Observations of (1×1) LEED patterns suggest that under the combined time and temperature conditions present during the laser annealing process the atoms in the outermost layer or layers are not able to organize after the regrowth of the molten region, into the ordered geometric arrangements corresponding to the reconstructed surface.

C. Structural Model for Laser-Annealed Si(111)-(1×1)

The observation that a (1×1) LEED pattern is obtained from the (111) surface of Si after laser irradiation in a UHV environment, coupled with the fact that these surfaces are observed to be atomically clean after this treatment suggests that it offers the opportunity for investigating a clean semiconductor surface that exhibits no ordered lateral reconstruction. In particular, it should be possible to obtain information about surface relaxations. One way to investigate this question is to measure the intensities of the diffracted electron beams as a function of incident electron energy (I-V profiles). The experimentally measured profiles must then be compared to results obtained from fully converged dynamical LEED calculations assuming various structural models for the geometric arrangement in the outermost monolayers. A measure of the agreement between the experimental results and the predictions of the model calculations is provided by the R-factor (the lower the R-factor value, the better the agreement). A detailed LEED analysis for laser-annealed (111)-(1×1) surfaces of Si and Ge has been performed and the results for Si are discussed below (Zehner *et al.*, 1981).

A Si(111) surface that had been irradiated with the output of the laser at an energy density of ~2.0 J/cm^2 was used in these investigations. The intensities of the diffracted beams were measured as a function of electron energy using a Faraday cup operated as a retarding field analyzer. Data were obtained for all of the {10}, {01}, {20} and {02} beams, and three each of the {11} and {21} beams. Based on observations and conclusions drawn from previous studies, symmetrically equivalent beams were averaged to provide a data base containing six average profiles.

The experimental data base has been compared with the results obtained from fully converged dynamical LEED calculations. Details of these calculations can be found in Zehner *et al.*, 1981 and only the results will be summarized here. Comparison of profiles obtained from the dynamical LEED calculations to the measured I-V profiles suggests that the first interlayer spacing, d_{12}, is contracted by 25.5 ± 2.5% with respect to the bulk value and that the second interlayer spacing, d_{23}, is expanded 3.2 ± 1.5% with respect to the bulk values. Profiles calculated using these values are shown in Figure 31, which also contains the corresponding experimental profiles and single beam reliability factors, R, determined for each comparison. The six-beam R factor corresponding to Figure 31 is 0.115. This value indicates a very good agreement between calculated and experimental profiles in a conventional LEED analysis, and suggests that the proposed structural model is highly probable. Furthermore, this R value is significantly lower than any reported value obtained in a LEED analysis of any semiconductor surface. The changes in interlayer spacings determined from this analysis correspond to nearest-neighbor bond length changes of −0.058 and +0.075Å.

It should be noted that the interpretation of results obtained in a recent study of the reversible high-temperature (~1150K) transition of the (7 × 7) to a (1×1) structure has led to the conclusion that a disordered surface exists above the transition temperature (Bennett *et al.*, 1981). Although the results of the LEED analysis presented above suggest a highly probable structural model for the laser-irradiated surface, only ordered structures could be tested in this analysis. Since the laser-irradiated surface is known to undergo an irreversible transition from (1×1) to (7 × 7) at ~700K, a study of this transition should aid in distinguishing between differing models for each surface structure.

D. Electronic Structure of Si(111) - (1×1)

An examination of the electronic structure in the surface region of the laser-annealed Si(111) surface is of interest in view of the results of the LEED analysis just discussed. Theoretical

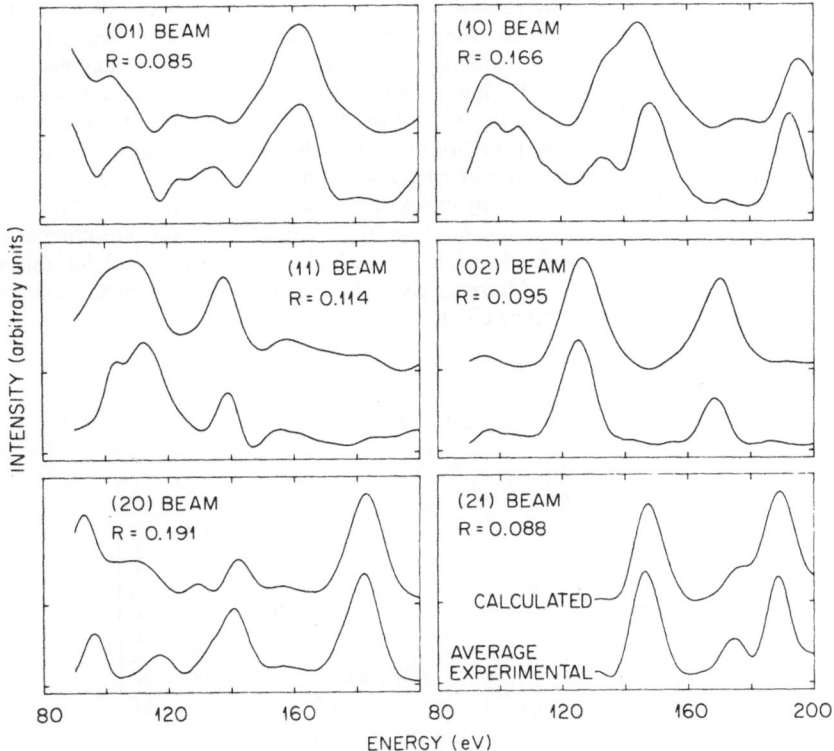

Figure 31   A comparison of the averaged experimental I-V profiles with calculated results for $\Delta d_{12} = -25.5\%$ and $\Delta d_{23} = 3.2\%$. TFA potential, $\theta_D = 550K$, $V_{0i} = 4.5$ eV. From Zehner *et al.*, 1981.

one-electron band structure calculations (Eastman, 1980) predict that such a surface would be metallic, with a half-filled band of dangling bond states at the Fermi energy E_F, and would be very different from that observed for the annealed Si(111)-(7 × 7) surface as well as for the cleaved Si(111)-(2 × 1) surface. It is assumed that surface states are bandlike in all of these calculations.

Angle-resolved and angle-integrated photoemission studies of both valence band surface states and surface core-level shifts for the laser-annealed Si(111)-(1×1) surface prepared in the same manner as was done for the LEED study and for a Si(111)-(7 × 7) surface, prepared by thermally annealing the (1×1) surface, have been performed (Zehner *et al.*, 1981a). The measurements were made using the display-type spectrometer at the synchrotron radiation source Tantalus I.

In Figure 32, angle-integrated photoemission spectra are presented for a laser-annealed (1×1) surface and for a (7 × 7) surface prepared by thermally annealing the (1×1) surface to ~1175K. For the (1×1) surface, the dashed line shows the spectrum obtained after a hydrogen exposure (~500L of activated H) which resulted in about a saturation monolayer coverage of H (same normalization). The difference between the solid curve and the dashed curve within ~3 eV of E_F represents surface state emission. Two predominant surface state features are seen (i) at −1.8 eV and (ii) at −0.85 eV (relative to E_F) on both the (1×1) and

(7 × 7) surfaces. Although these surface state features exhibit different angular emission distributions, the results obtained from both surfaces are identical. For the (7 × 7) surface, a third weaker feature (iii) is seen at the Fermi level E_F which is at 0.5 ± 0.05 eV above the valence band maximum E_v. It exhibits a well-defined hexagonal angular emission distribution that is peaked at the Brillouin zone boundary of a (2 × 2) surface unit cell. These states at E_F, which correspond to ~3% of the total emission intensity from the surface states within ~3 eV of E_F, appear to originate via the (7 × 7) reconstruction from the −0.85 eV states which show a corresponding reduced intensity for the (7 × 7) surface relative to the (1×1) surface. The absence of any emission in the gap at E_F for the (1×1) surface is not in accord with predictions of one-electron-band calculations for an unreconstructed surface and therefore appears to be inconsistent with the structural model for this surface determined in the LEED analysis. Alternatively, calculations neglecting intrasite correlations may not be adequate to predict surface energy bands.

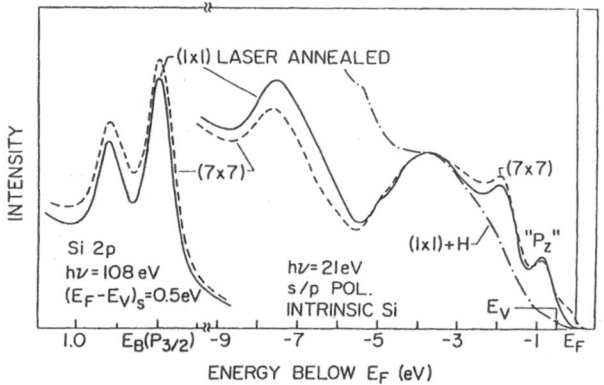

Figure 32   Angle integrated photoemission spectra for the valence bands and 2p core levels of laser-annealed Si(111)-(1×1) and thermally annealed Si(111)-(7 × 7) surfaces. Two prominent surface-state levels are seen for both, i.e., p_z-like levels at −0.85 eV and levels at −1.8 eV. From Zehner et al., 1981a.

Angle-integrated photoemission spectra of the Si $2p_{3/2}$ core-level for the (1×1) and (7 × 7) surfaces are shown in Figure 33 (solid lines) for a photon energy $h\nu = 120$ eV, an energy for which the spectra are surface-sensitive (escape depth 5.4Å) with about one-half the emission intensity corresponding to the outer double layer of Si surface atoms. To obtain these spectra the total 2p core-level spectra are decomposed into similarly shaped $2p_{1/2}$ and $2p_{3/2}$ contributions and then a $2p_{3/2}$ spectrum for a "bulk" contribution corresponding to the layers below the outer double layer is subtracted. The resulting curves (dashed lines in Figure 33) show the spectral distributions of surface core levels for the outer two surface layers (one double layer). The bulk line positions were found to be the same (within ±20 meV) for both surfaces, i.e., both have the same band bending at the surface ($E_F − E_v = 0.5 ± 0.05$ eV).

As seen in Figure 33, $2p_{3/2}$ surface core-level spectra (dashed curves) for the (1×1) and (7 × 7) surfaces are very similar and differ greatly from the spectra for the Si(111)-H(1×1). Both show characteristic low-binding-energy peaks, with the (1×1) surface showing a peak at −0.8 eV relative to the bulk with an intensity corresponding ~(1/4 ± 1/12) of a surface layer of atoms and the (7 × 7) surface showing a peak at −0.7 eV with an intensity corresponding to ~(1/6 ± 1/12) of a surface layer of atoms. Both surface spectra also show core levels on the high-binding-energy side of the bulk line at about the same position.

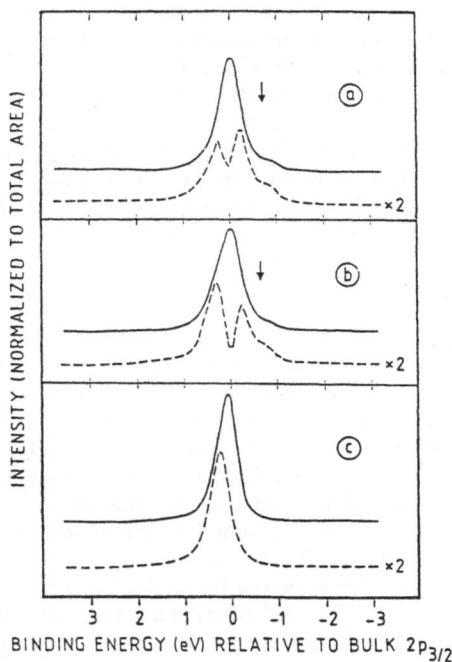

Figure 33 Si $2p_{3/2}$ core-level spectra for (a) Si(111)-(1×1), (b) Si(111)-(7 × 7) and (c) Si(111)-H(1×1). $h\nu = 120$ eV. The contribution due to the outer double layer of surface atoms is shown by dashed lines. From Zehner et al., 1981a.

The strong similarity of the valence band surface states and surface core-level spectra for the laser-annealed (1×1) surface and thermally-annealed (7×7) surface indicate that these surfaces have very similar local bonding geometries and differ mainly in long-range order involving geometrical rearrangements that are only a perturbation of the average local bonding geometry. An interesting question then involves the LEED analysis (Zehner et al., 1981) which gives such a good agreement with data using a model (1×1) geometry that appears to be different from that needed to describe the surface electronic structure. One possible explanation is that LEED is not particularly sensitive to long-range disorder if it is present on the (1×1) surface. Another explanation is that photoemission can rule out the relaxed, ordered (1×1) geometry only if the surface states are bandlike as assumed in one-electron-band calculations. However, correlation effects might be very important for these narrow surface levels. Theoretical proposals (Duke et al., 1981; Del Sole et al., 1981) have been put forth recently that would make the photoemission data from the Si(111)-(1×1) surface consistent with the unreconstructed relaxed surface predicted by the LEED analysis. In these models it is assumed that strong correlations dominate the surface state band structure. This idea is, in part, substantiated by experimental results obtained in a study on the cleaved Si(111)-(2 × 1) surface (Himpsel et al., 1981) which show the existence of two dangling bond states. Both theories predict a low-temperature antiferromagnetic ground state and downward dispersion of the dangling bond along $\Gamma - J$ at low temperature, and these predictions should be tested experimentally.

Two other photoemission studies of laser-annealed Si(111)-(1×1) using different annealing conditions (McKinley *et al.*, 1980; Chabal *et al.*, 1980) have been reported and in agreement with the above results find no occupied states in the gap. In one of these studies (Chabal *et al.*, 1980), as a consequence of the observation of weak half-order beams and photoemission spectra similar to those obtained from a Si(111)-(2 × 1), it is argued that the laser-annealed surface is buckled with no long-range order but with a short-range (2 × 1) reconstruction. It is suggested that different laser annealing conditions (depth of melt, regrowth velocity) can result in different surface structures. If this is true, then laser annealing with different annealing parameters may aid in determining the driving mechanisms for the reconstructions observed using conventional preparation procedures.

E. Surface Electronic Properties of Highly-Doped Si

Previous investigations have shown that both implanted B and As occupy substitutional sites in silicon subsequent to ion implantation and laser annealing even when their concentrations significantly exceed equilibrium solubility limits (White *et al.*, 1980). This observation suggests that these techniques may be used to alter or tailor the electronic structure at the surface and in the near surface region. To examine this possibility, photoemission techniques, as described in the previous section, have been used to investigate highly degenerate n-type Si(111)-(1×1) surfaces as a function of As concentration up to $\sim 5 \times 10^{21}/cm^3$ (~ 10 at.%) and degenerate p-type Si(111)-(1×1) as a function of B concentration up to $\sim 1 \times 10^{21}/cm^3$ (~ 2 at.%) (Eastman *et al.*, 1981). These concentrations considerably exceed equilibrium solubility limits. The samples were prepared by implanting As (100 keV, $7 \times 10^{16}/cm^2$ and $1 \times 10^{16}/cm^2$) and B (35 keV, $2 \times 10^{16}/cm^2$) and then *in situ* laser annealing. The maximum doping concentrations are about 10 and 3 times the concentrations of electrically active As and B achievable by conventional techniques, respectively. All laser-annealed Si(111) surfaces exhibited a (1×1) LEED pattern.

Angle-integrated photoemission spectra for the valence bands and $2p$ core levels are presented in Figures 34 and 35, for intrinsic Si(111)-(1×1), degenerate n-type As-doped (4 and 7 at.%) Si(111)-(1×1) and degenerate p-type B-doped (1 at.%) Si(111)-(1×1) surfaces. In Figure 34, spectra are normalized to constant total emission within 5 eV of E_F, and energies are given relative to the valence-band maximum at the surface (E_v^s). The Fermi-level position relative to the valence-band maximum at the surface ($E_F - E_v^s$) was previously determined to be 0.5 eV for intrinsic Si(111)-(1×1) (Zehner *et al.*, 1981a). E_F is seen to shift markedly with doping, i.e., from 0.25 eV above E_v^s for the B-doped sample to the conduction-band minimum $E_c^s = 1.1$ eV for the 7% As-doped sample. For intrinsic Si(111)-(1×1) in Figure 34, the dashed-dotted line shows the effect of an adsorbed monolayer of H, which is to remove the two predominant surface-state levels at 0.4 and 1.3 eV below E_v^s, as discussed in Section D. Relative to intrinsic Si, for highly degenerate (~ 1 at.%) B doping, the two states are unaltered and the principal changes are that E_F moves down by 0.25 eV and the surface becomes metallic.

More dramatic effects are seen with As doping. At 4 at.% As doping, the surface states have become significantly altered, while E_F has increased by 0.1 eV relative to the intrinsic Si. At this concentration, the upper "spz-like" dangling bond state has become much weaker and shifted upward in energy by 0.3 eV, the lower -1.4 eV state has increased significantly in intensity but is unshifted, and the surface has become metallic with new states near E_F. As the doping is further increased from 4 to 7 at.%, E_F rapidly shifts and becomes pinned at the conduction band minimum E_c^s. Also, the upper sp_z-like surface-state continues to diminish in intensity so as to be nearly imperceptible by 7 at.% doping, and the lower surface state becomes extremely intense. The conduction-band minima (Δ_{min}) near X become

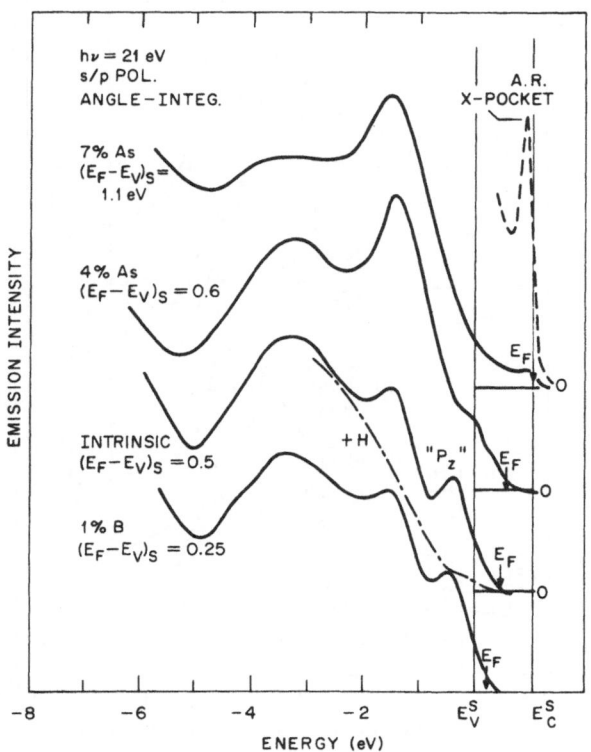

Figure 34   Photoemission spectra (PDOS) for the valence bands of laser-annealed (111)-(1×1) surfaces of intrinsic and highly doped Si. The levels near −0.4 and −1.3 eV are due to surface states. E_v^s, E_c^s and E_F denote the valence-band maximum, conduction-band minimum and Fermi-level positions at the surface. From Eastman *et al.*, 1981.

occupied, and emission from these minima is observed as intense elliptical lobes in angle-resolved photoemission spectra (dotted line labeled AR in Figure 34).

The Si $2p$ core-level spectra shown in Figure 35 were obtained using $h\nu = 108$ eV (dashed lines) and $h\nu = 120$ eV (solid lines) which provide "bulk-sensitive" and "surface-sensitive" spectra, respectively (Himpsel *et al.*, 1980). S and B denote surface and bulk $2p$ levels. For intrinsic Si and 7 at.% As-doped Si surfaces, the $2p$ levels are the same for both $h\nu = 108$ eV and $h\nu = 120$ eV; i.e., the "surface" S and "bulk" B $2p_{3/2}$ levels coincide in binding energy. This is indicative of the "flat-band" condition at the surface as shown by the schematic diagrams on the left of Figure 35. (CB and VB denote the conduction-band and valence-band edges.) A finding of significant interest is that at an As doping level of 7% or greater, E_F shifts across the entire gap from 0.5 eV for intrinsic (1×1) Si to the conduction band minimum at 1.1 eV. By depositing a thin Au film on this surface it was possible to show via Si $2p$ core-level measurements (not shown) that E_F remained unchanged (within ~50 meV). Thus a "zero-barrier-height" Schottky barrier was formed. For 4 at.% As and 1 at.% B doping levels the bands are bent upward and downward in the surface region, respectively. The broadening and smearing of the spectra when compared to that obtained for intrinsic Si reflect the distribution in energy of $2p$ core levels in the sampled region.

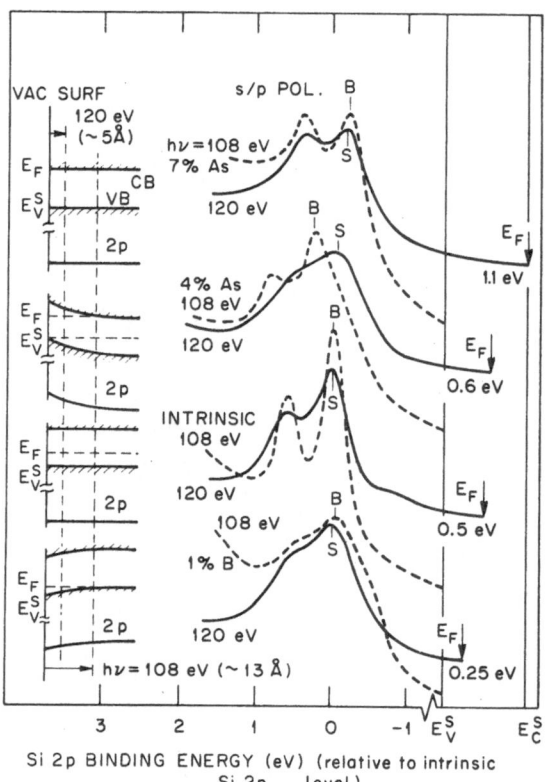

Figure 35   Angle-integrated photoemission spectra for the Si $2p$ core levels of the four Si surfaces depicted in Figure 34. The zero binding-energy corresponds to the Si $2p_{3/2}$ level for bulk Si(111)-(1×1). The schematic diagrams on the left depict band bendings and energy-level positions; the escape depths are $\ell \sim 5$Å for $h\nu = 120$ eV and ~ 13Å for $h\nu = 108$ eV. From Eastman *et al.*, 1981.

F. Summary of Surface Properties

It has been shown that the properties of the surface region of Si crystals can be altered by laser annealing in a UHV environment. The production of atomically-clean surfaces in processing times $\ll 1$ sec and the restoration of geometric order to the surface region of a damaged crystal indicate that this technique provides an alternative approach to conventional surface preparation procedures. When combined with ion implantation, laser annealing can be used to tailor both the geometric lattice (interatomic spacings) and electronic structure in the surface region.

These observations indicate that laser annealing of surfaces in UHV has potential as a tool for both surface science and practical application. Although it has been tested extensively only on Si crystals, the results obtained suggest that it will be applicable to a wide range of materials. Additional investigations concerned with the surface properties of materials irradiated with the laser as well as the changes that occur by varying the annealing conditions (regrowth velocity, depth of melt) need to be performed in order to characterize and understand this processing technique more completely.

REFERENCES

Appleton, B. R., and Celler, G. K. editors (1982). "Laser and Electron — Beam Interactions with Solids," North-Holland, N.Y.

Auston, D. H., Golovchenko, J. A., Smith, P. R., Surko, C. M. and Venkatesan, T. N. C. (1978). Appl. Phys. Lett. *33*, 538.

Aziz, M. J. (1982). J. Appl. Phys. *53*, 1158.

Baeri, P., Campisano, S. U., Foti, G. and Rimini, E. (1978). J. Appl. Phys. *50*, 788.

Baeri, P., Campisano, S. U., Grimaldi, M. G. and Rimini, E. (1980). *Laser and Electron Beam Processing of Materials*, ed. by C. W. White and P. S. Peercy, Academic Press, New York, p. 130.

Baeri, P., Poate, J. M., Campisano, S. U., Foti, G., Rimini, E. and Cullis, A. G., (1981). Appl. Phys. Lett. *37*, 912.

Baeri, P., Foti, G., Poate, J. M., Campisano, S. U. and Cullis, A. G. (1981a). Appl. Phys. Lett. *38*, 800.

Baeri, P. and Campisano, S. U. (1982). *Laser Annealing of Semiconductors*, ed. by J. M. Poate and J. W. Mayer, Academic Press, New York, Chapter 4.

Baeri, P. (1982). *Laser and Electron Beam Interactions with Solids*, ed. by B. R. Appleton and G. K. Celler, North Holland, New York, p. 151.

Baeri, P. and Barbarino, A. E. (1982). Private communication.

Baker, J. C. and Cahn, J. W. (1969). Acta. Metall. *17*, 575.

Bennett, P. A. and Webb, M. B. (1981). Surf. Sci. *104*, 74; (1981). J. Vac. Sci. Technol. *18*, 847.

Cahn, J. W., Coriell, S. R. and Boettinger, W. J. (1980). *Laser and Electron Beam Processing of Materials*, ed. by C. W. White and P. S. Peercy, p. 89. Academic Press, New York.

Campisano, S. U., Baeri, P., Grimaldi, M. G., Foti, G. and Rimini, E. (1980), J. Appl. Phys. *51*, 3968.

Campisano, S. U., Rimini, E., Baeri, P. and Foti, G. (1980a). Appl. Phys. Lett. *37*, 170.

Campisano, S. U., Foti, G., Baeri, P., Grimaldi, M. G. and Rimini, E. (1980b). Appl. Phys. Lett. *37*, 719.

Campisano, S. U., Baumgart, H. and Rozgonyi, G. (1982). Appl. Phys. Lett., submitted for publication.

Chabal, Y. J., Rowe, J. E. and Zehner, D. A. (1981). Phys. Rev. Lett. *46*, 600.

Cline, H. E. (1982). J. Appl. Phys. (in press).

Cullis, A. G. and Katz, L. E. (1974). Phil. Mag. *30*, 1419.

Cullis, A. G., Poate, J. M. and Celler, G. K. (1979). *Laser-Solid Interactions and Laser Processing-1978*, ed. by S. D. Ferris, H. J. Leamy and J. M. Poate, American Institute of Physics, New York, p. 311.

Cullis, A. G., Webber, H. C., Poate, J. M. and Simons, A. L. (1980). Appl. Phys. Lett. *36*, 320.

Cullis, A. G., Webber, H. C. Poate, J. M. and Chew, N. G. (1980a). J. Microsec. *118*, 41.

Cullis, A. G., Hurle, D. T. J., Webber, H. C., Chew, N. G., Poate, J. M., Baeri, P. and Foti, G. (1981). Appl. Phys. Lett. *38*, 642.

Cullis, A. G., Series, R., Webber, H. C. and Chew, N. G. (1981a). *Semiconductor Silicon 1981*, ed. by H. R. Huff, R. J. Kriegler and Y. Takeishi, Electrochemical Society, New Jersey, p. 518.

Cullis, A. G., Webber, H. C. and Chew, N. G. (1982). *Laser and Electron Beam Interactions with Solids*, ed. by B. R. Appleton and G. K. Celler, North Holland, p. 131.

Del Sole, R. and Chadi, D. J. (1981). Phys. Rev. B *24*, 7431.

Duke, C. B. and Ford, W. K. (1981). Surf. Sci. *111*, L685.

Eastman, D. E. (1980). J. Vac. Sci. Technol. *17*, 492.

Eastman, D. E., Heimann, P., Himpsel, F. J., Reihl, B., Zehner, D. M. and White, C. W. (1981). Phys. Rev. B *24*, 3647.

Ferris, S. D., Leamy, H. J. and Poate, J. M., editors (1979). "Laser-Solid Interactions and Laser Processing — 1978," American Institute of Physics, N.Y.

Foti, G. (1981). Nucl. Instrum. Meth. *182/183*, 573.

Galvin, G. J., Thompson, M. O., Mayer, J. W., Hammond, R. B., Paulter, N. and Peercy, P. S. (1982). Phys. Rev. Lett. *48*, 33.

Gibbons, J. F., Hess, L. D. and Sigmon, T. W., editors (1981). "Laser and Electron-Beam Solid Interactions and Materials Processing," North-Holland, N.Y.

Himpsel, F. J., Heimann, P., Chiang, T. C. and Eastman, D. E. (1980). Phys. Rev. Lett. *45*, 1112.

Himpsel, F. J., Heimann, P. and Eastman, D. E. (1981). Phys. Rev. B *24*, 2003.

Jackson, K. A., Gilmer, G. H. and Leamy, H. J. (1980). *Laser and Electron Beam Processing of Materials*, ed. by C. W. White and P. S. Peercy, p. 104. Academic Press, New York.

Jindall, B. K. and Tiller, W. A. (1968). J. Chem. Phys. *49*, 4632.

Larson, B. C., White, C. W. and Appleton, B. R. (1978). Appl. Phys. Lett. *32*, 801.

Liu, P. L., Bloembergen, N. and Hodgson, R. T. (1979). Appl. Phys. Lett. *34*, 864.

McKinley, A., Parlce, A. W., Hughes, G. J., Fryar, J. and Williams, R. H. (1980). J. Phys. D: Appl. Phys. *13*, 138.

Morehead, F. (1980). *Laser and Electron Beam Processing of Materials*, ed. by C. W. White and P. S. Peercy, Academic Press, New York, p. 143.

Mullins, W. W. and Sekerka, R. F. (1964). J. Appl. Phys. *35*, 444.

Narayan, J., Young, R. T. and White, C. W. (1978). J. Appl. Phys. *49*, 3912.

Narayan, J. (1979). Appl. Phys. Lett. *34*, 312.

Narayan, J. (1981). J. Appl. Phys. *52*, 1289.

Nozaki, T., Yaturugi, Y., Akiyama, N., Endo, Y. and Makide, Y. (1974). J. Radioanal. Chem. *19*, 109.

Poate, J. M., Leamy, H. J., Sheng, T. T. and Celler, G. K. (1978). Appl. Phys. Lett. *33*, 918.

Saris, F. W. (1981). Private communication.

Sekerka, R. F. (1965). J. Appl. Phys. *36*, 264.

Stuck, R., Fogarassy, E., Grob, J. J. and Siffert, P. (1980). Appl. Phys. *23*, 15.

Tiller, W. A., Jackson, K. A., Rutter, J. W. and Chalmer, B. (1953). Acta Metall. *1*, 428.

Turnbull, D. (1980). J. de Physique C4, 109.

Trumbore, F. (1960). Bell Syst. Tech. Jour. *39*, 205.

van Gurp, G. J., Eggermont, G. E. J., Tamminga, Y., Stacy, W. T. and Gijsbers, J. R. M. (1979). Appl. Phys. Lett. *35*, 273.

Wang, J. C., Wood, R. F. and Pronko, P. P. (1978). Appl. Phys. Lett. *33*, 455.

Westendorp, J. F. M., Wang, Z. and Saris, F. W. (1981). *Laser and Electron Beam Interactions with Solids*, ed. by B. R. Appleton and G. K. Celler, North Holland, New York, p. 255.

White, C. W., Narayan, J. and Young, R. T. (1979). Science *204*, 461.

White, C. W., Pronko, P. P., Wilson, S. R., Appleton, B. R., Narayan, J. and Young, R. T. (1979a). J. Appl. Phys. *50*, 3261.

White, C. W. and Peercy, P. S. editors (1980). "Laser and Electron Beam Processing of Materials," Academic Press, N.Y.

White, C. W., Wilson, S. R., Appleton, B. R. and Young, F. W., Jr., (1980). J. Appl. Phys. *51*, 738.

White, C. W., Appleton, B. R., Stritzker, B., Zehner, D. M. and Wilson, S. R. (1981). *Laser and Electron-Beam Solid Interactions and Materials Processing*, ed. by J. F. Gibbons, L. D. Hess and T. W. Sigmon, p. 59, North Holland Publishing Co., Inc., New York.

White, C. W., Appleton, B. R. and Wilson, S. R. (1982). *Laser Annealing of Semiconductors*, ed. by J. M. Poate and J. W. Mayer, Academic Press, New York, Chapter 5.

Williams, J. S. and Elliman, R. G. (1981). Nucl. Instrum. Meth. *182/183*, 389.

Wood, R. F., Kirkpatrick, J. R. and Giles, G. E. (1981). Phys. Rev. B *23*, 5555.

Wood, R. F. (1982). Phys. Rev. B, in press.

Yoffa, E. J. (1980). *Laser and Electron Beam Processing of Materials*, ed. by C. W. White and P. S. Peercy, Academic Press, New York, p. 59.

Zehner, D. M., White, C. W. and Ownby, G. W. (1980). Surf. Sci. *92*, L67.

Zehner, D. M., White, C. W. and Ownby, G. W. (1980a). Appl. Phys. Lett. *37*, 456.

Zehner, D. M., White, C. W. and Ownby, G. W. (1980b). Appl. Phys. Lett. *36*, 56.

Zehner, D. M., White, C. W. and Ownby, G. W. (1980c). *Laser and Electron Beam Processing of Materials*, ed. by C. W. White and P. S. Peercy, Academic Press, New York, p. 201.

Zehner, D. M., Noonan, J. R., Davis, H. L. and White, C. W. (1981). J. Vac. Sci. Technol. *18*, 852.

Zehner, D. M., White, C. W., Heimann, P., Reihl, B., Himpsel, F. J. and Eastman, D. E. (1981a). Phys. Rev. B *24*, 4875.

SOLID PHASE RECRYSTALLIZATION PROCESSES IN SILICON

J. S. WILLIAMS

Department of Communication and Electronic Engineering
Royal Melbourne Institute of Technology, Melbourne 3000, Australia

CONTRIBUTORS: S. U. Campisano, A. G. Cullis, S. S. Lau, J. W. Mayer, D. Turnbull and C. W. White

I. INTRODUCTION

Ion implantation into crystalline silicon typically produces an amorphous near-surface layer which is metastable and recrystallizes on subsequent heat treatment. The recrystallization process usually proceeds epitaxially on the underlying crystalline silicon substrate via either liquid or solid phase processes depending on the annealing procedure adopted. The previous chapter treated the annealing behavior during intense and rapid (nsec) laser or electron beam irradiations which induce local melting of the near-surface layers and recrystallization via liquid phase epitaxy. Intriguing impurity redistribution, segregation and supersaturation effects are observed to result from this annealing process. The magnitude of the observed effects is dependent upon the velocity of the liquid-solid interface (in the range of 1 m/sec) during the ultra-rapid resolidification time. The present chapter outlines the annealing behavior during near-surface solid phase crystallization of silicon, where similarly intriguing impurity movement and supersaturation effects are observed, although the recrystallization rate can be as much as ten orders of magnitude slower (1Å/sec) than during ultra-rapid liquid phase epitaxial growth.

Solid phase epitaxial growth (SPEG) of silicon is conveniently achieved in a furnace at temperatures of >500°C (Mayer et al., 1968; Csepregi et al., 1975). Alternatively, solid phase recrystallization can be induced in a much shorter time scale using scanning CW laser and electron beams (Gat and Gibbons, 1978; Williams et al., 1978; Regolini et al., 1979a; McMahon and Ahmed, 1979) or by employing strip heaters, high intensity arc lamps, solar energy and incoherent light sources (Fan et al., 1981; Gat, 1981; Lau et al., 1979; Fulks et al., 1981; Harrison et al., 1981). The laser and e-beam irradiations can locally raise the surface temperature >900°C for times of the order of milliseconds, whereas the latter heating methods essentially provide heating of the entire wafer to temperatures above 600°C for times in the range of 1 to 100 seconds. Although solid phase annealing can be induced over wide time and temperature ranges using the various transient annealing methods, the basic crystallization mechanisms and impurity redistribution processes are essentially similar. This chapter concentrates specifically on the surface crystallization behavior in silicon induced by furnace annealing. However, relevant results which have been obtained using transient annealing methods will be cited where appropriate.

II. OVERVIEW OF EXPERIMENTAL OBSERVATIONS

The kinetics of SPEG are influenced by several parameters, including the underlying substrate orientation and the type and concentration of implant species. However, work at Caltech has indicated that the growth kinetics of Si^+ ion implanted <100> oriented silicon are particularly simple (Csepregi et al., 1975). For this impurity-free situation, the planar amorphous-crystalline interface is observed to move towards the surface with a uniform velocity and a well defined activation energy of 2.35 eV. For furnace annealing at 550°C, the interface velocity is 1.5Å/sec and the regrown epitaxial layer is relatively defect-free (Lau et al., 1980).

When substrate orientations other than <100> are employed and implant species other than Si^+ are used to produce amorphous silicon, the regrowth kinetics and the nature of the regrown layers are often considerably more complex. In addition, the implanted impurity atoms may well redistribute prior to or during the regrowth process. Impurity segregation and precipitation effects at high implant concentrations can further influence the growth kinetics and crystalline quality. Furthermore, for annealing conditions in which the diffusion length of the implanted impurity is small (~1Å) during the time of regrowth, it is possible to incorporate metastable concentrations of impurities onto silicon lattice sites during SPEG. These various regrowth processes are considered, in turn, in the following sections.

A. Regrowth-Impurity Interactions

1. Low-dose Implantation

The strong dependence of epitaxial regrowth rate on the orientation of the underlying substrate is illustrated in Figure 1 (Csepregi et al., 1978). This shows regrown layer thickness vs annealing time at 550°C for <100>, <110> and <111> oriented silicon

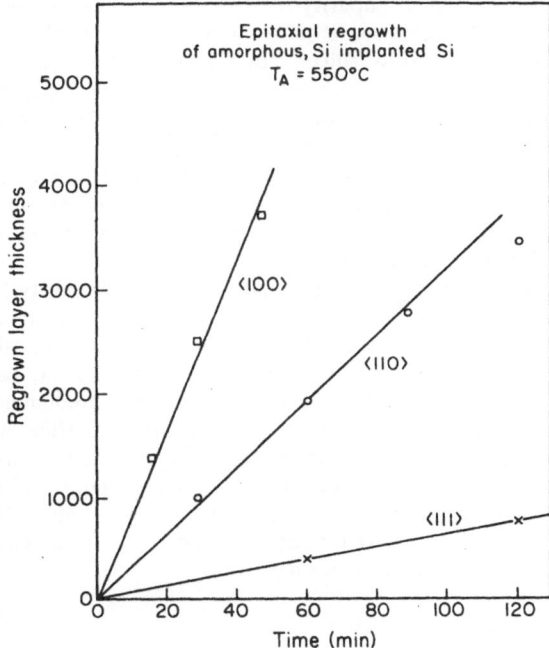

Figure 1   Regrown layer thickness versus time of anneal at 550°C for <100>, <110> and <111> - oriented silicon wafers that had been implanted with ^{28}Si ions at LN_2 temperatures. (After Csepregi et al., 1978.)

substrates initially implanted with Si^+ ions to produce an impurity-free amorphous surface layer ~4000Å in thickness. All orientations exhibit the same activation energy (2.35 eV) but <100> samples regrow about 3 times faster than <110> samples and about 25 times faster than the initial growth rate for <111> substrates. The growth of <111> oriented silicon is non-linear with annealing time and exhibits pronounced twinning along (111) planes, the twins occupying about 30 to 40 percent of the overall volume of the regrown layer (Lau et al., 1980).

Detailed regrowth studies at Caltech have shown that small concentrations of implanted impurities (≤1 atom percent) can profoundly increase (e.g., for B,P,As) or decrease (e.g., for O,N,C) the regrowth rate of <100> oriented silicon substrates (Csepregi et al., 1977; Kennedy et al., 1977). This behavior is illustrated in Figure 2 for P^+ and O^+ implants into <100> silicon. The regrowth rates have been normalized to that under impurity-free conditions. Clearly, the regrowth rate increases with implant concentration for phosphorus and decreases with implant concentration for oxygen. Further observations (Kennedy et al., 1977; Csepregi et al., 1977) have indicated that B^+ implanted samples result in the strongest increase in regrowth rate (about 2.5 times that of P^+ and As^+ at the same concentration) and exhibit an activation energy for regrowth slightly less than that of impurity-free silicon. In addition, the regrowth rate appears to be dominated by the localized impurity concentration near the growth interface.

The above data indicate that electrically active shallow level dopants in silicon, whether n- or p-type impurities, enhance the regrowth rate. More recent experiments indicate that multiple implantation of both n- and p-type dopants have a compensation-like effect, whereby the regrowth rate returns to that of impurity-free silicon (Suni *et al.*, 1982a). This effect is illustrated in Figure 3 for the combination of As^+ and B^+ implants (Figure 3a) and As^+ and P^+ implants (Figure 3b) into <100> silicon. In Figure 3a, the regrowth rate is slow (and close to the impurity-free case) in regions where the boron and arsenic concentrations are nearly equal. In contrast, Figure 3b shows that the regrowth rate is enhanced for multiple P^+ and As^+ implants (both n-type) compared with a single As^+ implant, indicated by the dashed curve. Furthermore, implantation of low concentrations of Ge^+ into <100> silicon (no doping effect) has been shown to have no significant effect on the regrowth rate (Mezey *et al.*, 1981). Models which attempt to explain the effect of dopants on the regrowth rate are considered later, in Section III.

As mentioned earlier, the quality of epitaxial growth for impurity-free <111> silicon substrates is poor, with the regrown layer containing a high density of twins and other defects. The introduction of small concentrations of low solubility impurities, such as the rare gases and Pb (Williams and Grant, 1976; Williams *et al.*, 1977), can result in severe disruption to epitaxial growth and ultimately to the predominance of alternate (polycrystalline) growth processes. Such effects have been attributed to nucleation of polycrystalline growth at implant precipitates or aggregates within the amorphous layer. These processes are more typical in the higher implant dose regime and will be described in more detail in the following section. In contrast to <111> substrates, good crystalline quality of epitaxially regrown layers is usually obtained for <100> silicon implanted with low impurity concentrations (<1 atom percent).

2. High-dose Implantation

As the implanted concentration of impurities increases beyond the 1 atom % levels in the previous section, the regrowth behavior becomes decidedly more complex. The trend for the regrowth rate to increase (almost linearly) with implant concentration (as illustrated in Figure 3 for P^+ implants) does not continue to high concentrations. Recent evidence (Williams and Elliman, 1980; Campisano and Barbarino, 1981) suggests that the regrowth rate reaches a maximum and then slows down appreciably at concentrations close to or exceeding the equilibrium solubility limit. For example, the regrowth behavior is shown in Figure 4 for 10^{16} As/cm^2 implanted at 50 keV in <100> silicon (Williams and Elliman, 1980). The high resolution channeling spectra in Figure 4a illustrate the isothermal anneal behavior at 505°C for various anneal times. Figure 4b plots the local regrowth rate (extracted from Figure 4a) as the growth interface proceeds through the arsenic concentration profile. Clearly, the regrowth rate is shown to exceed that of impurity-free silicon (shown dashed for reference) for the low arsenic concentrations ($\leqslant 1.5 \times 10^{21}/cm^3$) in the tail of the distribution. However, as the As concentration increases beyond $\sim 1.5 \times 10^{21}/cm^3$, the regrowth rate is observed to slow down rapidly and to ultimately reach a value over an order of magnitude below that of pure silicon at a concentration of 2.5×10^{21} As/cm^3. It is interesting to note that the equilibrium solid solubility limit for As in silicon is $\sim 1.5 \times 10^{21}/cm^3$ (Trumbore, 1960).

A further example of retardation in regrowth rate at high implant concentrations is shown in Figure 5 for Te implanted <100> silicon (Campisano and Barbarino, 1981). Compared with the impurity-free regrowth rate, low concentrations of Te, below $\sim 7 \times 10^{19}/cm^3$, increase the growth rate, whereas higher concentrations retard the regrowth.

In addition to modifying the regrowth rate, the type and concentration of the implant species can have a considerable influence on the crystalline quality of the regrown layer. For example, poor quality regrowth is illustrated in Figure 6, where channeling spectra indicate that SPEG is completely inhibited for 600°C annealing of a <100> sample implanted with

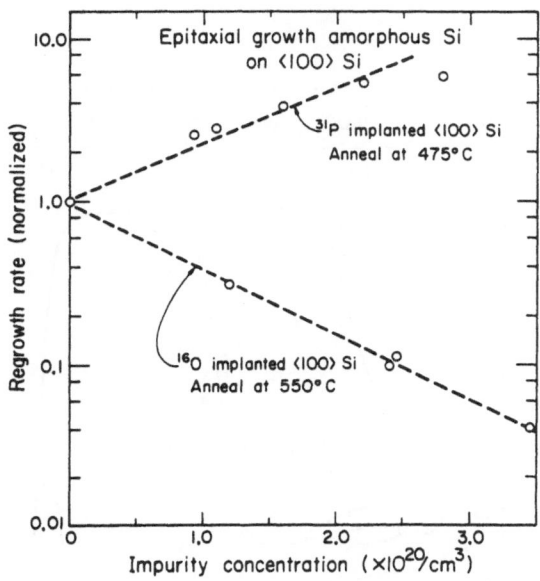

Figure 2   Epitaxial growth rate versus impurity concentration for <100> silicon implanted with[31]P and [16]O ions. The growth rate in the impurity free portion of the sample was found to be 1.5Å/sec at 550°C and 0.054Å/sec at 475°C. (After Lau *et al.*, 1980.)

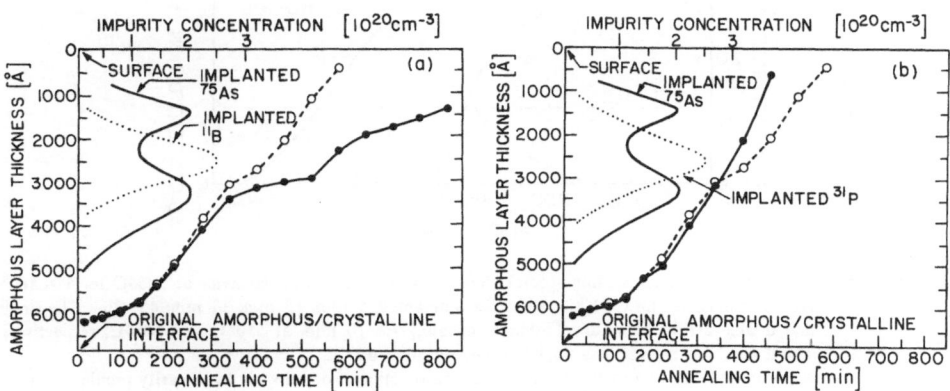

Figure 3   Amorphous layer thickness versus annealing time at 475°C for (a) [75]As and [11]B implanted, and (b) [75]As and ^{31}P implanted <100> silicon, with calculated impurity profiles superimposed on the data. The dashed line shows the regrowth characteristics of a reference sample implanted with [75]As only. (After Suni *et al.*, 1982a.)

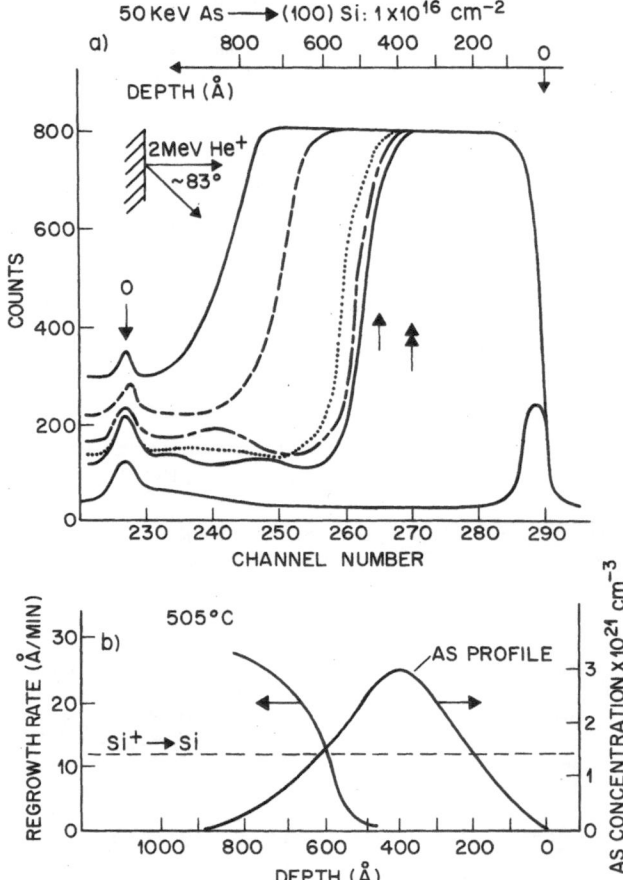

Figure 4   (a)   Silicon portion of channeling spectra showing isothermal anneal behavior at 505°C for 1×10^{16}
As/cm^2 implanted into <100> silicon: as-implanted; 5 min; 15 min; 30 min; 45 min. The single
and double arrows refer to additional annealing for 10 min. at 550°C and 565°C, respectively.
The dashed line represents a final 10 min. anneal at 600°C.

            (b)   Amorphous layer regrowth rate at 505°C as a function of depth with As impurity profile shown for
comparison. The dashed line represents the Si$^+$ → Si (impurity-free) regrowth rate at 505°C.
(Data from Williams and Elliman, 1980.)

2×10^{15} Cu/cm^2 at 70 keV (Campisano *et al.*, 1980a). In this case, polycrystalline regrowth takes place in the surface layers. Similar regrowth behavior has been observed for moderate dose implants ($\sim 1 \times 10^{15}$/cm^2) into <100> silicon of other low solubility species as Ag$^+$, Pb$^+$ and several of the rare gas ions (Campisano *et al.*, 1980b; Williams *et al.*, 1977; Williams and Grant, 1976; Revesz *et al.*, 1978; Cullis *et al.*, 1978). These results can be compared with the annealing behavior for much higher doses of high solubility species, such as 10^{16}As/cm^2 implanted <100> silicon, as shown in Figure 4, where complete epitaxy is obtained after 600°C thermal treatment (see dashed spectrum). However, an increase in the As$^+$ implant dose to 10^{17}/cm^2 in <100> silicon (peak concentration $\sim$28 atom percent) has been observed to result in poor quality epitaxial growth during 600°C annealing (Williams and Elliman, 1981). This dose-dependence of regrowth quality is also typically observed with <111> substrates, where poor quality SPEG (containing twins and stacking faults) can result at rather low impurity concentrations (Christodoulides *et al.*, 1978). Clearly, the quality of the regrown layer and the onset of polycrystalline growth for both <111> and <100> substrates can be correlated with (i) the slow down in epitaxial regrowth rate as the implant concentration increases and, (ii) with the equilibrium solubility limit of the implant species in silicon.

A suggested mechanism for poor quality SPEG leading to polycrystalline nucleation is illustrated schematically in Figure 7. For moderate and high dose implants, the silicon is amorphized to a depth well beyond the peak in the implant distribution. During annealing, epitaxial regrowth proceeds in the region of the profile of low implant concentrations (c.f. the initial epitaxial growth evident from Figure 6) but is increasingly impeded as the growth interface encompasses high concentration regions. The slow down in regrowth rate may well be attributable either to local strain or to impurity precipitation effects within the implant layer, processes which are discussed more fully in Section III. As epitaxy is retarded, polycrystalline nucleation within the amorphous layer may well dominate the latter stages of recrystallization. Polycrystalline nucleation is to be favored in cases where: (i) low solubility, fast diffusing species such as rare gases, Cu or Ag precipitate in the amorphous phase *prior to* regrowth, thus providing suitable nucleation sites for competing silicon growth processes, or (ii) implant segregation and precipitation processes occur *during* regrowth, particularly at concentrations exceeding the solubility limit. Evidence for the former process has been obtained from TEM analysis of rare gas implants into <100> silicon (Revesz *et al.*, 1978; Wittmer *et al.*, 1978), where gas bubbles have been observed within the amorphous surface layer prior to recrystallization. The observation of gas bubbles is illustrated in the TEM micrographs of Figure 8 (Cullis *et al.*, 1978). The possibility of such precipitation processes which can lead to polycrystalline nucleation during epitaxial growth is discussed more fully in the following section.

B. Impurity Segregation, Precipitation and Redistribution

The redistribution of ion implanted impurities in silicon can take place via several distinct processes. In this section, these redistribution processes are distinguished, for convenience, in terms of when they occur during annealing: (i) in amorphous silicon *prior* to recrystallization, (ii) *during* the amorphous-to-crystalline phase change, and (iii) in crystalline silicon *following* the recrystallization process.

1. Redistribution Prior to Recrystallization

In the previous section, it was shown that the segregation and precipitation of implanted impurities in amorphous silicon could constitute a barrier towards subsequent epitaxial growth (Figure. 6-8). It is of relevance in this section to briefly examine the possible ways in which

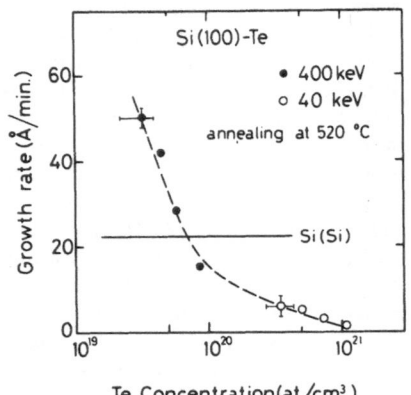

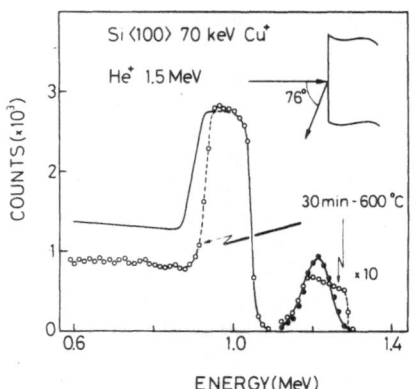

Figure 5   Regrowth rate as a function of Te concentration for Te⁺ implanted <100> silicon annealed at 520°C. The solid line indicates the impurity-free regrowth rate. (After Campisano and Barbarino, 1981.)

Figure 6   Channeling analysis of 2×10^{15} Cu/cm² implanted into <100> silicon and annealed at 600°C for 300 min. Solid curve indicates the aligned, as-implanted spectrum. (After Campisano et al., 1980a.)

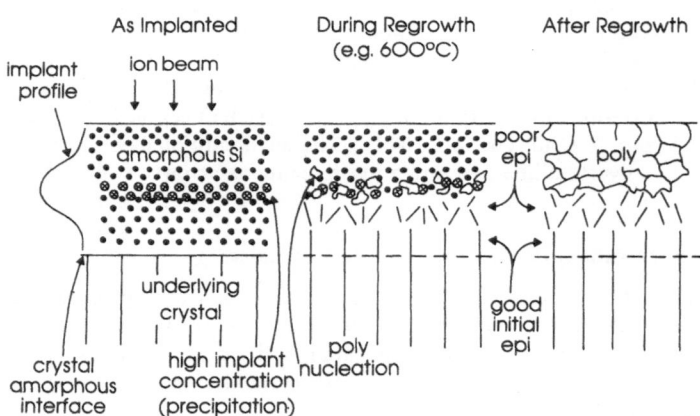

Figure 7   Schematic illustration of impeded SPEG and the initiation of alternate (polycrystalline) growth processes for high implant concentrations.

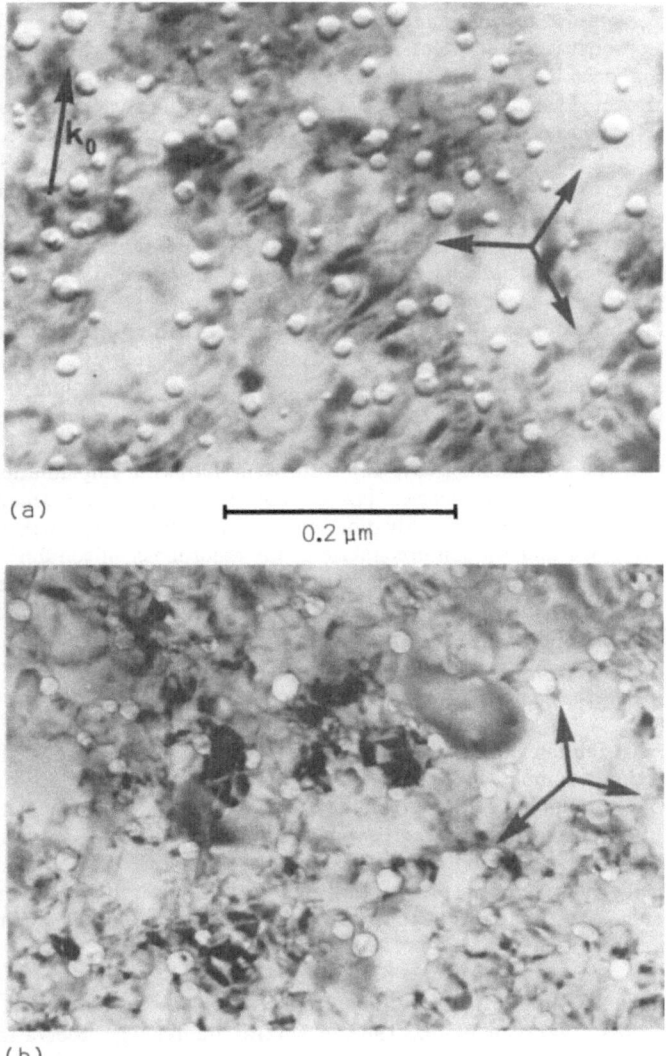

(a)

0.2 μm

(b)

Figure 8   Ne$^+$ ion implanted and annealed (111) Si layer: TEM micrograph showing faceted bubbles; (a) near [111] pole (<110> type directions indicated), topographical contrast; (b) [110] pole (<111> - type directions indicated by short arrow, [100] - type directions by long arrow), underfocus. (After Cullis *et al.*, 1978.)

impurities can redistribute and segregate in amorphous silicon prior to recrystallization both during the implantation process itself and during annealing at temperatures and for times insufficient to induce silicon recrystallization.

The various possibilities for redistribution, segregation and precipitation during implantation are examined in considerable detail in Chapters 8 and 9. In the brief treatment here, distinction is made between processes involving: (a) recoil and cascade effects, whereby previously implanted atoms are redistributed by knock-on or within the collision cascade of subsequently implanted ions, and (b) radiation-assisted diffusion processes involving, in particular, the generation of mobile defects during bombardment and associated solute motion under a concentration gradient or other (e.g., chemical) driving forces. The former process, of itself, leads merely to the modification of the implanted concentration profile: locally, the impurity atoms remain randomly distributed within the host matrix rather than segregating or forming precipitates. However, the latter radiation-assisted-migration process can lead to appreciable segregation and precipitation of elemental or more complex second phases. Furthermore, under the influence of a chemical driving force, solid state reactions and near-surface compound formation may result, as can occur for silicide-forming species such as Pt and Ni implanted into silicon. Such processes are particularly relevant to ion beam mixing, and are discussed in Chapter 9. As illustrated in Figure. 6-8, the agglomerates and second phase precipitates which result from radiation-assisted migration can influence the nature of subsequent silicon regrowth.

Precipitation in the amorphous phase can also occur following implantation, but prior to recrystallization, if the implant species has a migration energy (in amorphous silicon) appreciably lower than the 2.35 eV activation energy for epitaxial regrowth. In such cases, fast diffusers such as Cu and Ag can precipitate in the amorphous silicon during annealing at 550-600°C in times 10 to 100 times shorter than the time taken for completion of SPEG (Campisano *et al.*, 1980a,b). Thus, epitaxy can be retarded by precipitation which occurs ahead of the advancing epitaxial growth front.

2. Interface Segregation During Regrowth

An intriguing redistribution phenomenon occurring during epitaxial regrowth is illustrated in Figure 9. For this example (1×10^{15} In/cm^2 implanted <100> silicon), isothermal annealing at 555°C results in both progressive epitaxial growth towards the surface, as shown in Figure 9a, and accompanying segregation and "push out" of In impurity at the amorphous-crystal interface during epitaxy, as shown in Figure 9b (Williams and Elliman, 1981; 1982a). Following complete recrystallization, part of the implanted In remains within bulk Si and part has accumulated at the near-surface, ahead of the advancing amorphous-crystal boundary.

This "push out" phenomenon has also been observed for As, Sb, In, Pb, Tl and Bi implants into <100> silicon (Williams and Elliman, 1981, 1982a; Williams and Short, 1981; Fletcher *et al.*, 1981; Narayan *et al.*, 1981). Several interesting observations relating to this effect have been reported (Williams and Elliman, 1982a). (i) Interface segregation and "push out" of the implant species has only been observed for slow diffusing species which have not precipitated prior to epitaxy. Such species may be characterized by a diffusion length for atomic motion which is ≤1Å in the time taken for complete epitaxial recrystallization. (ii) "Push out" is not observed below a critical impurity concentration level. Consequently, the phenomenon is not representative of conventional segregation processes which are characterized by a concentration-independent segregation coefficient. The process is not, therefore, analogous to zone migration at a moving liquid-solid interface, as described in the previous chapter. (iii) The critical concentration at which the onset of "push out" is observed

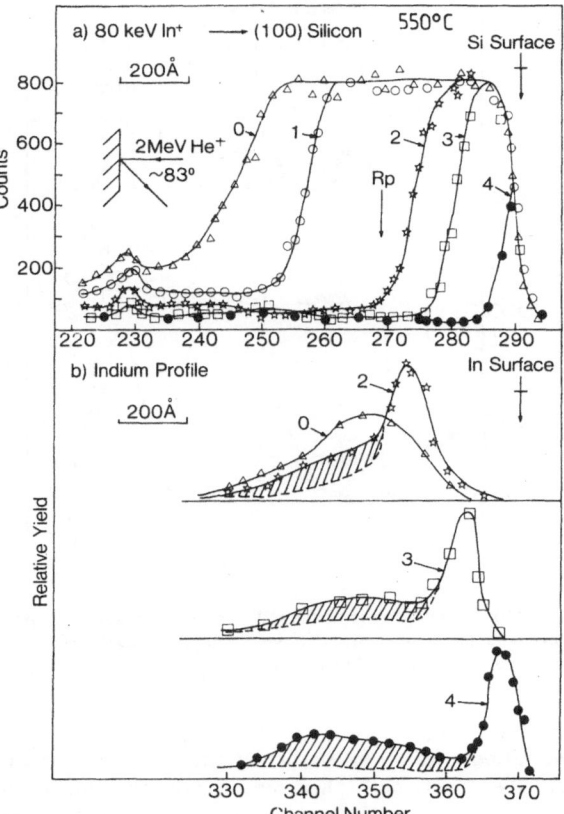

Figure 9   Channeling spectra showing epitaxial regrowth and correlated implant redistribution for 80 keV, 10^{15} In/cm^2 implanted into <100> silicon and annealed isothermally at 550°C for 0 mins (0); 3 mins (1); 10 mins (2); 15 mins (3); 30 mins (4).
(a)   Silicon portion of <100> aligned spectra.
(b)   Indium random profiles (solid curves) and <100> aligned profiles (dashed curved) corresponding to the spectra in (a). (After Williams and Elliman, 1981.)

is species dependent and, moreover, is always significantly greater than the maximum retrograde solid solubility for the particular impurity in silicon. The solubility aspects are discussed more fully in Section IIC. (iv) The "push out" phenomenon is observed at implant concentrations where the SPEG process is inhibited. For example, as the critical concentration is approached, the regrowth rate is observed to significantly slow down (Williams and Elliman, 1980, 1982a). Moreover, for concentrations significantly above the onset of "push out", polycrystalline regrowth often ensues (Williams and Elliman, 1981). (v) The maximum concentration of impurity which has been observed to segregate at a moving amorphous-crystalline interface is ~1 monolayer. For In and Pb impurities, concentrations in excess of 1 monolayer appear to completely retard epitaxial growth and initiate polycrystalline recovery, presumably by local impurity precipitation at the amorphous-crystalline interface.

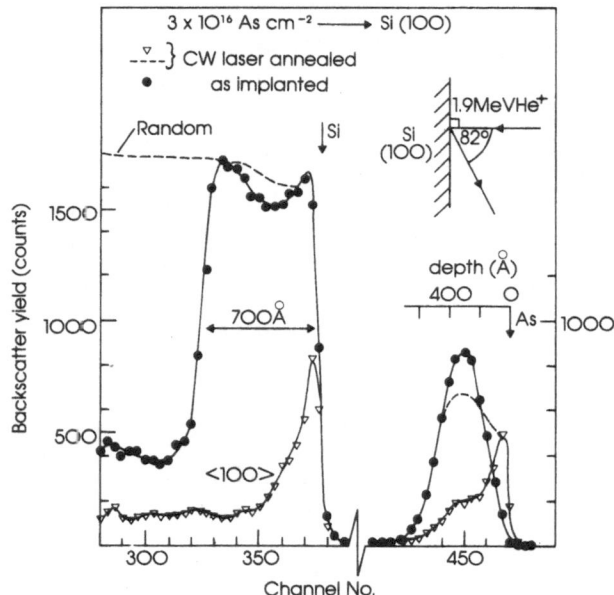

Figure 10   Random and <100> aligned spectra for CW Ar ion laser annealed, As⁺ ion ion implanted (100) silicon, indicating profile redistribution during solid phase regrowth. (Laser spot 40 μm, 8W power and 15 msec dwell time.) (Data from Williams *et al.*, 1979.)

The above observations refer, in particular, to furnace annealing at temperatures $\leqslant 600°C$. However, the "push out" phenomenon appears also to be observable during transient solid phase annealing at considerably higher temperatures. For example, Figure 10 shows arsenic redistribution during CW Ar ion laser annealing of 3×10^{16} As/cm² implanted <100> silicon at ~900°C for 15 msec.

C. Redistribution and Precipitation in Crystalline Silicon

Following recrystallization at temperatures $\leqslant 550°C$, impurity redistribution and precipitation effects may take place under suitable conditions within the recrystallized layer. The particular redistribution or precipitation process depends upon the nature and quality of the recrystallized layer. Two distinct situations can arise. Firstly, recrystallization may result in good quality SPEG with no redistribution or precipitation of the implant species during regrowth. Subsequent heat treatment at higher temperatures and/or for longer times may induce impurity migration via normal diffusion processes. As a consequence, the often observed profile broadening may occur for high (>900°C) anneal temperatures (Dearnaley *et al.*, 1973) and/or precipitation of the implanted impurity may result if the impurity diffusion length is greater than the mean separation of impurity atoms. Local precipitation has often been observed to result from high temperature ($\geqslant 1000°C$) processing of high dose B, P and As implanted silicon (Dearnaley *et al.*, 1973). However, more recently, the early stages of implant precipitation have been observed at somewhat lower anneal temperatures (600-900°C) for Bi (Campisano *et al.*, 1980a) and Sb (Fletcher *et al.*, 1981; Pogany, 1982) implanted <100> silicon. The TEM results in Figure 11 illustrate the situation in the Sb implanted silicon system. Figure 11a shows the formation of dislocation loops and rod-like defects at 850°C for 3×10^{15} Sb/cm^2 implanted <100> silicon. These defects are not evident at lower temperatures and appear to be associated with the initial stages of Sb precipitation, which can be clearly identified from diffraction patterns (not shown). Figure 11b shows the result of 1000°C annealing of 1×10^{16} Sb/cm^2 implanted <100> silicon in which regular Sb precipitates are observed: these are oriented with respect to the silicon lattice, as indicated by Sb spots in the accompanying diffraction pattern. Interestingly, annealing 1×10^{15} Sb/cm^2 implanted silicon up to 1000°C does not produce precipitation or any significant generation of extended defects. This concentration-dependent effect is related to equilibrium solubility limits and is pursued further in Section IIC.

A second situation amenable to implant redistribution following recrystallization is that in which the regrown layer contains a multitude of extended defects. In this case, the defects may act as nucleation sites for impurity precipitation and/or constitute fast diffusion paths for the impurity to redistribute throughout the recrystallized layer. Figure 12 illustrates redistribution of Pb in <111> silicon via the latter process. Following recrystallization at 565°C for 40 mins to produce a polycrystalline layer, the Pb implanted profile has not changed. However, annealing for longer times at 565°C has resulted in progressive Pb redistribution and segregation at the near-surface, presumably arising from rapid grain-boundary-assisted outdiffusion (Christodoulides *et al.*, 1980). Such redistribution effects are most typically observed in poorly recrystallized <111> silicon (Christodoulides *et al.*, 1978; Blood *et al.*, 1979) at implant concentrations exceeding about 1×10^{15}/cm^2.

Finally, implant redistribution and precipitation may proceed via several cooperative processes during annealing of a particular implant system. For example, in Figure 13 the channeling spectra illustrate both "push out" of Pb at the crystal-amorphous interface *during* the initial stages of epitaxial regrowth at 505°C, and further redistribution, most probably via grain-boundary-assisted outdiffusion, *following* polycrystalline regrowth of the near-surface region at 565°C (Williams and Elliman, 1981).

D. Supersaturated Solid Solutions

Impurities can be implanted into solids at concentrations which greatly exceed maximum equilibrium values. However, in silicon, subsequent annealing to reconstitute the crystalline lattice structure may lead to impurity precipitation effects and to a return to near-equilibrium solubility conditions. Indeed, in the previous sections, it was suggested that the onset of retarded regrowth rate, poor quality epitaxial growth, and concentration-dependent redistribution and precipitation effects may well be related to solid solubility limits for the implanted impurity in silicon. In terms of regrowth and redistribution, it has already been shown that a clear distinction can be made between fast and slow diffusing impurities in silicon. Similarly for solid solubility, one may expect differences between fast diffusers, which can precipitate ahead of the advancing crystal-amorphous interface during epitaxy, and slow diffusers, which may remain randomly dispersed as the recrystallization front sweeps to the surface to incorporate them into the crystalline lattice. In the latter case, supersaturated solid solutions might be expected to accompany SPEG. However, subsequent higher temperature and/or longer time annealing may induce precipitation within crystalline silicon and effect a return to equilibrium solubility when the impurity atoms become mobile in the lattice. In this section, both the formation and stability of SPEG-induced supersaturated solid solutions are examined.

1. Formation

Early studies of ion implantation and furnace annealing in silicon (see, for example, Picraux *et al.*, 1969; Mayer *et al.*, 1970; Gyulai *et al.*, 1971; Sigurd and Bjorkquist, 1972;

(a)

(b)

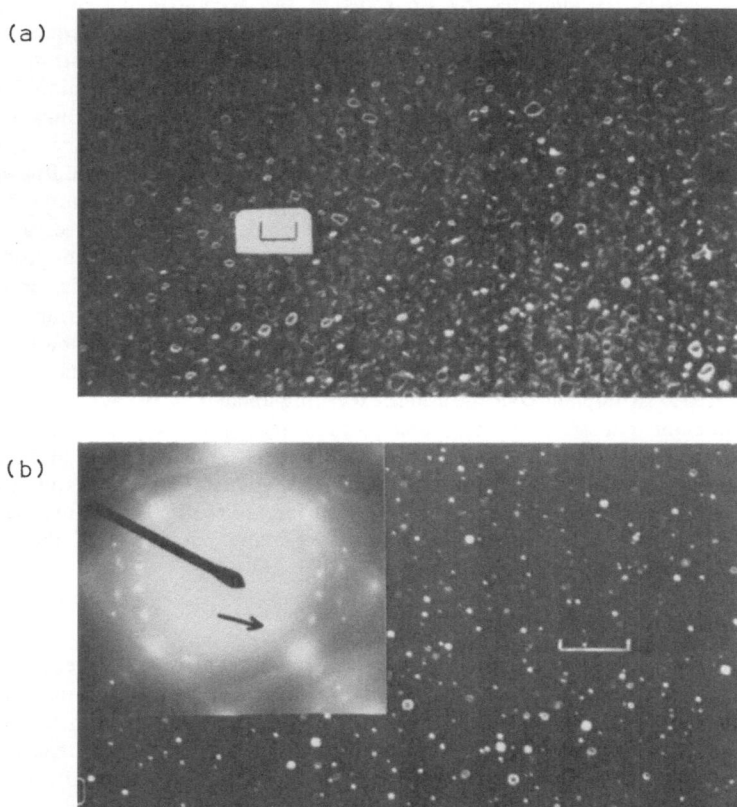

Figure 11   (a)   TEM dark field (440) micrograph showing the initial stages of precipitation and associated loop
                  formation for 3×10^{15} Sb/cm^2 implanted (100) Si annealed at 850°C for 30 min.   Scale marker
                  is 1000Å.
            (b)   TEM diffraction pattern and associated dark field micrograph (imaging the arrowed Sb spot) for
                  1000°C annealing of 1×10^{16} Sb/cm^2 implanted into (100) silicon. Regular Sb precipitates are
                  shown, which are oriented with respect to the Si host lattice. Scale marker is 1000Å. (After
                  Pogany, 1982.)

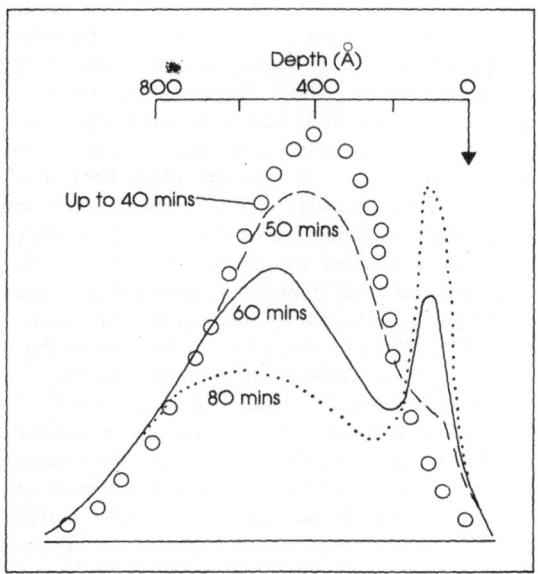

Figure 12   Pb profiles showing grain-boundary-assisted outdiffusion *following* polycrystalline regrowth for annealing of 2×10^{15} Pb/cm^2 implanted <111> silicon at 565˚C.   (After Christodoulides *et al.*, 1980.)

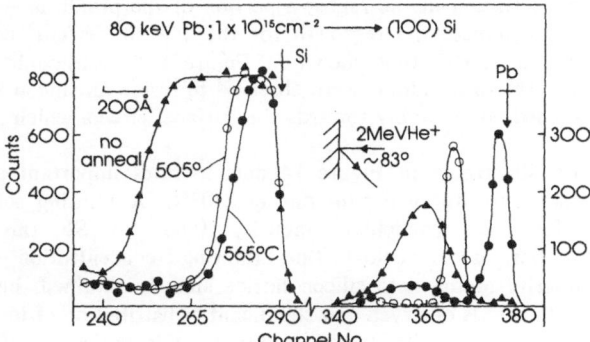

Figure 13   Channeling spectra illustrating two distinct redistribution processes for annealing of 1×10^{15} Pb/cm^2 implanted <100> silicon.   Annealing to 505˚C to induce partial SPEG facilitates "push out" of Pb and segregation at the amorphous-crystal interface.   Further annealing at 565˚C forms a polycrystalline surface layer through which the segregated Pb outdiffuses to build up at the surface.   (Data taken from Christodoulides *et al.*, 1980.)

Dearnaley *et al.*, 1973) gave clear indications of the possibility of forming supersaturated solid solutions via solid phase recrystallization. Such studies often employed channeling techniques to determine the substitutional concentrations for group III, IV, V and VI impurities in silicon. These impurities comprise those which have low diffusivity in silicon and, thus, would be expected to remain randomly dispersed during epitaxial growth. More recent examinations (Williams *et al.*, 1977; Blood *et al.*, 1979; Williams *et al.*, 1980; Josquin and Tamminga, 1978) have indicated that maximum measurable substitutional solubilities appear to depend on implant species, substrate orientation and annealing schedule. More detailed measurements (Williams and Elliman, 1980, 1981, 1982a; Campisano *et al.*, 1980a,b; Fletcher *et al.*, 1981) have clearly identified that highest solubilities are obtained for (100) silicon and furnace anneal temperatures $\leqslant 600°C$. In addition, maximum measured solubility limits appear to be controlled by several correlated factors, including impurity diffusivity, atom size, regrowth rate and impurity redistribution. Some of these more recent observations are illustrated below. The backscattering and channeling spectra of Figure 14 (Williams and Elliman, 1982a) illustrate some typical features of the Sb-implanted <100> silicon system. Figure 14a indicates that good epitaxial recovery was obtained following annealing of 5×10^{15} Sb/cm^2 implanted <100> silicon at 580°C. From a comparison of random and aligned Sb yields in Figure 14a, together with more detailed angular scans (Williams and Short, 1981), greater than 97% of Sb atoms are estimated to reside on silicon lattice sites. This provides a peak substitutional Sb concentration of $\simeq 1.3 \times 10^{21}$/cm^3, at least an order of magnitude above the equilibrium solubility maximum for Sb in silicon (Trumbore, 1960). Also, within the ~ 40Å depth resolution in Figure 14a, no redistribution of Sb has occurred during annealing. Significantly, all implant doses below 5×10^{15} Sb/cm^2 exhibited good epitaxy, near 100% Sb substitutionality and no redistribution.

The effect of increased Sb implant dose is shown in Figure 14b, where a mere factor of two increase in dose has resulted in a dramatic reduction in regrowth rate for the same anneal conditions as in Figure 14a. A further 60 min anneal at 580°C (not shown) does not remove the residual ~ 400Å thick amorphous layer indicated in Figure 14b. The maximum Sb substitutional concentration in the regrown portion of the profile is $\sim 1.3 \times 10^{21}$/cm^3, as before. Moreover, to complete the regrowth for the 1×10^{16}Sb/cm^2 case, annealing was carried out at 650 and 700°C (not shown in Figure 14). Although reasonable quality epitaxy resulted, only 10% of Sb atoms were observed to reside on silicon lattice sites and the Sb profile had broadened considerably towards the surface, effects which suggest appreciable Sb precipitation.

The Sb-Si system illustrated in Figure 14 demonstrates important general features of impurity trapping onto substitutional sites during SPEG. A limiting soluble concentration appears to exist for each particular impurity (and for Sb this concentration is $\sim 1.3 \times 10^{21}$/cm^3). For doses below this limiting concentration, almost complete incorporation of impurity atoms onto silicon lattice sites is obtained, but, above it, severe retardation of regrowth rate is observed and significant redistribution of implant is obtained.

The "push out" phenomenon illustrated in Figure 9 is indicative of the redistribution which can take place during regrowth for concentrations just above the maximum measured substitutional concentration. For the In-Si system shown in Figure 9, the onset of "push out" establishes a limiting soluble concentration of 5×10^{19}/cm^3 (compared with the maximum equilibrium limit of 8×10^{17}/cm^3). Thus, the "push out" process appears to control the maximum concentration of impurity which can be incorporated onto silicon lattice sites during SPEG of low diffusivity impurities in silicon.

The metastable solid solubility effects described above have also been observed by other workers for Sb and In implanted silicon (Fletcher *et al.*, 1981). Furthermore, similar results have been obtained for As, Pb, Tl, Bi and Te implants into <100> silicon (Williams and

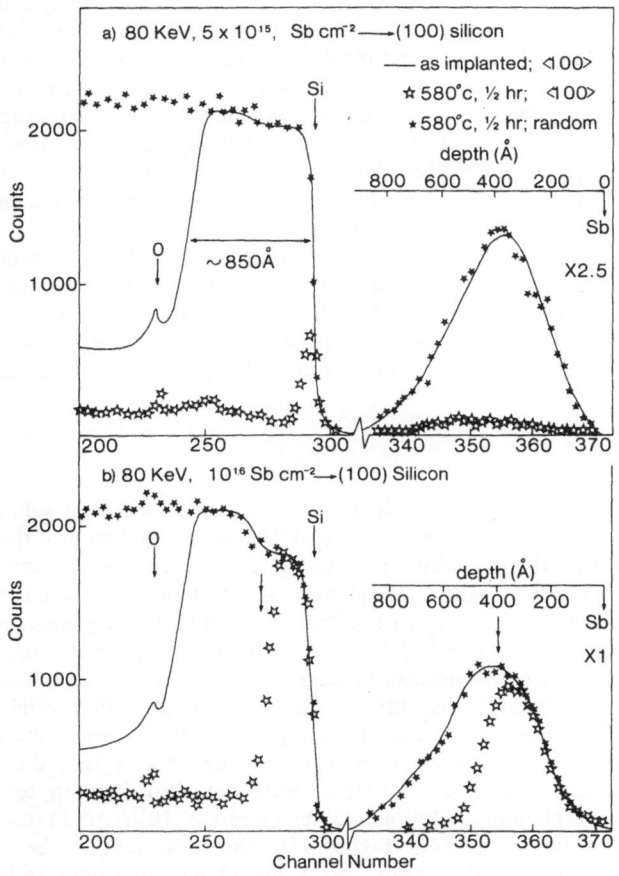

Figure 14   High resolution random and <100> aligned spectra from Sb implanted (100) silicon before and following annealing at 580°C for 30 min.
(a)   A dose of 5×10^{15} Sb/cm^2, which exhibits good regrowth and high substitutionality.
(b)   A dose of 1×10^{16} Sb/cm^2, which exhibits retarded regrowth. The extent of epitaxy is indicated on the Sb profile by the double-arrows. (After Williams and Elliman, 1982a.)

Elliman, 1981; Williams and Short, 1981; Campisano et al., 1980a,b). For all these impurities, the measured solubility was observed to substantially exceed that reported in the literature (Trumbore, 1960; Hanson, 1958) for equilibrium solid solutions.

Supersaturated solid solutions can also be formed in the solid phase by transient annealing at temperatures considerably in excess of 600°C. For example, metastable As concentrations in <100> silicon have been observed following scanned CW laser and electron beam annealing at >900°C in the time scale of milliseconds (Regolini et al., 1979b; Lietoila et al., 1979). In addition, annealing in a time scale of one to several seconds using strip heaters (Lietoila et al., 1981) and incoherent light sources (Harrison et al., 1981) can also result in substitutional concentrations in excess of solid solubility for As, P and Sb implants into <100> silicon. However, the maximum soluble concentrations achievable from the higher temperature transient annealing methods appear to be below those achievable via low temperature furnace annealing (Williams and Short, 1981). Nevertheless, the "push out" process (as shown in Figure 10) still appears to control the maximum solid solubility which can be obtained under transient anneal conditions.

Although the formation processes for supersaturated solid solutions are decidedly different when induced by solid or liquid phase epitaxy, it is somewhat intriguing that the magnitude of the maximum attainable solid solubility, for many of the impurities, is similar for both liquid and solid phase processes. These comparisons were presented in the previous chapter: possible analogies between the different formation processes are discussed further in Section IIIC.

2. Stability

Having formed supersaturated solid solutions via furnace or transient solid phase annealing methods, it is of interest to examine their stability during subsequent thermal processing. Figure 15 illustrates the stability of the Bi-Si system (Campisano et al., 1980a; Campisano et al., 1981). Figure 15a shows the isochronal annealing behavior of the substitutional Bi fraction for various Bi implant doses, all of which produce supersaturated Bi solid solutions following SPEG at 550°C. A clear dose-dependence is observed, whereby the higher doses exhibit decreased substitutionality at lower anneal temperatures. Based upon this behavior, it is speculated that the metastable Bi phase will precipitate if the mean diffusion length of the impurity at the annealing temperature exceeds the average impurity distance (Campisano et al., 1980a,b). It is interesting to note that the substitutional Bi solubility substantially exceeds $2 \times 10^{19}/cm^3$ following annealing up to 925°C and this compares with the maximum equilibrium solubility of $8 \times 10^{17}/cm^3$ (Trumbore, 1960). As illustrated in the phase diagram in Figure 15b, this behavior can be represented by a maximum non-equilibrium soluble component α' for 30 min annealing, to be compared with the equilibrium phase α, obtained after infinite time (Campisano, 1981).

Studies of the metastable Sb-Si system (Williams and Short, 1981, 1982; Fletcher et al., 1981; Pogany, 1982) have shown dose-dependent behavior similar to the Bi-Si system and have identified interesting microstructural effects relating to precipitation, as illustrated earlier in Figure 11. This precipitation has been observed to correlate well with the reduction in Sb substitutionality, as measured by channeling, and with electrical activity, obtained from resistivity measurements (Williams and Short, 1981, 1982). The correlation between Sb substitutionality and electrical activity is illustrated in Figure 16. Figure 16b indicates that 30 min isochronal annealing up to 1000°C reduces the substitutionality for implant doses exceeding $1 \times 10^{15}/cm^2$, corresponding to a soluble concentration of $2.6 \times 10^{20}/cm^3$ which is still a factor of 3 in excess of the maximum equilibrium value (Trumbore, 1960). Except for the lower temperature ($\leqslant 700°C$) measurements for the

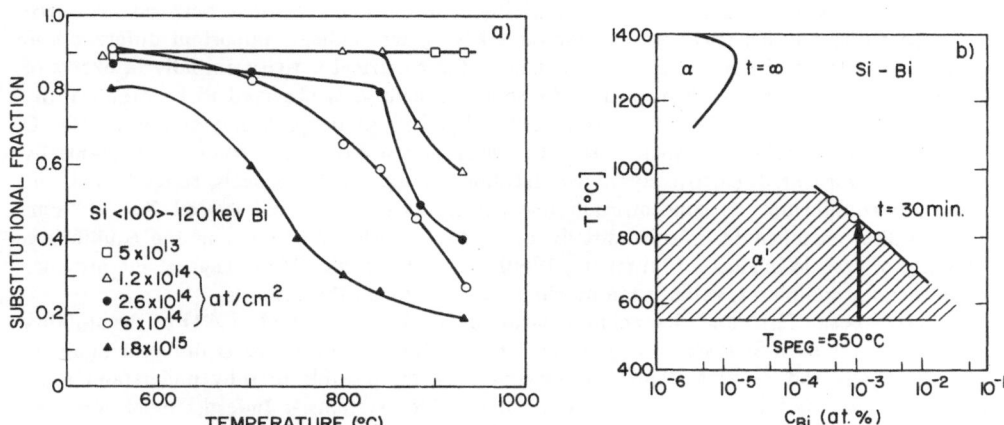

Figure 15   (a)   Substitutional Bi fraction as a function of temperature for 30 min isochronal annealing steps. All samples were first annealed to obtain a supersaturated solid solution. (After Campisano *et al.*, 1980a.)

(b)   The Bi-Si phase diagram showing the equilibrium solid solution α, and the metastable phase α' following annealing for $t = 30$ min (data taken from (a)). (After Campisano, 1981.)

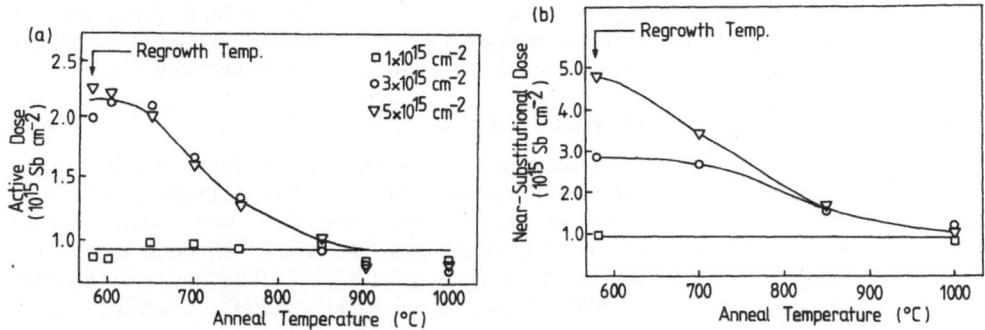

Figure 16   (a)   Active dose, as obtained from resistivity measurements, and

(b)   near-substitutional dose, as obtained from channeling measurements, plotted as a function of anneal temperature (10 to 30 min annealing) for metastable Sb-Si solid solutions formed at 580°C.

5×10^{15} Sb/cm^2 implant, the electrical activity in Figure 16a correlates well with the Sb substitutional concentrations.

Metastable As concentrations, observed following transient solid phase annealing of <100> silicon (Lietoila *et al.*, 1979, 1981; Regolini *et al.*, 1979) have also shown instability during subsequent thermal processing. However, correlations between measured substitutionality and electrical activity for the As-Si system indicate important differences in comparison with the Sb-Si system. For example, the electrical activity, initially in excess of 10^{21}/cm^3 and consistent with measured As substitutionality, is observed to be dramatically reduced following rather short time anneals (<30 mins) at temperatures as low as 600°C (Lietoila *et al.*, 1981). In such cases, the near-substitutional As fraction substantially exceeds the measured electrically active fraction. Longer-time anneals suggest that the maximum equilibrium As concentration (electrically active) does not exceed 3×10^{20}/cm^3 which, rather surprisingly, is substantially below the reported maximum in the equilibrium solubility of arsenic in silicon (Trumbore, 1960). Lietoila *et al.* (1981) suggest the presence of As-vacancy complexes, rather than precipitation, to explain their results.

Finally, recent TEM and channeling measurements (Fletcher *et al.*, 1981) of the stability of the metastable In-Si system suggest that the precipitation behavior is different again to both the Sb-Si and As-Si systems. At present it is not possible to fully understand these differences in stability for the various metastable systems. Indeed, more detailed examinations of microstructure and correlations with both substitutionality and electrical activity are needed for metastable concentrations of all p- and n- type dopants in silicon. In this regard, supersaturated solutions formed by liquid phase epitaxy where the regrown layers are extended-defect-free (Chapter 4), should be examined for comparison with those formed by SPEG, where grown-in defects may play a role in the subsequent precipitation processes.

III. REGROWTH MODELS

In order to account for the various regrowth observations outlined in Section II, one ideally requires a comprehensive atomistic model for the moving amorphous-crystal interface during the SPEG process. Although no complete model exists, this section briefly reviews the current understanding of the crystal-amorphous interface in silicon and outlines recently suggested models which have been put forward to explain the observed regrowth effects.

A. The Amorphous-Crystalline Interface: Impurity-free Regrowth

The regrowth kinetics of the SPEG process in impurity-free <100> silicon exhibit a single, well-defined activation energy, with the amorphous-crystal boundary maintaining a laterally uniform velocity as it moves toward the surface. Consistent with thermodynamic considerations in covalently bonded solids (Spaepen and Turnbull, 1979), these observations suggest that the growth process is controlled by interfacial bond breaking and rearrangement rather than long range transport of silicon atoms. Based on this concept, the regrowth rate v, along the <100> direction, is given by (Lau *et al.*, 1980):

$$v = a \, v_0 \exp(- \Delta E_a / kT),$$

where a is the atomic spacing along the <100> direction, v_0 is the jump frequency, ΔE_a is the measured activation energy, k is Boltzmann's constant, and T is the temperature in degrees Kelvin. Using this approach, however, provides a calculated growth velocity about two orders of magnitude slower than the experimental values. This, therefore, suggests that cooperative bond breaking processes may be important during SPEG.

Cooperative or multiple bond breaking and rearrangement processes can be visualized in terms of a structural model for the amorphous-crystalline interface proposed by Spaepen (1978). This model asserts that the interfacial free energy will be minimized in situations where no dangling bonds are present at the interface, a unique nearest neighbor distance is preserved, and bond angle distortion, responsible for the major part of the interfacial free energy, is minimized. It has been proposed (Spaepen and Turnbull, 1979, 1982) that crystal growth in this system could be effected by reconstruction of the network in the wake of a single broken bond running along a ledge. A large number of atoms sites could be reconstructed on the crystalline phase by such a moving defect. This process is illustrated in Figure 17 for the advancement of a [110] ledge on (111) silicon planes. The breaking of one bond (e.g., DF) is sufficient to rearrange the five- and seven-fold rings, characteristic of the amorphous phase, into chain-type six-fold rings which make up the crystalline (111) planes (Spaepen and Turnbull, 1979). As shown, this process can allow the crystallization of several atoms, and thus increase the regrowth velocity substantially above that calculated from single (uncorrelated) bond-breaking processes.

The interface model and the associated growth process envisaged by Spaepen (1978) can be used to provide a qualitative explanation for the observed orientation dependence of regrowth rate in impurity-free silicon. As discussed by Spaepen and Turnbull (1982), interface-controlled growth ensures that the crystal is bounded by planes normal to the directions of slowest growth, which are usually planes on which atomic packing is most dense. In silicon, this suggests that regrowth occurs via a (111) layering mechanism in which the crystal would be bounded at the amorphous-crystal interface by the most dense (111) planes. Thus, the amorphous-crystal interface for substrate orientations deviating from <111> may be expected to break up into terraces which expose (111) planes. One may then envisage the growth process to be governed by the motion of [110] ledges, as the densest directions bounding {111} terraces (Spaepen and Turnbull, 1982). The growth situation illustrated in Figure 17 depicts this process. Based upon such a preferred growth direction, the dependence of growth rate on substrate orientation essentially becomes a geometrical argument, where the regrowth time will be equivalent to the time required to recrystallize a row of atoms along (111) planes between the original amorphous-crystal interface and the top surface. This suggests (Spaepen and Turnbull, 1982) that the regrowth rate (or, more precisely, the fraction of growth sites or ledges) will increase as sin θ, where θ is the angle of inclination of the substrate normal to <111> orientations. This prediction provides rather good qualitative agreement with the observed orientation dependence of regrowth rate (Csepregi et al., 1978).

Other models to account for the orientation dependence have been previously reviewed by Lau (1978), and are summarized as follows: (1) A geometric model has been proposed by Csepregi et al. (1978) in which it is assumed that the amorphous-crystalline interface is not reconstructed and that crystallization takes place on interface sites where at least two nearest neighbors are in crystalline positions. This requirement leads to an atom growth sequence which follows <211> directions on (111) planes. Thus, similar to the model of Spaepen and Turnbull (1979), the orientation dependence becomes essentially a geometric argument where the predicted trends are in qualitative agreement with the experimental observations (Csepregi et al., 1978). (2) Seshan and EerNisse (1978) have suggested a stress relaxation model to account for the large differences between regrowth rate on <100> and <111> oriented silicon. In this model, it is argued that recrystallization reduces the observed biaxial stress in ion implanted layers. It is suggested that stress relief occurs via the motion of Shockley partial dislocations ($b = 1/6$ [211]) under the influence of the resolved shear stresses on the (111) planes. Frank partial dislocations with $b = 1/3$ [111] are assumed to be nucleated, upon heating, at the amorphous-crystalline interface by the aggregation of point

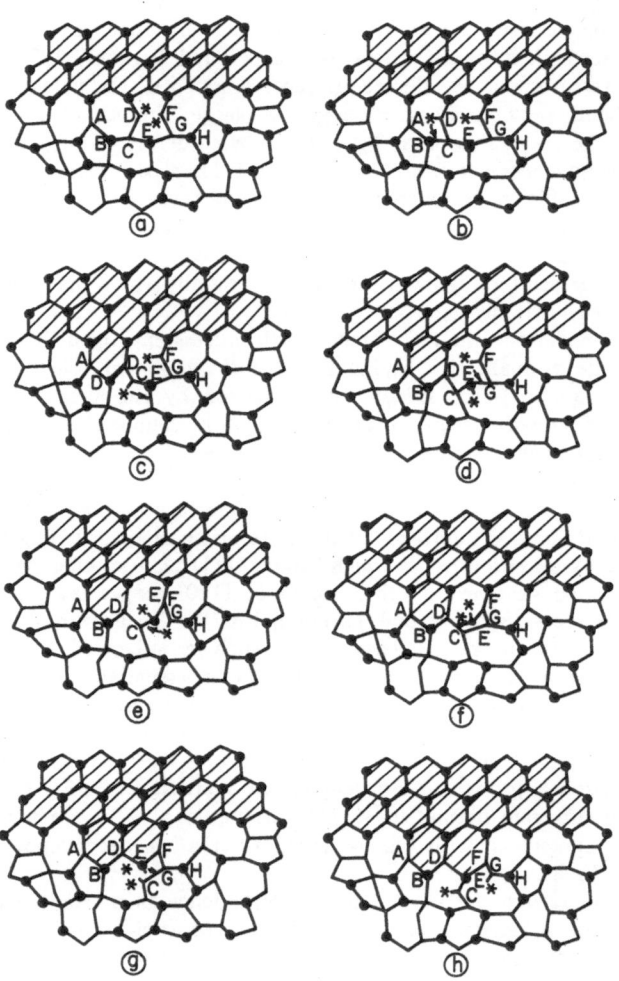

Figure 17   Successive steps in the crystallization of an amorphous tetrahedrally coordinated silicon structure by the advancement of a [110] ledge alone a (111) terrace. The upper portion of each figure represents the crystalline phase and the lower portion the amorphous phase. The breaking of an initial bond (DF) is shown to effect reconstruction of several atoms to the crystalline phase. (After Spaepen and Turnbull, 1979.)

defects in the crystalline region. Higher shear stresses on (100) and (110) substrates, in contrast to (111), lead to the formation of the required Shockley partial dislocations, which provide abundant nucleation sites for growth since they intersect both (100) and (110) surfaces. Although this model predicts better growth for (100) and (110) orientations, compared with (111) substrates, it implies similar growth rates for both the (100) and (110) orientations, in sharp disagreement with experiments (Csepregi et al., 1978).

None of the regrowth models described above addresses the necessary nucleation processes for the development of crystalline growth sites at the amorphous-crystalline interface. Indeed, in order to explain the regrowth-rate effects and orientation dependence in a more quantitative manner, it will be necessary to further develop an understanding of the processes which control nucleation of growth sites. This aspect is particularly relevant to an explanation of the dramatic effects of small impurity concentrations on regrowth rate, where, in their present form, the models presented in this section are decidedly inadequate.

B. Low Dose Impurity Effects

The first accounts describing the effect of small impurity concentrations on epitaxial regrowth rate (Kennedy et al., 1977; Csepregi et al., 1977; Lau et al., 1980) suggested that doping could lead to several processes which could initiate the observed changes in regrowth rate. For example, it was suggested that shallow donors (P, As) or acceptors (B) may influence bond breaking processes at the interface via lower impurity-silicon bond energies. Furthermore, since doping produces a shift in the Fermi level, electronic effects may result in enhanced generation and diffusivity of those defects which constitute growth sites during SPEG. Indeed, the more recently observed compensation-like effect on regrowth (Suni et al., 1982) strongly suggests that such electrical processes could have a major influence on the regrowth rate. However, a major result to be dealt with is that the regrowth rate can be altered by more than two orders of magnitude by an impurity concentration as low as 0.5 atomic percent located at the growth interface. In this regard, cooperative bond breaking phenomena (as discussed earlier) may be important in providing multi-atom reordering at the interface in the vicinity of an impurity atom or an impurity-generated defect. In this section, some recently proposed mechanisms to account for the observed low dose impurity effects are examined.

1. Strain Models

Several authors have suggested that lattice strain, arising from the mismatch of atomic size between the impurity and lattice atoms, may enhance the regrowth rate. For example, Suni et al., (1982) point out that the accumulation of local strains surrounding an impurity atom may develop into a macroscopic strain in the lattice, thus providing a driving force for migration. Such a process may enhance regrowth by facilitating atom motion, notably silicon interstitials, across the interface from amorphous to crystalline silicon. In the previous section, it was suggested (Seshan and EerNisse, 1978; Lau, 1978) that differences in lateral strain could significantly influence impurity-free regrowth along different substrate directions via generation of Shockley partial dislocations at the interface. Thus, impurity-size-induced strains may be expected to similarly aid the regrowth process.

Campisano (1982) has noted an interesting relationship between regrowth rate, at constant impurity concentration, and the absolute covalent radius difference between the particular impurity and silicon $|r-r_{Si}|$. This correlation is shown in Figure 18 for dopant concentrations of 0.125 atom percent: the trend suggests a linear relationship between strain (which is proportional to $|r-r_{Si}|$) and regrowth rate. In addition, Campisano (1982) notes that, at low impurity concentrations, the regrowth rate increases linearly with dose (as

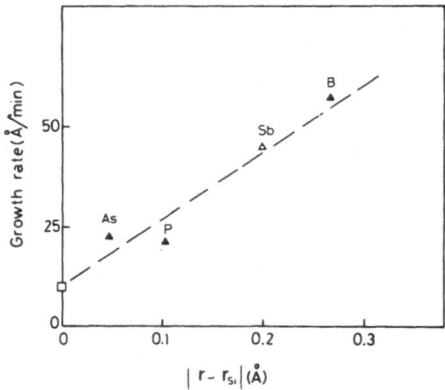

Figure 18   Experimental growth rate of ion implanted silicon <100>, measured at 500°C and linearly extrapolated
to a dopant concentration of 0.125 atom percent, as a function of the absolute difference in covalent
radius between the tetrahedrally coordinated impurity and silicon, $|r - r_{Si}|$. (After Campisano, 1982).

illustrated by the phosphorus implant data shown earlier in Figure 2) and speculates that
impurity-induced regrowth-rate effects result from strain-enhanced recrystallization for
impurity concentrations below the limit of solid solubility. Moreover, above solid solubility
limits, Campisano (1982) suggests (as indicated by regrowth measurements for C, O, N and
Te) that the regrowth speed should decrease.

Although a mechanism of stress-enhanced regrowth rate is apparently consistent with
observed regrowth behavior for single implants, it is difficult to envisage such a model
explaining the compensation effect. Based solely on a strain model, one might expect
"compensation" only when the integrated (compressive) strain of impurities smaller than
silicon is just balanced by the integrated (tensile) strain of larger impurities. This is not
consistent with the dual boron and arsenic implant data of Suni et al. (1982) (illustrated in
Figure 3) where equal concentrations of boron $(r-r_{Si} = -0.29\text{Å})$ and arsenic
$(r-r_{Si} = +0.01)$ result in a regrowth rate similar to impurity-free silicon, even though the
compressive strain component is dominant.

2. Defect Generation Models

In terms of the Spaepen model of the growth process, it was intimated earlier that regrowth
at the amorphous-crystal interface may be controlled by defect nucleation processes which
initiate bond breaking and allow crystallization to proceed along [110] ledges bounding (111)
terraces. Thus, the rate limiting process in epitaxial growth may well be the availability of
favorable growth sites on ledges. In this regard, it is interesting to examine possible defect
configurations which could constitute growth sites at the interface and, further, to establish
the dependence of this defect concentration on implanted impurity concentration.

Suni et al. (1982a,b) note that doping of silicon both enhances the self diffusivity
(Fairfield and Masters, 1967) and results in a proportional increase in the concentration of
charged defects (Shaw, 1975). It is suggested that the availability of charged defects at the
interface (e.g., vacancies) may control bond breaking and hence the regrowth rate (Walser
and Bene, 1978; Suni et al., 1982a,b). Thus, the regrowth rate may be expected to increase
with the concentration of (charged) impurities. Although precise details of such a model
have not been established, the notion of charged vacancy generation by doping (presumably

initiated by Fermi level changes) can explain, at least qualitatively, the doping and compensation-like effects on regrowth rate.

A more detailed model to describe the SPEG process, which is also based on electronic structure effects and charged defect generation, is that of Williams and Elliman (1982b). In this case, the suggested mechanism of regrowth draws upon an analogy between the effect of doping on dislocation motion in crystalline silicon and impurity-enhanced epitaxial growth. The former process has been studied in some detail both experimentally (Patel and Chaudhari, 1966; Patel et al., 1976, 1977) and theoretically (Patel et al., 1976, 1977; Haasen, 1975; Hirsch, 1979, 1980, 1981). The effect of doping, for both n- and p- type impurities to concentrations of the order of $10^{19}/cm^3$, is to enhance measured dislocation velocities in silicon by in excess of two orders of magnitude. The scale of this effect is very similar to that observed for impurity-enhanced epitaxial regrowth rates, and hence the basis of an analogy between the two phenomena.

The effect of doping on dislocation velocity has been discussed in terms of a well established mechanism of dislocation motion by the formation and migration of kinks, and its dependence upon changes in the Fermi level (Hirsch, 1981). Although a complete understanding of the mechanism is lacking, several theories have been developed to explain the dependence of dislocation velocity on the Fermi level. These fall basically into two categories: (i) those in which kinks are assumed to have well defined donor and acceptor levels in the band gap and doping provides either an increase in the concentration of charged kinks or changes in the kink migration energy (Hirsch, 1979, 1981); (ii) those which assume that doping increases either the fraction of charged sites on dislocations (Patel and Testardi, 1977) or the fraction of charged dislocations (Haasen, 1975). Both approaches predict an increase in dislocation velocity for n- and p-type doping of silicon.

In the regrowth model of Williams and Elliman (1982b), the growth sites at the amorphous-crystalline interface are envisaged as kink-like steps on [$\bar{1}10$] ledges, as illustrated in the simplified schematic representation in Figure 19. It is proposed that the SPEG process is controlled by the motion of these kink-like growth sites BB' alone [$\bar{1}10$] ledges AA'. In a manner analogous to dislocation motion, doping may enhance the epitaxial regrowth velocity by either increasing the concentration of charged kinks or reducing their migration energy. Figure 19 shows regrowth via kink motion on the upper ledge AA': this process may be envisaged as a cooperative process in which a moving kink recrystallizes many atoms before annihilation. The lower ledge illustrates pinning of kinks at a "strong" bond between certain impurity atoms and silicon. The solid lines joining the row of ledge atoms in Figure 19 are not meant to represent the true bonding situation: atoms at kink sites may be either fully coordinated or possess dangling bonds, similar to dislocation models (Hirsch, 1981).

The above model has several appealing features: (i) it can account for compensation-like regrowth effects, (ii) it is based upon a strong analogy with dislocation motion in silicon, where the doping effects are better understood, and (iii) it can readily incorporate the basic details of the Spaepen model (Spaepen and Turnbull, 1979) for the structure of the crystalline-amorphous interface. Nevertheless, much theoretical and experimental work is needed to establish both the details and the validity of such a regrowth mechanism.

C. High Dose Impurity Effects

As outlined in Section II, several intriguing processes can be observed during SPEG at high impurity concentrations for implant species which are effectively immobile during silicon recrystallization. These effects, notably retarded regrowth rate, "push out" of impurity at the moving amorphous-crystal interface, formation of metastable solid solutions and the onset of alternate (polycrystalline) growth processes, typically occur at implant concentrations which substantially exceed the maximum equilibrium solid solubility limits for the particular

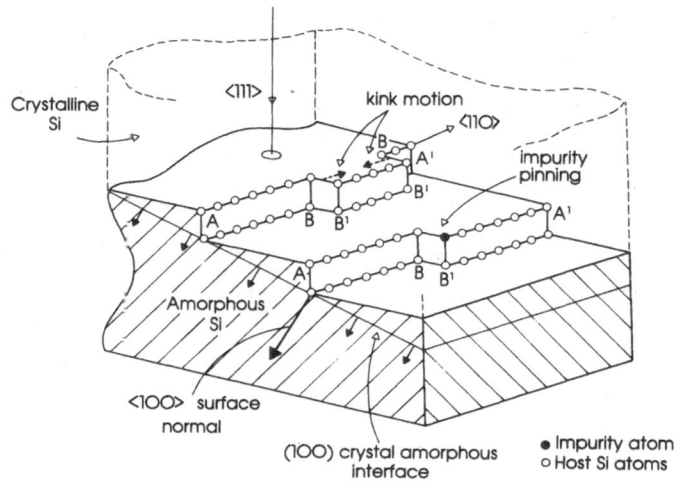

Figure 19   Simplified representation of the SPEG process for regrowth of (100) silicon in terms of kink (BB')
generation and motion along [110] ledges (AA') which connect (111) terraces. The upper cutaway
portion of the figure, bounded by the dashed lines, represents the crystalline phase, whereas the lower
hatched portion represents amorphous silicon. The lines connecting the indicated ledge atoms are not
meant to be representative of the true bonding situation. The possibility of impurity-pinning of kinks via
strong bonds with silicon is also illustrated.

impurity in silicon.   In this section, recently suggested mechanisms, put forward to account
for the observed high dose impurity effects, are reviewed.

Campisano *et al.* (1980a,b) have suggested that the mechanisms of formation of
supersaturated solid solutions during liquid phase epitaxial growth (LPEG) or SPEG are
analogous.   The concept is that slow diffusing impurities in solid silicon may be trapped at a
moving interface on substitutional sites in great excess of the maximum solid solubility when
the impurity residence time at the interface is larger than the time to regrow one monolayer.
Such a solute trapping process (Baker and Cahn, 1969; Jackson *et al.*, 1980) would be
controlled by the diffusion coefficients in the two adjacent phases.   It is suggested that fast
diffusers may either inhibit SPEG (through precipitation and polycrystalline nucleation in
amorphous silicon) or be rejected to the surface during LPEG via zone migration at the
liquid-solid interface.   Although this model conveniently accounts for general differences
between trapping of fast and slow diffusing impurities in both SPEG and LPEG, it is not
entirely consistent with many of the specific observations during regrowth.   For example,
supersaturated solid solutions of Pb in Si can be formed by low temperature SPEG
(Christodoulides *et al.*, 1980; Williams and Elliman, 1981) but complete "zone refining" of
Pb is observed during LPEG (White, 1981).   Moreover, in terms of a solute trapping analogy
between SPEG and LPEG processes, one may have expected the "push out" phenomenon
during SPEG to be a conventional segregation process, attributable to diffusivity differences
in amorphous and crystalline silicon.   However, the "push out" phenomenon is clearly not
analogous to zone migration during LPEG.

Narayan *et al.* (1981), in discussing SPEG-induced "push out" for the In-Si system, note
that regrowth occurs at temperatures where indium is molten.   They suggest that indium
interface segregation and outdiffusion in amorphous silicon may result from an enhanced
liquid phase diffusion process.   However, observations of "push out" for Sb-Si, As-Si and Bi-Si
systems (Williams and Short, 1981), where the impurity melting point is not exceeded at the

regrowth temperature, are not consistent with such a model. Williams and Elliman (1982) have suggested that the formation of supersaturated solid solutions during SPEG may well originate from a solute trapping process, but that the maximum concentration of impurities which can be trapped on substitutional sites may be controlled by interface processes rather then by impurity diffusion in amorphous and crystalline silicon. A model, based on interfacial strain during SPEG, has been developed to account for the decreased regrowth rate and the "push out" phenomenon. It has been argued that size differences between the impurity and silicon atoms may well give rise to increased bond distortion and local strain at the amorphous-crystal interface when impurities are incorporated onto substitutional sites. The magnitude of the integrated strain would be expected to increase with increasing implant dose. It was speculated that high levels of interfacial stress/strain may (i) impede the rate of epitaxial growth, (ii) provide the driving force for "push out" of impurity into the less dense amorphous phase at the moving amorphous crystal interface, and, as a consequence, (iii) indirectly determine the maximum measured substitutional solubility limit during SPEG (Williams and Elliman, 1982; Williams and Short, 1981).

The incorporation of atoms larger than Si into the crystal lattice would be expected to induce a local tensile strain at the interface. The integrated effect of this strain component may well oppose recrystallization since the latter process necessitates a densification of the lattice of the order of 6%. This suggests that the magnitude of the *decrease* in epitaxial regrowth rate at high impurity concentrations should scale with both the *increase* in impurity concentration and the impurity size. Regrowth data for Te, In, Pb and As implants in <100> silicon (Campisano and Barbarino, 1981; Williams and Elliman, 1980, 1982) are indeed consistent with these expected trends, although more quantitative data and correlations with integrated strain would be needed to fully test the plausibility of a strain model. Alternatively, Campisano and Barbarino (1981) and Campisano (1982) note that regrowth rate is typically retarded at concentrations approaching the equilibrium solid solubility limit and suggest that precipitation processes may play a role in inhibiting the regrowth rate. However, in the absence of impurity migration for low diffusivity species at the regrowth temperature, this process is difficult to envisage unless stress-enhanced or other, similarly driven, segregation processes are operative in amorphous silicon prior to crystallization.

With regard to a strain model, a clear correlation has been found to exist (Williams and Short, 1981) between the measured metastable solid solubility limit obtained from SPEG and the tetrahedral covalent radius (Pauling, 1948) of the impurity. This relationship is illustrated in Figure 20, where large impurities such as Pb and In are observed to have a lower solubility limit ($\sim 1 \times 10^{20}$/cm^3) than those impurities with a size similar to that of Si, such as As and Ge, where the metastable solubility is high ($\sim 1 \times 10^{22}$/cm^3). Although such a correlation is consistent with an interfacial strain model, similar correlations have been observed between impurity size and solid solubility limit for both equilibrium solid solutions (Trumbore, 1960) and LPEG-induced solid solutions (White *et al.*, 1980). Thus, more detailed measurements are needed (in particular, correlations between integrated stress-strain and SPEG-induced solubility limits for a large number of impurities in silicon) to fully establish the validity of a strain model. Indeed, a most important question remains: why is the magnitude of the measured metastable solid solubility limit (for most impurities) almost identical for SPEG and LPEG processes even though the impurity segregation mechanisms appear to be decidedly different?

IV. SUMMARY AND CONCLUSIONS

Ion implantation of silicon provides a unique opportunity of studying solid phase epitaxial growth processes under a variety of important situations. Over the past few years, this aspect

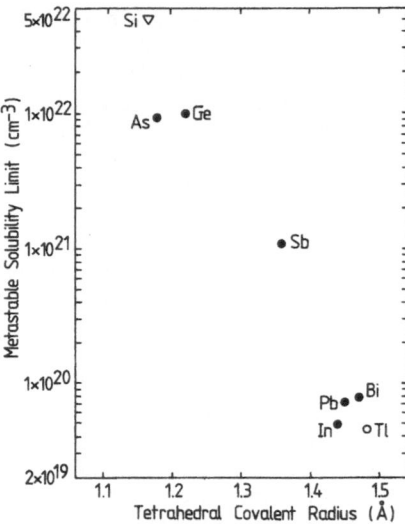

Figure 20   Plot of metastable solubility limit formed by SPEG plotted as a function of tetrahedral covalent radius for various implanted impurities in (100) silicon. (After Williams and Short, 1981.)

has been increasingly exploited to provide detailed measurements of the regrowth kinetics and, thus, allow crystallization models to be formulated. Firstly, the use of Si^+ implants into silicon has facilitated, without the complication of impurity-related effects, both the determination of the regrowth activation energy (2.35 eV) and the characterization of the strong dependence of regrowth rate on the substrate orientation. Such measurements have provided considerable insight into the growth processes on an atomic scale and have allowed structural models of the amorphous-crystalline interface to be examined and refined.

Secondly, the observations of impurity-regrowth interactions both at low and high concentrations have suggested that both electronic effects of doping and interfacial stress may play important roles in epitaxial growth. In particular, the enhancement of the regrowth rate by orders of magnitude for the introduction of only 0.5 atom percent of dopant atoms, together with the observed compensation-like effect, indicate that the regrowth velocity may be controlled by the generation of (charged) growth sites at the amorphous-crystalline interface. The similarity between observations of impurity-enhanced dislocation motion in silicon and the effects of impurities on regrowth rate have suggested an analogy between the two processes: epitaxial growth could, therefore, be envisaged to proceed via the generation and subsequent motion of kink-like defects on interfacial ledges and terraces. At high implant concentrations, the observations of retarded regrowth rate and impurity "push out" are more difficult to explain in terms of an atomistic regrowth model. In this case, it has been suggested that the build up of interfacial strain, resulting from size differences between the impurity and silicon atoms, may inhibit regrowth and constitute the driving force for interface segregation and "push out". However, at present, the proposed mechanisms to account for both low and high dose regrowth effects are somewhat speculative: further experimental and theoretical investigations are needed to fully clarify the regrowth process.

The introduction of impurities into silicon under the non-equilibrium conditions of ion bombardment allows impurity solubility, redistribution, segregation and precipitation processes to be studied prior to, during and following recrystallization. Several intriguing effects have been observed, most notably, the formation during SPEG of supersaturated solid solutions of low diffusivity impurity species in silicon. Investigations of the formation of such metastable solid solutions have indicated that the maximum concentration of impurity which can be incorporated onto silicon lattice sites during SPEG is limited by the onset of the "push out" process. It is interesting to note that the magnitude of the maximum soluble concentration is similar whether formed by solid phase or liquid phase epitaxy even though the formation mechanisms are decidedly different. For SPEG, it has been suggested that impurity size and attendant lattice strain may determine the maximum impurity concentration incorporated onto substitutional lattice sites.

Investigations of the stability of super-saturated solid solutions in silicon have indicated interesting differences between the various impurity-silicon systems in terms of the operative segregation and precipitation processes. These observations can provide valuable information on the early stages of impurity precipitation, including defect generation and associated impurity segregation. For example, arsenic appears to move off lattice sites into a vacancy-associated defect complex whereas antimony tends to segregate into regular precipitates which are oriented with respect to the surrounding silicon lattice. Differences between the stability and precipitation processes in these systems leads to significant differences in the levels of electrical activity compared with those expected from reported maximum equilibrium solid solubilities.

In contrast to the trapping of low diffusivity impurities on lattice sites during regrowth, fast diffusers such as Cu and Ag are observed to precipitate in the amorphous phase ahead of the advancing amorphous-crystal interface. As a result, epitaxy is inhibited and polycrystalline nucleation can ensue. In such cases, impurities may outdiffuse through the polycrystalline surface layer via grain-boundary-assisted migration processes.

Finally, apart from giving valuable insights into crystal growth phenomena and the formation and stability of metastable solid solutions, studies of solid phase epitaxial growth of ion implanted silicon have provided much important information to improve and extend the processing capabilities for the fabrication of advanced silicon devices. In this regard, more detailed electrical measurements to assess the properties of supersaturated solid solutions and their subsequent thermal stability would seem to constitute a fruitful avenue for future research in light of the fast approaching realization of very large scale integrated (VLSI) circuits.

REFERENCES

Baker, J. C. and Cahn, J. W. (1969), Acta Met. *17*, 575.
Blood, P., Brown, W. L. and Miller, G. L. (1979), J. Appl. Phys. *50*, 173.
Campisano, S. U., (1981), Private communication.
Campisano, S. U., (1982), Appl. Phys. Lett. (to be published).
Campisano, S. U., Rimini, E., Baeri, P. and Foti, G., (1980a), Appl. Phys. Lett. *37*, 170.
Campisano, S. U., Foti, G., Baeri, P., Grimaldi, M. G. and Rimini, E., (1980b), Appl. Phys. Lett. *37*, 719.
Campisano, S. U. and Barbarino, A. E., (1981), Applied Phys. *25*, 153.
Christodoulides, C. E., Baragiola, R. A., Chivers, D., Grant, W. A. and Williams, J. S., (1978), Rad. Eff. *36*, 73.
Christodoulides, C. E., Carter, G. and Williams, J. S., (1980), Rad. Eff. *48*, 91.
Csepregi, L., Mayer, J. W. and Sigmon, T. W. (1975), Phys. Lett. *54A*, 157.
Csepregi, L., Kennedy, E. F., Gallagher, T. J., Mayer, J. W. and Sigmon, T. W., (1977), J. Appl. Phys. *48*, 4234.
Csepregi, L., Kennedy, E. F., Mayer, J. W. and Sigmon, T. W., (1978), J. Appl. Phys. *49*, 3906.
Cullis, A. G., Seidel, T. E. and Meek, R. L., (1978), J. Appl. Phys. *49*, 5188.
Dearnaley, G., Freeman, J. H., Nelson, R. S., and Stephen, J., (1973), "Ion Implantation," North-Holland, Amsterdam.
Fairfield, J. M. and Masters, B. J., (1967), J. Appl. Phys. *38*, 3148.

Fan, J. C. C., Geis, M. W. and Tsaur, B. Y., (1981), Appl. Phys. Lett. *38*, 365.

Fletcher, J., Narayan, J. and Holland, O. W., (1981), Inst. Phys. Conf. Ser. *60*, 295.

Fulks, R. T., Russo, C. J., Hanley, P. R. and Kamins, T. I., (1981), Appl. Phys. Lett. *39*, 150.

Gat, A. and Gibbons, J. G., (1978), Appl. Phys. Lett. *32*, 142.

Gat, A., (1981), IEEE Elect. Dev. Lett. *2*, 85.

Gyulai, J., Pashley, R. D., Meyer, O. and Mayer, J. W., (1971), Rad. Eff. *7*, 17.

Haasen, P., (1975), Phys. Stat. Sol. (a) *28*, 145.

Hansen, M., (1958), "Constitution of Binary Alloys," McGraw Hill, New York.

Harrison, H. B., Grigg, M., Short, K. T., Williams, J. S. and Zylewicz, A., (1981), Symposium A, Boston MRS meeting (to be published).

Hirsch, P. B. (1979), J. de Phys. *40*, C6-117.

Hirsch, P. B., (1980), J. Microsc. *118*, 3.

Hirsch, P. B., (1981), in "Defects in Semiconductors," p. 257, (J. Narayan and T. Y. Tan, eds.), North Holland, New York.

Jackson, K. A., Gilmer, G. H. and Leamy, H. J., (1980), in "Laser and Electron Beam Processing of Materials," p. 104, (C. W. White and P. S. Peercy, eds.), Academic Press, New York.

Josquin, W. J. M. J. and Tamminga, Y., (1978), Appl. Phys. *15*, 73.

Kennedy, E. F., Csepregi, L., Mayer, J. W. and Sigmon, T. W., (1977), J. Appl. Phys. *48*, 4241.

Lau, S. S., (1978), J. Vac. Sci. Technol. *15*, 1656.

Lau, S. S., Von Allmen, M., Golecki, I., Nicolet, M. A., Kennedy, E. F. and Tseng, W. F., (1979), Appl. Phys. Lett. *35*, 327.

Lau, S. S., Tseng, W. F. and Mayer, J. W., (1980), in "Handbook on Semiconductors," (S. P. Keller, Ed.), Vol. 3, Ch. 7.

Lietoila, A., Gibbons, J. F., Magee, T. J., Peng, J. and Hong, J. D., (1979), Appl. Phys. Lett. *35*, 532.

Lietoila, A., Gold, R. B., Gibbons, J. F., Sigmon, T. W., Scovell, P. D. and Young, J. M., (1981), J. Appl. Phys. *52*, 230.

McMahon, R. A. and Ahmed, H., (1979), in "Laser and Electron Beam Processing of Electronic Materials," (G. L. Anderson, G. K. Celler and G. A. Rozgonyi, eds.), Electrochem. Soc., Princeton.

Mayer, J. W., Eriksson, L., Picraux, S. T. and Davies, J. A., (1968), Can. J. Phys. *45*, 663.

Mayer, J. W., Eriksson, L. and Davies, J. A., (1970), "Ion Implantation in Semiconductors," Academic Press, New York.

Mezey, G., Matteson, S. M. and Gyulai, J., (1981), in "Ion Beam Modification of Materials," Pt. II, p. 587, (R. E. Benenson, E. N. Kaufmann, G. L. Miller and W. W. Scholz, eds.), North Holland, Amsterdam.

Narayan, J., Fletcher, J. and Holland, O. W., (1981), Symposium E, Boston MRS Meeting (to be published).

Patel, J. P. and Chaudhari, A. R., (1966), Phys. Rev. *143*, 601.

Patel, J. P. and Testardi, L. R., (1977), Appl. Phys. Lett. *30*, 3.

Patel, J. R., Testardi, L. R. and Freeland, P. E., (1976), Phys. Rev. *B13*, 3548.

Patel, J. R., Testardi, L. and Freeland, P. E., (1977), Phys. Rev. *B15*, 4121.

Pauling, L., (1948), "The Nature of the Chemical Bond," p. 179, Cornell University Press, Ithaca, New York.

Pogany, A. P., (1982), to be published.

Picraux, S. T., Johansson, N. G. E. and Mayer, J. W., (1969), in "Semiconductor Silicon," p. 442, Electrochem. Soc., New York.

Regolini, J. L., Gibbons, J. F., Sigmon, T. W. and Pease, R. F. W., (1979a), Appl. Phys. Lett. *34*, 410.

Regolini, J. L., Sigmon, T. W. and Gibbons, J. F., (1979b), Appl. Phys. Lett. *35*, 114.

Revesz, P., Wittmer, M., Roth, J. and Mayer, J. W., (1978), J. Appl. Phys. *49*, 5199.

Seshan, K. and EerNisse, E. P., (1978), Appl. Phys. Lett. *33*, 21.

Shaw, D., (1975), Phys. Stat. Sol. (a). *72*, 11.

Sigurd, D. and Bjorkquist, K., (1972), Rad. Eff. *17*, 209.

Spaepen, F., (1978), Acta Met. *26*, 1167.

Spaepen, F. and Turnbull, D., (1979), in "Laser-Solid Interactions and Laser Processing - 1978," p. 73, (S. D. Ferris, H. J. Leamy and J. M. Poate, eds.), AIP Conf. Ser. *50*, New York.

Spaepen, F. and Turnbull, D., (1982), in "Laser and Electron Beam Processing of Semiconductor Structures," Chapter 3, (J. M. Poate and J. W. Mayer, eds.), Academic Press, New York.

Suni, I., Goltz, G., Grimaldi, M. G., Nicolet, M. A, and Lau, S. S, (1982a), Appl. Phys. Lett. (in press).

Suni, I., Goltz, G., Nicolet, M. A. and Lau, S. S., (1982b), Thin Solid Films (in press).

Walser, R. M. and Bene, R. W., (1978), in "Thin Film Phenomena - Interfaces and Interactions," p. 284, (J. E. E. Baglin and J. M. Poate, eds.), Electrochem. Soc., Princeton.

White, C. W., (1981), private communication.

White, C. W., Wilson, S. R. and Appleton, B. R., (1980), J. Appl. Phys. *57*, 738.

Williams, J. S., (1979), in "Laser and Electron Beam Processing of Electronic Materials," p. 249, (G. L. Anderson, G. K. Celler and G. A. Rozgonyi, eds.), Electrochem. Soc., Princeton.

Williams, J. S. and Grant, W. A., (1976), in "Application of Ion Beams to Materials," (G. Carter, J. S. Colligon and W. A. Grant, eds.), Inst. Phys. Conf. Ser. *28*, 31.

Williams, J. S., Christodoulides, C. E., Grant, W. A., Andrew, R., Brawn, J. R. and Booth, M., (1977), Rad. Effects *32*, 55.

Williams, J. S., Brown, W. L., Leamy, H. J., Poate, J. M., Rodgers, J. W., Rousseau, D., Rozgonyi, G. A., Shelnutt, J. A. and Sheng, T. T., (1978), Appl. Phys. Lett. *33*, 542.

Williams, J. S., Brown, W. L. and Poate, J. M., (1979), in "Laser Solid Interactions and Laser Processing -1978," p. 399, (S. D. Ferris, H. J. Leamy and J. M. Poate, eds.), AIP Conf. Ser. *50*, New York.

Williams, J. S. and Elliman, R. G., (1980), Appl. Phys. Lett. *37*, 829.

Williams, J. S., Christodoulides, C. E. and Grant, W. A., (1980), Rad. Eff. *48*, 78.

Williams, J. S. and Elliman, R. G., (1981), Nucl. Instr. Meth. *183*, 758.

Williams, J. S. and Short, K. T., (1981), Symposium E. Boston MRS meeting (to be published).

Williams, J. S. and Elliman, R. G., (1982a), Appl. Phys. Lett. (in press).

Williams, J. S. and Elliman, R. G., (1982b), to be published.

Williams, J. S. and Short, K. T., (1982), Appl. Phys. Lett. (in press).

Wittmer, M., Roth, J., Revesz, P. and Mayer, J. W., (1978), J. Appl. Phys. *49*, 5207.

CHAPTER 6

METASTABLE ALLOYS AND SUPERCONDUCTIVITY

B. STRITZKER

Institut für Festkörperforschung, Kernforschungsanlage Jülich, D-5170 Jülich, West Germany

CONTRIBUTORS: H. Bernas, W. Buckel, S. S. Lau, G. Linker, J. S. Williams and P. Zieman

I. INTRODUCTION:

During the last ten years considerable interest has grown in the production of metastable alloys. The reason for this development is the need for new materials with special properties which cannot be met using standard equilibrium growth methods. In addition many applications require this special material only in the near surface region, whereas the bulk properties of the material should not be affected. In order to meet these different requirements, the basic properties of such metastable alloys have to be well understood. This was a challenge for many different groups to investigate metastable materials using different production and detection methods. Thus a lot of knowledge about metastable alloys has been accumulated in the last years. Out of many reviews on this topic only a few shall be cited. The selection is based on the different production methods for these materials.

A large number of examples of vapor quenched materials is given in the review by Bergmann (1976) where the importance of these materials for superconductivity is stressed. A recent review by Güntherodt et al. (1980) summarizes the electronic properties of metallic glasses produced by liquid quenching. The production of metastable alloys by ion implantation was studied intensively by Kaufmann et al. (1981). They implanted many different ions into Be. The resulting lattice sites of these metastable substitutional or interstitial alloys can be well described by a Miedema plot. Poate (1980) has given a review on metastable surface alloys produced not only by ion implantation, but also by pulsed laser and electron beam irradiation. Metastable Si-based alloys as achieved by laser annealing are treated in Chapter 4. In the following paper the different ways of producing metastable alloys will be emphasized.

The technological interest in these metastable alloys arose from quite different areas because of their unique properties, e.g., with respect to corrosion resistance and mechanical strength. These surface properties are reviewed by Gilman (1980) and their technological importance is pointed out. In addition metastable metallic glasses possess very interesting magnetic properties combined with mechanical hardness, an often very desirable combination. Moreover, metastable alloys often have very interesting superconducting properties (Bergmann, 1976; Johnson, 1980). Historically, there has been a long standing motivation in developing new materials with improved superconducting transition temperatures, critical magnetic fields and critical currents. Thus a lot of experimental and theoretical knowledge has been gathered for several decades. This review will therefore be mainly restricted to superconducting metastable alloys and the different methods of production. In addition the study of superconducting alloys gives intrinsic information about the state of the metastable phases, as will be discussed below.

Non-equilibrium, metastable phases are interesting for superconducting purposes due to metastability with respect to two different properties: 1) Equilibrium solubility limits can be highly exceeded, i.e., the region where a special phase exists can be substantially extended. This advantage of ion implantation is used in a variety of experiments in superconductivity (Meyer, 1980b; Stritzker, 1978). 2) New non-equilibrium crystal structures or even amorphous phases can be achieved in a metastable phase.

Both kinds of metastability i.e., metastability with respect to composition or with respect to structure are naturally correlated depending on the production method. For instance low temperature ion implantation will both exceed solubility limits and introduce lattice disorder. Both effects are desirable in superconductivity, since variation of the composition within the same crystal structure allows the variation of the electronic density of states at the Fermi energy, $N(0)$, and a change of the crystal structure changes the phonon spectrum of the system, $F(\omega)$. Thus it is possible to alter both the important parameters $N(0)$ and $F(\omega)$ which determine the superconducting properties. This was the reason for the surprising result that the disordered metastable system with the highest known T_c of $\sim 17K$, consists of three elements Pd, Cu and H, each of them non-superconducting in the pure phase (Stritzker, 1974).

Since amorphous metallic phases belong to an interesting class of materials which will be discussed in detail in this chapter, the stability considerations for amorphous systems reviewed by Cotteril (1976) and Duwez (1979) will be briefly discussed. The most widely accepted model to describe amorphous phases is the dense-random-packing of hard spheres. This model, first suggested by Bernal (1960) for liquids, was the basis for further computer models. The dense-random-packing model gives a good description for amorphous phases containing only one kind of atom. To date there are only two metals known, namely Ga and Bi, which can be quenched as pure materials into an amorphous state.

The overwhelming majority of metallic amorphous systems contain at least two kinds of atoms of different atomic size. Normally at least 10 at .% of another atomic species is required to stabilize the amorphous state. In the Bernal model this stabilization is explained by a sterical argument, that means filling the holes in a dense-random-packed host material with the stabilizing, smaller kind of atoms. However, this consideration does not take into account the important chemical nature of the different atomic species which is a very important feature to stabilization. For instance atomic species such as semi-metals with a somewhat covalent (i.e., anisotropic) type of bonding, can stabilize the amorphous phase much more effectively than metal atoms with isotropic metallic bonding. So it is very important to emphasize that in general a random packing of atoms of different sizes cannot explain the properties of an amorphous phase. In many metallic glasses produced by liquid quenching, the short range order as determined by the special kind of chemical bonding influences the physical properties (Cargill, 1975).

For the determination of the structure of metastable phases the standard investigation techniques like x-ray and electron diffraction have normally been applied. Since very often only thin surface layers are produced, the resulting metastable surface alloys have to be investigated by glancing x-ray methods such as the Reed camera or Seemann-Bohlin arrangement. In addition Rutherford backscattering yields detailed information about the composition of the metastable phase (Chu et al. 1978). For single crystal hosts, channeling techniques can provide information both about the lattice location of the alloying element and the amount of lattice disorder introduced during the preparation of the metastable phase.

The following will emphasize the point that studies of superconducting alloys can also yield much information about the microscopic state of the metastable phase. The measurement of the superconducting properties of a new metastable phase also entail determination of the electronic properties in the normal conducting state. The determination of the normal state resistance behavior, i.e., temperature dependence as well as residual resistivity, and the superconducting properties give easily interpretable information about different metastable alloys.

A superconductor is characterized by its transition temperature, T_c, below which the repulsive Coulomb interaction, μ^x, between conduction electrons is overwhelmed by the attractive electron-phonon interaction, λ (McMillan, 1968):

$$T_c \propto <\omega> \left[\exp - \frac{1-\lambda}{\lambda - \mu^x} \right] \qquad (1)$$

with the phonon energy ω. The electron-phonon interaction determining T_c is given by

$$\lambda = 2 \int \frac{\alpha_s^2(\omega) F(\omega)}{\omega} \, d\omega = \frac{N(0) \cdot <I^2>}{M <\omega^2>} \tag{2}$$

with $\alpha_s^2(\omega) F(\omega)$ =   electron-phonon interaction parameter times phonon density of states

         $N(0)$ =   electronic density of states at the Fermi energy

         I^2 =   electron-phonon matrix element

         M =   atomic mass

From the equivalency in equation (2) it can be seen that T_c or λ can be explained in two different ways in a more phononic or a more electronic picture. Due to the electron-phonon interaction both descriptions are equivalent and it depends on the special system which description is more adequate. In order to obtain high values for T_c or λ one needs either weak phonon modes or high $N(0)$. The important electron-phonon interaction, the parameter $\alpha_s^2 F$ can be determined by superconducting tunneling experiments.

Additional important information about the basic electronic conductance mechanism can be obtained from the normal conducting state just by measuring the sample resistivity (ρ) versus temperature:

i) Metallic behavior where resistivity shows the following behavior:

$$\rho = \rho_0 + \rho(T) \tag{3}$$

with ρ_0 = residual resistivity, determined by scattering of conductance electrons with impurities or imperfections. The resistivity $\rho(T)$ is due to electron-phonon scattering which increases with increasing temperature. For simple metals $\rho(T)$ is proportional to T at higher temperatures and from more general considerations:

$$\rho(T) = \frac{m}{e^2} \frac{1}{N(0)} \int \frac{\alpha_\rho^2(\omega) F(\omega)}{\sinh(\omega/T)} \, d\omega \tag{4}$$

with electron mass m and elementary charge e. Thus by measuring simply $\rho(T)$ it is possible to deduce $\alpha_\rho^2 F$ (Gorska et al. 1977). However, it should be noted that $\alpha_\rho^2 F$ in (4) is already averaged with respect to the electron moments. That means that there is, in principle, a difference between $\alpha_s^2 F$ in (2) and $\alpha_\rho^2 F$ in (4) which turns out, in practice, not to be of much importance (Bernas and Nedellec, 1981).

ii) Metallic glasses and many amorphous systems show very little temperature dependence of the resistivity or very often even a negative temperature coefficient, i.e., a slight increase of resistivity with decreasing temperature. The different models explaining this effect are reviewed by Güntherodt et al. (1980).

iii) Semiconductors show an exponential increase of the resistivity with decreasing temperature since the number of free carriers is determined by a thermal activation process $n \alpha \sum_\alpha e^{-\frac{E_a}{kT}}$ with different activation energies E_a due to intrinsic and impurity properties. Lowering of the temperature means a rapid decrease of n and thus a steep increase of resistance.

In conclusion the electronic and superconducting properties of a metastable phase provide important information about its intrinsic properties in addition to structural investigations.

II. NON-EQUILIBRIUM METHODS:

In the following section the different non-equilibrium methods used for the production of amorphous and other metastable alloys will be described. These methods comprise vapor quenching, liquid quenching, laser quenching as well as ion irradiation. The different quench rates and their dependence on various parameters will be discussed.

A. Vapor Quenching

Figure 1 shows the schematic of an apparatus used for vapor quenching. The desired material, which can be a pure material or an alloy with components of similar evaporation pressure, is evaporated with an electron gun or a resistance heated boat. Two different, independently controlled vapor sources should be used if one wants to quench alloys consisting of components with different vapor pressures. The vaporized atoms are condensed onto a substrate opposite the source. In the following we will discuss only examples where the substrate holder provides substrate cooling to liquid Helium temperatures. At 4K the atoms are not able to rearrange over larger distances. Thus highly disordered or even amorphous solid phases can be achieved. Typical layer thicknesses are less than 1000Å. It is desirable that the vapor quenching experiments take place at pressures better than 10^{-8} Torr otherwise impurity effects can predominate. The quenching rate connected with the rapid cooling of the vaporized atoms hitting the cold substrate surface can be estimated to be about 10^{14} K/sec. This quenching rate can be reduced by an increase of the substrate temperature during the condensation of the vaporized atoms. In addition the quenching rate depends not only on the substrate temperature but also on the special material properties, such as melting point and thermal conductivity.

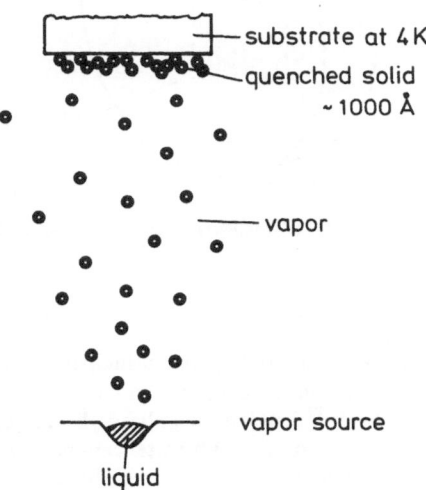

Figure 1   Schematic representation of Vapor Quenching.

B. Liquid Quenching

The material to be quenched is melted within a glass tube by high frequency heating under an argon atmosphere or *in vacuo*. Then the liquid is forced by the Ar-pressure out of a capillary at the bottom of the glass tube. The resulting jet of liquid material solidifies at the surface of a water cooled, rotating copper disk (see Figure 2) resulting in about 20 μm thick ribbons of several millimeters width and of several meters length. With a similar technique, where droplets of the desired material are rapidly quenched between a water cooled piston-anvil arrangement, 25 μm thick splats with about 2 cm diameter can be produced. The quenching rate involved in both processes can be estimated to about 10^6 K/sec. The quenching rate can be influenced by the temperature and the velocity of either the rotating disk or the piston anvil arrangement.

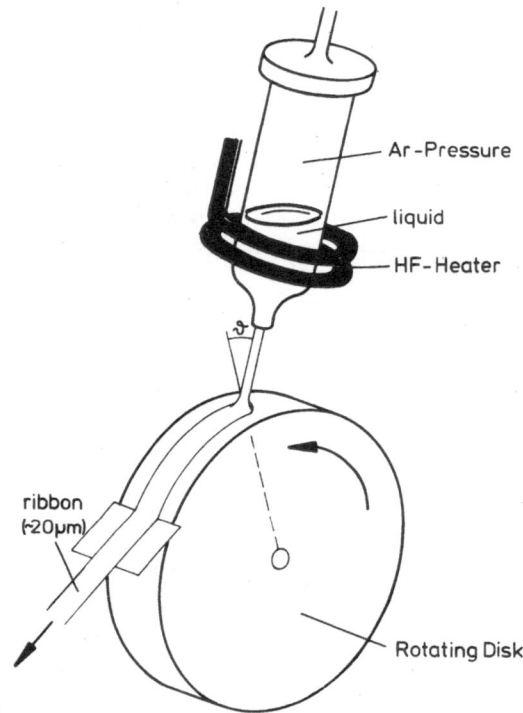

Figure 2   Schematic representation of Liquid Quenching.

C. Ion Implantation And Irradiation

Figure 3a and Figure 3b show schematically the principal arrangements necessary for ion implantation and ion irradiation. In both cases ion beams of the order of 100 keV and fluences in the range of 10^{14} to 10^{17} ions/cm^2 bombard the target which is mounted onto a substrate holder. The temperature of the substrate can be varied between 2K and room temperature or even above room temperature. The material of interest is evaporated as a thin solid film on top of the substrate. In the implantation case the accelerated ions come to rest within the thin film thus altering the composition. In contrast the composition of the film is not changed during the irradiation because the ions penetrate through the evaporated material and stick inside the substrate. Both ion implantation and irradiation may result in a change of the structure. This can be interpreted by different models in terms of the introduction of lattice defects, radiation enhanced diffusion or rapid quenching of a locally heated collision cascade region. The quenching rate of the latter "thermal spike" model can be influenced by the substrate temperature, the local energy due to the incoming ions as well as by parameters of the sample material. A detailed discussion of the different processes is given in Chapter 7.

D. Laser Quenching

The principle configuration is shown in Figure 4. A thin film ($\sim$1000Å thick) of the desired material has been evaporated onto a substrate which has negligible absorption to the applied laser light. The temperature of the substrate can be varied over a wide range (i.e., 4K - 600K). The thin film can be melted with a high power laser pulse. The temperature of the substrate itself is not significantly raised because it does not couple to the laser light. Shortly after the laser pulse terminates, the molten layer starts to resolidify rapidly from the boundary at the colder substrate. The resolidification is very fast resulting in quenching rates of about 10^9 K/sec. The quenching rate depends not only on internal parameters of the material but also on laser conditions and substrate temperature. A detailed consideration of this topic is given in Chapter 2.

III. INTERNAL ENERGY CONSIDERATIONS

In order to discuss different metastable states, a simple model will be introduced. This so-called Ostwald step rule (Ostwald, 1897) has been used to describe the old vapor quenching results (Buckel and Hilsch, 1954).

The Ostwald step rules states the following. After condensation of a vapor, it is necessary for the condensate to pass consecutively through all possible high temperature phases until the equilibrium phase is reached. This is shown schematically as the solid line in Figure 5 where the internal energy of the system (without any T-dependence) is plotted versus temperature. Starting from the high internal energy of the vapor phase at high temperatures the internal energy is reduced during a slow cooldown passing through the liquid phase and then through the different high temperature phases, i.e., HTI and HTII, until the equilibrium phase is reached. By means of sufficiently rapid cooling it should be possible to quench in some of these intermediate phases at low temperatures. For instance the experimental results suggest that the amorphous phases obtained by vapor quenching onto substrates at 4K can be described as frozen liquids. The liquid phase on the other hand is the high temperature phase following the vapor phase.

In order to demonstrate more clearly the application of the Ostwald step rule, a very fast quenching process is included in the schematic phase diagram of Figure 5 as dashed lines. The vapor phase is cooled down extremely rapidly to very low temperatures as indicated by an arrow. Then an amorphous metastable state can be stabilized. The properties of this amorphous phase suggest that it is identical to a frozen liquid phase. This means that in our simple phase diagram in Figure 5, the value of the internal energy (without T dependence) of the amorphous phase is equal to that of the liquid phase. There are several experimental facts which support this analogy (review by Bergmann, 1976): 1) The diffraction patterns for electron or x-ray diffraction are nearly identical for the amorphous and liquid phases. 2) The residual resistivities are very similar as can be seen from Table I. 3) Hall resistivities are the same to within 10%.

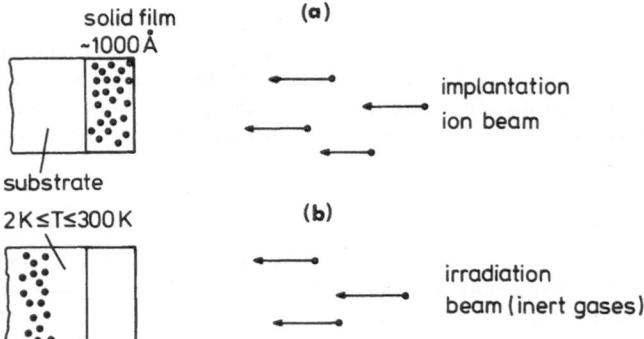

Figure 3   Schematic representation of Ion Implantation and Irradiation.

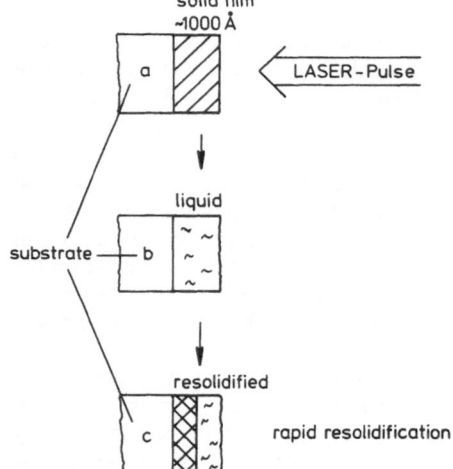

Figure 4   Schematic representation of Laser Quenching.

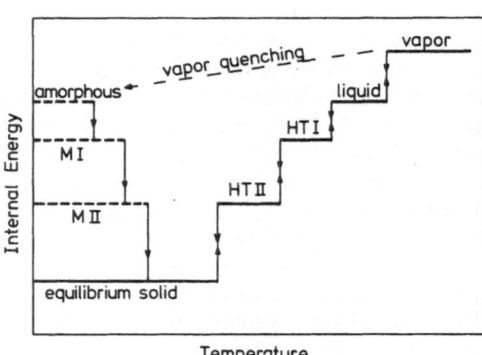

Figure 5   Schematic representation of the internal energy as a function of temperature.

TABLE I

Comparison of residual resistivities of liquid and amorphous phases
(Bergmann, 1976)

LIQUID		AMORPHOUS	
Metal	ρ_L $\{10^{-4}\Omega cm\}$	ρ_S	Alloy
Bi	1.3	1.6	Bi
Ga	.26	.29	Ga
Pb	.95	.78	$Pb_{.75}Bi_{.25}$
Sn	.48	.47	$Sn_{.86}Cu_{.14}$
In	.33	.33	$In_{.85}Sb_{.15}$
Ti	.73	—	—

By means of careful annealing of this amorphous phase it is possible in many cases to pass consecutively through other metastable phases which have their analogues in intermediate high temperature phases. Thus the internal energy can be reduced in a stepwise fashion until the equilibrium phase is reached. Such metastable phases are shown in Figure 5 as MI and MII in analogy to the high temperature phases HTI and HTII with the corresponding internal energy values. The structures of MI and MII are equivalent to HTI and HTII.

IV. VAPOR QUENCHING; COMPARISON WITH OTHER NON-EQUILIBRIUM TECHNIQUES

A. Ga

In the following the Ostwald step rule will be demonstrated for Ga. Gallium is the only pure metal besides Bi which can be vapor quenched into a liquid-like amorphous state (Buckel and Hilsch, 1954). In general, amorphous metals need about 10 at.% of an alloy partner to stabilize the amorphous state. The stability criteria have already been discussed in the introduction.

Figure 6 shows the resistance of vapor quenched Ga versus temperature. The resulting amorphous phase has both a high $T_c = 8.5K$ and a high residual resistivity $\rho_0 = 29\ \mu\Omega cm$. Annealing to more than 15K results in an irreversible decrease of the resistance. The amorphous phase has transformed into β—Ga with $T_c = 6.3K$ and a low resistivity of $\rho_0 = 3\ \mu\Omega cm$. This crystalline metastable β-phase has similar structured and electronic properties to a high pressure phase of Ga. Above 60K this metastable phase transforms into the equilibrium α-phase with a low $T_c = 1.07K$ and a medium resistivity of $12\ \mu\Omega cm$. This example demonstrates that the measurement of the electric resistivity provides easy evidence not only for phase transitions but also for the characterization of the different phases by the different values for $d\rho/dT$. After the structure of the different phases has been determined, for instance by electron diffraction in the case of Ga (Buckel, 1954), it is only necessary to measure the resistivity as function of temperature to identify the various phases. Figure 7 transfers the experimental results of Fig. 6 into the simple internal energy model. Vapor quenching results in the amorphous, i.e., the frozen liquidlike phase, which transforms at 15K into the metastable β-Ga with lower internal energy. The β-phase transforms at 60K into the equilibrium α-phase which has the lowest T_c value. At this point a common property of the non-transition metals should be emphasized: the liquid-like amorphous phase has a substantially higher T_c than the crystalline equilibrium phase. This can be understood by equation (2). Amorphization implies, in general, a substantially less densely packed structure because of the many internal holes and imperfections. Thus many atoms can oscillate more freely, so $F(\omega)$ is smeared and shifted to lower energies, ω, as compared to the crystalline material. In addition the electron-phonon coupling, $\alpha^2(\omega)$ is also enlarged. Since λ is determined by the low energy part of $\alpha^2(\omega)F(\omega)$, this means that the amorphous phase has a higher λ and thus a higher T_c.

In a recent unpublished experiment, Görlach, Hitzfeld, Ziemann and Buckel (1981) studied the influence of ion irradiation in Ga thin-film targets at 4K. They irradiated all three Ga phases with Ar^+-ions of 275 keV penetrating through the Ga films. Starting with α-Ga they achieved an immediate transformation into the amorphous phase as demonstrated in Figure 8 by the steep increase of both T_c and ρ. T_c and ρ saturate at the values typical for amorphous Ga. Starting from an amorphous Ga film, low-temperature ion irradiation did not change the amorphous phase. Also β-Ga could be transformed into amorphous Ga with Ar^+ doses of the order of 10^{16}. The authors believe that in this latter case oxygen-impurities are homogeneously distributed throughout the Ga-film. Thus the amorphous phase is stabilized and does not immediately transform back into the surrounding metastable β-phase, where the nearest neighbor coordination is very similar to that in the amorphous phase. On the other hand α-Ga and amorphous Ga have very different nearest neighbor coordinations. Thus it is much easier to transform α-Ga into amorphous Ga. These results mean that ion irradiation at low temperatures produces the same amorphous phase as vapor quenching. This can be easily understood by the assumptions of the thermal spike model: the high energy density in a cascade, as produced by a heavy ion, leads to a locally restricted, "molten" region within a cold matrix. The high temperature gradients to the cold matrix, due to the small molten region, lead to extremely high cooling rates comparable to that of vapor quenching ($\sim 10^{14}$ K/sec). Details of the thermal spike model are discussed in Chapter 7. The validity of this model in explaining the Ga results is underlined by the fact, that Görlach et al. could not produce the amorphous Ga phase by He-irradiation which does not produce extended cascades.

The transformation from α- and β-Ga into amorphous Ga by low temperature ion irradiation is indicated by the thick solid arrows at 4K in the simple phase diagram in Figure 7. This description does not include the intermediate high temperature state within the thermal spike region.

B. Be, In, Ge-Cu

In the following, mainly systems of non-transition metals will be discussed where the general rule applies that lattice disorder enhances T_c because $\alpha^2(\omega)F(\omega)$ is more enhanced for low energies (ω) (Bergmann, 1976). An interesting material in this respect is Be, which in the crystalline state only becomes superconducting at $T_c = 0.026$K. In the highly disordered, vapor quenched state T_c is enhanced to about 9K (Comberg et al., 1975). This means an enhancement of more than a factor 300 by lattice disorder. It is still not clear if this vapor

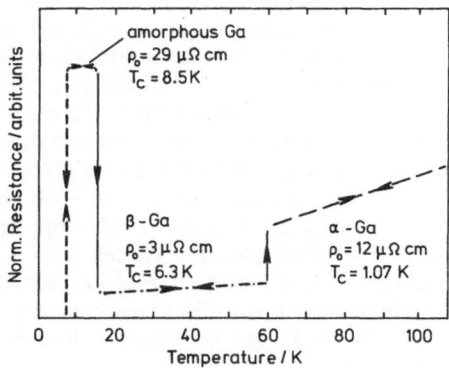

Figure 6   Normalized resistance of vapor quenched Ga as a function of temperature (Buckel and Hilsch, 1954).

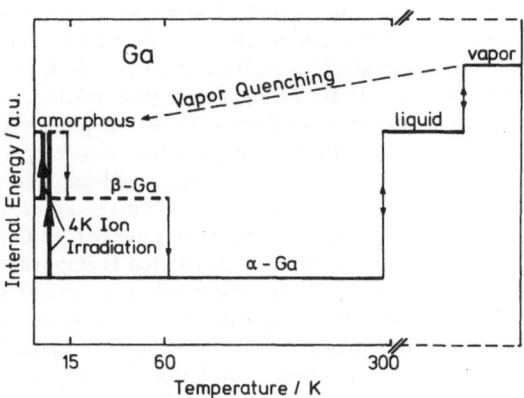

Figure 7   Schematic representation of the internal energy of Ga as a function of temperature.

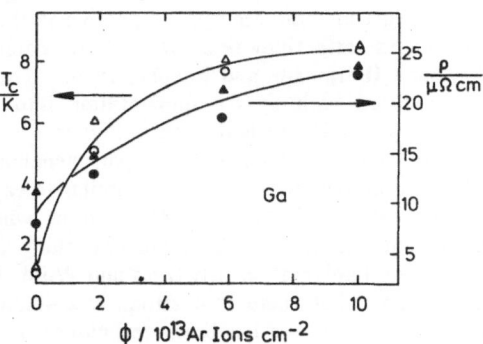

Figure 8 T_c and ρ_0 of α-Ga versus dose of irradiating Ar ions (at 4K) indicating the transformation into amorphous Ga (Görlach et al., 1981).

quenched phase is liquid-like amorphous or micro-crystalline. There were different attempts to compare this large effect on T_c by vapor quenching with the method of low temperature ion irradiation or implantation. Figure 9 shows the results obtained on a Be-film which had been evaporated on quartz substrates at room temperature (Stritzker and Ewert, 1978). The as-condensed films were not superconducting above 1.0K whereas a T_c of 1.1K was achieved after irradiation with only 2×10^{13} Kr$^+$/cm^2. With increasing fluence T_c increases to about 2.5K and stays nearly constant between 10^{15} to 6×10^{16} Kr$^+$/cm^2. Then both T_c and the resistance increase steeply. T_c reaches a maximum value of ~6K which is below the value of ~9K for the vapor quenched material. Klein et al. (1976) implanted O into Be films at 4K and achieved a T_c as high as 9K. However, they started from a very disordered material with a T_c ~7K. Both experiments show that the use of ion techniques at low temperatures results in Be-phases comparable to that obtained by vapor quenching.

Indium is another example where T_c can be enhanced by lattice disorder. Heim et al. (1978) compared both vapor quenching and low temperature ion irradiation. The result is shown in Figure 10 where T_c is plotted versus concentration of implanted In$^+$ ions into In films held at 2K. They studied In films in two different initial states i) annealed In with $T_c = 3.6$K (dashed curve) ii) vapor quenched, microcrystalline In with $T_c = 4.35$K (solid curve). Low-temperature In implantation results in a continuous increase of T_c of the as-annealed film over the whole concentration regime. At 1.2 at.% In implantation, T_c reaches the value of vapor quenched In. Then T_c increases even further with implantation dose. Also the T_c of the vapor quenched film increases with low temperature implantation after an initial T_c-depression. The result in both cases implies that low-temperature ion implantation produces a more disordered phase than vapor quenching and implantation can be considered to have a higher quenching rate. In a recent paper Hofmann et al. (1981) found different fluences of Ar had different effects on both ρ and T_c of annealed In films. The higher fluence regime is explained by the stabilizing effect of redistributed O impurities.

The Ge-Cu system represents another example where the mechanism of ion implantation at 4K can be compared with vapor quenching. If Ge vapor is condensed onto a substrate at room temperature an amorphous phase is achieved with respect to long range order. However, the nearest neighbor coordination is still that of the Ge crystal (diamond type). That is the reason that this "amorphous" phase still shows a semiconducting resistance behavior. In contrast, the vapor quenched amorphous Ge phase which needs about 20 at.% Cu for stabilization behaves metallic and becomes superconducting at 3.3K for Ge$_{50}$Cu$_{50}$ (Stritzker and Wühl, 1971). This metallic amorphous phase has the coordination of liquid Ge, i.e., it is in the frozen liquid state (Nowak et al., 1978). Each Ge atom gives 4 free electrons to the conduction band (Bergmann and Koepke, 1976). This liquid-like amorphous state of Ge can also be produced both by Cu implantation into Ge at 4K and by Kr irradiation of Ge$_{80}$Cu$_{20}$ films kept at 4K. In both cases a higher $T_c = 3.7$K is achieved for a lower Cu concentration of only ~20 at.% compared to vapor quenching (Stritzker and Wühl, 1976). Again the ion techniques at low temperatures are similar to vapor quenching. Similar observations have been made for the Si-Au or Si-Cu systems where a superconducting, liquid-like amorphous phase could be obtained by vapor quenching (Möckel and Baumann, 1980) and by low temperature ion implantation (Stritzker and Wühl, 1976).

These different examples show that there is a strong resemblance between the phases obtained by vapor quenching and by heavy ion beam techniques. It depends on the special material which of the two techniques is more favorable for superconductivity, i.e., for a higher degree of lattice disorder. Up to now there is only one example, Pd, where both non-equilibrium techniques give different results. Vapor-quenched Pd has a high initial resistivity and is not superconducting. Annealing of this Pd results in a decreasing lattice disorder and

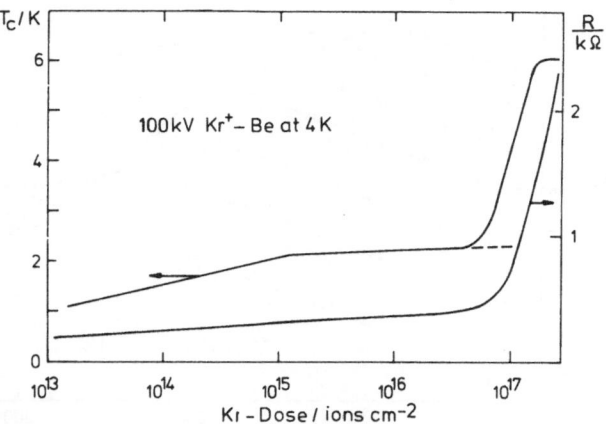

Figure 9 T_c and resistance R of Be film versus dose of irradiating Kr-ions (at 4K).

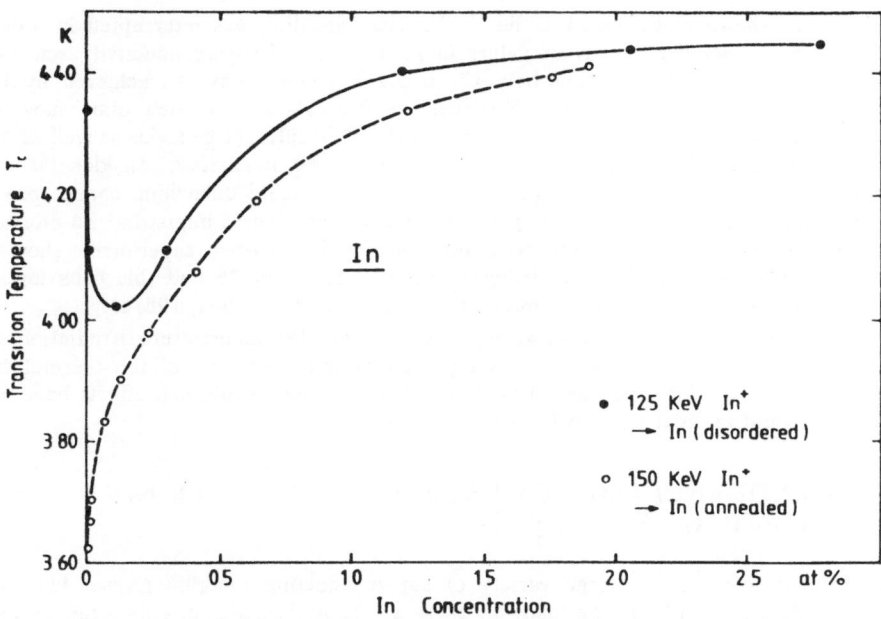

Figure 10 T_c of initially disordered (solid line) and annealed (dashed line) In films as a function of low temperature implanted In concentration (Heim *et al.*, 1978).

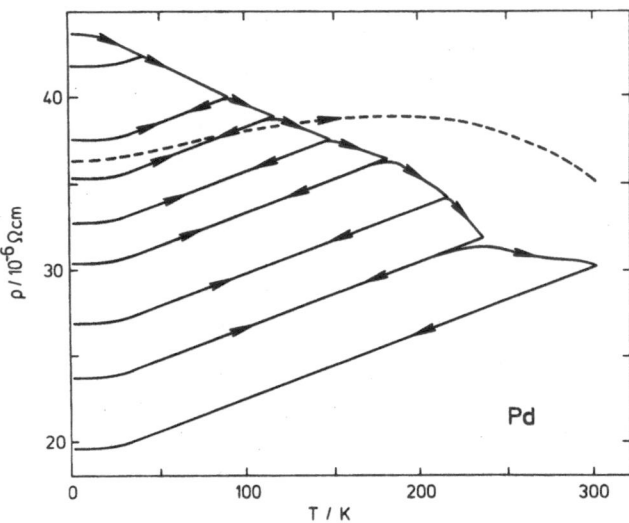

Figure 11   Resistivity ρ of vapor quenched 400Å thick Pd-films as a function of annealing temperature for as-condensed Pd (solid curve) and after 4K He$^+$ irradiation into the superconducting state (dashed curve).

ρ decreases as shown in Figure 11 (solid line). The annealing was interrupted in order to look for superconductivity at various values of ρ. No sign of superconductivity was found above 0.1K. However, superconductivity ($T_c \leqslant 3.2$K) could always be achieved by He$^+$-irradiation of such Pd films at 4K (Stritzker, 1979). These irradiated films show very different annealing curves (dashed line in Figure 11). This different behavior as well as other indications led to the following interpretation. Only the ion irradiation can kick Pd atoms into interstitial positions within the Pd lattice. Vapor or liquid quenching cannot produce interstitial atoms because they are energetically unfavorable. These interstitial Pd atoms are thought to be the reason for the observed superconductivity. Recent experiments show that the susceptibility during irradiation decreases about a factor of 25 and this behavior is in agreement with the assumption of Pd interstitials (Meyer and Stritzker, 1981b).

The preceding experiments have strongly hinted that low-temperature irradiation with heavy ions can be understood by the assumption of rapid quenching of the thermal spike region. However, the Pd experiment has shown that additional collisional effects have to be taken into account in order to describe the ion techniques.

V. LIQUID QUENCHING; COMPARISON WITH OTHER NON-EQUILIBRIUM TECHNIQUES

In the last five years a large variety of superconducting metallic glasses have been produced (Johnson, 1978). In the following there will be a discussion of only a few examples which have been investigated by ion beam techniques.

A. Zr-Cu

Zr-Cu metallic glasses are found to be superconducting. Although the x-ray diffraction pattern looks "amorphous" it is interesting to see if low temperature ion techniques has any effect on T_c and ρ of these metallic glass. Table II shows the resulting increases in both T_c

and ρ for various Cu-concentrations after He-irradiation at 4K (Meyer, 1980). In all cases low temperature irradiation yields a positive influence, i.e., the system becomes even more amorphous due to the irradiation. Similar T_c values could be obtained by Cu-implantation into Zr. Figure 12 shows ΔT_c, the increase of T_c due to irradiation, as a function of annealing temperatures. In all cases ΔT_c vanishes on annealing to room temperature.

TABLE II

Increase of the transition temperature, $\Delta T_c/T_c$, and of the
resistivity $\Delta\rho/\rho$, of
ZrCu metallic glasses after 4K He$^+$ implantation (Meyer, 1980).

at % Cu	$\Delta\rho/\rho$	$\Delta T_c/T_c$
26	0.39	0.068
30	0.43	0.10
35	0.51	0.28
40	0.72	0.29
50	0.34	0.69

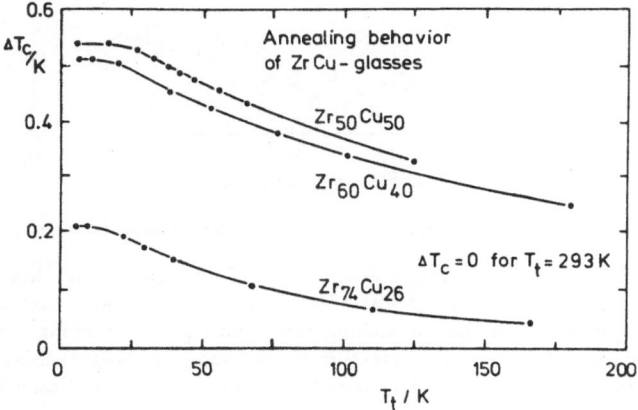

Figure 12   Reduction of the T_c increase, ΔT_c, induced by 4K He$^+$ irradiation as a function of annealing temperature T_t (Meyer, 1980).

B. (Ti, Zr, Hf)-Fe

Whereas it is easy to produce (Ti, Zr, Hf)$_{80}$ - Cu$_{20}$ metallic glasses it is difficult to prepare the corresponding Fe (instead of Cu) alloys by liquid quenching. It has not yet been possible to form Ti-Fe metallic glasses by liquid quenching. Due to the higher quenching rate of the heavy ion implantation at 4K, it was expected that it should be possible to also produce the Fe alloys in an amorphous state. The results are shown in Figure 13 where T_c is plotted as a function of the concentration of implanted Fe and Cu (for comparison). As can be seen all three Fe alloys can be produced and have a similar concentration dependence as the

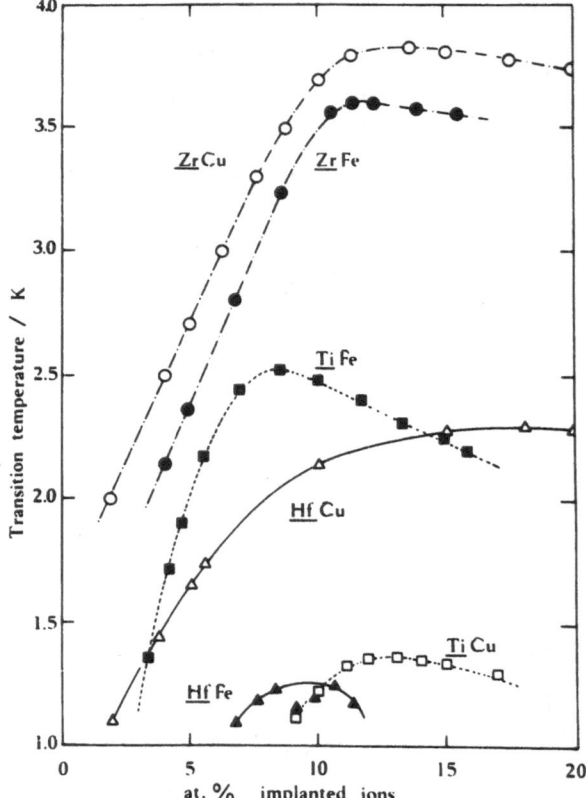

Figure 13 T_c for Ti, Zr and Hf as a function of low temperature implanted Cu or Fe concentrations.

corresponding Cu alloys. Ti-Fe, not producible by liquid quenching, has a substantially higher T_c value than Ti-Cu. These results (Meyer et al., 1981) show that ion implantation is indeed a technique with a higher quenching rate than liquid quenching. Ion implantation at 4K can produce a larger variety of metallic glasses than liquid quenching techniques. A detailed systematic study of (Ti, Zr, Hf) based - 3rd metal alloys will soon be published by Meyer and Stritzker.

C. $Pd_{80}Si_{20}$-H

The metallic glass $Pd_{80}Si_{20}$ is not a superconductor after preparation by liquid quenching. However it becomes superconducting at ~2.5K after H implantation at liquid He temperatures (Stritzker and Luo, 1979). The experiment was explained in analogy to the pure PdH system which is a superconductor at 9K (see the review on superconducting metal-H systems by Stritzker and Wühl, 1978). In a recent experiment the Orsay group (Bernas et al., 1980) showed that the appearance of superconductivity in $Pd_{80}Si_{20}$ is correlated with the doubling of the residual resistivity. In situ Rutherford backscattering and temperature cycling experiments during annealing indicate that the reduction in resistivity is not necessarily correlated with H release from the sample and that presumably H has stabilized another amorphous PdSi phase of higher resistivity. However, superconductivity only exists when H is in the sample, confirming the previous explanation for the PdH system.

All three examples show that ion implantation or irradiation can produce "more amorphous" states than the liquid quenching techniques. This is understandable since the estimated quenching rates are several orders of magnitude different (for example $\sim 10^{14}$ K/sec compared to $\sim 10^9$ K/sec).

VI. ION IRRADIATION AND IMPLANTATION; METHODS TO PRODUCE AMORPHOUS METALS

The production of amorphous materials by room temperature bombardment will briefly be discussed. As discussed in the introduction, amorphous materials need about 10 to 20 at.% of impurity atoms to stabilize an amorphous phase. This condition does not essentially change for ion implantation conditions. For instance 15-20 at.% metalloids (B, P, As, Sb) or 7-10 at.% Dy and Gd are required to amorphize ferromagnetic transition metals and alloys. Theoretically these amorphous phases can be understood by random packing models with a local minimum in the free energy. With respect to ion bombardment at room temperature there is an essential difference between a semiconductor and a metal. In the latter, with covalent directional bonds, point defects are not mobile in contrast to metals with non-directional bonds. Therefore one could expect to quench in collisional damage in a semiconductor whereas, in a metal, a dynamic recovery of collisional damage could occur. In a semiconductor a phase change, to the amorphous phase, should occur at quite low doses. However, in a metal a more and more defective crystal would be produced. Finally the metal lattice would collapse around an appropriate impurity with size comparable to the host atoms. Thus the amorphous phase would be stabilized.

With respect to superconductors, a detailed study has been performed by Linker (1981) for the bcc-superconductors Nb $(T_c = 9.2$K$)$ and Mo $(T_c = 0.92$K$)$. Both metals were implanted with different ions like N, Ne, P and S. The effect on T_c can be described in general by a decrease for Nb with a saturation value around 2K whereas T_c of Mo increases. The highest value of 8.4K is achieved for P implantation at 77K. An even higher value of 9.2K can be achieved by 4K implantation (Linker and Meyer, 1976). The samples were analyzed by detailed x-ray diffraction measurements. In Nb the defect structure is not much dependent on the ion species. The defect structure of Nb implanted with N can be described in terms of small static displacements of the host lattice atoms. A similar kind of structure is also observed in N-implanted Mo during the initial state of disorder (up to 10 at.% N). However, for higher N concentrations (about 20 at.%) partial amorphization of the Mo layers occurs. Amorphous Mo was also obtained for P and S implantation. Thus it is assumed that for high enough impurity concentrations (20-25 at.%, where x-ray measurements can no longer be performed) amorphization of Mo occurs during the ion implantation independent of the special ion species as long as a suitable "glass former" is used.

In a recent experiment Mendoza-Zelis et al. (1982) studied the influence of B implantation on the resistivity behavior of Pd films at 4K. It was known from previous experiments that Pd can be transformed into a superconductor $(T_c \simeq 3.8$K$)$ by B-implantation (Stritzker and Becker, 1975). Figure 14 shows the results of the recent experiments. The resistivity, ρ, is plotted versus implanted B concentration. After an initial increase ρ rises slowly with increasing B content up to the eutectic composition. With further increase of B concentration ρ increases steeply indicating amorphization. At about 33 at.% B the temperature dependence of the resistivity changes with $d\rho/dT < 0$. This negative temperature dependence, as in metallic glasses, is a strong indication for the amorphicity of these Pd-B alloys. In this region of $d\rho/dT < 0$, superconductivity starts and T_c is 3.7K at 55 at.% B. This is a clear indication that superconductivity is caused by the amorphicity of the Pd-B alloys as produced during the low temperature B implantation.

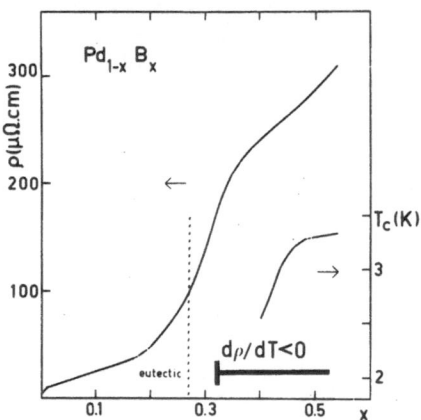

Figure 14   Resistivity ρ of Pd films as a function of low temperature implanted B concentration (Mendoza-Zelis *et al.*, 1980).

VII. TE-AU; A COMPARISON OF THE DIFFERENT NON-EQUILIBRIUM TECHNIQUES

In the following, the $Te_{1-x}Au_x$ system will be discussed in detail because many non-equilibrium techniques have been used in its formation. In addition $Te_{1-x}Au_x$ has interesting properties which make it attractive for an examination based on measurements of the electrical properties. The reason is that the different phases have completely different values of ρ and $d\rho/dT$.

In previous experiments, metastable $Te_{1-x}Au_x$ alloys had been produced both by liquid quenching (Luo and Klement, 1962, Tsui and Newkirk, 1969) and by vapor quenching (Krauss *et al.*, 1975) yielding the following phases:

1)   Amorphous $Te_{1-x}Au_x$ is produced by vapor quenching. This phase is semiconducting, i.e., ρ increases exponentially with decreasing temperature.

2)   Simple cubic $Te_{1-x}Au_x$ $(0.15 \leqslant x \leqslant 0.35)$ is produced by liquid quenching. This phase has a slight negative temperature coefficient of the resistivity $(d\rho/dT < 0)$ similar to metallic glass and becomes superconducting at about 2.3K.

3)   In contrast the equilibrium phase consists of a phase mixture of Te, Au and Te_2Au. This phase mixture behaves as metal or semiconductor depending on the composition, but is not superconducting.

Thus it is easy to distinguish between the structurally different metastable phases 1) and 2) and the stable phase 3) just by measuring ρ and $d\rho/dT$ without any structure determination. This property makes it easy to study the results of the two other non-equilibrium techniques, i.e., ion bombardment and laser quenching.

First the results of ion irradiation will be described (Meyer and Stritzker, 1981). Homogeneous Te-Au films ($\sim$1000Å thick) consisting of the equilibrium phase 3) were irradiated at 4K with He^+-ions penetrating totally through the sample. Figure 15 shows the resulting change of resistivity versus irradiation dose for $Te_{75}Au_{25}$. An initial steep increase of ρ due to the lattice disorder is followed by a smooth further increase of ρ. After irradiation with 3×10^{15} He^+/cm^2, the metallic character of phase 3) vanishes and the sample behaves like the semiconducting phase 1). After 6×10^{15} He^+/cm^2, ρ increases by 3

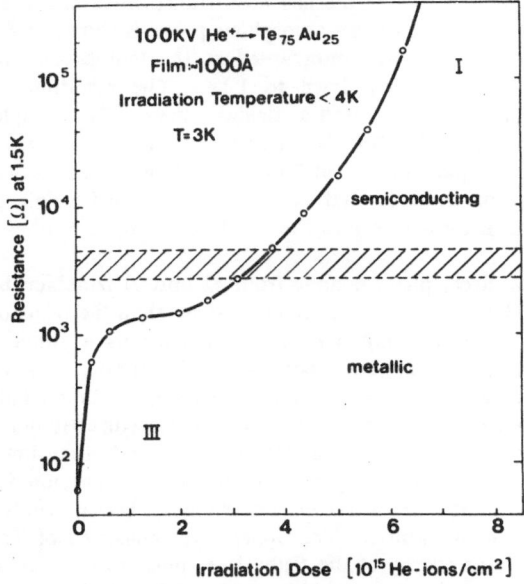

Figure 15 Resistance of a $Te_{75}Au_{25}$-film versus dose of irradiating He^+ ions (4K).

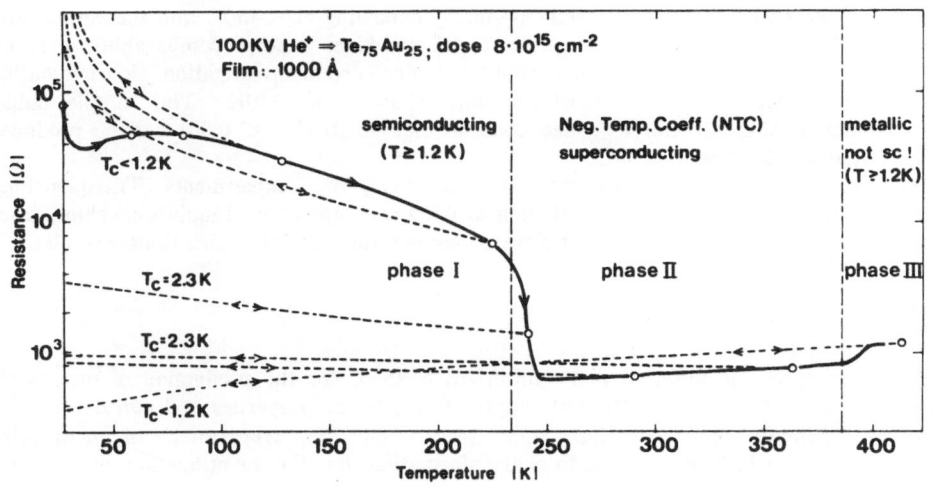

Figure 16 Annealing behavior of low temperature irradiated $Te_{75}Au_{25}$.

orders of magnitude. The whole sample has been transformed into the amorphous phase 1). The resulting resistance behavior of this amorphous phase is shown in Figure 16 as a function of temperature. The annealing was interrupted at the temperature indicated by the open points and the sample was cooled down to 1.1K. The resulting reversible temperature dependences of the resistance are plotted as dashed curves. The sample remains in phase 1) up to annealing temperatures of 230K. Then an irreversible drop of the resistance indicates a phase change which is completed at about 250K. The slight negative value of $d\rho/dT$ and the superconducting transition indicate that phase 2) has formed. This phase is stable up to about 400K where a transformation into the equilibrium metallic phase 3) ($d\rho/dT > 0$ and $\rho_3 > \rho_1$) occurs.

In order to see if a direct phase change from 3) into 2) is achievable, an irradiation was performed at 250K. This temperature is slightly higher than the transformation of 1) into 2). Figure 17 shows the resulting change of the resistance as a function of He dose. The resistance passes through a maximum, decreases, and saturates at a value below the initial resistance. This behavior ($\rho_2 < \rho_3$) and the occurrence of superconductivity at ~2.3K is convincing evidence that $\rho_2 < \rho_3$ phase 2) has been formed directly out of phase 3).

The two results obtained by ion bombardment suggest, that ion irradiation at 4K is comparable to vapor quenching (compare Section IV) whereas ion irradiation at 250K is similar to liquid quenching. In a recent experiment, Te-Au sandwich films were laser quenched at liquid He temperatures. The upper layer consisted of Te in order to ensure a better coupling of the laser energy of the Ruby laser-pulse to the Te-Au sandwich layers. A superconducting phase, i.e., phase 2), could be achieved by this laser quenching.

The various experimental results can be summarized in the schematic phase diagram for $Te_{1-x}Au_x$ based on the simple internal energy model (compare Section III) as shown in Figure 18. The equilibrium phases are indicated by solid lines, the metastable ones by dashed lines. The various processes can be described as follows. Vapor quenching produces the amorphous, superconducting phase 1). The analogy to a frozen liquid is indicated by the same internal energies for both liquid phase and phase 1). At 230K phase 1) transforms irreversibly into phase 2). Liquid quenching forms the simple cubic, superconducting phase 2) (for $0.15 \leqslant x \leqslant 0.35$) with lower internal energy than the liquid phase. (Presumably there exists a comparable high temperature or high pressure phase.) This metastable phase 2) transforms irreversibly during annealing at ~400K into the equilibrium phase 3). Laser quenching of samples at 4K produces results comparable to liquid quenching. The possible concentration regime is under current investigation. Ion irradiation at 4K forms phase 1) which transforms into phase 2) at 230K. This simple cubic, superconducting phase 2) (for extended concentrations $0.10 \leqslant x \leqslant 0.90$) can be produced directly by ion irradiation at 250K.

The following results can be extracted from the $Te_{1-x}Au_x$ - experiments. The quenching rates of vapor quenching and ion irradiation at 4K are comparable. Liquid quenching, laser quenching at 4K and ion irradiation at elevated temperatures of 250K give similar results.

VIII. CONCLUSION

In this chapter the different non-equilibrium methods for the production of metastable alloys have been discussed mainly with respect to electronic properties and correlation with structure. Primary emphasis has been placed on the preparation of metastable superconductors, which provide additional information for the identification of different phases.

Vapor quenching of metal atoms onto cold substrates (4K) can often result in amorphous phases closely resembling frozen liquid phases. The liquid phase can be imagined to occur

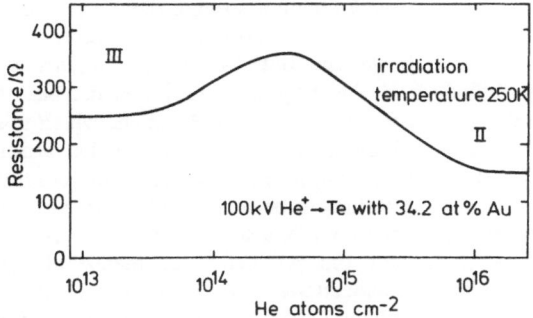

Figure 17   Resistance of a $Te_{66}Au_{34}$ film versus dose of irradiating He^+ ions (250K).

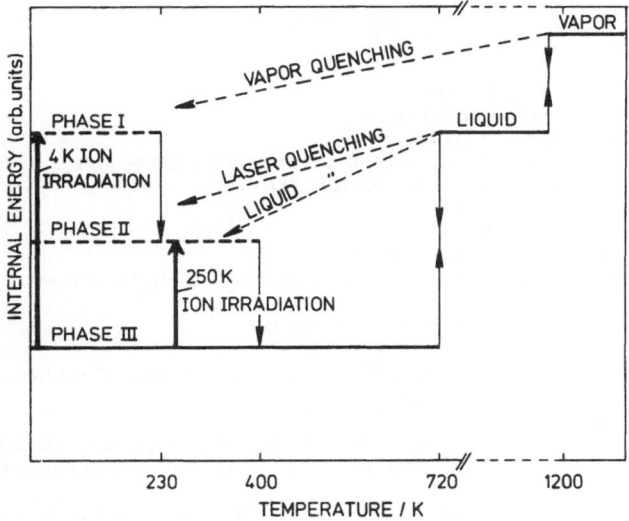

Figure 18   Schematic representation of the internal energy of Te-Au as a function of temperature.

immediately after condensation of the vapor with the result that its "structure" can be frozen in. In the same vein is the more commonly used method of liquid quenching, where a droplet of liquid is abruptly cooled to room temperature. However, it must be realized that the quenching rates will be several orders of magnitude slower than vapor quenching. Nevertheless, it is possible in many cases (when the stability criteria are fulfilled) to produce amorphous alloys which have no long-range order but which have, in general, short-range order more related to the crystalline phase. Laser quenching, where a liquid is also cooled very rapidly, yields results similar to liquid quenching.

The critical point of this discussion is the explanation of the results of ion bombardment. The quenching rate involved in this process can be influenced by the ion beam parameters and the target temperature. Most of the reported results can be understood in terms of a thermal spike model. The enormous energy of a heavy ion of several hundred keV is transferred to the lattice very locally within the small volume of the cascade. Comparison of the results of ion bombardment with the other liquid quenching methods suggests that the energy density within the thermal spike region is so high that there occurs some kind of liquid state which is very localized for an extremely short time. However, the collisional aspects of ion bombardment should not be forgotten. Initial changes of resistivities after very low dose implantations show the influence of the introduced radiation defects. In irradiated Pd, the occurrence of superconductivity can be explained by Pd atoms which are shot into interstitial positions. This collisional aspect becomes more important in ion beam mixing experiments where, in contrast to the irradiation experiments, considerable material transport occurs.

In summary there are many questions to be solved. More examples have to be studied in a more microscopic manner in order to obtain a better, detailed understanding. However, the examples presented here show that the production of metastable alloys by the various quenching methods can be mainly understood by thermodynamic considerations. The simple phase diagrams introduced in this chapter must be revised to account for the chemical nature of the systems. For detailed understanding, the real thermodynamic potentials must be calculated and thus the equation of state of the system under consideration.

REFERENCES

Bergmann, G. (1976), Physics Rep. *276*, 161.

Bergmann, G. and Koepke, R. (1976), private communication.

Bernal, J. D. (1960), Nature (London) *185*, 68.

Bernas, H., Travese, A., Zawislak, F. C., Chaumont, J. and Dumoulin, L. (1980), J. Physique *41*, C8-859.

Bernas, H. and Nedellec, P. (1981), Nucl. Instr. and Math., *183*, 845.

Bernas, H. *et al.*, (1981), to be published.

Buckel, W. (1954), Z. Physik, *238*, 136.

Buckel, W. and Hilsch, R. (1954), Z. Physik, *138*, 109.

Cargill, G. S. (1975), in "Solid State Physics," (H. Ehrenreich, F. Seitz and D. Turnbull, eds.) Academic Press, New York, *30*, 227.

Comberg, A., Ewert, S. and Wühl, H. (1975), Z. Physik, *B20*, 165.

Cotteril, R. M. J. (1976), American Scientist, *64*, 430.

Chu, W. K., Mayer, J. W. and Nicolet, M. A. (1978): "Backscattering Spectroscopy," Academic Press, New York.

Duwez, P. (1979), Proc. Indian Acad. Sci., *C2*, 117.

Gilman, J. J. (1980), J. Physique *41*, C8-811.

Görlach, U., Hitzfeld, M., Ziemann, P., Buckel, W. (1981), Verhandlungen der DPG 3/1981, p. 482, to be published in Z. Phys. I want to thank the authors for the permission to discuss their results prior to publication.

Gorska, A., Gorski, A. M., Igalson, J. I., Pindor, A. J. and Sniadower, L. (1977), 2nd Int. Conf. Hydrogen in Metals, Paris.

Güntherodt, H. J., Oelhafen, P., Lapka, R. *et al.*, (1980) J. Physique *41*, C8-381.

Heim, G., Bauriedl, W. and Buckel, W. (1978), J. Nucl. Mat., *72*, 263.

Hofmann, A., Ziemann, P. and Buckel, W. (1981), Nucl. Instr. Meth. *183*, 943.

Johnson, W. L. (1978), Caltech Report, CALT-822-104.

Johnson, W. L. (1980), J. Physique, *41*, C8-731.

Kaufmann, E. N. and Buene, L. (1981), Nucl. Instr. and Meth., *182*, 327.

Klein, J., Leger, A., Chaumont, J. and Bernas, H. (1976), unpublished.

Krauss, G., Müller, W. H.-G., Baumann, F. and Buckel, W. (1975), J. Less-Common Metals, *43*, 13.

Linker, G. and Meyer, O. (1976), Sol. State Comm., *20*, 695.

Linker, G. (1981), Nucl. Instr. Meth., *183*, 501.

Luo, H. L. and Klement, W. (1962), J. Chem. Phys., *36*, 1870.

McMillan, W. L. (1968), Phys. Rev., *167*, 331.

Mendoza-Zelis, L., Bernas, H., Traverse, A. and Chaumont, J. (1982), in "Metastable Materials Formation by Ion Implantation," (S. T. Picraux and W. J. Choyke, eds.), North Holland, New York, p. 223.

Meyer, J. D. (1980), J. Physique, *41*, C8-762.

Meyer, O. (1980b), in "Treatise on Materials Science and Technology," (J. K. Hirvonen, ed.) p. 415, Academic Press, New York.

Meyer, J. D., Ochmann, F. and Stritzker, B. (1981), Solid State Comm. *39*, 419.

Meyer, J. D. and Stritzker, B. (1981), Nucl. Instr. Meth., *183*, 965.

Meyer, J. D. and Stritzker, B. (1981b), to be published.

Möckel, P. and Baumann, F. (1980), Phys. Stat. Sol. (a) *57*, 585.

Nowak, H. J., Leitz, H. and Buckel, W. (1978), Phys. Stat. Sol. (a), *49*, 73.

Poate, J. M. (1980), in "Laser and Electron Beam Processing of Materials," (C. W. White and P. S. Peercy, eds.) p. 691. Academic Press, New York.

Ostwald, W. (1897), Z. Physik Chem, *22*, 289.

Stritzker, B. (1974), Z. Physik, *268*, 261.

Stritzker, B. (1978), J. Nucl. Mat., *72*, 256.

Stritzker, B. (1979), Phys. Rev. Lett., *42*, 1769 and Inst. Phys. Conf. Ser., *55*, 529.

Stritzker, B. and Becker, J. (1975), Phys. Lett., *51A*, 147.

Stritzker, B. and Ewert, S. (1978), to be published.

Stritzker, B. and Luo, H. L. (1979), Solid State Comm., *29*, 811.

Stritzker, B. and Wühl, H. (1971), Z. Physik, *243*, 361.

Stritzker, B. and Wühl, H. (1976), Z. Physik B, *24*, 367.

Stritzker, B. and Wühl, H. (1978), in "Hydrogen in Metals II," (G. Alefeld and J. Völkl, eds.), p. 243, Springer Verlag, Berlin.

Tsuei, C. C. and Newkirk, L. R. (1969), Phys. Rev., *183*, 619.

CHAPTER 7

COLLISION CASCADES AND SPIKE EFFECTS

J. A. DAVIES

**Atomic Energy of Canada Limited Research Corporation,
Chalk River, Ontario, Canada**

CONTRIBUTORS: J. Bøttiger, W. Hofer and U. Littmark

I. INTRODUCTION

In a typical pulsed-laser or electron beam annealing or alloying study, the mean deposited energy density, θ, (corresponding to a 1-2 J/cm^2 pulse distributed over ~0.5 μm depth) is of the order of 1-2 eV/atom and the quenching time t_q for the resulting molten zone is ~10^{-7} seconds. There are at least three ways in which *ion* beams may be used to achieve comparable energy densities and quench rates (Table I):

(i) High-intensity pulsed ion sources are now available (e.g., at Cornell) which can deposit several J/cm^2 of 200 keV H$^+$ or Ba$^+$ within an individual pulse of ~10^{-7} sec duration. The range of such ions is less than 1 μm and hence θ and t_q should be directly comparable to the laser annealing case, as Baglin (IBM) and the Cornell group (Baglin, 1981) have already clearly demonstrated.

(ii) *Scanned Microbeam*. An alternative way of achieving equivalent θ and t_q values, suggested several years ago by Lindhard (1976), is to scan with a sufficiently intense microbeam that the required θ value is achieved during a single sweep across the surface. A beam of 1 MeV Sb_2^+ ions, focussed down to a 3-μm spot, would deposit the required 1-2 eV/atom in $\leq 10^{-7}$ sec. Hence, a sweep rate of 30 m/sec should be sufficient to attain the required θ level. To date, no experimental attempt to test this suggestion has yet been made, although the required beam parameters are probably within the scope of current microbeam technology. Some groups have used much lower energy densities plus continuous (repetitive) scanning to achieve implantation self-annealing, but this produces annealing conditions that are more closely analogous to the CW-mode of laser annealing.

(iii) Under certain ion beam conditions (high atomic number and low beam energy), the energy density within a *single* collision cascade may attain or even exceed the 1-2 eV/atom level. However, the radius r of such a collision cascade is extremely small (typically $\sim 10^{-2}$ μm or less) and the corresponding quenching rate ($t_q = r^2/4D$, where D is the thermal diffusivity) would be many orders of magnitude faster than in the pulsed (or scanned) high current density cases. The quenching rate may even be significantly faster than the rate of electron/phonon coupling, so that (depending on the nature of the collision cascade) one can have a spike consisting of "hot" atoms and "cold" electrons, or vice versa. Consequently, markedly different types of annealing behavior may be anticipated and normal thermal conductivity treatments may no longer be quantitatively valid.

The first two methods of using ion beams to achieve high θ values produce conditions that are quite analogous to the pulsed laser annealing behavior reported elsewhere in this book and hence are not discussed further here. Instead, the present chapter concentrates entirely on

TABLE I

High Energy Density Processing

Process	Pulse Length (sec)	Depth (μm)	Energy Density eV/atom	Quench Time (sec)
Pulsed laser (or electron) anneal	10^{-8}	0.5	1-2	10^{-7}
Pulsed Ion Beam (Cornell)	10^{-7}	0.1-1	1-2	10^{-7}
Scanned Microbeam	10^{-7}	0.5	1-2	10^{-7}
Individual Collision Cascade	10^{-14}	0.01	$<10^{-3}$ up to >10	10^{-12}

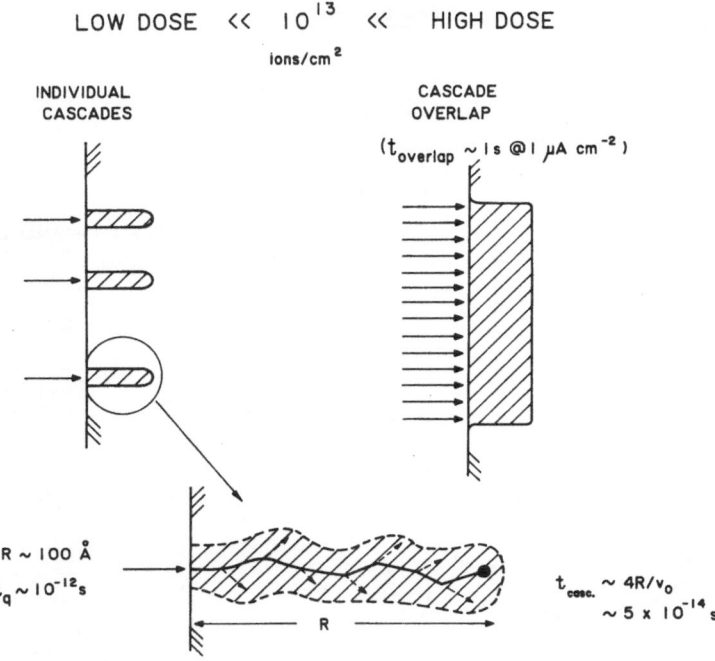

Figure 1:   Collision cascade schematic, indicating the typical dimensions and corresponding time scales involved.

the third method: i.e., we discuss in some detail the question of achieving high θ within a single collision cascade and review some of the experimental evidence for materials modification resulting from such thermal spikes. At the outset, we shall consider some general features of collision cascades in order to provide a basic framework for the subsequent treatment of the high-density cascade regime; this basic framework also serves as an introduction to other cascade related phenomena such as ion beam mixing (Chapter 9), sputtering (Chapter 10) and solute redistribution effects (Chapter 8).

II. COLLISION CASCADE CONCEPTS

One important concept is the distinction between the individual cascade regime and the cascade overlap regime (Figure 1). At a current density of ~1 μA/cm^2, successive cascades of ~20Å radius will occur in a given region of surface at roughly one second intervals, whereas the quenching time (t_q) of such a cascade is typically only a few picoseconds. Hence, in the "energy spike" context, cascade overlap effects should always be completely negligible. Even at the extremely high current densities (300 A/cm^2) of the Cornell pulsed ion source, the characteristic overlap time is still several orders of magnitude longer than t_q and hence, before the arrival of another cascade, the deposited energy density becomes dissipated over a volume much larger than cascade dimensions. On the other hand, in the more general context of radiation damage and ion-beam mixing, where point defect migration processes often play a prominent role, cascade overlap effects must always be considered, except when the total dose is small ($<10^{13}$ ions/cm^2).

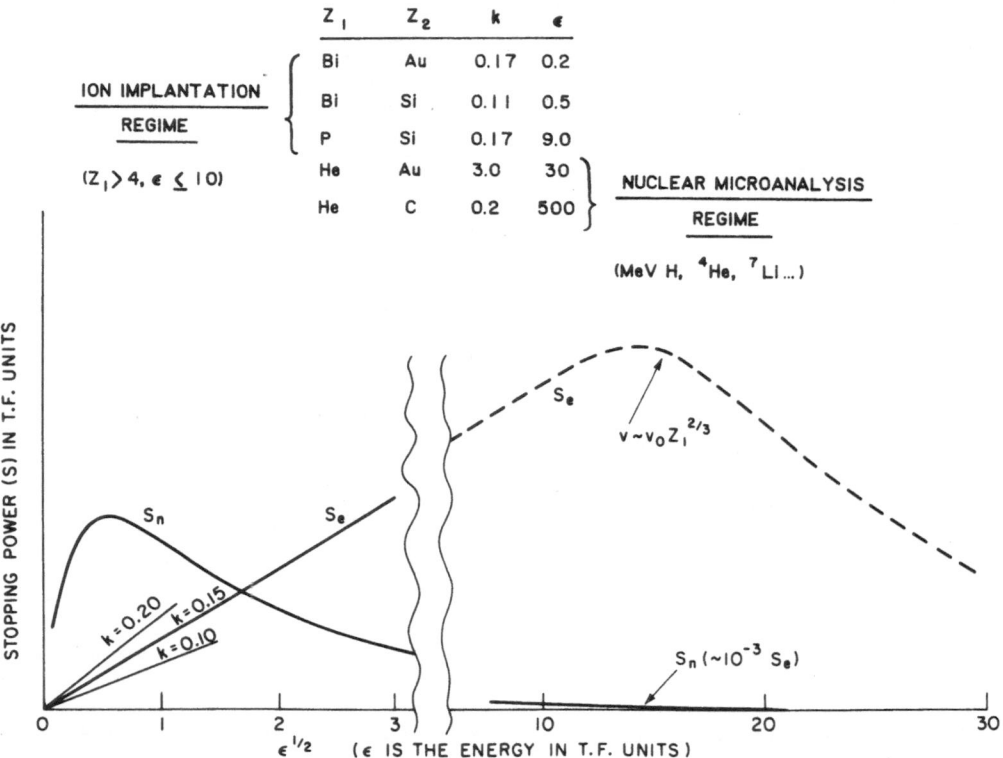

SOME TYPICAL 500 keV CASES

Z_1	Z_2	k	ϵ
Bi	Au	0.17	0.2
Bi	Si	0.11	0.5
P	Si	0.17	9.0
He	Au	3.0	30
He	C	0.2	500

ION IMPLANTATION
REGIME

$(Z_1 > 4, \epsilon \lesssim 10)$

NUCLEAR MICROANALYSIS
REGIME

(MeV H, ^{4}He, ^{7}Li ...)

STOPPING POWER (S) IN T.F. UNITS

S_n

S_e

S_e

$v \sim v_0 Z_1^{2/3}$

k = 0.20
k = 0.15
k = 0.10

$S_n (\sim 10^{-3} S_e)$

0   1   2   3   10   20   30

$\epsilon^{1/2}$ (ϵ IS THE ENERGY IN T.F. UNITS)

Figure 2:   Stopping power versus energy in dimensionless units (based on Lindhard *et al.*, 1963a).

Figure 2 provides a convenient overview of the main energy loss processes (nuclear stopping, S_n, and electronic stopping, S_e), over the entire range of interest in ion bombardment studies. The abscissa scale, ϵ, is the energy expressed in Lindhard's dimensionless Thomas-Fermi (T.F.) units (Lindhard *et al.*, 1963a): i.e., it is the ratio of the T.F. screening length, a, to the distance of closest approach b in an unscreened collision (where

$$a = 0.8853 a_0 \, [Z_1^{2/3} + Z_2^{2/3}]^{1/2}; \; b = Z_1 Z_2 e^2 \, (M_1 + M_2)/M_2 E;$$

a_0 is the Bohr radius, e is the electronic charge and E is the initial energy of the incident ion).

The advantages of such an energy scaling are twofold. Firstly, it enables S_n (and associated quantities such as multiple scattering, sputtering and damage production) to be expressed in terms of a semi-universal curve for all combinations of Z_1 and Z_2, as shown in Figure 2. Secondly, it permits a clean separation of almost all ion bombardment studies into two widely different regimes: namely,

(i) *Ion Implantation Regime* - $\epsilon \lesssim 10$ (and $Z_1 > 4$). This consists of relatively slow moving heavy ions, with velocities much smaller than the T.F. velocity, $v_0 Z_1^{2/3}$, and it is the regime in which S_n and S_e both contribute significantly to the slowing down. Consequently, the theoretical treatment here is more complex and less accurate than in regime (ii). Regime (i) covers the energy region in which sputtering and damage cascade effects are dominant; hence, sputter-etching and ion-beam mixing fall exclusively within this domain. It is also the regime of main interest in almost all ion implantation studies.

(ii) *Nuclear Microanalysis Regime* - $\epsilon \gg 10$ (and low Z_1). In this energy region, S_n has fallen to a negligible and approximately constant fraction ($\sim10^{-3}$) of S_e. Hence, the ion's trajectory is almost linear, sputtering and collision cascade effects are generally small, and a quantitative theoretical framework exists for accurately predicting the energy loss and channeling behavior. This is the energy region widely used for quantitative nuclear microanalysis methods such as Rutherford backscattering and nuclear reaction analyses. The maximum total rate of energy deposition (i.e., $S_n + S_e$) is comparable in both regimes and under appropriate conditions, as we shall see later, this may produce observable "spike" effects. However, these spike effects should exhibit at least one major difference between the two regimes. At very large ϵ (regime (ii)), the energy deposition process is almost exclusively electronic excitation (S_e) and the resulting spike therefore consists initially of "hot" electrons and relatively "cold" atoms; if the subsequent electron/phonon coupling rate (approximately $\sim10^{-11}$ sec) is sufficiently slow compared to thermal diffusivity, then the atoms within the spike would remain relatively cold. In regime (i), on the other hand, both S_n and S_e are always substantial and hence the initial spike conditions produce "hot" atoms and "hot" electrons simultaneously. Note that S_e never becomes negligible compared to S_n, even at extremely ϵ low values.

Before leaving Figure 2, it should be noted that S_e does not exhibit the property of T.F. scaling; hence, unlike S_n, it cannot be described by a single universal curve. However, at low energies, S_e increases linearly with $\epsilon^{1/2}$ and, as long as $Z_1 \geq Z_2$, the slope k falls within the fairly narrow range, 0.14 ± 0.03. The magnitude of ϵ and k for some typical 500-keV ions has been included in Figure 2 to illustrate this point.

III. DEPOSITED ENERGY DENSITY, θ

The deposited energy density θ within each individual cascade is a strong function of Z_1, Z_2, E and target density. It can vary by at least four orders of magnitude: i.e., from $<10^{-3}$ eV/atom to >10 eV/atom. Quantitative estimates of θ may be obtained by extending Lindhard's stopping power description of the primary ion (Figure 2) to include the recoiling atoms as well. Again, it is convenient (Lindhard *et al.*, 1963b) to separate the energy-loss mechanisms into two types: nuclear (recoil) loss $\nu(E)$ and electronic excitation $\eta(E)$, with $\nu(E) + \eta(E) = E_{ion}$. Note that, although $\nu(E)$ and S_n are closely related quantities, $\nu(E)$ is always ($\sim20-30\%$) *smaller* than the integrated nuclear stopping power $(\int S_n)$ value of the primary ion; this difference arises because the energetic recoils created along the primary ion's path lose some of their kinetic energy by electronic excitation in subsequent collisions. Winterbon *et al.* (1970) have developed analytical procedures, based on the transport equation, that quite adequately predict the mean dimensions (depth, transverse and longitudinal straggling) of the collision cascade; comprehensive tabulations are found for example in Winterbon (1975). An alternative procedure is to use Monte Carlo techniques to simulate individual collision cascades and then average the results from several hundred cascades.

Figure 3 shows the excellent agreement between the analytical and Monte Carlo predictions, indicating that one can predict with some confidence the average dimensions of the collision cascade. In many applications, one is interested in the displaced atom distribution resulting from such a cascade; this requires additional assumptions to be made concerning the threshold displacement energy (E_d) and the probability of subsequent vacancy/interstitial recombination occurring within the cascade. The usual procedure (the Kinchin-Pease method) is to assume that only those recoils receiving an energy greater than E_d become displaced. This leads to a fairly simple relationship (Sigmund, 1969) between $\nu(E)$ and the number N_d of displacements produced within the cascade: namely,

$$N_d = \frac{0.4\nu(E)}{E_d} \qquad\qquad (1)$$

However, in the present energy spike context, we are primarily concerned with the depth distribution of $\nu(E)$ and so can avoid these complications.

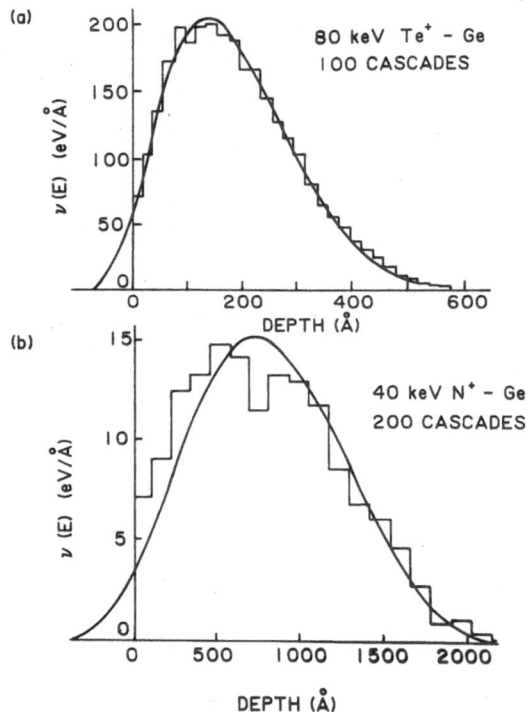

Figure 3:   Comparison of the deposited energy distributions, $\nu(E)$, derived from analytical (solid curve) and Monte Carlo (histogram) methods. (Adapted from Walker and Thompson, 1978). Note the large change in depth scale between the upper and lower Figures, indicating that the heavier ion (Te^+) is producing a much larger θ than the N^+ cascade.

In estimating the appropriate value θ for an individual cascade, it is extremely important to distinguish between the *average* depth distribution of $\nu(E)$ resulting from a large number of cascades (as shown in Figure 3) and the distribution produced within an individual cascade. The latter quantity can best be obtained by Monte Carlo simulation; two examples are shown in Figure 4, with the corresponding average distribution included for comparison. When the incident ion is heavier than the substrate (e.g., the Bi^+ into Si case), the deposited energy distribution within the individual cascade is seen to be distributed fairly uniformly throughout the $\nu(E)$ envelope. In this case, the average $\nu(E)$ distribution provides a reasonable approximation of the individual cascade behavior. However, at the other extreme of mass ratio (e.g., N^+ into Si), we see that the individual cascade occupies only an extremely small fraction of the corresponding $\nu(E)$ envelope and hence θ in each individual cascade will be much larger than that calculated from the volume of the overall $\nu(E)$ envelope.

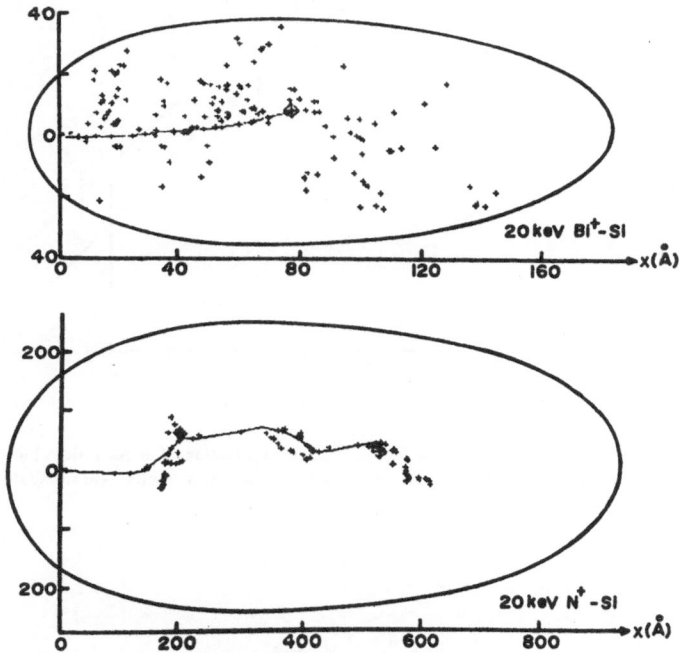

Figure 4:   Relationship between individual (Monte Carlo) cascade dimensions and the average distribution obtained from the transport equation. The elliptical curves represent the $\nu(E)$ contour at 10% of $\nu(E)_{max}$. The primary ion trajectory is indicated by a continuous line and those cascade atoms receiving more than E_d are denoted by +. (Adapted from Walker and Thompson, 1978).

The magnitude of this volume correction factor (V_R) has been evaluated by Walker and Thompson (1978) as a function of mass ratio and their results in Si and Ge are shown in Figure 5. For the Bi^+ into Si case of Figure 4 $(M_2/M_1 \approx 0.13)$, the appropriate volume reduction $(\delta_{corr})^3$ is ~ 0.7 (i.e., not much less than unity), but for the N^+ into Si case the corresponding reduction is almost a factor of 100. Obviously, for low-mass ions, the calculated value of the deposited energy density θ within each cascade can be increased enormously by this effect; in such cases, there may also be considerable local variations in θ from one region of the cascade to another, i.e., so-called sub-cascades.

Taking the above considerations into account and using power-law approximations of the Thomas-Fermi potential to simplify the calculations, Sigmund (1974) and Winterbon (1981) have provided a convenient set of graphs and computer tabulations from which θ for the individual cascade (plus other cascade parameters of interest) may be evaluated for all combinations of Z_1, Z_2 and E. For example, within the $m \approx 1/3$ power law regime (i.e., for $\epsilon \lesssim 0.2$), Sigmund (1974) has shown that $\theta \approx G_3 N^2/E$ where N is the target density in

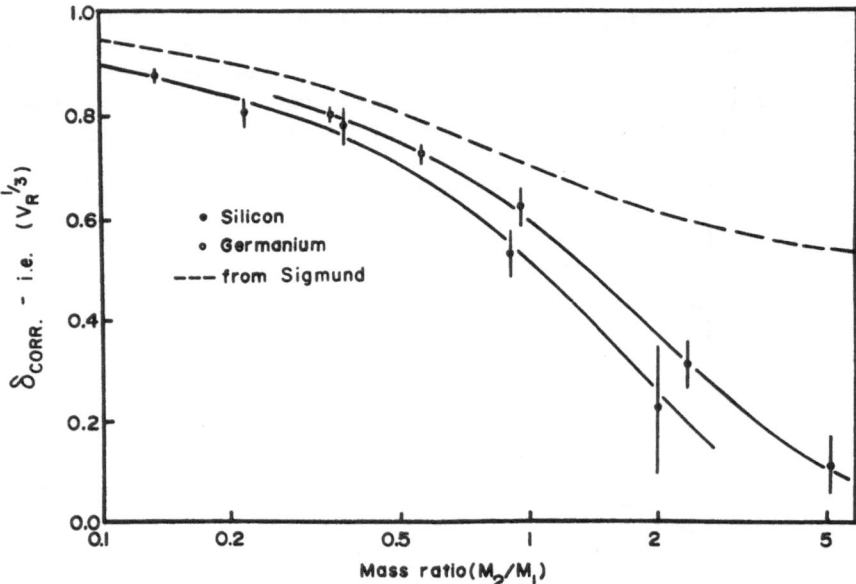

Figure 5:   One-dimensional cascade correction factor, δ_{corr}, as a function of mass ratio. The dashed curve was obtained analytically by Sigmund; the points are Monte Carlo results. (From Walker and Thompson, 1978).

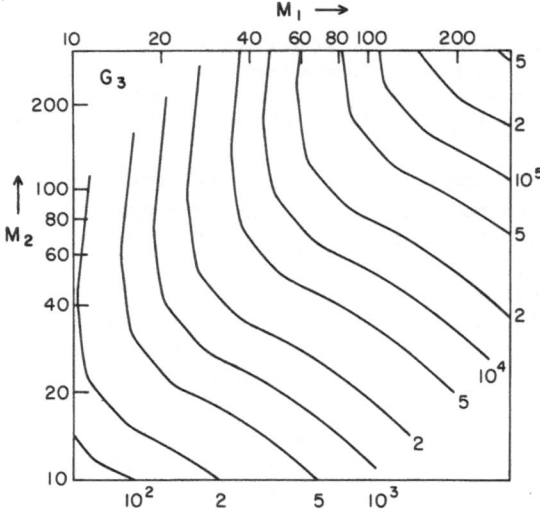

Figure 6:   Contour plots of G_3 (eV Å^6 keV) as a function of M_1 and M_2. (Taken from Sigmund, 1974.)

atoms/Å^3, E is the incident energy in keV and G_3 depends only on M_1 and M_2 (as shown in Figure 6). Within the $m = 1/2$ regime (valid for ϵ values between ~0.2 and 5), θ exhibits an even stronger E dependence, i.e., $\theta = G_2 N^2 / E^2$.

Table II gives some representative values of θ, illustrating its dependence on atomic density and Z_1 and also the strong reciprocal dependence on beam energy. To further illustrate this very strong energy dependence, Sigmund (1974) has taken a typical medium mass case (Te^+ into Ag) and has evaluated the mean deposited energy density as a function of the incident energy (Figure 7). Also included in Figure 7 is Sigmund's estimate of the subsequent quenching rate, based on a high pressure kinetic gas model for estimating the appropriate thermal conductivity value in such a "hot", short-lived spike.

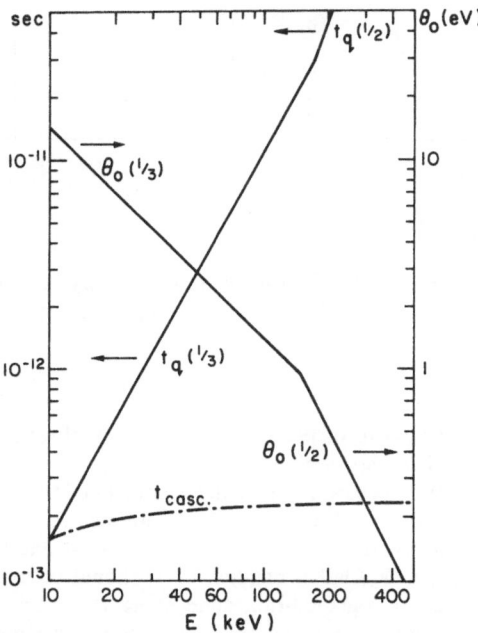

Figure 7: Initial deposited energy density θ_0 versus incident energy E for Ag bombarded with Te^+ ions: for the $m = 1/2$ and $m = 1/3$ regimes t_{casc} is the cascade propagation time (Figure 1), t_q is the time required to quench θ_0 to half of its initial value, and the values in brackets denote the $m = 1/2$ and $m = 1/3$ regimes. (Taken from Sigmund, 1974.)

Provided Z_1 and Z_2 are almost equal, it is possible to develop quite simple scaling laws for estimating θ: for example, in the $m = 1/3$ power law approximation, the ion range expressed in g/cm^2 depends only weakly on Z (namely $R \propto Z^{1/6}$); the resulting expression for θ is then directly proportional to ($\rho^2 Z^{1/2}/E$) where ρ is the target density in g/cm^3. Evidence for this strong dependence on target density is clearly seen in Table II.

In this section, we have considered at some length the problem of estimating θ as a function of various experimental parameters and have seen that under appropriate conditions (i.e., low E, high Z_1 and Z_2) values considerably in excess of 1-2 eV/atom may be achieved throughout the central region of a collision cascade.

TABLE II

Dependence of θ on Energy, Density
and Atomic Number

E (keV)	Z_1	Z_2	ρ (g/cm^3)	θ (eV/atom)
100	Si	Si		0.001
10	Si	Si	2.5	0.1
10	Au	Si		1.0
10	Au	Au	19.3	25.0
10	Au	Bi	9.8	6.0

IV. ION-BEAM MIXING

In each collision cascade, the deposited nuclear recoil energy $\nu(E)$ creates a large number of displaced atoms and the net displacement resulting from this recoil motion is termed ballistic (or cascade) mixing.

A. Individual Cascade Regime

Within a single cascade, the *mean* displacement distance is always insignificantly small: for example Andersen (1979) has shown that each Frenkel pair has an average displacement range R_{recoil} of only ~6–10 Å; consequently, since the ratio of the number (N_d) of displaced atoms (see equation 1) to the total number N_v of atoms within the central core of the cascade is much less than unity, the mean atomic displacement due to ballistic mixing within a single cascade (i.e., $R_{recoil} \times N_d/N_v$) is negligible. It should be noted however that cascade propagation tends to produce a vacancy-rich zone along the track of each energetic recoil atom, with a corresponding interstitial-rich zone towards the perimeter of each subcascade. Hence, from a radiation damage perspective, sufficient mass transport to produce clusters such as vacancy or interstitial loops can occur within a single cascade. Also, after the cascade propagation has ended, there may be considerable diffusive motion resulting from the high concentration of point defects created within the cascade envelope. Such radiation-enhanced diffusion effects depend strongly on such factors as target temperature and dose rate (flux) and are discussed further in Chapters 8 and 10.

At sufficiently high θ values to produce significant spike effects, one might also expect rapid diffusive mixing to occur throughout the hot spike volume. However, the subsequent quenching rate (Section VI) is usually so rapid that in an individual cascade the diffusion length $(D_l t)^{1/2}$ would not exceed a few Å, even for a completely molten spike.

B. Cascade Overlap Regime

So far, we have considered only the mixing that occurs within an individual cascade. However, the doses used in ion-beam mixing (Chapter 9) and sputtering (Chapter 10) are clearly in the cascade-overlap regime of Figure 1. Even at a dose of 10^{16} ions/cm^2, the implanted region receives at least $\sim 10^3$ successive overlapping cascades and, under such conditions, the cumulative effects of ballistic mixing or of pseudo-molten spikes is no longer negligible: for example, simple relationships have been developed (Tsaur et al., 1979), showing that the effective broadening, $(D_I t)^{1/2}$ of a sharp interface due to ballistic mixing effects increases approximately as the square root of the bombardment dose.

Extensive theoretical studies of the predicted magnitude of such ballistic mixing under typical sputtering and ion-beam mixing conditions have appeared in the literature (Littmark, 1980a, 1980b; Sigmund, 1980, 1981; Gras-Marti, 1981). Hence, we will only summarize here the general conclusions to be drawn from these studies.

Since all target atoms within the ion range participate in the ballistic mixing process, it is only the relative motion of marker atoms versus neighboring host atoms that can result in an observable change in the mean depth distribution. Furthermore, at sufficiently high fluences for which all atoms are significantly displaced, simple ballistic mixing calculations show that considerable target density fluctuations would be created by prolonged bombardment. Due to the higher probability of recoil generation at the cascade centers than at the boundaries, there is primarily a dilution of target atoms at depths were the $d\nu(E)/dx$ distribution has its maximum value and a corresponding density increase at lower and deeper depths (see Figure 8). Superimposed on these density variations is a strong dilution near the surface due to sputter ejection and an increase near the mean range of the bombarding ions, since these extra atoms must also be accommodated in the lattice. These density fluctuation effects are far too large for the lattice to accommodate and hence some sort of density relaxation correction must be applied in order to obtain a realistic *depth* scale. One simple model, proposed by Littmark (1980a), is to assume that the lattice relaxes to a constant density, as shown in Figure 8. In reality, the lattice probably does not fully relax to its original (constant) density value; consequently, the depth scale expressed in distance units (Å) depends on the specific relaxation model chosen. It should be emphasized however that, as long as the depth scale is expressed in areal density units (atoms/cm^2), it is essentially unaffected by such relaxation effects and is therefore always valid.

Sputter profiling experiments generally involve a rather low energy beam in which the ion range is shorter than the initial depth of the marker atoms. In such a case, ballistic mixing of host and marker atoms moves the residual peak position of the marker to progressively shallower depths than predicted by sputtering alone, as shown in Figure 9. At the same time, a small but very penetrating tail is also produced in the marker profile. The corresponding sputtering yield of the marker atoms versus the depth removed is shown in Figure 10. Note that, although the peak value has clearly been shifted to a much shallower depth than the original 200 Å position, the overall *mean* depth has been almost unaffected by this ballistic mixing.

Ion-beam mixing experiments, on the other hand, require much higher energy ion beams, whose range is considerably longer than the depth of the interface across which mixing is being studied. In such cases, ballistic mixing shifts the interface position to progressively larger depths than predicted by sputtering alone, as is shown schematically in Figure 11.

A significant non-linear effect at the fluences used in ion beam mixing has been observed by Matteson et al. (1981) in their molecular ion bombardments of 30 Å Pt layers on Si. Using beams of As^+, As_2^+ and As_3^+ at a constant atom dose (5×10^{15} As atoms/cm^2) and a constant velocity (120 keV/atom), they found the amount of mixing to increase strongly with the number of atoms in the molecular ion, thus clearly establishing the failure of the linear collision cascade model to adequately describe the mixing behavior of high θ cascades. In this context, it should be noted that a 200-400 keV Xe^+ beam (typical of many ion beams mixing studies discussed in Chapter 9) produces a θ value comparable to that of the 240 keV As_2^+ beam used in this study (Matteson et al., 1981). Hence, at the doses used in ion beam

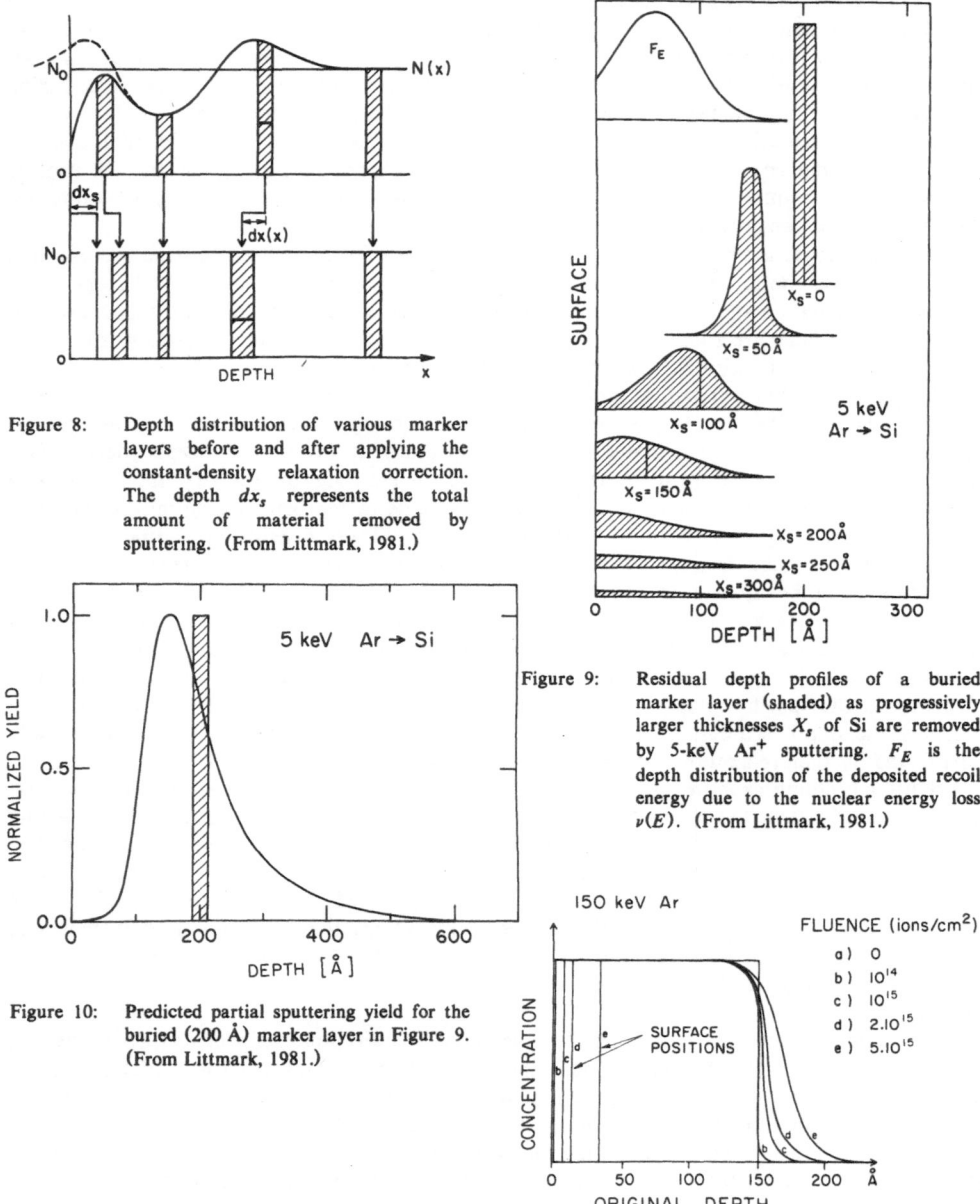

Figure 8: Depth distribution of various marker layers before and after applying the constant-density relaxation correction. The depth dx_s represents the total amount of material removed by sputtering. (From Littmark, 1981.)

Figure 9: Residual depth profiles of a buried marker layer (shaded) as progressively larger thicknesses X_s of Si are removed by 5-keV Ar$^+$ sputtering. F_E is the depth distribution of the deposited recoil energy due to the nuclear energy loss $\nu(E)$. (From Littmark, 1981.)

Figure 10: Predicted partial sputtering yield for the buried (200 Å) marker layer in Figure 9. (From Littmark, 1981.)

Figure 11: Schematic representation of the collisional mixing change in interface position as a function of the total dose (fluence) of 150-keV Ar$^+$ ions. In each case, the depth scale has been converted to the original scale by adding a correction for the amount of sputtered material. Experimentally this is observable by comparison with a deep-lying, undisturbed marker. (From Littmark, 1981.)

mixing, spike effects may frequently play a significant role, especially in high-Z targets such as Au and Pt.

It should be emphasized that, in this section, we have considered only those effects resulting directly from the collision cascade process: i.e., ballistic mixing and (at high θ) energy spike effects. However, in many systems, additional and often much larger mixing contributions may result from post-cascade effects such as radiation-enhanced diffusion. The importance of these temperature-dependent effects is discussed further in the next three chapters (8-10).

V. EXPERIMENTAL EXAMPLES OF SPIKE EFFECTS

In Section III, we saw that appropriate conditions of low energy and high atomic number can produce θ values of several eV/atom (or even higher) within the central core of a collision cascade. Before discussing the difficult question of the subsequent rapid quenching processes involved in such small (10-100 Å) cascades, let us first look at some experimental examples in which spike effects appear to be playing a significant role. The concept of a thermal spike is a fairly old and controversial topic, and over the past 25 years, a wide variety of experimental results have been attributed to it. In many cases, subsequent work showed that other explanations were involved. In recent years, however, rather compelling evidence of significant spike effects has been found in at least three widely different types of experimental study.

A. Implantation Damage in Semiconductors

In ion-implanted semiconductors such as Si and Ge, the observed damage level (N_d) at low temperature can be almost an order of magnitude *greater* than that predicted from linear collision cascade theory, especially for high-Z ions such as Tl (Figure 12). Note that the high-energy portion of each curve bends over to approximately the same slope as the N_{kp} line, indicating that the rate of damage creation at high energy agrees well with collision cascade theory and that the enhanced value of N_d is entirely due to the low-energy portion of each ion track, where the deposited energy density becomes extremely high. The reciprocal slope ($d\nu(E)/dN_d$) of the curve in Figure 12 is a measure of the effective energy required to create each displaced atom. Obviously, at high $\nu(E)$, this must correspond closely to the displacement energy E_d (14 eV in Si), since all the curves become approximately parallel to the N_{kp} line. But, at low $\nu(E)$ and high Z_1 (i.e., high θ), the reciprocal slope is seen to fall rapidly; in the case of Tl, the limiting slope eventually approaches the heat of melting value (0.7 eV) for Si.

In a closely related electron microscopy study, Howe and Rainville (1981) found that at sufficiently high θ (i.e., ≥ 0.5 eV/atom) each implanted ion produced an observable amorphous zone whose radius is in excellent agreement with that derived from the RBS measurement of N_d. This strongly suggests that each high density cascade is initially a molten zone which subsequently quenches rapidly enough to remain amorphous.

In this context it should be noted that the solidification rate (due to the extremely rapid quenching of the small cascade volume) is of the order of 100 Å in $\sim 10^{-12}$ sec i.e., 10^3-10^4 m/sec. As discussed elsewhere in this volume (Chapters 3,4), the rate of epitaxial regrowth in a typical pulsed laser annealing experiment is only ~ 5 sec. Furthermore, there are theoretical reasons for believing that the maximum epitaxial regrowth rate in Si cannot exceed ~ 20 m/sec. Hence, the extremely rapid quenching of a collision cascade spike would be expected to produce a non-epitaxial (i.e., amorphous) region in Si or Ge.

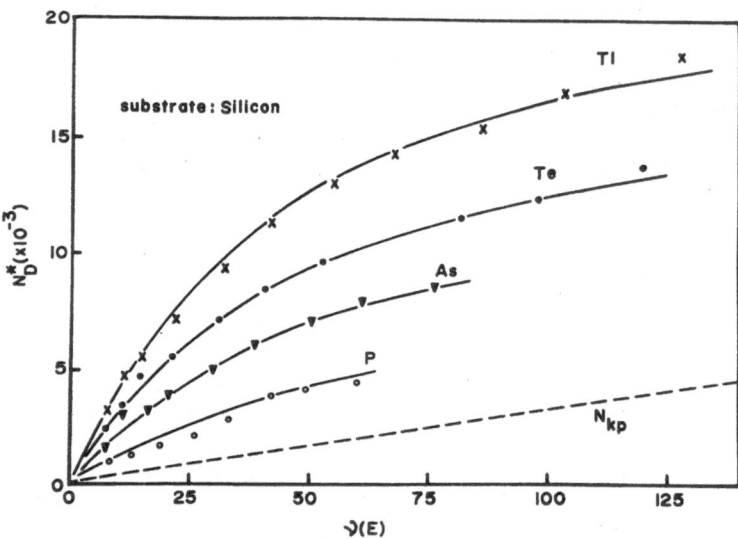

Figure 12: The observed total number N_d^* of displaced atoms per cascade versus $\nu(E)$ in silicon at 35K, as determined by the RBS/channeling technique (Thompson and Walker, 1978). The dashed line N_{kp} is the value predicted by linear cascade theory and the Kinchin-Pease relation (equation 1).

B. Sputtering of Metals

Simple collision cascade theory predicts that the observed sputtering yield should increase linearly with the deposited energy density (θ_s) in the surface region of the cascade. Strong deviations from this linear dependence have been reported, firstly by Andersen and Bay (1973) in a series of comparisons of diatomic versus equal-velocity monatomic ion bombardments (e.g., 50-keV Te_2^+ versus 25-keV Te^+) and more recently by Thompson (1981) in an extensive sputtering study of Ag, Au and Pt targets (Figure 13). Two clearly resolved regions of behavior are observed. At low energy densities, the predicted linear dependence from collision cascade theory agrees well with the experimental data. But, at higher θ_s values, the data in each case exhibit a sharp transition to a much stronger than linear dependence, indicating the onset of some sort of collective process such as a thermal spike.

One of the more puzzling and controversial early investigations of possible spike effects was the study by Nelson (1965) of the temperature dependence of the sputtering yield in various metals. In each case, as the melting point was approached, he observed a sharp (exponential) increase in sputtering yield (Figure 14) which he attributed to some sort of spike-induced evaporation process. An increase in substrate temperature would certainly reduce somewhat the energy/atom required to achieve a given θ_s value; hence, the size of the resulting spike and therefore the sputtering yield should both increase with substrate temperature. However, Nelson's observed temperature effects are far too large and too sudden to be explained by such a simple mechanism. The integrated heat capacity over the temperature ranges in Figure 14 is so small (compared to the latent heats of fusion and evaporation) that one would only predict at most a few % increase in the thermal sputtering component. Furthermore, for the spike dimensions and temperatures proposed by Nelson, his experimentally observed yield values are many orders of magnitude greater than the maximum possible spike evaporation rate.

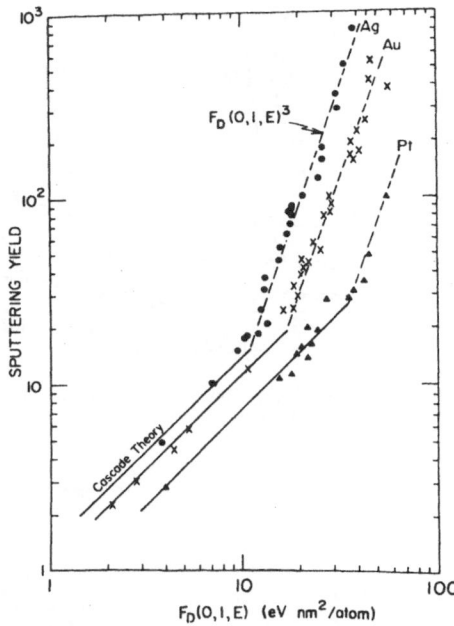

Figure 13: Sputtering yields for Ag, Au and Pt bombarded with various monatomic and polyatomic ions as a function of the surface deposited energy density F_D (i.e., θ_s) obtained by Monte Carlo calculations. (From Thompson, 1981 - note that in his paper the abscissa contains a scaling error of 10^4 which at the author's request we have now corrected (Thompson, 1982)).

Hofer (1981), in a similar study of the temperature dependence of the sputtering yield of Ag by low-energy Ar^+ and Xe^+, has recently obtained an almost identical curve to that in Figure 14. However, he finds that this enhanced sputtering at $T \geq 900$ K persists even when the ion beam is turned off, indicating that, for Ag at least, the anomalous temperature dependence has nothing to do with collision cascades or thermal spikes, but is merely the onset of significant thermal evaporation from the entire target surface. A quick inspection of the various threshold temperatures in Figure 14 shows that the corresponding metal vapor pressures all exceed ~10^{-4} torr; hence, continuous thermal evaporation may well be the dominant process in every case studied by Nelson (1965). A quick inspection of the various threshold temperatures in Figure 14 shows that the corresponding metal vapor pressures all exceed ~10^{-4} torr; hence, continuous thermal evaporation may well be the dominant process in every case studied by Nelson (1965).

C. Frozen Gas Erosion by MeV Ions

Our third and final example of spike effects involves a completely different ϵ-regime, namely the sputtering of frozen gases by MeV low-Z ions such as $^4He^+$. Here, the ratio of $\nu(E)/\eta(E)$ is extremely small (~10^{-3}), as shown in Figure 2. Hence the sputtering yield due to atomic collision processes is usually negligible, as can be seen by extrapolation of the curves of Figure 15. In metals, the energy deposited via electronic excitation ($\eta(E)$) is rapidly dissipated throughout the lattice and so does not contribute significantly to any spike effects. However, in certain non-metallic materials such as frozen gases, ice, and UF_4, very

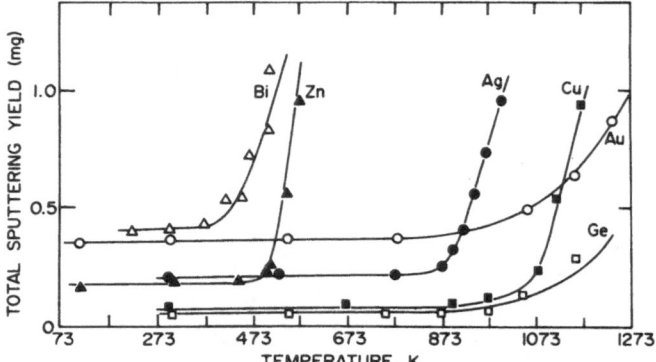

Figure 14: Temperature dependence of the total sputtering yield (weight loss) for several metal targets bombarded with 45-keV Xe^+ to a fluence of $2.9 \times 10^{16}/cm^2$. (From Nelson, 1965.)

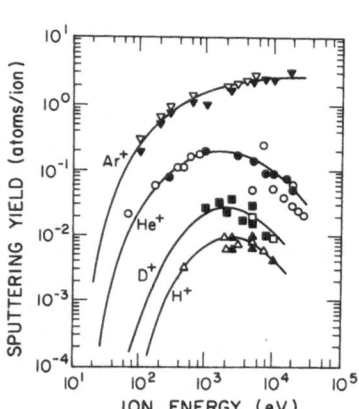

Figure 15: Sputtering yields versus ion energy for various low-Z ions on stainless steel.

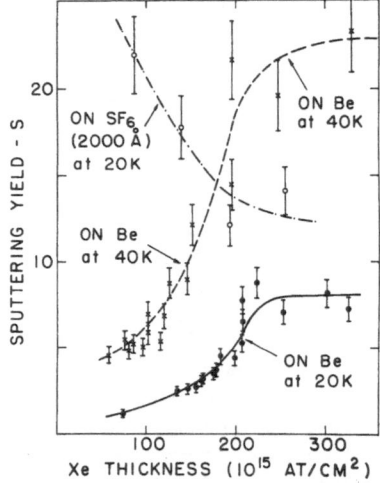

Figure 16: Thickness and substrate dependence of the sputtering yields S for 1.0 MeV He into Xe. (From Ollerhead et al., 1980.)

large sputtering yields are observed (Table III) and these yields generally increase approximately as the *square* of the total deposited energy density. Presumably, in these electrically insulating materials, the $\eta(E)$ contribution is not so rapidly dissipated as in metals and may perhaps contribute to some sort of spike effects.

In the case of the rare gases (Xe, Kr and Ar), the observed yield exhibits a strong dependence on the thickness of the frozen gas and also on the thermal conductivity of the underlying substrate, as can be seen in Figure 16. This behavior is interpreted by Ollerhead et al. (1980) as strong evidence that in these cases the sputtering is due to some sort of thermal spike process. The observed enhancement in Xe sputtering yield on raising the substrate temperature from 25K to 40K (Figure 16) is puzzlingly large for a spike effect,

TABLE III

Sputtering Yields for MeV ^{4}He Ions

Target (~20K)	S (atoms/ion)	Reference
Xe	10	Ollerhead *et al.* (Chalk River)
Kr	20	Ollerhead *et al.* (Chalk River)
Ar	100	Bøttiger *et al.* (Aarhus)
Ice (H_2O)	10	Brown *et al.* (Bell Labs)
CO_2	100	Brown *et al.* (Bell Labs)
N_2	300	Pirronello *et al.* (Catania)
UF_4 (4 MeV $^{16}O^+$)	4	Griffith *et al.* (Cal. Tech.)
Various metals	10^{-3}	

for the same argument as we advanced previously in relation to Figure 14. However, a more careful study of the Ar temperature dependence reported by Bøttiger *et al.* (1981) shows that the observed sputtering yield is completely independent of temperature (Figure 17) until one reaches a threshold temperature of ~24K for Ar, (~50K for Xe), above which thermal evaporation losses during the time of the experiment become significant. Even at low T, Bøttiger observed a significant increase in the sputtering yield at high current density, indicating that beam heating effects can raise the temperature of the entire frozen Ar film above the 24K threshold for evaporation losses to become significant. A similar beam heating effect may have been responsible for the anomalous temperature dependence in Figure 16, since considerably higher beam currents were involved in this earlier work.

From the examples presented above, we see that large non-linear spike effects can be observed under widely different ion bombardment conditions.

VI. THERMAL SPIKE CONCEPTS - AND COMPLICATIONS

Even when θ within the collision cascade far exceeds the characteristic threshold for some specific process (e.g., melting or evaporation), there are still several criteria to be met for a simple thermal spike concept to have quantitative significance. Firstly, the cascade dimensions must be large enough for statistical dynamics to apply: for example, the diameter of a 10 keV Au^+ into Au cascade is only ~20 Å and hence would contain less than 500 target atoms. A second closely-related requirement to this is that the thermal gradient across the spike should be reasonably small. In the above Au^+ case, where the initial value of θ (~25 eV/atom) is roughly 10^3 greater than that of the surrounding substrate, the initial "thermal" gradient would be enormous, i.e., almost a factor of 10 per lattice spacing. Finally, the quenching time constant $t_q = R^2/4D$ (where R is the ion range and D is the thermal diffusivity of the target) must be larger than the time required for electron/phonon coupling (i.e., $\geq 10^{-11}$ sec); otherwise, the spike will require two different "temperatures" to describe it

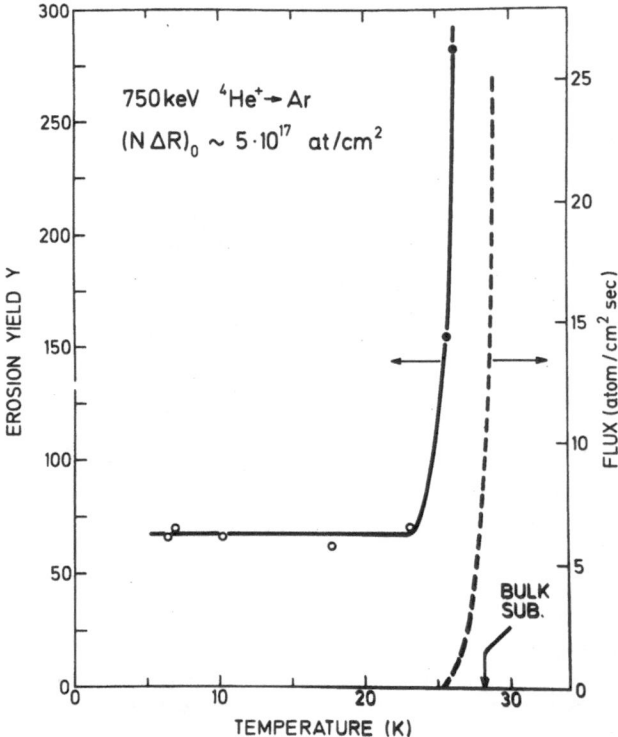

Figure 17: Temperature dependence of the erosion yield Y of Ar by 750 keV ^{4}He$^+$ ions. (From Bøttiger, 1981.)

properly: one for the electron system, and a quite different one for the atomic motion. Under such circumstances, the choice of an appropriate D value to describe the "thermal diffusivity" presents a severe problem.

A comprehensive review of the theoretical basis for treating these high density cascade effects, including the relative merits of the various spike models (thermal spike, displacement spike, shockwave and ionization spike) has recently been prepared by Thompson (1981). Generally speaking, quantitative treatment in terms of simple thermal diffusivity is not possible, as the above criteria are rarely fulfilled within the small dimensions of a typical collision cascade. However, to give some insight into the problems involved, Winterbon (1981) assumes the validity of Boltzman statistics and bulk thermal diffusivity concepts and uses them to calculate the magnitude of the various spike parameters (spike size, temperature gradient and quenching rate) for each of the experimental systems in Section V. The results are summarized in Tables IV and V.

In the case of the enhanced implantation damage in semiconductors (Table IV), we may establish a reasonable upper limit to the size of the "molten spike" by dividing the total available upper limit recoil energy $\nu(E)$ by the enthalpy of melting ΔH (eV/atom). Note that these ΔH values include the integrated heat capacity from the substrate temperature up to the melting point. The resulting spike radius (r_{max}) agrees remarkably well with the size (r_{exptl}) of the amorphous zones measured by electron microscopy (Howe and Rainville, 1981). In both Si and Ge, if we use standard thermal diffusivity data, the estimated magnitude of

TABLE IV

Non-Linear Cascade Damage Parameters

	Si	Ge
Ion Beam	30 keV	30 keV
	Bi_2^+	Bi_2^+
$\nu(E)/E$	0.8	0.8
ΔH (eV/atom)	0.7	0.6
D (cm²/sec)	0.23	0.15
r_{max} (Å)	56	65
t_q (10^{-12} sec)	0.4	0.7
dT/dx (K/Å)	30	20
r_{exptl} (Å)	26	33

the quenching time (t_q) to reduce θ to half its initial value is less than 10^{-12} sec. This is probably far too short for electron phonon equilibration to occur; however, since most of the energy is deposited directly into atomic motion, t_q is still sufficiently long to create a random liquid-like distribution of atoms. It should be emphasized, nevertheless, that 10^{-12} sec is far too short for significant diffusive mixing to occur within the spike volume since even for a hot liquid, the quantity $(D_l t_q)^{1/2}$ would not exceed 1-10 Å.

In treating enhanced sputtering (Table V), we estimate the approximate surface area of the spike by dividing the initial rate of energy loss (dE/Ndx) of the incident beam by the enthalpy of melting ΔH. These ΔH values include the integrated heat capacity from the substrate temperature to that at which the vapor pressure becomes ~1 atmosphere. For MeV ion beams, where the main component of dE/dx is electronic excitation, this estimate of r_{max} is far too large unless some mechanism exists for converting electronic excitation into thermal motion more rapidly than t_q. In the case of Xe (and the other inert gas targets), the estimated value for t_q (~10^{-10} sec) is large enough that electron phonon equilibrium (~10^{-11} sec) probably could occur. It is interesting to note that these inert gas targets are the only materials in which the calculated evaporation yield ($S_{thermal}$) from the spike is comparable in magnitude to the observed sputtering yield. For all the other cases in Table V, t_q is too short and the thermal gradients are so large that simple thermal diffusivity concepts would certainly not apply. Nevertheless, the experimental sputtering yields are all several orders of magnitude greater than the values ($S_{cascade}$) predicted by linear collision cascade theory, indicating that spike effects of some sort must be dominant in every case.

It is intriguing to note that simple thermal spike considerations would predict the sputtering yield in Xe to be at least a factor of 10^3 greater than that in ice or in UF_4, and yet

TABLE V

Non-Linear Sputtering Cascade Parameters

Target	Au	Xe	H_2O (ice)	UF_4
Ion Beam	90 keV Sb_3^+	1 MeV He^+	1 MeV He^+	4 MeV $^{16}O^+$
$\nu(E)/E$	0.8	0.005	0.003	0.004
ΔH (eV/atom)	1	0.06	0.2	2.0
D (cm^2/sec)	0.7	0.002	0.01	0.01
r_{max} (Å)	80	100	30	20
t_q (10^{-12} sec)	0.3	150	2	1
dT/dx ($K/Å$)	100	1	10	200
$S_{cascade}$ (atoms/ion)	40	0.02	0.001	0.01
$S_{thermal}$ ($\propto \pi r^2 \cdot t_q$)	0.01	20	0.01	<0.005
$S_{observed}$ (atoms/ion)	600	10	10	4

the measured S values are almost identical (Table V). Clearly, there must be some drastically different and more energy-efficient spike mechanism involved in the latter two materials: perhaps a coulomb "explosion" mechanism due to the high ionization density along the incident ion trajectory is playing a role (Brown *et al.*, 1980).

For metal targets such as Au, there is a further complication in selecting an appropriate thermal diffusivity value, D. Since t_q is far too short for electron phonon equilibration to occur, and since $\nu(E)$ is the dominant energy deposition process in this case, we should probably replace the metallic thermal diffusivity value of 0.7 cm^2/sec by a D value more typical of an insulator (i.e., $\sim 10^{-2}$ cm^2/sec). This would increase considerably the estimated spike contribution to the sputtering yield, i.e., ~ 1 atom/ion, but this is still almost a factor of 10^3 less than $S_{observed}$.

VII. CONCLUSIONS

Despite the difficulty in developing a quantitatively reliable basis for estimating spike phenomena, there are still several important conclusions to be drawn.

(i) Spike effects do exist - and are widespread whenever the deposited energy density θ within the cascade approaches the eV/atom level.

(ii) The quenching time t_q of the spike is sufficiently long for crystalline random-liquid transitions to occur and also (in some cases) for significant evaporation (i.e., sputtering).

(iii) A t_q value of $\sim 10^{-12}$ sec is far too short for significant diffusive mixing to occur.

(iv) The choice of thermal diffusivity (D) is a key problem because t_q is usually shorter than the equilibrium time between atomic and electronic temperatures.

(v) Theoretical estimates are not yet of much quantitative value.

Finally, we should emphasize that the present chapter is primarily concerned with one specific energy density regime: namely, the spike effects that occur in high deposited-energy-density cascades where linear collision-cascade theory is no longer applicable. At lower energy densities, the situation is in much better shape in that quantitative estimates of implantation damage, sputtering and ballistic mixing effects can usually be obtained directly from the well-established linear collision cascade theory.

REFERENCES

Andersen, H. H., and Bay, H. L., (1973) Rad. Eff. *19*, 139.

Andersen, H. H., (1979) Appl. Phys. *18*, 131.

Baglin, J. (1981), oral contribution at the Trevi Institute on Surface Modification and Alloying.

Bøttiger, J., (1981) oral contribution at the Trevi Institute on Surface Modification and Alloying.

Brown, W. L., Augustyniak, W. M., Brody, E., Cooper, B., Lanzerotti, L. J., Ramirez, A., Evatt, R. and Johnson, R. E., (1980) Nucl. Instr. Meth. *170*, 321.

Gras-Marti, A. and Sigmund, P., (1981) Nucl. Instr. Meth. *180*, 211.

Griffith, J. E., Weller, R. A., Seiberling, L. E., and Tombrello, T. A. (1980) Rad. Eff. *51*, 223.

Hofer, W., (1981) oral contribution at the Trevi Institute on Surface Modification and Alloying.

Howe, L. M., and Rainville, M. H., (1981) Nucl. Instr. Meth. *182/183*, 143.

Lindhard, J., Scharff, M., and Schiott, H. E., (1963a) Kgl. Danske Vid. Selsk. Mat. fys. Medd. *33*, No. 14.

Lindhard, J., Nielsen, V., Scharff, M. and Thompsen, P. V., (1963b) Kgl. Danske Vid. Selsk. Mat. fys. Medd. *33*, No. 10.

Lindhard, J., (1976) private communication.

Littmark, U. and Hofer, W. O., (1980a) Nucl. Instr. Meth. *168*, 329.

Littmark, U. and Hofer, W. O., (1980b) Nucl. Instr. Meth. *170*, 177.

Littmark, U., (1981), oral contribution at the Trevi Institute on Surface Modification and Alloying.

Matteson, S., Paine, B. M., Grimaldi, M. G., Mazey, G. and Nicolet, M. A., (1981) Nucl. Instr. Meth. *182/183*, 43.

Nelson, R. S., (1965) Phil. Mag. *11*, 291.

Ollerhead, R. W., Bøttiger, J., Davies, J. A., L'Ecuyer, J., Haugen, H. H., and Matsunami, N., (1980) Rad. Eff. *49*, 203.

Pirronello, V., Strazzulla, G., Foti, G. and Rimini, E. (1981) Nucl. Instr. Meth. *182/183*, 315.

Sigmund, P., (1969) Appl. Phys. Lett. *14*, 114.

Sigmund, P., (1974) Appl. Phys. Lett. *25*, 169 (+ an erratum correction in *27*, 52).

Sigmund, P. and Gras-Marti, A., (1980) Nucl. Instr. Meth. *168*, 389.

Sigmund, P. and Gras-Marti, A., (1981) Nucl. Instr. Meth. *182/183*, 25.

Thompson, D. A., and Walker, R. S., (1978) Rad. Eff. *36*, 91.

Thompson, D. A., (1981) Rad. Effects *56*, 105.

Thompson, D. A., (1982) private communication.

Tsaur, B. Y., Matteson, S., Chapman, G., Liau, Z. L. and Nicolet, M. A., (1979) Appl. Phys. Lett. *35*, 825.

Walker, R. S. and Thompson, D. A., (1978) Rad. Effects *37*, 113.

Winterbon, K. B., Sigmund, P. and Sanders, J. B. (1970) Kgl. Danske Vid. Selsk. Mat. fys. Medd. *37*, No. 14.

Winterbon, K. B., (1975) Ion Implantation Range and Energy Distributions, Vol. 2, Low Energy (Plenum Press, New York).

Winterbon, K. B., (1981) 20-p. computer printout, private communication.

CHAPTER 8

SOLUTE REDISTRIBUTION AND PRECIPITATE STABILITY: POINT-DEFECT MEDIATED EFFECTS

A. D. MARWICK

IBM Research Center, Yorktown Heights, New York

CONTRIBUTOR: H. Wiedersich

I. INTRODUCTION

Ion implantation or bombardment carried out to modify a surface will cause radiation damage. Point-defects are produced when atoms are displaced from their sites in a crystalline solid by energetic collisions, and the point defects can diffuse if the temperature is sufficiently high. Defect migration allows atomic diffusion. In this chapter the consequences of the atomic diffusion due to the migration of irradiation-produced point defects will be explored.

The important difference between defect-mediated processes (i.e., those in which migration of point-defects plays an important part) and those caused by direct atomic displacement (such as displacement mixing) is that the former are thermally activated. Therefore, thermodynamic concepts are useful, even though they must be applied with caution. Although an alloy irradiated at an elevated temperature is very far from its equilibrium state, the evidence so far indicates that the phases observed are those known from the equilibrium phase diagram (Williams and Titchmarsh, 1979). However, the phases observed may not be those expected for the mean sample composition and the irradiation temperature, because solute drift, discussed below, changes the composition of small volumes within the material.

II. MECHANISMS

This section will review the processes by which solute transport occurs under irradiation. The solute may have been introduced by implantation, or be native to the alloy being irradiated. The irradiation may be done deliberately to introduce damage, but damage also inevitably accompanies ion-implantation. Much surface processing is done at ambient or only moderately elevated temperatures; indeed the ability to modify surface properties without high-temperature processing may be an important advantage of ion implantation. Unfortunately most of the existing body of knowledge of the metallurgical changes brought about by irradiation damage relates to high temperatures at which power-producing nuclear reactors operate. This regime is in many ways simpler than the moderate temperature regime, and certainly much more is known about it. Nevertheless it is still an active topic of research within the nuclear technology field. A recent workshop discussed these applications in detail (Gatlinburg, 1979).

A. Diffusion Under Irradiation

1. Radiation-Enhanced Diffusion

Vacancy-interstitial pairs are generated by displacement of target atoms. The target atoms may have been struck directly by the projectile, or by another target atom set in motion in a collision cascade. The rate at which vacancy-interstitial pairs are formed is the damage rate K, given by the Kinchin-Pease formula in units of displacements per atom per second (dpa/sec):

$$K = \frac{0.4}{NE_d} \phi \frac{\partial E_{dam}}{\partial x}$$ (0)

where N is the atomic density, E_d is the displacement energy, ϕ is the flux of bombarding particles, and $\partial E_{dam}/\partial x$ the damage energy deposited into the target per unit path length of the projectile. Codes for calculating K for ion bombardment of various structures exist, although unfortunately they do not usually embody all existing theoretical knowledge of damage and penetration phenomena. It should be noted that the displacement energy E_d is *not* the same as the displacement threshold. It is essentially a phenomenological parameter which determines the slope of a curve of K vs. $\partial E_{dam}/\partial x$. A value of 40 eV for Fe and Fe-based alloys has been adopted by the radiation-damage community (Norgett *et al.*, 1975).

Vacancy-interstitial pairs produced in collision cascades tend to be generated in clusters and it is thought that their recombination is thereby enhanced. Therefore the displacement rate K probably overestimates the production rate of defects which eventually diffuse over significant distances. Low temperature measurements of the Frenkel-pair generation rate (Averback *et al.*, 1978; Kirk and Greenwood, 1979; Coltman *et al.*, 1981) indicate that 30% or less of Frenkel pairs survive annealing in the last stages of collision cascades. Even fewer may survive at high irradiation temperatures.

  a. Diffusion in pure metals. Once generated, point defects can diffuse at a rate determined by the target temperature. Interstitialcies in pure materials are usually much more mobile than vacancies, although in the Ag-Zn alloy studied by Hillairet (see Hillairet, 1978) interstitials are only about as mobile as vacancies. Mobile defects can be annihilated by recombination (e.g., a vacancy with an interstitialcy), by diffusion to sinks (surfaces, grain boundaries, dislocations, etc.) or by clustering. The mean concentration of defects in a material may be calculated approximately from rate equations. For a simple pure material in which clustering of defects can be neglected, the concentrations of vacancies and interstitialcies are governed by the following equations:

$$\frac{\partial C_\nu}{\partial t} = K - RC_\nu C_i - k_\nu^2 D_\nu C_\nu \tag{1}$$

$$\frac{\partial C_i}{\partial t} = K - RC_\nu C_i - k_i^2 D_i C_i \tag{2}$$

Here R is the recombination coefficient, k^2 the sink density and D the diffusion coefficient of the defect. Definitions of these quantities can be found elsewhere (Marwick, 1981). Suffice it to say that the above equations can be readily solved to give point-defect concentrations, and therefore the self-diffusion coefficient for the irradiated metal. To this should be added the diffusion coefficient due to thermally generated vacancies. The concentration of thermally generated interstitialcies is negligible in metals.

  The total radiation-enhanced self-diffusion coefficient given approximately by

$$D_{\text{irr}} = D_\nu C_\nu + D_i C_i, \tag{3}$$

(see Sizmann (1978) for more detail) is plotted in Figure 1 for the case of Ni. Also shown is the diffusion coefficient due to thermally-generated vacancies. The enhancement of diffusivity due to irradiation can be many orders of magnitude. In this plot steady-state values of the defect concentration have been used, and the saturation of vacancy concentration at low temperature has been neglected. These and other complications are discussed in following sections of this chapter.

  The self-diffusion coefficient D_{irr} given by equation (3) exhibits well-known scaling behavior once defect concentrations have reached steady state. The variation of D_{irr} with temperature or dose rate depends on which of the two loss mechanisms embodied in equations (1) and (2) is dominating. Where loss to fixed sinks (or any other first-order process) predominates, D_{irr} is independent of temperature and scales as the first power of the ion flux. Therefore, diffusion lengths are independent of ion flux. At lower temperatures defect concentrations are higher, and recombination may be the dominant loss mechanism, in which case the radiation-enhanced diffusion coefficient is temperature dependent, and is proportional to the square-root of the ion flux. The different dependences on temperature and flux of the diffusion coefficient in these two regimes can be used to distinguish them from one another, in principle.

  From equations (1) and (2) it may be readily shown that fixed sinks dominate defect loss when their density is above a critical value k_c^2 given by

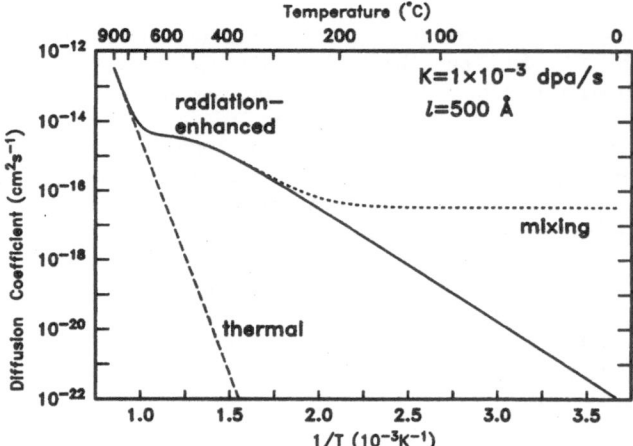

Figure 1:   The mean radiation-enhanced diffusion coefficient in a 500 Å thick foil of Ni at a damage rate typical of
ion bombardment, calculated from the analytical solution to the rate equations due to Lam *et al.*
(1974). A sink density of $10^{10}/cm^2$ was assumed. The curve shown represents a somewhat idealized
picture, in that saturation of the vacancy concentration and time-dependence of the defect concentrations
at low temperature have been neglected. The thermal diffusion coefficient and the diffusion coefficient
due to ballistic mixing (see text) are also shown for comparison.

$$k_c^2 \sim (K/d_r a^2)^{1/2} \tag{4}$$

where a is the edge length of the unit cell. For example, in Al at 100°C and a damage rate
of 10^{-2} dpa/sec, a grain size of about 100 Å would be needed if grain boundaries were to
dominate the loss of point-defects.

The curves in Figure 1 represent a somewhat idealized situation and are not strictly
applicable to any but very carefully designed self-diffusion experiments. Among the
complicating factors met with in more practical cases are trapping of point-defects at solute
atoms in alloys, the non-achievement of steady state in low temperature irradiations (both
discussed in the next section), clustering of point-defects, and time-dependent or
temperature-dependent sink concentrations. The latter arise when (as is often the case) the
ion beam generates significant numbers of defect sinks, whose density may well be
temperature dependent because of nucleation or annealing during irradiation. Therefore a
temperature-dependent diffusion coefficient may be observed even where most point-defects
are lost to sinks, in contrast to the predictions of the simplest theory.

b. Diffusion in alloys.   Since surface modification with ion beams often involves alloying,
or bombardment of alloys, it would be useful to be able to predict solute diffusion coefficients
in irradiated alloys with reasonable accuracy. Unfortunately a host of processes which
complicate the picture presented so far are known to occur in alloys. Some of these are
outlined in this section.

An important effect in alloys is interaction of defects with solute atoms. In general, these
interactions reduce the radiation-enhanced diffusion coefficient by hindering the migration of
defects and therefore enhancing recombination of vacancies and interstitials. However, the
opposite effect is possible: some solutes enhance the rate of vacancy migration (Le Claire,
1978). In dilute alloys, solute-defect interactions can be regarded as a perturbation of pure-
metal behavior. In a concentrated alloy, on the other hand, a different approach is required.

The two cases will be considered in turn.

(i) *Dilute Alloys.* Perturbation of Cr diffusion in dilute implanted Ni-Cr and Ni-Cr-Si alloys by the presence of Si in the latter was studied by Piller and Marwick (1979). They found that the diffusion of Cr under He ion bombardment was greatly reduced by the presence of 0.33 at. %Si in the Ni-Cr, and they concluded that Si atoms in solution were trapping point-defects, enhancing recombination, and lowering the concentration of free defects available to cause Cr diffusion.

Piller and Marwick modelled the effects of static (i.e., not mobile) traps on radiation-enhanced diffusion by calculating the diffusion coefficient given in Eq. (3) in the presence of 0.3 at.% of traps, with varying binding energies. The damage rate was 3×10^{-4} dpa/sec, in their experiments. The calculation followed the time-dependence of the defect concentrations, and the diffusion coefficient at the end of 1800 sec of irradiation was derived. Results are shown in Figure 2.

The diffusion coefficients calculated from the static trap model have a different dependence on temperature from those calculated for a pure material at steady state (Figure 1) because of two factors: trapping of defects at solute atoms, which is particularly effective at moderate-to-high temperatures; and saturation of the vacancy concentration at low temperatures, which leads to constant diffusivities by interstitialcy diffusion.

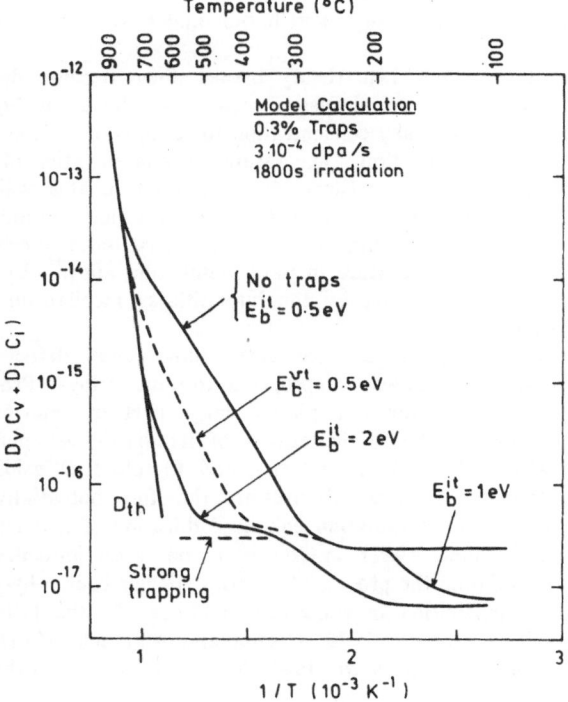

Figure 2:   Model calculations of self-diffusion coefficients in an irradiated dilute alloy containing 0.3 at.% of immobile traps for different trap binding energies (Piller and Marwick 1979). The diffusion coefficient is reduced when trapping is sufficiently strong, because the concentrations of free point-defects are reduced.

The effectiveness of traps in reducing diffusion depends on whether they trap vacancies or interstitials. As can be seen from Figure 2, interstitial traps with binding energies less than about 2 eV have little effect at moderate temperature. On the other hand, even weak vacancy traps reduce D_{irr} considerably. With high trap binding energies or large damage rates, the traps fill up with trapped defects until they reach 50% occupancy (Brailsford and Bullough, 1978). At this stage recombination at the traps is a first order process, and the radiation-enhanced diffusion coefficient becomes

$$D_{irr} = \frac{1}{3} a^2 K / C_u \tag{5}$$

where a is the cubic unit edge length, and C_u is the solute (trap) concentration. This limit is plotted in Figure 2 as the line marked "strong trapping."

Although the calculation just described assumed that the traps were immobile, this is probably not so in most cases where the traps are solute atoms, though it applies where the traps are ordered zones, phase boundaries or solute-atom clusters. Atomic traps associated with a defect are expected to be mobile; the solute-atom/point-defect pair is called a complex. If the complex is mobile, the solute atom diffuses by diffusion of the complex. A diffusion coefficient for Si in a Ni-Si alloy has been measured by Potter et al. (1979). The speed of diffusion of the complex affects the efficiency of trapping. A model of the interaction of trapping and solute diffusion in complexes which gives good qualitative agreement with experiment has been given by Johnson and Lam (1976). The diffusion and breakup of complexes is one mechanism of solute redistribution under irradiation, described in a later section.

At lower temperatures simple rate theory breaks down because defect concentrations saturate due to clustering, and the diffusion coefficients shown in Figure 2 are mainly determined by the saturation vacancy concentration for temperatures less than about 250°C, under the conditions of the calculation. The saturation concentration of vacancies is about 0.1% (Birtcher and Blewitt, 1978). Since C_v is limited in this way the interstitialcy concentration is enhanced, because otherwise most interstitialcies would be annihilated by recombination with vacancies. Self-diffusion by the interstitialcy mechanism is therefore expected to dominate at low temperature in pure metals and alloys where solute atoms can diffuse by this mechanism. Diffusion by the interstitialcy mechanism in dilute alloys is treated theoretically by Barbu (1980).

(ii) Concentrated Alloys. In more concentrated alloys, solute defect interactions cannot be treated in the pairwise approximation appropriate to dilute alloys. Processes like trapping at solute atoms and ordered zones can play a large part in determining the diffusion coefficients. In multi-phase systems, the effects of irradiation on phase stability partly determine the composition of the bulk material, and therefore diffusion coefficients. The kinetics of the evolution of concentrated mixtures are, therefore, potentially very complex.

While a few measurements of radiation enhanced diffusion of a fourth solute in ternary Fe-Cr-Ni alloys exist (Piller 1981), most work on radiation-enhanced diffusion in concentrated alloys has followed the process of ordering, or has been a by-product of attempts to understand solute redistribution in concentrated alloys. In the latter case, theoretical (Marwick, 1978; Wiedersich et al., 1979; Horton and Marwick, 1981) and experimental work (Turner et al., 1980; Wagner et al., 1981) have concentrated on the redistribution.

2. Ballistic Mixing

Another important phenomenon in low or moderate temperature ion-implantation is that of "displacement" or "cascade" mixing described in more detail in Chapter 7. An initial estimate of the diffusion coefficient due to ballistic mixing was that of Andersen (1979):

$$D_{\text{displ}} = 0.067 \ (F_D/N) \ \phi \ (\frac{\bar{r}^2}{E_d}), \tag{6}$$

where F_D is the energy deposited into elastic collisions, ϕ is the ion flux, and $\bar{r}^2/E_d$ is a measure of the efficiency of mixing. Tsaur *et al.* (1979) found that this constant was 70 Å^2/eV for thin Pt markers in Si, and 200 Å^2/eV for Si markers in Pt. With literature values for E_d, these measurements imply values of $\bar{r}^2$ which are much larger than the 100 Å^2 or so envisaged by Andersen. An estimate of the diffusion coefficient due to displacement mixing is plotted in Figure 1, using $\bar{r}^2/E_d$ of 50 Å^2/eV, with E_d of 40 eV. The displacement mixing diffusion coefficient is much smaller than the radiation-enhanced diffusion coefficient, except at low temperature.

The term "low temperature", meaning the temperature regime in which diffusion of radiation-induced point-defects is sluggish, can be rendered more precise with a knowledge of the vacancy migration energy. This quantity is shown in Table I, along with the temperature at which the diffusion length for vacancies is 200 Å in a 20 minute anneal. This temperature has been selected to give a measure of the temperature at which thermally activated diffusion of irradiation-produced defects could contribute appreciably to the evolution of compositional or constitutional gradients in an irradiated material. The materials for which such effects might be expected at or near room temperature include Ag, Al, Au and Cu.

TABLE I

Vacancy migration energy, E_v, and temperature for a vacancy diffusion length of 200Å in a 20 min. anneal, for selected metals.

Material	$E_v{}^{(a)}$ (eV)	T_{200}(K)
Al	0.62	245
Ag	0.66	260
Au	0.83	327
Cu	0.72	284
Mo	1.3	513
Fe	1.32[(b)]	521
Ni	1.38	544
Pt	1.42	560
W	1.7	671

(a) Values from Baluffi (1978), except where indicated.
(b) Kiritani *et al.* (1979).

3. Point Defect Processes Within Cascades

Although ballistic displacement mixing is intrinsically athermal, present knowledge of collision cascades allows some discussion of possibly thermally activated processes in cascade mixing. Here the term "cascade mixing" refers to matter-transport processes occurring as a direct consequence of the concentration of displacement events into a small volume.

It is known that vacancy loops can form within collision cascades, and that the probability of formation is larger in dense cascades. (Jenkins *et al.*, 1978) The agglomeration of vacancies implies the occurrence of atomic motion. We can estimate the magnitude of the resulting diffusion coefficient by noting that if the mean diffusion length within a cascade volume is ℓ, then the bulk diffusion rate given the occurrence of cascades of volume V at a rate r per unit volume.sec is

$$D_{casc} = \ell^2 V r \tag{7}$$

The cascade occurrence rate r is related to the damage rate K by $r = NK/n_d$, where n_d is the number of Frenkel pairs produced per cascade. The number of atoms within the cascade volume is given by NV and we can guess that this might be about ten times the number actually permanently displaced. If ℓ is about 1Å (which is consistent with the fact that disordered zones form within cascades in ordered alloys), we then find

$$D_{casc} \cong 10^{-15} K \text{ cm}^2/\text{sec} \tag{8}$$

which is of the same order as pure ballistic mixing. This is a very rough estimate.

An important point is that vacancy movement in a cascade may be affected by the bulk lattice temperatures. The vacancy movement occurs during "cooling" of the cascade, as is shown by Guinan and Kinney's (1981) molecular-dynamics calculations. (Their results also account for the reduction in the number of point-defects surviving in the wake of a collision cascade). Atomic motion due to this mechanism will be additional to the prompt mixing caused by ballistic transport.

B. Redistribution Mechanisms

Encounters between point-defects and solute atoms allow the latter to diffuse and also induce solute redistribution if there is a gradient of point-defect concentration (equivalently: if there is a point-defect flux). Two extremes of behavior can be distinguished: the formation of mobile complexes between solute atoms and defects, and purely transient encounters. These extremes will be considered in turn. It should be borne in mind that the behavior of any solute will lie somewhere between the two extremes.

1. Redistribution by Formation of Mobile Complexes

If it is energetically favorable for a point-defect/solute atom complex to form, and if the resulting complex is mobile at the irradiation temperature, then solute redistribution can occur. Net transport of solute occurs because point-defect concentration is high. The complex, being able to diffuse, then undergoes a random walk. Since the binding energy is finite, it will eventually break up. If this happens where the point-defect concentration is low, the solute atom has a correspondingly smaller chance of again being rendered mobile by an encounter with a point defect. In this way, solute atoms are transported from defect-rich regions to regions where the defect concentration is low.

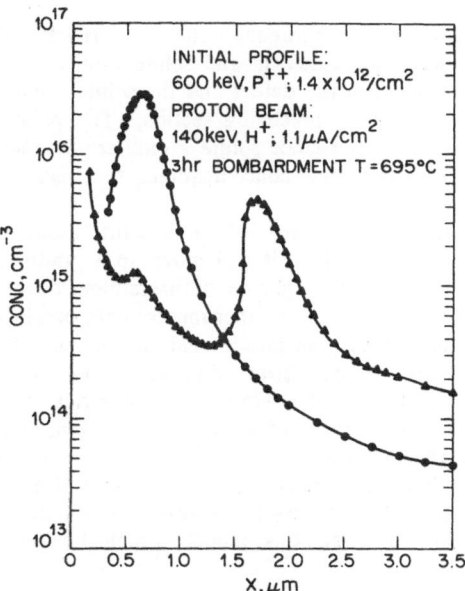

Figure 3: The development of the concentration profile of a 600 keV P implant due to proton bombardment at 695°C (Akutagawa *et al.*, 1979). Phosphorous atoms have been transported out of the defect-rich region near the end of the protons' range at a depth of 1.3 microns by forming mobile complexes with point-defects produced by irradiation.

An example of such behavior is shown in Figure 3. In this experiment a P implant in Si was irradiated with protons at 695°C. After irradiation, P was depleted near the end of the proton range, where most of the point-defects were generated. Both the vacancy (Morikawa *et al.*, 1980; Fair, 1981) and the interstitial (Gosele and Strunk, 1979) have been postulated to be the defect which forms the complex with the P. Many examples of similar redistribution in metals are known (see the review by Rehn, 1981).

In cubic metals, interstitialcies (which are only produced in significant numbers by irradiation) have a dumb-bell structure. One possible form of interstitialcy-solute complex is the mixed dumb-bell, which is thought to form with undersized solutes (Johnson and Lam, 1976; Dederichs *et al.*, 1978). Molecular dynamics studies of the structure of mixed dumb-bells have recently been reported by Lam *et al.*, (1980). Such a complex can migrate by a sequence of jumps and rotations. Little is known about the actual mechanism, except in a very few cases such as the Al-Fe alloys studied by Rehn *et al.* (1978). Even less is known about interstitialcy migration in concentrated alloys or intermetallic compounds.

The solute-vacancy complex is easier to visualize. It migrates by a mixture of exchanges between the solute atom and its neighboring vacancy, and jumps of the vacancy between sites which are nearest-neighbors to the solute. The calculations of Gupta *et al.* (Gupta and Lam, 1979; Gupta, 1980) show that rather large binding energies (greater than 1 eV) are possible.

In many cases, solute redistribution is observed but it is not known which point-defect species is responsible. This may not matter in practice, since qualitative predictions can often be made without knowing the defect type.

2. Redistribution by the Inverse Kirkendall Effect

Solute redistribution by the inverse Kirkendall effect (Marwick, 1978) takes place because of transient encounters between point-defects and solute atoms. The encounters are transient when there is no binding between the defect and the solute atom - therefore this case is the opposite to that considered in the preceding section. The point defect flux in an alloy is accompanied by an atom flux, and since solute atoms are in general present in this flux in different proportions from their total concentrations, they are redistributed in the regions where the defect flux occurs.

The case of vacancies is the best studied. If a solute atom is a faster vacancy diffuser than the other components of an alloy, it will move up a gradient of vacancy concentration (Marwick and Piller, 1979). Conversely, slow diffusers move down.

As an example, Figure 4 shows some experimental data obtained by Turner *et al.* (1980) from measurements of the composition profile near the surface of Fe-Cr-Ni alloys irradiated with 3 MeV Ni ions at high temperatures. Also shown are composition profiles calculated with an inverse Kirkendall effect model (Horton and Marwick 1981). In the calculations the vacancy flux was taken to be the cause of solute redistribution, and therefore the low vacancy diffusion coefficient of the Ni atoms accounts for their enrichment on the surface sink (Marwick, 1978). (Similar enrichment would be caused if the Ni were enriched in the interstitial flux (Wiedersich, 1979)). As the Figure shows, only moderately good agreement with the data was achieved using this model, which however contained no adjustable parameters.

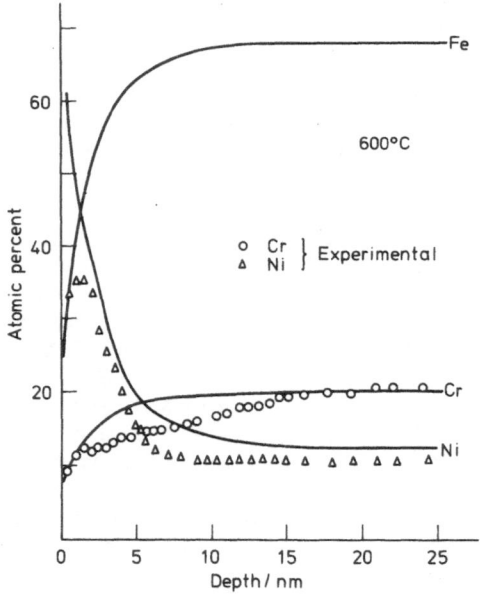

Figure 4:   Near surface composition of an Fe-Cr-Ni alloy as determined by Turner *et al.* (1980) following irradiation with 3 MeV Ni ions to a damage dose of 6 dpa. Also shown are theoretical calculations by Horton and Marwick (1981) using an inverse Kirkendall effect model.

C. Point Defect Gradients

Whether solute redistribution takes place via solute/defect complexes, or by the inverse Kirkendall effect, the driving force is a difference in point-defect concentration from place to place within an irradiated solid. In this section we discuss the origin and magnitude of such differences. We will first discuss gradients of *production,* then gradients produced by the presence of *sinks.* It is necessary to bear in mind that although solute redistribution in one dimension may be noted where a depth profiling technique is used, redistribution in three dimensions around internal sinks occurs simultaneously.

1. Gradients of Production

If the point-defect production rate varies from place to place within the irradiated solid, then the steady-state defect concentration will likewise tend to vary. The mean damage production rate varies over the range of the bombarding particle, and this variation is significant if the range is comparable to the thickness of the region of interest or the diffusion length of point-defects. An example is shown schematically in Figure 5a. Here a projectile with a relatively short range (say 200 Å) produces point defects whose mean diffusion length before annihilation is of the same order as the ion's range. Therefore, many defects are lost by diffusion to the surface, and the defect gradients tend to be steep. Note also that the peak of the point-defect distribution lies beyond the projectile's range, and that the defect-rich zone extends well beyond the damage. Calculations for a specific case may be found in Marwick and Piller (1980).

If the projectile's range is large compared to diffusion lengths then Figure 5b will apply. Here the projectile range is perhaps some thousands of Angstroms. Point-defects are of course produced over the whole range of this heavy ion, and the production rate peaks at a depth slightly smaller than the ion's projected range. The point-defect profile is essentially a diffusion-broadened version of the damage profile, except that near the surface sink there is a region depleted in defects. The defect concentration gradient at this point is usually sharper than those at the end of the projectile's range, and therefore causes more redistribution.

2. Gradients Around Sinks

Sinks are parts of the structure of a solid which absorb and emit point defects. Examples are dislocations, grain boundaries and surfaces. In thermal equilibrium emission and absorption of defects occur at equal rates, but under irradiation the point-defect concentration is raised far above the equilibrium value and the sinks mostly absorb. The defect concentration in their immediate vicinity is held at or near the low thermal equilibrium if they are perfect or near-perfect sinks. Therefore a gradient of point-defect concentration is maintained near the sink.

A fairly simple and tractable case very relevant to ion-beam modification of materials is that of the gradients near a surface. In fact we can be even more specific and discuss loss to the surfaces of a thin layer, since this is a common experimental situation. Analytic solutions of the point defect profile for such a case were published by Lam *et al.* (1974).

Figure 6 shows the vacancy flux through the surface of a 200 Å thick foil, as a fraction of the mean vacancy concentration in the foil. This quantity is a measure of the amount of solute redistribution to be expected. We have plotted it as a function of temperature for a damage rate of 10^{-3} dpa/sec. The plot shows an important feature. The maximum defect flux occurs at a relatively high temperature. At low temperatures most defects recombine,

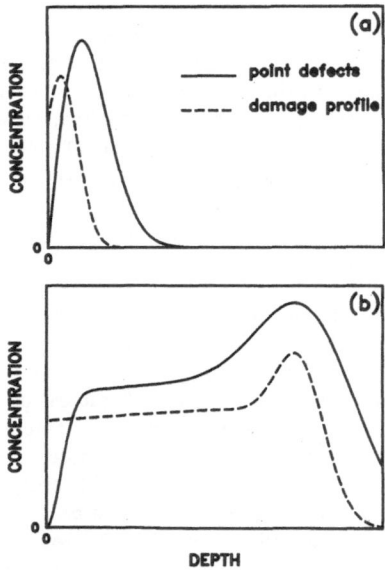

Figure 5:   Schematic illustrations of the relationship of damage profiles to the resulting point-defect distributions when point-defects are mobile. (a) Low energy. The range of the ions is less than or comparable to the diffusion length of the point-defects. (b) High energy. The steepest gradient in the defect profile is at the surface.

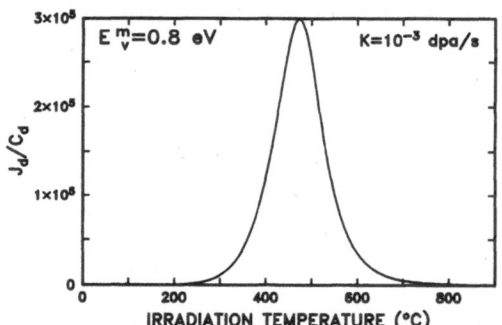

Figure 6:   Calculated vacancy flux through the surface of a thin foil of Al under irradiation. The greatest flux occurs at 475°C, and solute redistribution is greatest at that temperature.

and therefore few flow through the foil surface. At high temperature, on the other hand, almost all the radiation-induced defects flow to the surface, but their numbers are dwarfed by the large numbers of thermally generated vacancies which allow thermal diffusion of solute atoms, thus smoothing out any solute gradients which might otherwise occur. The net result is that the greatest solute redistribution usually occurs only within a rather narrow temperature range.

That is not to say, however, that solute redistribution never takes place around room temperature. Solute transport by solute/interstitialcy complexes seems to allow redistribution in some materials at room temperature, for example in Ni-Si alloys (Piller and Marwick, 1978). The reason is probably that at low temperature the interstitialcy concentration can be anomalously high as already discussed. This promotes the formation of solute-interstitialcy complexes.

III. REDISTRIBUTION EFFECTS

Having surveyed the theory of solute redistribution, in this section we examine some manifestations of these effects in ion-bombardment situations. We will emphasize ion-bombardment experiments because of their application to materials modification. However, it should be realized that much of the original interest in solute redistribution came from its effects in neutron-irradiated steels used in fusion and fission reactors. Therefore, much of the literature on the subject is oriented towards nuclear technology. Nevertheless, most experimental studies have been done with ion beams, and these experiments are of interest to the materials-modification community. The main difference in the two fields is the emphasis on high-temperature applications in the reactor-oriented work.

A. Low Energy

Our first example of solute redistribution induced by ion bombardment is one in which radiation-enhanced diffusion, rather than solute redistribution, turns out to be an important effect. In the Ni-Cu alloy studied by Rehn et al. (Rehn et al. 1979; Rehn and Wiedersich, 1980) bombardment with 5 keV Ar ions at elevated temperatures produces a competition between radiation-enhanced Gibbsian adsorption of Cu to the bombarded surface, and preferential removal of Cu by sputtering. This is an example of radiation-enhanced diffusion kinetically enabling a process which would otherwise be too slow at the irradiation temperature to be significant. Figure 7 shows measured depth profiles of Ni concentration following 5 keV bombardment at high temperature. These profiles can be explained (Lam and Wiedersich, 1981) as being a result of a combination of radiation-enhanced diffusion, Gibbsian adsorption, and preferential sputtering. Although Gibbsian adsorption tends to enrich Cu on the surface, Cu is preferentially sputtered off. Therefore the sub-surface region becomes depleted in Cu. The depletion propagates through the layer in which radiation-enhanced diffusion is significant, which is much thicker than the range of the projectiles (see Figure 5a). Thus the amount and the spatial extent of the composition modification are increased by the presence of radiation-induced point-defects.

Modelling of this system by Lam and Wiedersich (1981) has clarified the relationship between the processes operating. In addition to those mentioned, displacement mixing and radiation-induced segregation are found to play a role. Figure 8 shows calculated steady-state depth profiles of Cu content. The concentration on the surface is determined by preferential sputtering, while the immediate sub-surface layer is depleted in Cu to a greater extent at higher temperature, because of the tendency for radiation-enhanced diffusion coefficients to be higher at higher temperature (see Figure 1).

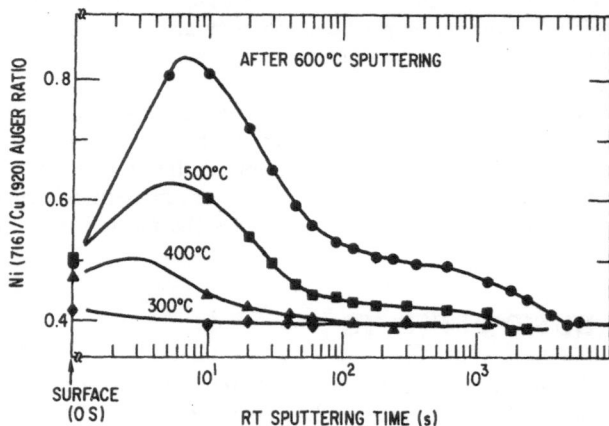

Figure 7:   Peak-to-peak ratios of the Ni (716 eV) and Cu (920 eV) Auger transitions measured after sputtering
             for various periods of time at room temperature (Rehn *et al.*, 1979). The sputtering rate was
             ~6 Å/sec. Each specimen had previously been sputtered for 2 h at the indicated temperature,
             corresponding to the removal of ~4.3 microns of material.

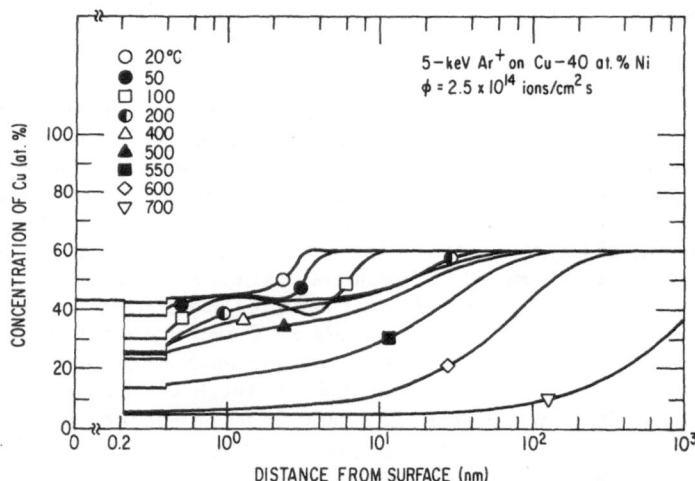

Figure 8:   Calculated profiles of Cu concentration in the Cu-40% Ni alloy bombarded with 5 keV Ar ions at
             various temperatures (Lam and Wiedersich 1981).

Experiments which demonstrate point-defect migration effects on the distribution of solute near a surface are those which follow the diffusion of an implanted solute as it is irradiated with self-ions of the host. Figure 9 shows an example (Hobbs and Marwick, 1981). The experimental conditions correspond to Figure 5a, the projected range of the Ni being 180 Å. It was found that the Mn implant, which is thought to redistribute by the inverse Kirkendall effect in Ni, drifted to greater depths and broadened under irradiation. This is easily understood as a combination of drift and radiation-enhanced diffusion. After a dose of 4.8×10^{15} ions/cm^2 (14 dpa) the solute profile takes on a final steady-state shape. This occurs when solute fluxes due to solute gradients are exactly balanced by solute fluxes due to vacancy gradients (Marwick and Piller, 1980). The final solute distribution is therefore determined by the point-defect distribution. Note that for this reason the Mn profile does not continue to broaden indefinitely when bombarded by low-energy ions.

Several consequences of similar behavior have been explored for different solutes in Ni in a series of experiments. For example, in high-temperature implants the radiation damage produced by the bombarding particle can modify its distribution (Marwick and Piller, 1980). Figure 10 shows an example, comparing normalized high- and low-temperature implants of Mn into Ni at 500°C. In the high-temperature implantation the final Mn distribution is believed to reflect the underlying point-defect profile. Note that because the implant concentration on the surface is reduced by the drift of the Mn atoms against the vacancy flux, saturation of the amount implanted is probably postponed, and the saturation concentration increased, although this has not been checked experimentally.

A similar effect of solute redistribution occurs during recoil implantation of Al into Ni at elevated temperature (Marwick and Piller 1981) (Figure 11). This has the effect of increasing the penetration of the Al implant. The Al is initially introduced into the Ni by recoil implantation, in other words by displacement mixing at the Al/Ni interface. Although the great majority of implanted Al atoms come to rest at a very shallow depth, < 10 Å, they are immediately affected by the defect currents generated by the bombarding ion, and "sucked down" into the bulk of the material. Since they subsequently can undergo radiation-enhanced diffusion, they can penetrate to a considerable depth. This effect was shown to be most marked at higher temperature, as expected given that vacancy fluxes are then higher.

B. High Energy Ion Implantation

There are a number of important differences between low and high energy ion bombardment which are related to the distribution of defect production relative to the surface. The maximum of the defect production moves from the surface to well into the interior (of order microns deep), and the number of defects produced per incoming ion increases from a few to $\sim 10^3$ as the energy of the ion is increased from the keV into the MeV range. At low energies sputtering is important relative to diffusion enhanced by radiation or effects driven by defect fluxes, but is nearly negligible at high energies. Radiation-enhanced diffusion peaks close to the surface at low energy (Figure 5a) and tails off toward the interior as defects, that diffuse out from the production zone, are eliminated by recombination and sink annihilation. At high energy (Figure 5a), an additional region of high diffusivity exists between the damage peak and the surface. Long range defect fluxes originate close to the surface and are mainly directed toward the interior, for low energy bombardment. At high energy, long range defect fluxes from the peak damage zone toward the surface become important also; segregation to internal sinks within the damage zone can drastically alter the microstructure, in particular precipitating phases which may be induced or redistributed. Precipitate stability will be discussed in the next section. Furthermore, during high energy

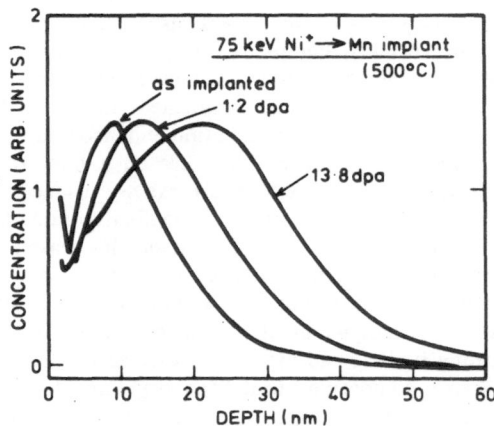

Figure 9:   Redistribution of a $2.2 \times 10^{14}/cm^2$ Mn implant in Ni as a result of irradiation with 75 keV Ni ions at 500°C (Hobbs and Marwick, 1981). The implant profile changes by a combination of radiation-enhanced diffusion and solute drift in the vacancy gradients set up by the irradiation. 1 dpa corresponds to 3.4×10^{14} ions/cm^2 of 75 keV.

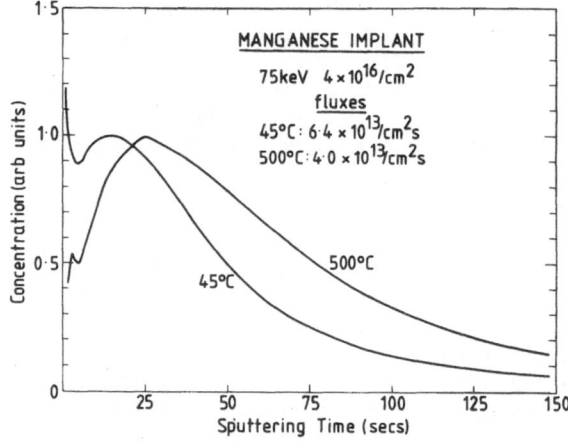

Figure 10: Normalized depth profiles of low and high temperature Mn implants in Ni, showing the redistribution of Mn brought about by the irradiation damage during the implant at elevated temperature. The implant doses were 6.4 and $4.0 \times 10^{13}/cm^2$ at low and high temperatures respectively (Marwick and Piller, 1980).

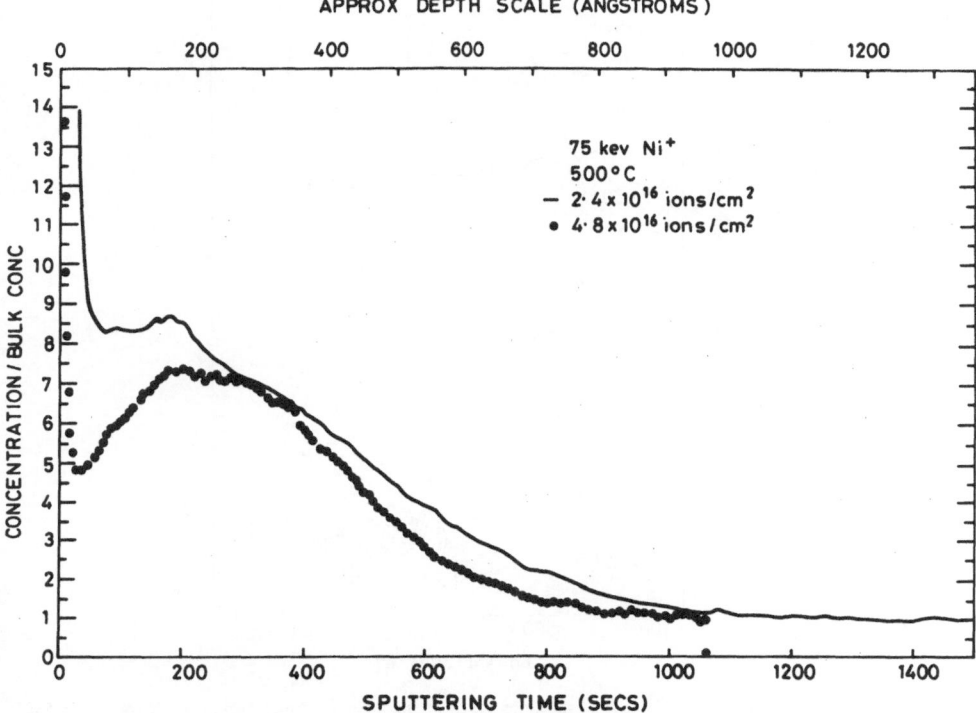

Figure 11: Aluminum concentration profiles within a Ni surface following bombardment of a 100 Å overlayer with 75 keV Ni ions at 500°C to doses of 2.4 and 4.8 × 10¹⁶ ions/cm². At the lower dose some of the overlayer remains, but this is subsequently sputtered off, leaving recoil-implanted Al whose distribution is determined by defect-gradient effects (Marwick and Piller, 1981).

bombardment at elevated temperatures a near-surface region exists in which the compositional and microstructural development is dominated by defect fluxes to the surface.

Recently, detailed studies on the kinetics of solute redistribution within the surface layer have been made in Ni-Si alloys by Rehn *et al.* (1981b). They investigated the growth of Ni_3Si films during bombardment of Ni-17.7 at.% Si alloys with 2 MeV He and Li ions at temperatures in the range of 345 to 650°C. A portion of the backscattered ions were used to determine the film thickness *in situ* as a function of time (or dose) as described by Rehn *et al.* (1980) and Averback *et al.* (1981b). The radiation-induced segregation of Si to the surface causes the precipitation of continuous surface films of Ni_3Si not only in two-phase Ni-Si alloys ≤10 at.% Si but also solid solutions of Si in Ni (Potter *et al.*, (1977); Robrock and Okamoto, (1979)). It appears likely that the undersized Si atoms are transported predominantly by tightly bound Si-interstitial complexes (Okamoto *et al.*, 1981). However, Si-vacancy complexes may also contribute to the segregation. Recent calculations by Gupta (1980) suggest a binding energy of 0.8 eV between a vacancy and a Si atom in Ni.

The growth of the Ni_3Si layer during 2 MeV He ion bombardment at 530°C as measured by Rutherford backscattering is illustrated in Figure 12. The thickness of the Ni_3Si layer is plotted in Figure 13 as a function of the square root of dose for a number of temperatures (Rehn *et al.*, 1981b). The parabolic growth of the film indicates strongly that the growth

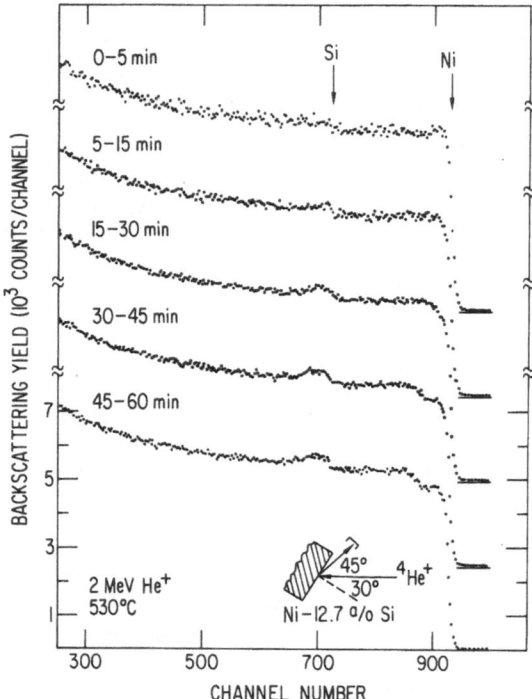

Figure 12:   RBS spectra of a Ni-12.7 at.% Si specimen taken at the indicated times during bombardment at 530°C
with 2 MeV He ions at a calculated dose rate of 3.1×10^{-4} dpa/sec (Rehn *et al.*, 1981).

rate is limited by a diffusion process. As shown in Figure 14, the parabolic growth rate
constant A (slopes of straight lines in Figure 13) can be represented by a straight line with
an apparent activation energy of 0.3 eV on an Arrhenius plot from about 350 to 550°C
(Rehn, 1981). A proportionality to the square root of dose of the amount of an alloying
component transported by defect fluxes to the surface has also been observed for the solid
solution alloy system Ni-Cu (Rehn *et al.*, 1981a; Wagner *et al.*, 1981). Thus, parabolic
kinetics of radiation-induced segregation are not restricted to the growth of a precipitate layer
on a surface.

Parabolic growth kinetics can result from a variety of diffusion-limited transport processes,
such as slow transport of defects through the precipitate phase, transport of solute through a
severely solute depleted region, or long range transport from the bulk of the alloy (Wiedersich
and Lam 1982). Here, we will illustrate parabolic kinetics resulting from transport of tightly
bound Si-interstitial complexes through a Si-depleted region of thickness X_d to the growing
Ni_3Si layer of thickness X_{Ni_3Si}. The interstitial flux to the surface incorporates every Si atom
arriving at the boundary because of the tight-binding assumption. The depleted layer is
characterized by the condition $0 \leqslant C_{Si} \cong C_{Si}^c \leqslant C_i$, i.e., the Si concentration present in the
depleted layer is low enough that all Si atoms are in the form of complexes. Furthermore,
the boundary condition at the precipitate interface is $C_{Si}^c \cong 0$ and at the depleted-
layer/matrix interface $C_{Si}^c \cong C_i$ where C_i is the steady-state interstitial concentration in the
bulk of the alloy. Thus, the flux of Si to the precipitate is given by

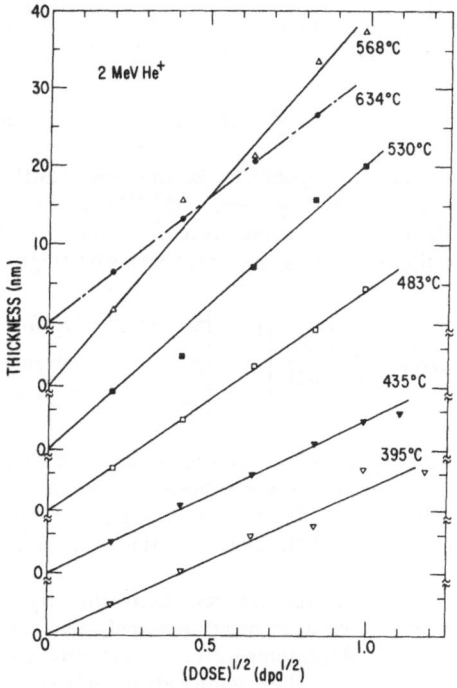

Figure 13: Plot of the Ni₃Si coating thickness as a function of the square-root of dose during bombardment at several temperatures with 2 MeV He ions at a calculated dose rate of 3.1×10^{-4} dpa/sec. The linear dependence indicates diffusion-limited growth kinetics (Rehn *et al.*, 1981b).

$$J_{\text{Si}} = - D^c_{\text{Si}} \left(\frac{dC^c_{\text{Si}}}{dX} \right) \cong - D^c_{\text{Si}} \left(\frac{C_i}{X_d} \right) \tag{9}$$

where D^c_{Si} is the diffusion coefficient of the Si-interstitial complexes. Because of solute conservation

$$X_{\text{Ni}_3\text{Si}} C_{\text{Ni}_3\text{Si}} = X_d C^0_{\text{Si}} \tag{10}$$

ignoring the small amount of solute in the depleted zone and transitions on either side of it. $C_{\text{Ni}_3\text{Si}}$ and C^0_{Si} are the Si-concentrations in the precipitate and the bulk of the alloy, respectively. The growth rate of the precipitate layer is given by

$$\frac{dX_{\text{Ni}_3\text{Si}}}{dt} = \frac{-J_{\text{Si}}}{C_{\text{Ni}_3\text{Si}}} \cong \frac{D^c_{\text{Si}} C_i C^0_{\text{Si}}}{C^2_{\text{Ni}_3\text{Si}} X_{\text{Ni}_3\text{Si}}}. \tag{11}$$

Integration yields the precipitate layer thickness as a function of time or of dose, $\phi = Kt$, respectively:

$$X_{\text{Ni}_3\text{Si}} = \left[\frac{2C_{\text{Si}}^0}{C_{\text{Ni}_3\text{Si}}^2} \cdot D_{\text{Si}}^{\varepsilon} C_i \cdot t \right]^{1/2} = \left[\frac{2C_{\text{Si}}^0}{C_{\text{Ni}_3\text{Si}}^2} \cdot \frac{D_{\text{Si}}^{\varepsilon} C_i}{K} \cdot \phi \right]^{1/2}, \tag{12}$$

i.e., the layer growth is proportional to $t^{1/2}$, or at constant displacement rate, K, proportional to $\phi^{1/2}$.

The temperature and dose-rate dependence of the film growth is contained in the factor $(D_{\text{Si}} C_i / K)^{1/2}$. Conventional rate theory (see, e.g., Wiedersich, 1972) yields some simple approximations for this factor in the defect recombination (low temperature, T), the sink (intermediate T), and for the thermal vacancy concentration (high T) dominated regimes,

$$(D_{\text{Si}}^{\varepsilon} C_i / K)^{1/2} \cong \left(\frac{b^2}{6\Omega} \right)^{1/2} \cdot \begin{cases} (\nu_v / Ka)^{1/4} & \text{low } T \\ p^{-1/2} & \text{intermediate } T \\ (ac_v^{th})^{-1/2} & \text{high } T. \end{cases} \tag{13}$$

Here, b is the interatomic distance, Ω is the atomic volume, ν_v is the vacancy jump frequency, a is a constant related to the recombination volume, p^{-1} is the average number of jumps a defect performs before annihilating at a sink, and c_v^{th} is the atomic fraction of vacancies at thermal equilibrium. These approximations are valid when the defects with the lowest mobility are vacancies.

These approximations show that the apparent activation energy of the rate constant for film growth should be 1/4 of the vacancy *migration* energy at low temperature, and $-1/2$ of the vacancy *formation* energy at high temperature. Both predictions are consistent with the measurements shown in Figure 14. The temperature independent region at intermediate temperatures is missing in Figure 14 presumably because the sink density was too low to dominate defect annihilation at any temperature. The rate constant should be reduced proportional to the 1/4 power of the defect production rate in the low temperature defect recombination regime, and be independent of K at high temperatures. Again, the experimental findings, Figure 14, are in reasonable agreement with the predictions (Rehn, 1981). Despite the good agreement of the experimental results with the simple model described, it is not assured that this model is in fact the correct one to be applied to the growth of Ni$_3$Si layers on Ni-12.7 at.% Si. Models based on other rate-limiting solute transport processes give closely similar relations between film thickness and dose to those derived here (Wiedersich and Lam, 1982).

The growth of Ni$_3$Si layers has also been used to obtain information about the effects of ion mass or average cascade size on the production of defects available for long-range atom transport (Averback *et al.*, 1982a; Rehn, 1981). Growth constants were measured for bombardment with 2 MeV He^4, 2 MeV Li^7, 3 MeV Ni^{58} and 3.25 MeV Kr^{84} with the beam current densities adjusted to give the same calculated total defect production rates. The rate constants show a significant reduction with increasing ion mass (Figure 14). The reduction has been interpreted as resulting both from increasing recombination within larger, denser cascades and from loss of freely migrating defects at sinks provided by defect clusters formed in dense cascades (Rehn, 1981).

In another study performed with high-energy ions, Marwick and Piller (1981) studied the enhancement of recoil implantation of Al into Ni with 200 keV N ions. The depth profile measured by Marwick and Piller after N irradiation is shown in Figure 15. The shape of the depth profile results from the operation of four processes. First, Al is believed to be introduced into the Ni in the first instance by recoil implantation. Next, it is swept out of the

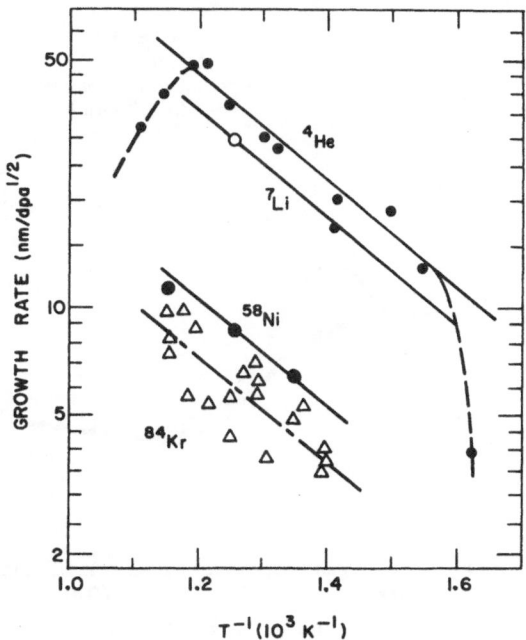

Figure 14: Arrhenius plot of the coating growth rate constants measured for a Ni-12.7 at.% Si alloy. The calculated displacement rates for the different ions were kept approximately constant at 3×10^{-4} dpa/sec (Rehn *et al.*, 1981b).

shallow surface layer in which it comes to rest by the inverse Kirkendall effect induced by the flux of vacancies to the surface. Once away from the surface, the Al atoms undergo radiation-enhanced diffusion in the defect-rich layer which is roughly as thick as the N range (Figure 5b applies). Finally the Al atoms are confined within this layer by the action of the defect flux caused by the gradient of production at the end of the N's range. In fact the peak of Al concentration at around 2000 Å depth is due to the peak in vacancy concentration around the damage peak at the end of the N ion's range. This technique of recoil implantation at elevated temperature has the advantage that dopants can be implanted to considerable depths, with the use of simple gas ion accelerators. A similar technique was used by Baumvol *et al.* (1980) to introduce Sn into Fe. Significant reductions in wear were observed. Dearnaley and Goode (1981) have coined the term "bombardment diffusion" to describe the introduction of a solute into a surface in this way.

Finally, it is worth mentioning that surface enrichment can occur not only in solid solutions but also in ion bombarded *compounds*. Wagner *et al.* (1982) found that when Ni_3Si was bombarded with MeV Ar, Li or Ni ions at a dose rate of 10^{-3} dpa/sec at temperatures around 520°C a surface layer enriched beyond stoichiometry could be detected by Auger spectroscopy. The layer was found to be 20 to 50 Å thick. At the highest doses studied the Si concentration in this layer reached 50%. The mechanism of enrichment remains obscure. However, radiation-induced defects are known to migrate within Ni_3Si because dislocation loops form within the ordered phase in this material during high-temperature electron irradiation (Wagner *et al.*, 1982). Presumably the migration of defects can cause chemical redistribution in this compound, as well as in metallic solid solutions.

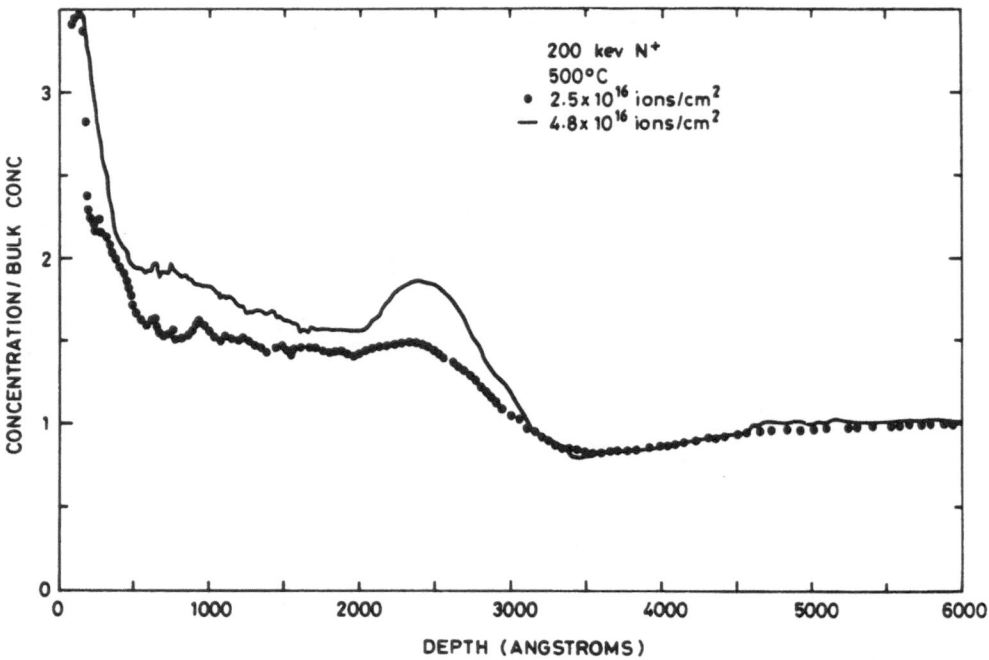

Figure 15:   Depth profiles of Al recoil-implanted into Ni by 200 keV N ion bombardment at 500°C (Marwick and Piller, 1981). The Al overlayer was initially 100 Å thick. Movement of pre-existing Al solute into the defect-rich region near the end of the nitrogen's range causes the dip in Al concentration near 3500 Å. Recoil-implanted Al is diffusing from the surface.

Some general considerations related to the mechanism of thermal diffusion in metal silicides have been put forward recently by Peterson *et al.* (1982), and these may be useful in interpreting the results of irradiation experiments.

IV. PHASE STABILITY UNDER IRRADIATION

The stability of certain phases under irradiation depends on such factors as radiation-enhanced diffusion, which enhances solute mobility, displacement mixing, which tends to homogenize the target, and solute redistribution in defect gradients, which changes the composition of small volumes of material. In general, all these mechanisms operate at once. For convenience we distinguish two regimes: high temperature, where displacement mixing plays only a small part; and low temperature where it may dominate.

A. Precipitate Stability in the High Temperature Regime

The high temperature regime is characterized by high defect and solute mobility, and by appreciable solute redistribution by defect currents. Because solute atoms diffuse so fast by defect-mediated processes, displacement mixing plays only a secondary part. Rapid radiation-enhanced diffusion increases the rate of processes such as phase transformation or Ostwald ripening which occur without irradiation. Despite the greatly enhanced diffusion rate under irradiation, interfacial processes appear to be fast enough for local equilibrium to be maintained (Marwick, 1981). The largest effect on high-temperature precipitate stability appears to be due to changes in local solute concentrations under irradiation.

Solute redistribution is driven by defect fluxes, as we have seen, and therefore occurs near sinks. Potter (1978) found that γ' precipitates of Ni_3Si in a thin foil of Ni-12.8 at.% Al alloy formed a thin layer at the position of maximum defect concentration in the foil. Aluminum is enriched at such a position because it moves in the opposite direction to a vacancy flux, by the inverse Kirkendall effect. The precipitate layer was not in the center of the foil, because the rate of damage production in the foil was not uniform throughout its thickness, and therefore the peak defect concentration was not in the centre of the foil.

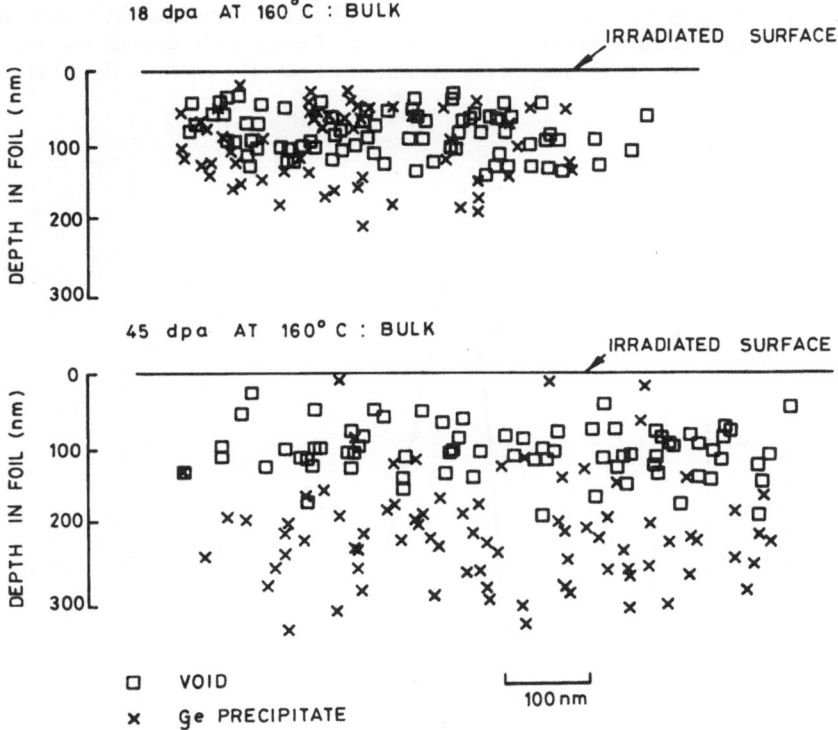

Figure 16: Two-dimensional plots of the depth distribution of irradiation-induced voids and Ge precipitates in an Al-0.75 at.% Ge alloy irradiated with 100 keV Al ions to the peak damage doses indicated (Rusbridge, 1981). Voids and precipitates both nucleate in the defect-rich layer near the damage peak at 750Å. At higher dose, Ge precipitates redistribute but voids remain.

Redistribution due to a gradient of production of point defects in a thick sample was observed by Rusbridge (1981) in an Al(Ge) alloy. Irradiation with 200 keV Al ions was found to produce a layer of voids near the damage peak, as expected. At the same depth a layer of Ge precipitates appeared (Figure 16). The precipitates only nucleated readily in defect-rich regions of this alloy because they need an excess of vacancies to grow (Bertram et al., 1978). Therefore the low-dose microstructure was expected. At higher doses the void layer remained in the same place but the precipitates were redistributed as shown in Figure 16b. this was explained when the Ge concentration profile was measured by sputter profiling and SIMS (Figure 17). It was found that at this higher dose Ge had been depleted in the defect-rich layer. Because of the many limitations of sputter profiling, the Ge concentration in the dip in Figure 17 could not be given accurately. However, since the the Ge precipitates dissolved, it must have been less than the solubility limit of Ge in Al, approximately 0.2 at.% at the irradiation temperature. The Ge removed from this layer reappeared as excess Ge concentration at shallower and greater depths, as seen in Figure 17. The surface peak in Ge concentration is accounted for by the few large precipitates marked in Figure 16, whose mean size was much greater than that of the precipitates in the bulk. This increased size may be due to a combination of their stress-free growth environment, and ripening by rapid surface diffusion.

Apparently similar redistribution in an experiment aimed at ion-beam mixing Ge into Al has been described by Lau *et al.* (1981) (Figure 18). They found that Xe bombardment at 100°C led to an anomalous Ge concentration profile. Compared to thermal treatment at the same depth, there was a large peak in Ge concentration near the center of the Al layer with a

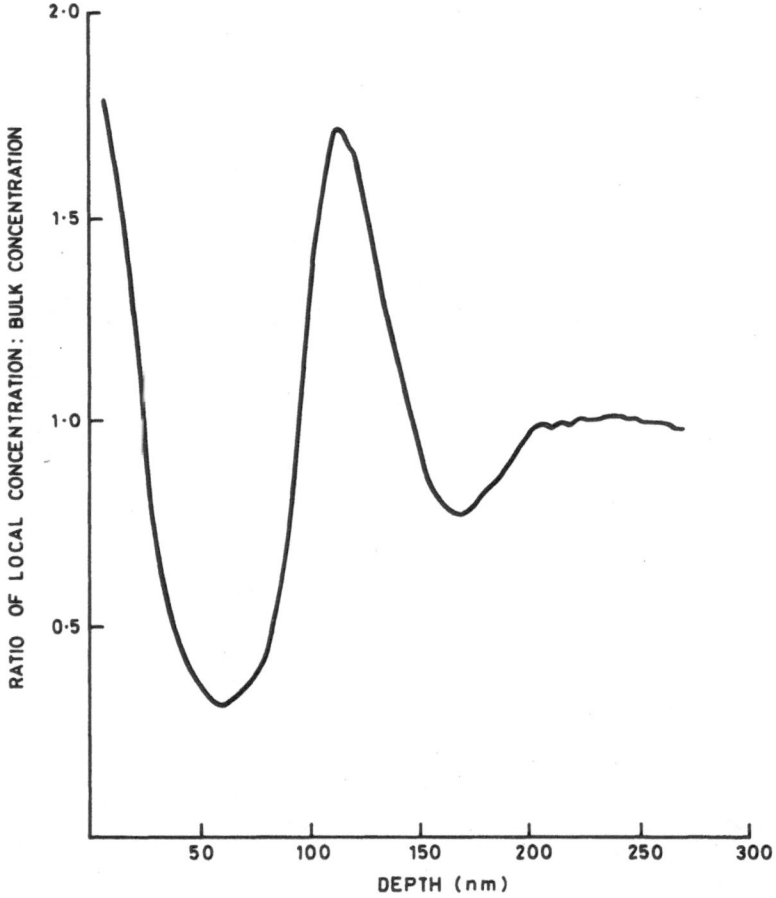

Figure 17: SIMS profile of Ge concentration in the high-dose specimen of Figure 16, showing depletion of Ge in the defect-rich layer (Rusbridge, 1981).

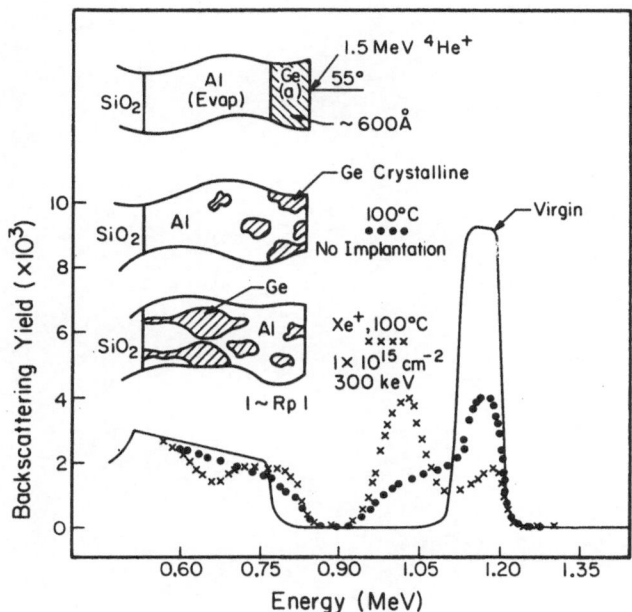

Figure 18: RBS profiles of Ge-Al thin-film structures (Lau *et al.*, 1981). Bombardment with Xe ions at 100°C produces a Ge distribution different from that resulting from thermal annealing. The difference probably results from the same process noted by Rusbridge (1981), namely a combination of radiation-enhanced diffusion, and solute drift in the defect gradients produced by the ion-irradiation.

dip at somewhat lesser depth. This can be understood in the same terms as Rusbridge's experiment, if the peak of the point-defect concentration lies at a depth between the mean projected range, R_p, of the projectiles and the surface, at the position of the dip observed in the Ge concentration.

The examples given so far have concerned macroscopic solute redistribution. Since defect sinks also occur within irradiated materials, as well as at their surfaces, solute redistribution at a fine scale can also occur at these sinks. Such redistribution can have repercussions for the stability of precipitates at or near the sinks. For example, if a solute is enriched at a sink, the solubility limit may be exceeded and a second phase appear in what was previously a solid solution. The best studied cases are that of Si in Ni (see, e.g., Robrock and Okamoto, 1979) and austenitic stainless steels (Williams and Titchmarsh, 1981). Other cases of solute enrichment at sinks include Ge in Ni (Barbu, 1979), and Be in Cu (Bartels *et al.*, 1979). More examples are cited by Wiedersich (1980) and Rehn (1981).

When a solute is depleted at sinks, the resulting increase in solute concentration elsewhere can also lead to anomalous precipitation. For example Little and Stow (1979) found that precipitates of the Cr rich α' phase appeared in Fe- 10% Cr alloy for which the equilibrium solubility at the irradiation temperature was 14%. Figure 19 shows the microstructure in a specimen which was neutron irradiated in a fast reactor to 30 dpa at 420°C. Depleted zones around voids are seen. Another example of depletion of Cr at sinks, this time grain boundaries in Fe-5 at.% Cr and Fe-13 at.% Cr has been measured by Takahashi *et al.* (1981) using high-resolution TEM compositional analysis. Chromium presumably redistributes by the inverse Kirkendall effect in these alloys.

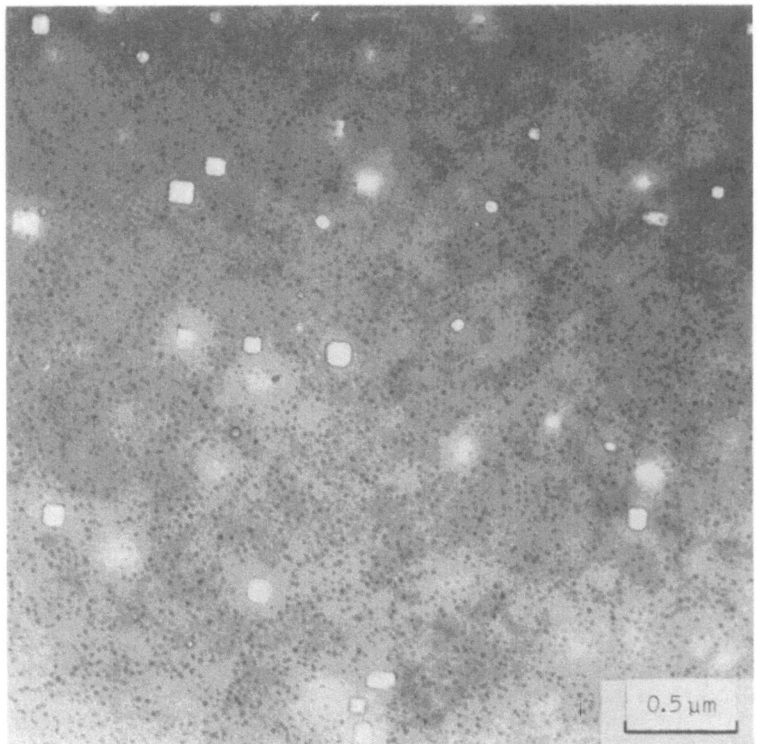

Figure 19:   Voids and α chromium-rich precipitates in Fe-10 wt.% Cr alloy neutron irradiated to a damage dose of
30 dpa at 420°C (Little and Stow, 1979). Precipitate-free zones have formed around voids as a result
of Cr depletion. Cr drifts away from voids because of the defect gradient near these sinks.

B. Precipitate Stability in the Low Temperature Regime

"Low" temperatures are characterized by relatively low defect, especially vacancy, mobility.
Solute diffusivity is also low, and therefore solute transport by displacement mixing can be a
significant contribution to the total. As already pointed out, displacement mixing may be
described by the mathematics of diffusion theory and characterized by a diffusion coefficient,
but it is a non-equilibrium process which cares nothing for the differences of free energy
within the target. It simply tends to homogenize and disorder, and would ultimately reduce
heterogeneous targets to uniform mixtures.

Examples of such a trend in its early stages are afforded by the work of Jones (1978) on
thoria precipitates in steel, and Rusbridge (1981) on Ge precipitates in Al. In the latter work
Ge precipitates, formed in Al(Ge) alloys by ion irradiation at 200°C, were subjected to
further irradiation at liquid N temperature, then warmed to room temperature for
examination by TEM. It was found that the original precipitates were changed in
appearance, having been partly replaced by a cloud of smaller precipitates. These had re-
nucleated in the Ge-rich zone surrounding the original Ge precipitates, which had been partly
destroyed by displacement mixing.

In Al, solute is appreciably mobile at room temperature and precipitates can re-nucleate, but in an alloy where diffusion is slower, precipitates bombarded at room temperature may simply disappear (Nelson *et al.*, 1972). Where they have been there are presumably zones of high solute concentration, but this has not yet been confirmed by measurement.

V. CONCLUSIONS AND FUTURE TRENDS FOR SURFACE PROCESSING

Surface modification can take two forms: change of composition and change of structure. The two are of course not mutually exclusive. Exploitation of point-defect mediated effects can help to reach a desired end in either area. Conversely, an appreciation of solute diffusion and redistribution is necessary to interpret the complex phenomena met with in ion-beam modification work.

In many materials irradiation or implantation at elevated temperature is necessary if solute diffusivity is to be increased, and if solute redistribution is to be used to alter implant profiles or to modify sub-surface precipitation by solute redistribution. In work done to date only a few instances of target temperatures being deliberately raised for these purposes have been reported. However temperature rises no doubt occur in many implantations where high beam currents are used.

In the future there will be more deliberate control of target temperature. Low target temperatures will be used to promote mixing and to take the target away from equilibrium phases. High target temperatures will be used to promote solute diffusion and redistribution. A combination of low and high temperature treatments may also be found advantageous to achieve a desired solute profile.

Also to be expected in future work is a trend away from using rare-gas ions for inducing radiation damage or mixing. The inert gas, being insoluble, tends to precipitate as bubbles. This inhomogeneity in the target may be undesirable. Soluble or chemically active ions such as N are widely used to modify wear properties of surfaces, but are not very damaging because of their low mass. Where high levels of damage are required the use of self-ions has much to recommend it, although the complexity of the ion-sources required is a serious disadvantage. Where a heavy dopant ion is to be implanted, the damage produced by this beam may be turned to good advantage.

Research in at least two areas is needed to back up future application of ion beams to surface processing: investigation of radiation enhanced diffusion in compounds like metal silicides; and a better understanding of atomic migration by interstitialcy mechanisms below the vacancy migration temperature.

Atomic migration in compounds must occur during the phase transformations often observed under bombardment (Mayer, 1981). Little or no work on radiation-enhanced diffusion in such materials seems to have been done, however, nor is data on point-defect migration energies available in most cases. However, such information is essential if the kinetics of phase transformations under irradiation are to be quantitatively understood.

Atomic migration below the vacancy migration energy, presumably by an interstitialcy mechanism, has been observed in Cu-Ni alloys by Poershke and Wollenburger (1980) and in Fe-Cr-Ni alloys by Mantl *et al.* (1979). Where anything other than homogenization of the target occurs during irradiation below the vacancy migration temperature, the interstitialcy mechanism of solute diffusion is presumably effective. More understanding in this area is again desirable.

Ion bombardment can modify the depth distribution of precipitating phases, and even their existence. By proper choice of bombarding ion energy and irradiation temperature the possibility exists that the precipitate density may be deliberately altered in a chosen depth range. For example, low temperature irradiation may destroy precipitates (though perhaps leaving behind small solute-enriched zones). Alternatively the solute distribution may be deliberately altered by a high temperature irradiation to deplete or enrich the precipitate density. Thus the near surface microstructure may be tailored for specific purposes, even without doping the surface (if a self-ion is used). Given the present state of knowledge the choice of conditions for doing this must be determined empirically except where the system being used has already received intensive study, and only very few have.

REFERENCES

Akutagawa, W., Dunlap, H. L., Hart, R. and Marsh, O. J. (1979), J. Appl. Phys. *50*, 777.

Andersen, H. H. (1979), Appl. Phys. *18*, 131.

Averback, R. S., Benedeck, R. and Merkle, K. L. (1978), Phys. Rev. B *18*, 4156.

Averback, R. S., Rehn, L. E., Okamoto, P. R. and Cook, R. E. (1981a) Nucl. Instr. Methods *182 & 183*, 79.

Averback, R. S., Rehn, L. E., Wiedersich, H. and Cook, R. E. (1981b) in "Phase Stability Under Irradiation," J. R. Holland, L. K. Mansur and D. I. Potter, eds. (AIME, New York, NY), p. 101.

Baluffi, R. W. (1978), J. Nucl. Mat. *69 & 70*, 240 and references therein.

Barbu, A. (1979), proc. conf. on irradiation behavior of metallic materials for fast reactor components, J. Poirier and J. M. Dupouy (eds), (CEA, Paris), p. 57.

Barbu, A. (1980), Acta Met *28*, 499.

Bartels, A., Dworschak, F., Meurer, H. P., Abromet, C. and Wollenberger, H. (1979), J. Nucl. Mat. *83*, 24.

Baumvol, I., Jr., Watkins, R. E. J., Longworth, G. and Dearnaley, G. (1980), in "Low Energy Ion Beams 1980," I. H. Wilson and K. G. Stevens, eds., (Conf. Ser. No. 54, Institute of Physics, London), p. 201.

Bertram, K., Minter, F. J., Hudson, J. A. and Russel, K. C. (1978), J. Nucl. Mat. *75*, 42.

Birtcher, R. C. and Blewitt, T. H. (1978), J. Nucl. Mat. *69 & 70*, 783.

Brailsford, A. D. and Bullough, R. J. (1978), J. Nucl. Mat. *69 & 70*, 434.

Coltman, Jr., R. R., Klabunde, C. E. and Williams, J. M. (1981), J. Nucl. Mat. *99*, 284.

Dearnaley, G. and Goode, P. D. (1981), Nucl. Instr. Methods *189*, 117.

Dederichs, P. H., Lehmann, C., Schober, H. R., Scholz, A. and Zeller, R. (1978), J. Nucl. Mat. *69*, 176.

Fair, R. B. (1981), J. Appl. Phys. *51*, 5828.

Gatlinburg (1979). Workshop on solute segregation in metals, Gatlinburg 1979, published in J. Nucl. Mat. vols. 69 & 70.

Gosele, U. and Strunk, H. (1979), Appl. Phys. *20*, 265.

Guinan, M. W. and Kinney, J. H. (1981), presented at the second topical meeting on fusion-reactor materials, Seattle 1981. To be published in J. Nucl. Mat.

Gupta, R. P. (1980), Phys. Rev. *B22*, 5900.

Gupta, R. P. and Lam, N. Q. (1979), Scripta Met *13*, 1005.

Hillairet, J. (1978), Nuclear Mat. *69 & 70*, 776.

Hobbs, J. E. and Marwick, A. D. (1981), Rad. Effects Lett. *58*, 83.

Horton, M. E. and Marwick, A. D. (1981), Proc. conf. on Atomic Collisions in Solids Lyons, to be published in Nucl. Instr. Methods.

Jenkins, M. L., English, C. A. and Eyre, B. L. (1978), Phil. Mag. *38A*, 97.

Johnson, R. A. and Lam, N. Q. (1976), Phys. Rev. *B13*, 4364.

Jones, R. H. (1978), J. Nucl. Mat. *74*, 163.

Kiritani, M., Takata, H., Morijama, K. and Fujita, F. E. (1979), Phil. Mag. *40A*, 779.

Kirk, M. A. and Greenwood, L. R. (1979), J. Nucl. Mat. *80*, 159.

Lam, N. Q., Rothman, S. J. and Sizmann, R. (1974), Rad Effects *23*, 53.

Lam, N. Q., Doan, N. V. and Adda, Y. (1980), J. Phys. F. *10*, 2359.

Lam, N. Q. and Wiedersich, H. (1981), submitted to Rad. Effects.

Lau, S. S., Tsaur, B. Y., von Allmen, M., Mayer, J. W., Stritzker, B., White, C. W., and Appleton, B. (1981), Nucl. Instr. Methods *182 & 183*, 97.

LeClaire, A. D. (1978), J. Nucl. Mat. *69 & 70*, 70.

Little, E. A. and Stow, D. A. (1979), J. Nucl. Mat. *87*, 25.

Mantl, S., Sharma, B. D. and Antesberger, G. (1979), Phil. Mag. *39A*, 389.

Marwick, A. D. (1978), J. Phys. F. *8*, 1849.

Marwick, A. D. (1981), Nucl. Instr. Methods *182 & 183*, 827.

Marwick, A. D. and Piller, R. C. (1979), J. Nucl. Mat. *83*, 35.

Marwick, A. D. and Piller, R. C. (1980), Rad. Effects *47*, 195.

Marwick, A. D. and Piller, R. C. (1981), Nucl. Instr. Methods *182 & 183*, 121.

Mayer, J. (1981), Nucl. Instr. Methods *182 & 183*, 1.

Morikawa, Y., Yamamoto, K. and Nagami, K. (1980), Appl. Phys. Lett. *36*, 997.

Nelson, R. S., Hudson, J. A. and Mazey, D. J. (1972), J. Nucl. Mat. *44*, 318.

Norgett, M. J., Robinson, M. T. and Torrens, I. M. (1975), Nucl. Eng. Design *33*, 50.

Okamoto, P. R., Rehn, L. E., Averback, R. S., Robrock, K. H. and Wiedersich, H. (1981), Yamada Conference V on Point Defects and Defect Interactions in Metals, Kyoto, Japan, in press.

Petersson, C. S., Baglin, J. E. E., Dempsey, J. J., d'Heurle, F. M. and La Placa, S. J. (1982), submitted to J. Appl. Phys.

Piller, R. C. (1981), private communication.

Piller, R. C. and Marwick, A. D. (1978), J. Nucl. Mat. *71*, 309.

Piller, R. C. and Marwick, A. D. (1979), J. Nucl. Mat. *83*, 42.

Poershke, R. and Wollenburger, H. (1980), Rad. Effects *49*, 225.

Potter, D. I. (1978), Rad. Effects *35*, 115.

Potter, D. I., Rehn, L. E., Okamoto, P. R. and Wiedersich, H. (1977), Scripta Met. *11*, 1095.

Potter, D. I. and Wiedersich, H. (1979), J. Nucl. Mat. *83*, 208.

Rehn, L. E. (1981), Symposium on Metastable Material Formation by Ion Implantation, Materials Research Society, Boston, in press.

Rehn, L. E. and Wiedersich, H. (1980), Thin Solid Films, *73*, 139.

Rehn, L. E. Robrock, K. H., and Jacques, H. (1978), J. Phys. F *8*, 1835.

Rehn, L. E., Danyluk, S. and Wiedersich, H. (1979), Phys. Rev. Letts. *13*, 1764.

Rehn, L. E., Averback, R. S. and Okamoto, P. R. (1980), "Characterization of near-surface region in irradiated Ni$_3$Si alloys," Las Vegas, 1980.

Rehn, L. E., Wagner W. and Wiedersich, H. (1981a), Scripta Met. *15*, 683.

Rehn, L. E., Wagner, W., Okamoto, P. R. and Wiedersich, H. (1981b), "In-situ Ni$_3$Si surface film growth in Ni-12.7 at.% Si alloys," to be published.

Robrock, K. H. and Okamoto, P. (1979), proc. conf. on irradiation behavior of metallic materials for fast reactor components, J. Poirier and J. M. Dupouy (eds), (CEA, Paris) p. 57.

Rusbridge, K. (1981), Nucl. Instr. Methods *182 & 183*, 521.

Sizmann, R. (1978), J. Nucl. Mat. *69 & 70*, 386.

Takahashi, H., Ohnuki, S. and Takeyama, T. (1981), "Radiation-induced segregation at internal sinks in electron-irradiated binary alloys," presented at the second topical meeting on fusion-reactor materials, Seattle 1981. To be published in J. Nucl. Mat.

Tsaur, B. Y., Matteson, S., Chapman, G., Liau, Z. L. and Nicolet, M. A. (1979), Appl. Phys. Lett. *35*, 825.

Turner, A. P. L., Nolfi, F. V. and Sethi, V. (1980), in "Phase Stability During Irradiation," J. R. Holland and D. I. Potter, eds., (Met. Soc. AIME).

Wagner, W., Rehn, L. E. and Wiedersich, H. (1981), Phil. Mag., in press.

Wagner, W., Rehn, L. E. and Wiedersich, H. (1982), to be submitted for publication.

Wiedersich, H. (1972), Rad. Effects *12*, 111.

Wiedersich, H., Okamoto, P. R. and Lam, N. Q. (1979), J. Nucl. Mat. *83*, 98.

Wiedersich, H. and Okamoto, P. R. (1980), in "Phase Stability During Irradiation," J. R. Holland, L. K. Mansur and D. I. Potter, eds., (Met. Soc. AIME).

Wiedersich, H. and Lam, N. Q. (1982), in "Phase Transformations and Solute Redistribution in Alloys during Irradiation," F. V. Nolfi, ed. (Applied Science Publishers, Barkin, U.K.).

Williams, T. M. and Titchmarsh, J. M. (1981), J. Nucl. Mat. *98*, 223.

CHAPTER 9

ION BEAM MIXING

J. W. MAYER

Department of Materials Science, Cornell University, Ithaca, New York

S. S. LAU

**Department of Electrical Engineering and Computer Sciences,
University of California, San Diego, California**

CONTRIBUTORS: J. A. Davies, U. Littmark, A. Marwick, F. Saris and
H. Wiedersich

I. INTRODUCTION

The field of ion beam mixing studies the compositional and structural changes of a two or multiple component system under the influence of ion irradiation. Due to these changes, material properties of the system may be modified in ways sometimes difficult to achieve by conventional methods. From a collisional point of view, ion beam mixing effects result from particle-solid interactions. The phenomena become more complex when thermodynamical forces are involved, and these forces are sometimes modified by the presence of high concentrations of defects in the system.

Ion beam mixing (Tsaur *et al.*, 1979a) was developed to achieve ion beam modified materials with higher solute concentrations at lower ion doses than can be achieved with conventional high-dose implantation techniques. The concept, as shown in Figure 1, is to deposit a layer of material on the substrate, a Au film on Cu in this case, and then bombard the sample with ions that are sufficiently energetic so that the ion range exceeds the film thickness. The amount of Au that can be introduced into the Cu with ion-mixing greatly exceeds the maximum concentration that can be achieved by direct implantation where sputtering effects generally set the concentration limits. In other experiments (Mayer *et al.*, 1980) with single crystal Cu substrates it was shown that both techniques, direct implantation and ion beam mixing, could be used to form substitutional single-crystal solid solutions of Au or Pd in Cu. The initial experiments also demonstrated (Tsaur *et al.*, 1979b) that silicides could be formed by ion-induced reactions. Thus, from a materials modification viewpoint it was concluded that ion beam mixing was an attractive alternate to direct implantation when the alloying species can be deposited as a thin film.

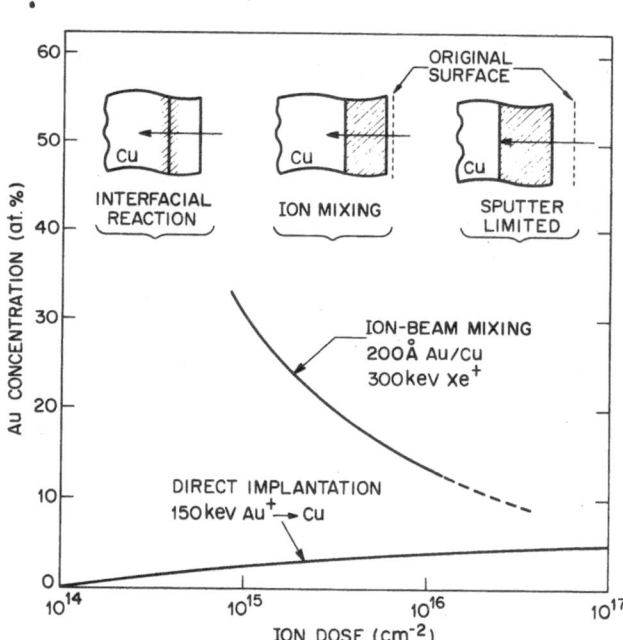

Figure 1:   Concentration versus dose for a Au film, 200 Å thick, ion-mixed with 300 keV Xe⁺ ions. The lower curve is for the direct implantation of 150 keV Au ions. (From Tsaur *et al.*, 1979a).

The precursors to the work on ion-beam mixing were studies of preferential sputtering of silicides where it was found that composition changes occurred over depths comparable to the penetration depth of the bombarding ions (Liau *et al.*, 1977; 1978a,b; Blank and Wittmaack, 1979; Wach and Wittmaack, 1981). Aside from the influence on sputter-depth-profiling (Liau *et al.*, 1979a), the work indicated that ion-induced reactions could be used for material modification (Liau and Mayer, 1980). However, from the viewpoint of sputtering not all materials respond to ion-induced reactions in as dramatic a fashion as silicides (see, for example, Chapter 10). To achieve intermixing in immiscible systems, one can deposit multiple, thin layers (Tsaur *et al.*, 1981a; Mayer *et al.*, 1981). As shown in Figure 2, one advantage of such an approach is that the composition of the mixed layer is established by the relative thicknesses of the deposited layers. This provides a "limited" supply system of fixed composition in contrast to the single layer approach where the concentration is dependent on the total dose of incident ions (Figure 1). As demonstrated by Tsaur *et al.* (1981a,b,c), ion mixing of multiple layers provides a direct method of examining metastable phase formation with ion beams. It provides yet another approach to the general field of metastable alloys Chapter 6).

In this chapter, we are primarily concerned with the mechanisms involved with ion-induced reactions rather than the nature of the mixed system. The applications of ion-beam mixing as a metallurgical tool depend upon our understanding of the dynamics of the process. At present, there is an understanding of the general trends based on a rather limited number

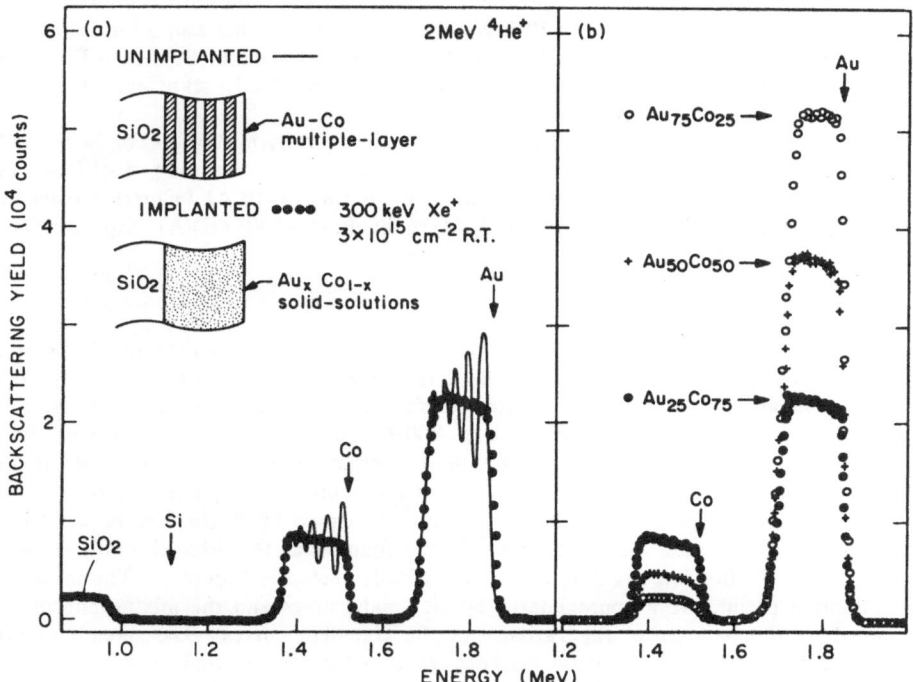

Figure 2: (a) Backscattering spectra showing the formation of a uniformly mixed Au-Co layer by ion-beam mixing of thin deposited Au and Co multiple layers. (b) Backscattering spectra of ion-mixed Au-Co films with compositions $Au_{75}Co_{25}$, $Au_{50}Co_{50}$, and $Au_{25}Co_{75}$.

of studies. The collision cascade (Chapter 7 and Thompson, 1981) and ballistic effects of recoils establish the lower bound on the amount of mixing. Another regime is established by defect migration during high temperature irradiation (Chapter 8). The compositional changes during ion bombardment and sputtering (Chapter 10) are certainly correlated with the composition established in ion beam mixing. These different approaches have not yet been tied together in a comprehensive framework. Consequently, we present an overview in the following sections that is based on specific examples. In the concluding sections we speculate on future trends.

II. MIXING EFFECTS IN COMPOUND FORMING SYSTEMS; Pt-SILICIDE

When the system under investigation is chemically unstable in its original compositional profile or structural state, interactions between different components may take place under steady-state thermal annealing conditions. For example, a compound is formed between A and B after annealing. It is, therefore, to be expected that the effects of ion beam mixing in these systems are not necessarily only those derived purely from collisions between different atoms. The effects of mixing in compound forming systems can be illustrated by using those observed in the Pt/Si system as examples.

The steady state thermal annealing behavior of a layer of Pt ($\leqslant 1000$ Å) on a Si substrate has been well established (Tu and Mayer, 1978; Ottaviani and Mayer, 1981). At temperatures of about 250°C, a compound of Pt_2Si forms at the Si/Pt interface. The compound continues to grow until the Pt is totally consumed. Further annealing leads to the formation of PtSi at the Si/Pt_2Si interface until the total conversion of Pt_2Si into PtSi. This configuration (Si/PtSi) then becomes stable against further steady state annealing up to temperatures near the melting point of the system.

It is, therefore, of interest to explore the reactions induced by ion irradiation in the Pt/Si system as compared to those reactions observed under steady state annealing conditions. Two types of sample configuration are of interest here: (i) A thin (~ 10 Å) Pt marker embedded in a matrix of amorphous Si, and (ii) A layer of Pt ($\sim 200-400$ Å) deposited on a crystalline Si substrate.

A. Thin Marker--Pt/Si System

The mixing effects of ion irradiation with thin markers were investigated by Tsaur et al. (1979c) and Matteson et al. (1981a). In the latter experiment, a thin Pt marker (~ 10 Å) was located about 200 Å below the matrix surface of amorphous (evaporated) Si. The amorphous Si with a total thickness of about 800 Å was deposited on a single crystal Si substrate. Inert gas ions were implanted into the sample with sufficient energies to give projected ranges of approximately 1100 Å. The temperatures of implantation ranged from ~80 to 523°K. The profiles of the Pt marker were measured by Rutherford backscattering techniques before and after ion irradiation. It was found that the original width of the Pt marker broadened after ion irradiation as schematically shown in Figure 3. The initial and post-irradiation profiles were approximated by Gaussian curves and the mixing effects were determined by the increase of the variance of the Gaussian curves. The spreadings after irradiation for a number of elemental markers embedded in Si appeared to be insensitive to the mass of the marker (Figure 4). This observation is strong evidence that the spreading is due primarily to interactions of the incoming ions with atoms in the matrix, and not with the marker atoms themselves. It is interesting to note that Pt and other silicide formers (Ni, Pd) exhibit non-Gaussian profiles after irradiation at high temperatures, whereas non-silicide formers (such as Ge, Sn, Sb and Au) exhibit Gaussian profiles up to ~523°K. Matteson *et al.* (1981a,b) treated the mixing effect in terms of a diffusional model where the variance of a Gaussian curve is expressed as the product of an effective diffusion coefficient and implantation time. This parameter, Dt (the square of the diffusion length) for a number of different thin markers has the following characteristics: (1) Dt is proportional to the ion dose, ϕ as shown in Figure 5. (2) Dt is relatively insensitive to the mass of the thin marker. (3) the ratio of Dt to ϕ (Dt/ϕ) is independent of temperature from ~80°K to ~300°K, and (4) Dt/ϕ is proportional to the nuclear stopping power, therefore, mixing efficiency is higher for heavy ions than that for light ions.

Based on these observations, and assuming a random flight model for ion beam mixing, Matteson *et al.* (1981b) then, relate the square of the diffusion length, Dt, to the mean square of the range of an impurity atom for a given recoil energy. According to this model, Dt is proportional to the ion dose, Dt is independent of temperature and dependent on the material properties through only their atomic masses, atomic numbers, and effective displacement energy of the marker atoms. These predictions are in good agreement with experimental results.

B. Thin Markers--General Comments

Spreading of impurity markers due to primary and secondary recoils within the collision cascade, often termed recoil mixing or ballistic mixing, has been treated by a number of authors: Andersen (1979), Littmark and Hofer (1980a,b), Sigmund and Gras-Marti (1981). For example, Sigmund and Gras-Marti treated theoretically the mixing phenomenon of dilute impurities in homogeneous matrices by ion bombardment from a collisional point of view. It was found that the spreading of the Pt impurity profiles in Si is consistent with the mechanism of relocation of matrix atoms under ion bombardment. The relocation of matrix atoms is primarily due to ion-matrix knock-on effects, and the impurity atoms spread as a result of the relative motion of impurity and matrix atoms.

Therefore, for dilute impurity atoms in a homogeneous matrix, we conclude that mixing of atoms under ion irradiation at room temperature and below is primarily due to interactions between incoming ions and the matrix atoms, and the interactions can be treated from a collisional point of view. The thin marker experiments provide an estimate of the lower limit or minimum amount of mixing that can occur in ion-beam induced reactions. With higher temperature irradiations where defect motion can predominate (see for example, Averback *et al.*, 1981 or Marwick in Chapter 8) the spreading of the marker can be increased due to radiation enhanced broadening.

For the purpose of materials modification, however, thin markers represent a dilute impurity system and we will not pursue the analysis of that system in further detail. We are primarily concerned with structures where large atomic concentrations can be introduced into the substrate as a result of ion bombardments.

C. Bilayer Configuration--Pt/Si System

As the impurity concentration increases in the matrix (as the marker increases in thickness), the mixing efficiency for the Pt/Si system is observed to increase. For example, at a dose of 2×10^{16} Xe ions/cm^2 at 300 keV, the effective diffusion length, $\sqrt{Dt}$, is 90 Å for a ~5 Å Pt marker, as compared to 180 Å for a 30 Å Pt marker in Si. As the Pt increases further to a bilayer configuration (a layer of Pt deposited on a Si substrate), an even thicker Pt$_2$Si phase is formed after a relatively low dose of ions.

For a sample with a bilayer configuration, the ion beam mixing phenomena are quite different from those observed for thin markers. Let us consider the experiments carried out

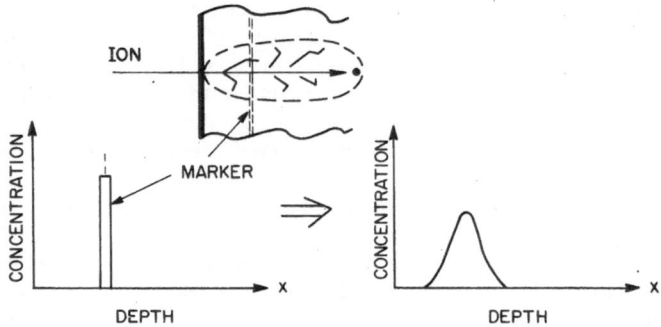

Figure 3:   Schematic of the depth distribution of a thin marker embedded in a matrix before and after ion irradiation.

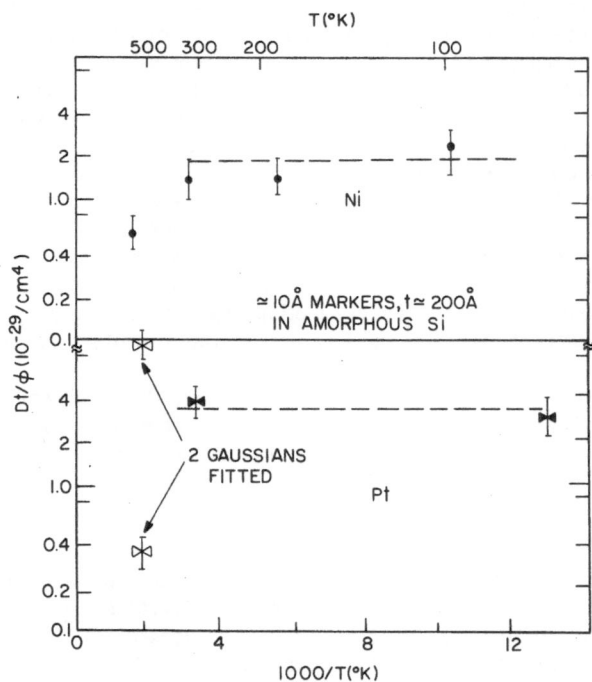

Figure 4:   The mixing parameter, Dt/ϕ, vs. 1000/T for marker layers at a depth of 200 Å in amorphous Si for a dose of $1 \times 10^{16}/cm^2$, 220 keV Kr ions (from Matteson *et al.*, 1981a).

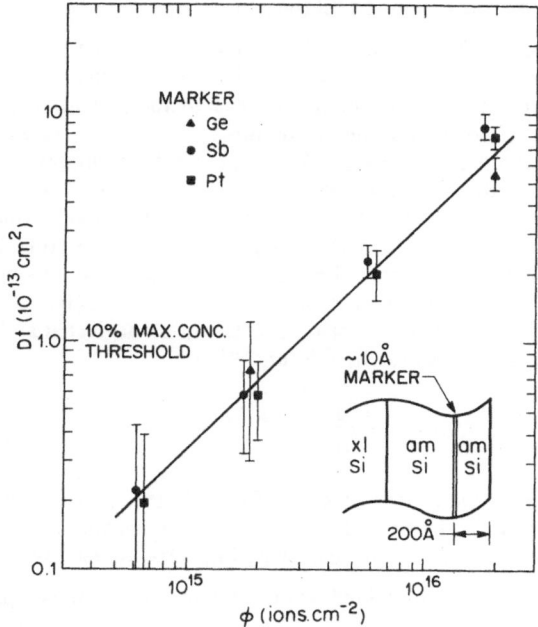

Figure 5:   The square of the effective diffusion length, Dt, vs. dose of 220 keV Kr ions incident at LN₂ temperature on markers in amorphous Si (from Matteson *et al.*, 1981a).

by Tsaur *et al.* (1979b) on the Pt/Si system. In this case, the samples consist of a layer of ~470 Å Pt deposited on <100> oriented Si substrates, and the effect of the ion beam was investigated after implanting Ar, Kr or Xe ions (300 keV) in the samples at room temperature. Mixing between Pt and Si was detected at very low Xe doses. For example, a mixed layer ~300 Å thick with a composition similar to Pt_2Si was obtained after only an implantation dose of 1×10^{14} Xe/cm². The thickness of the mixed layer increases with the square root of the total implanted dose (see Figure 6), until the Pt layer is consumed in the mixing reaction. The composition of the mixed layer remains relatively constant and close to Pt_2Si as mixing proceeds with increasing dose.

The reaction kinetics of ion beam induced Pt_2Si also depend on the mass of the incident ions as shown in Figure 6. For a given dose, the thickness ratio of induced Pt_2Si is 3:2.2:1 for Xe, Kr and Ar ions respectively. The thickness ratios correspond with the mass ratios of Xe, Kr and Ar ions (3.3:2.1:1), and the ratios (7.1:4:1) of nuclear energy loss in units of keV/μm for 300 keV Xe, Kr and Ar in a Pt film (Tsaur *et al.*, 1979b). This is only an estimate because of the change in energy loss across the metal/semiconductor interface (Averback *et al.*, 1982); however, we expect that the initial reaction rate should be proportional to the amount of energy dissipated within the interfacial region.

The presence of the silicide phase Pt_2Si was detected by x-ray diffraction experiments in the mixed layers. From a phase formation point of view, the growth of Pt_2Si induced by an ion beam is analogous to that induced by thermal annealing, in that the first phase formed is Pt_2Si and that the kinetics of growth follow a square root of time relationship.

D. Metastable Phase Formation--Pt/Si System

As mentioned previously, the thickness of the silicide layer increases with the implantation dose as the ion-induced reactions proceed. At doses higher than those required to consume all the unreacted metal, ion beam mixing can lead to the formation of more Si rich phases and/or amorphous layers of metal and Si mixtures. In the case of Pt/Si, the initial silicide formed by ion bombardment is Pt_2Si. After the total consumption of Pt to form Pt_2Si, further implantation leads to an amorphous Pt-Si mixed layer (Tsaur *et al.*, 1980a,b) and *not* to a PtSi layer, as observed with thermal annealing. As implantation proceeds, more Si is mixed into the amorphous layer and the composition of the mixture becomes richer in Si. The final composition of this mixture depends on the original Pt thickness, the range of the incident ions and the erosion of the sample surface by sputtering. If the mixing process is efficient (as in the case of Pt/Si) sputtering effects can be neglected and the final composition of mixed layer can be approximated by (Tsaur, 1980):

$$\frac{N_{Si}}{N_{Pt}} = \frac{C_{Si}(x_f - x_0)}{C_{Pt}x_0}$$

where C_{Si}, C_{Pt} are atomic densities of Si and Pt, x_0 = original thickness of Pt, x_f = penetration depth of implanted ions. The final composition of the mixed layer, therefore, increases in Si content with increasing implantation energy. As shown in Figure 7, for a Pt (240 Å)/Si sample implanted with Xe ions, the final composition ($\frac{N_{Si}}{N_{Pt}}$) of the amorphous layer is ~1.2 and ~1.8 for 150 keV and 250 keV ions, respectively.

The formation of disordered or amorphous layers by ion irradiation is well established (Schulson, 1979; Mayer *et al.*,1970). The point we stress is that ion beam mixing can form amorphous mixtures with compositions (more Si rich than PtSi in this case) that lie in interesting regions of the equilibrium phase diagram, Figure 8, where metastable phases may exist. The free energies of amorphous layers are generally higher than those of metastable crystalline phases. The amorphous alloy can easily transform into a metastable crystalline phase if such a phase exists and if atomic mobilities are provided. The formation of metastable phases in the Pt/Si system is demonstrated in the schematic diagram shown in Figure 9. The starting point for Figure 9 is a sample with a layer of PtSi (300 Å) formed by furnace heating on a <100> Si substrate. Implanting the sample with Xe^+ ions (300 keV) to various doses leads to the formation of amorphous layers with different compositions. A low temperature anneal (400°C, 30 min) causes the amorphous layers to transform into different metastable (Pt_2Si_3 and Pt_4Si_9) and stable (PtSi) phases. For example, a low dose of 4×10^{14}/cm² leads to the coexistence of Pt_2Si_3 (metastable) and PtSi (stable) phases after annealing. At 1×10^{15}/cm², the entire initial PtSi layer was uniformly mixed to form a $PtSi_{1.5}$ layer which transforms into a Pt_2Si_3 metastable crystalline phase. At even higher doses, the coexistence of two metastable phases (Pt_2Si_3 and Pt_4Si_9) or the formation of single Pt_4Si_9 metastable phase can be obtained. It is interesting to note that the compositions of Pt_2Si_3 and Pt_4Si_9 center around an eutectic composition of 67.5% Si in the Pt-Si phase diagram (Figure 8).

The crystal structure of Pt_2Si_3 has been identified to be hexagonal and decomposes into PtSi and Si at ~500°C (Tsaur *et al.*, 1980a,b). Low temperature resistivity measurements indicate that the crystalline Pt_2Si_3 phase exhibits a superconducting transition onset at ~4.2K and complete transition at ~3.6K.

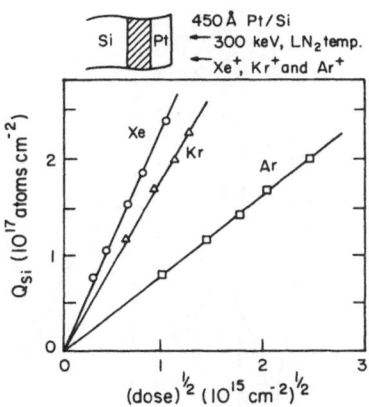

Figure 6: Average thickness of Pt$_2$Si as a function of dose$^{1/2}$ for 470 Å-thick Pt film on silicon substrates implanted with different ions (Ar, Kr and Xe). (From Tsaur *et al.*, 1979b.)

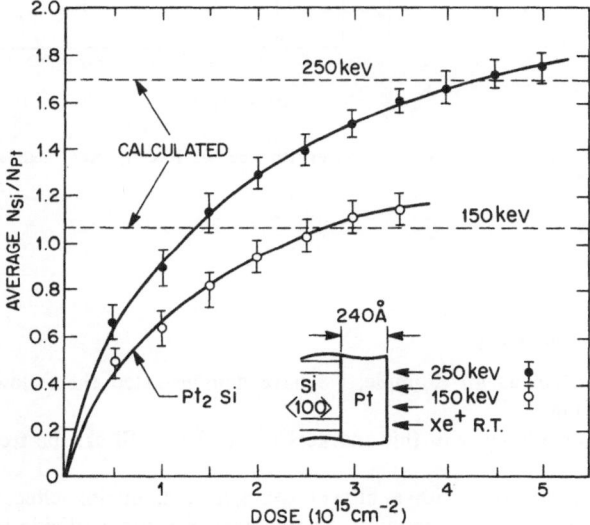

Figure 7: The measured N$_{Si}$/N$_{Pt}$ ratio as a function of ion dose for 240 Å Pt/Si samples implanted with Xe ions of 250 keV and 150 keV energies, respectively. The dashed lines are the calculated final compositions of the mixed layers with $x = x_f$, the ion penetration depth. (From Tsaur, 1980.)

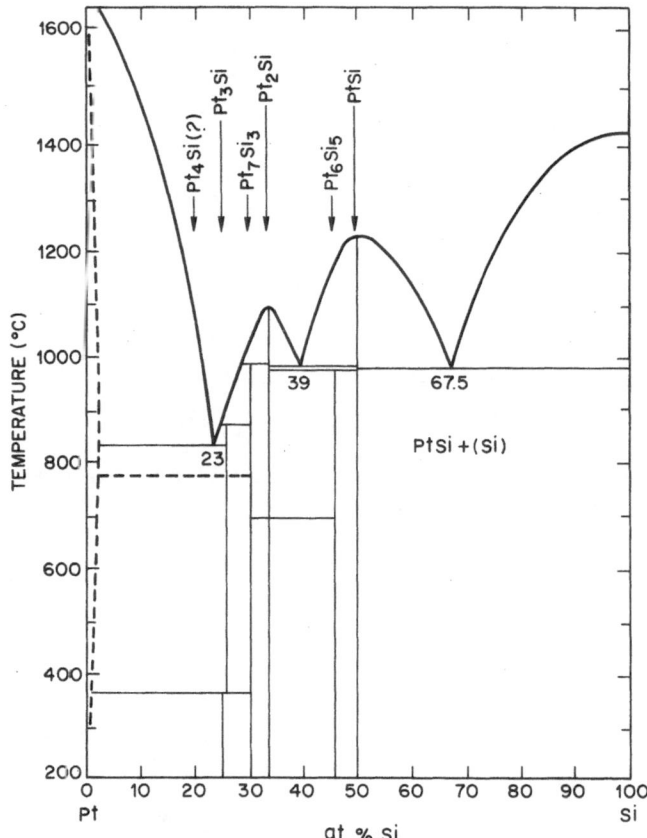

Figure 8:   Equilibrium Pt-Si phase diagram. The Pt-rich side has recently been modified but the Si-rich side remains unchanged.

E. Pt/Si System: Summary

Using the Pt/Si system as an example, we have demonstrated the following effects in ion beam induced reactions:

(1)   Ion beam mixing with thin markers embedded in Si can be treated by collisional theories.

(2)   For thicker layers with a bilayer configuration, mixing effects far exceed those predicted by collision models. In this case, a distinct silicide phase is formed by ion irradiation.

(3)   The formation of the first silicide phase induced by ion beam is analogous to that induced by thermal annealing.

(4)   Amorphous and metastable-crystalline phases can be obtained by ion beam mixing with relative ease.

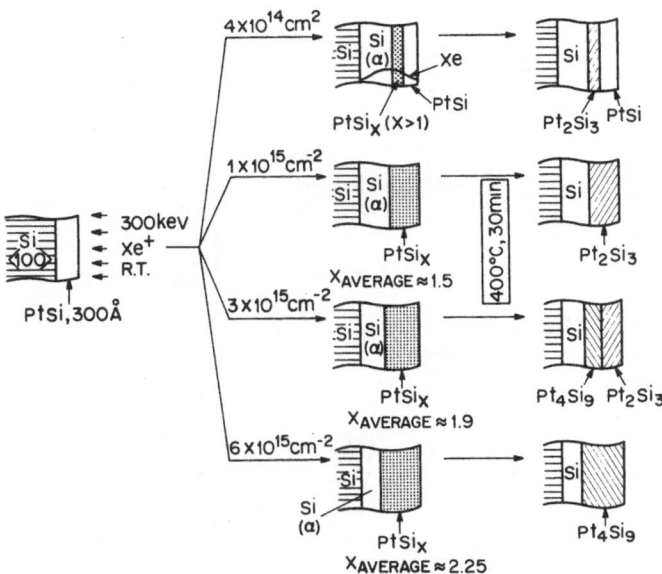

Figure 9:   Schematic diagram which shows the various Si-rich Pt-Si mixed layers produced by ion-beam mixing of a thin PtSi film on a Si substrate at different dose levels. Formation of the metastable phases Pt$_2$Si$_3$ and Pt$_4$Si$_9$ is achieved upon post-annealing.

III. MIXING EFFECTS IN SILICIDE FORMING SYSTEMS

Silicide forming systems are attractive for ion beam mixing studies because the thermal-annealing behavior of metal/Si structures has been thoroughly studied and because of the potential application of mixed layers in integrated circuit fabrication processes. Consequently, there have been numerous studies of silicide formation induced by ion bombardment in addition to the Pt/Si studies cited in Section II: Near-noble metals (Tsaur, 1980), including Pd (van der Weg et al., 1976); Chapman et al., 1979; Tsaur et al., 1979d) and Ni (Averback et al., 1982, Tsaur et al., 1979b); refractory metals (Wang et al., 1979, 1980; d'Heurle et al., 1980; and Tsai et al., 1980) including Mo (Chiang et al., 1981) and Nb (Kanayama et al., 1979 and Matteson et al., 1979), rare-earth metals (Tsaur and Hung, 1980) and even gold (Tsaur and Mayer, 1981) where metastable Au-Si crystalline phases were formed near the eutectic composition.

The general features of these studies were: 1) the same initial phases were formed with ion mixing as is found in thermal annealing (with the exception of the metastable Au$_5$Si$_2$ phase and hexagonal rather than tetragonal WSi$_2$), 2) elevated substrate temperatures are required to form well-defined phases in the refractory silicides which require annealing at temperatures of 450 to 600°C, and 3) the silicide thickness dependence on ion dose follows a (dose)$^{1/2}$ or linear dose dependence in systems which have (time)$^{1/2}$ or linear time dependence in furnace treatments. These results all suggest radiation enhanced diffusion with vacancy migration over the collision cascade dimensions in cases where efficient mixing and well-defined phases are found.

The formation of chromium-disilicide (Tsaur, 1980; Mayer et al., 1981) is an example of a linear dose dependence as indicated in the backscattering spectra shown in Figure 10.

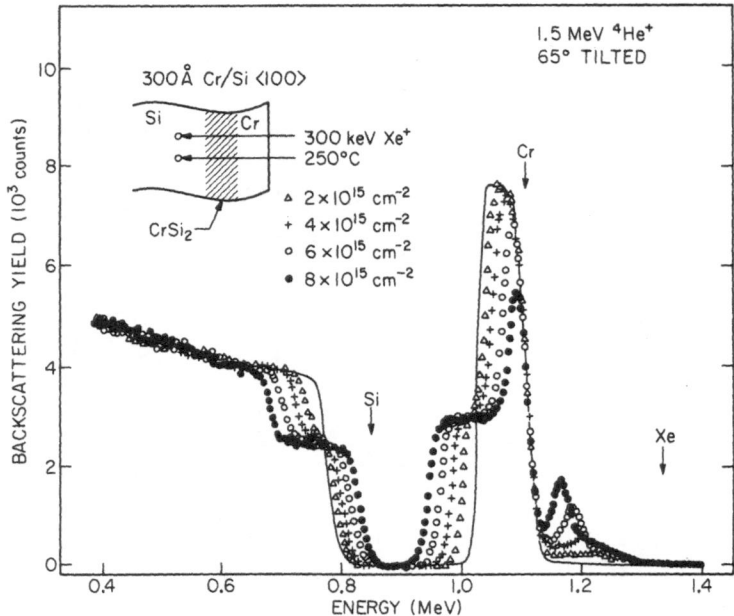

Figure 10: Backscattering spectra which show the layered growth of the CrSi$_2$ phase induced by ion-beam mixing. The thicknesses of CrSi$_2$ increase linearly with ion doses, which is similar to the kinetics of linear time dependence in normal thermal treatment. (From Mayer *et al.*, 1981.)

With ion-mixing at 250°C (as compared to 450°C with furnace annealing) the silicide composition is uniform and the thickness of the phase increases proportional to the dose. With irradiation at room temperature and below, the mixed layer is graded in composition. The temperature dependence of the amount of mixing for Cr and Nb silicide formation is shown in Figure 11. Below room temperature, the mixing is relatively temperature independent as would be suggested by ballistic or recoil mixing. In the thermally activated regime, the presence of well-defined phases show that point defects are at work; i.e., vacancy migration. At present there is no indication of long-range defect migration (beyond the extent of the collision cascade volume) since even in the early work (van der Weg *et al.*, 1976) mixing occurred in bilayer structures only when the ions penetrated past the metal/Si interface. We conclude that efficient mixing occurs only over collision-cascade dimensions.

IV. METAL/METAL BILAYER SYSTEMS

For systems composed of thin films of Al and the near-noble elements (Ni, Pd, Pt), intermetallic phases are formed by ion beam mixing (Mayer *et al.*, 1981). The rate of phase formation seems comparable to that found with silicides as shown in Figure 12; however, there are indications that the first phase formed by ion bombardment may differ from that formed by thermal reactions. We believe that there have not been sufficient experiments to verify the difference in the initial phases that are formed. The difference may be an artifact due to contamination at the interface between the two metal-films.

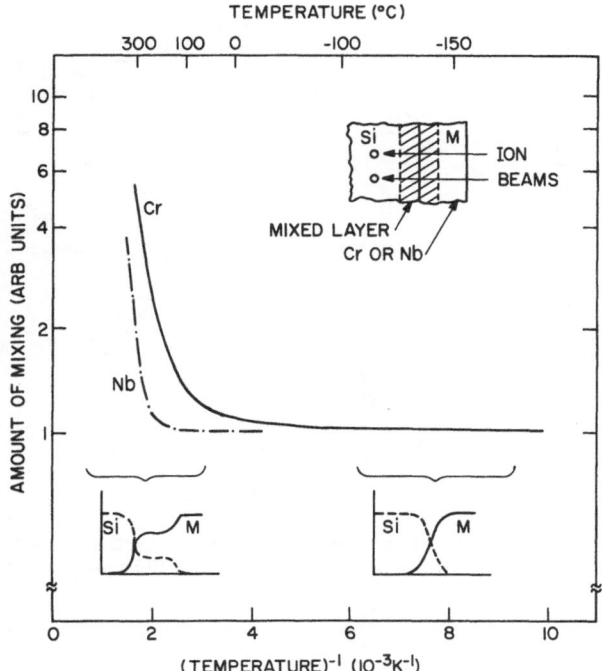

Figure 11: The amount of Si atoms contained in the Cr-Si and Nb-Si mixed layers versus reciprocal implantation temperature. Two distinct regimes are observed: at low temperatures the mixing is dominated by cascade mixing which is relatively insensitive to temperature; at high temperature, radiation-enhanced diffusion and chemical driving forces become dominant and the mixing increases drastically with temperature. (From Mayer *et al.*, 1980 and Matteson *et al.*, 1979.)

Based on the results obtained with silicides and Al intermetallics, we believe that ion beam mixing will lead to the formation of well-defined phases in systems whose equilibrium phase diagram shows the existence of phases with congruent melting points (diagrams similar to those shown in Figure 8 for the Pt-Si system). In addition we believe that the thermally-induced reactions can be used as a guide to estimating ion-induced reactions. Again, there is not sufficient experimental data to substantiate this speculation.

The behavior of Au/Ag thin film couples also shows that the amount of interdiffusion due to ion-mixing can be predicted from thermal data. Thermal treatments lead to more pronounced interdiffusion in polycrystalline Au/Ag film couples than with single-crystal, epitaxial couples due to the rapid diffusion paths offered by grain boundaries. The backscattering spectra shown in Figure 13 also indicate that interdiffusion is more pronounced in polycrystalline systems than in single crystal systems (Mayer *et al.*, 1982). The Au/Ag system also points out the difference between thermal- and ion-beam-induced reactions. In the ion beam mixed single crystal samples, metastable Au-Ag superlattice structures are formed whereas such phases are not found in thermal reactions.

The evidence to date indicates that ion-beam mixing is relatively sluggish in immiscible systems (that some intermixing occurs has been demonstrated in multi-layer systems (Tsaur *et al.*, 1981)). Wang *et al.* (1982) showed that ion mixing proceeded more efficiently in Au-Cu (see also Figure 1) than in the immiscible Au-W system.

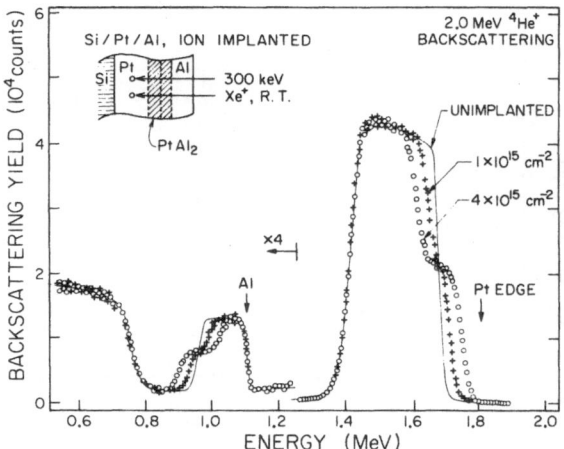

Figure 12: Backscattering spectra for Al-Pt thin-film samples that were implanted at RT with 300 keV Xe ions. The steps in the spectra correspond to the composition, $PtAl_2$ (from Mayer *et al.*, 1981).

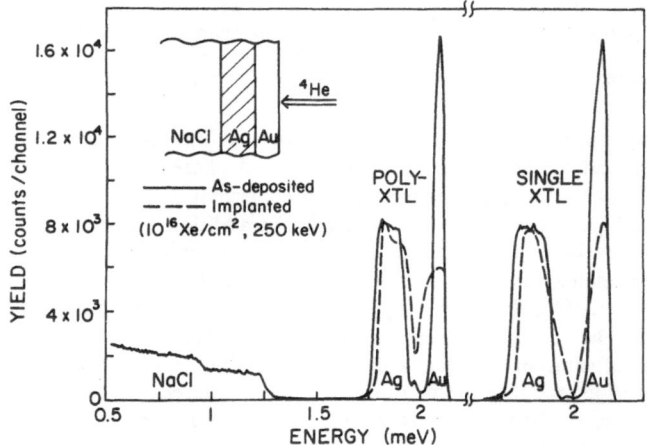

Figure 13: Backscattering spectra from single crystal and polycrystalline Au/Ag bilayers before and after a dose of 10^{16} Xe ion/cm^2 at 250 keV (from Mayer *et al.*, 1982).

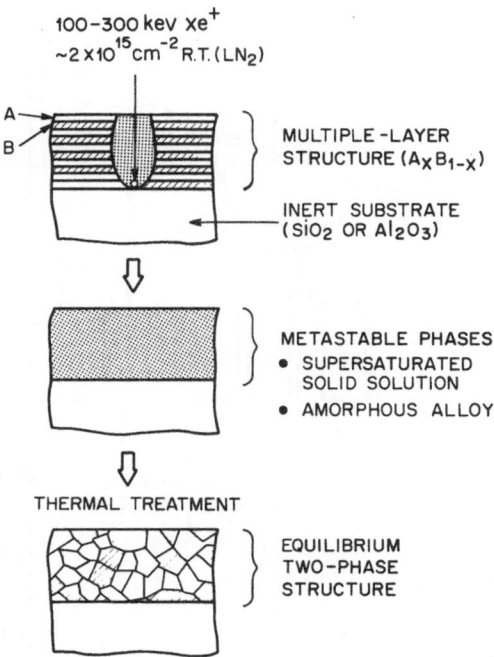

100–300 keV Xe^+
~$2 \times 10^{15} cm^{-2}$ R.T. (LN$_2$)

A
B

MULTIPLE-LAYER
STRUCTURE ($A_x B_{1-x}$)

INERT SUBSTRATE
(SiO_2 OR Al_2O_3)

METASTABLE PHASES
• SUPERSATURATED
  SOLID SOLUTION
• AMORPHOUS ALLOY

THERMAL TREATMENT

EQUILIBRIUM
TWO-PHASE
STRUCTURE

Figure 14: Schematic diagram showing the formation of metastable phases by ion-beam mixing. The metastable phase transforms back to the equilibrium state upon heat treatment.

V. MULTI-LAYER METAL/METAL SYSTEMS

The formation of metastable phases by ion-induced reactions in metal/metal systems can be conveniently studied in multiple-layer systems where the films are thin enough (~200 Å each) so that uniform mixing distributions can be forced even in immiscible systems. The flow of the experiments (Tsaur et al., 1981) is shown schematically in Figure 14. This approach generally leads to the formation of single-phase fcc alloys in Au-based systems (Tsaur et al., 1981b; Tsaur and Mäenpää, 1981). Even with relatively complex Au-based phase diagrams rather simple structures are formed with compositions determined by the ratios of initial film thicknesses. The combination of fcc (Au) and bcc (V) metals leads to fcc alloys formed with Au-rich compositions and bcc alloys formed with V-rich compositions (Figure 15).

From the standpoint of unraveling ion beam mixing mechanisms, multilayer systems are more useful in metastable phase than in diffusion and transport investigations. However there are strong indications that long range transport is involved; there is a pronounced increase in the grain size of the film. For the Au-Ni system (Tsaur and Mäenpää, 1981) shown in Figure 16, the grain size increased from about 200 Å to 1500 Å after irradiation.

We believe that increases of about 1000 Å in grain size due to ion-induced reactions is a clear indication of enhanced diffusion in which vacancy migration probably plays a role.

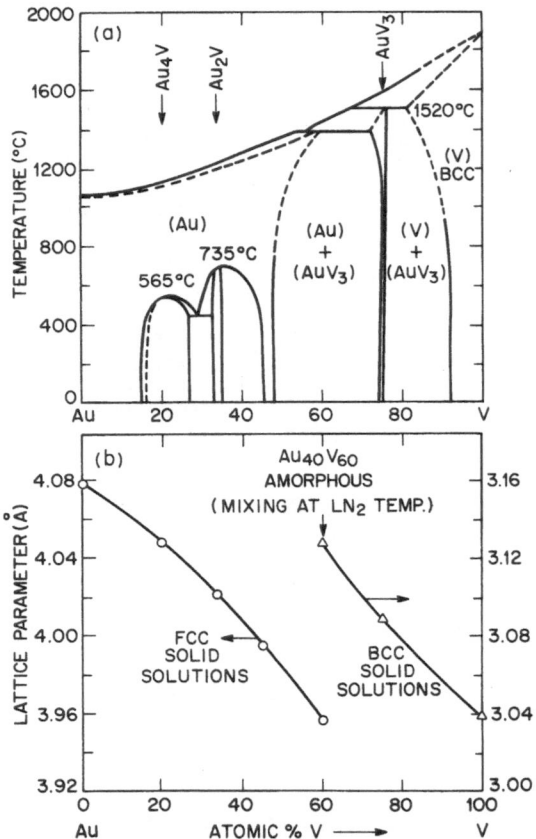

Figure 15: (a) Phase diagram of Au-V system, (b) lattice parameters of the ion-induced metastable phases vs. alloy composition. An amorphous phase is found for $Au_{40}V_{60}$ samples implanted at LN_2 temperature (from Tsaur *et al.*, 1981a).

VI. EUTECTIC SYSTEMS

Ion beam-mixing in eutectic systems has been reviewed by Lau *et al.* (1981). At present only a few systems have been investigated in detail. The Au-Si and Ag-Si systems are simple eutectics with quite different behavior. The Au-Si system not only exhibits strong diffusion effects during sputtering (Blank and Wittmaack, 1981) but also ion induced phase formation. The Ag-Si system on the other hand is relatively insensitive to ion mixing as compared to Au-Si (Lau *et al.*, 1981; Newcombe *et al.*, 1981). We believe that the different behavior is associated with the presence of metastable phases in the Au-Si system. So far, we have found strong interdiffusion in metal/metal and metal/Si systems only when phase formation occurs.

The Au-Si system with its fast interdiffusion rates also provides a good test of *de*-mixing at high ion beam doses. For prolonged irradiation at doses greater than that required to form the metastable phases (composition $\sim Au_7Si_3$), Si migrates to the surface and the metastable phases dissociates into Au and Si (Liu *et al.*, 1982). This behavior of de-mixing resembles

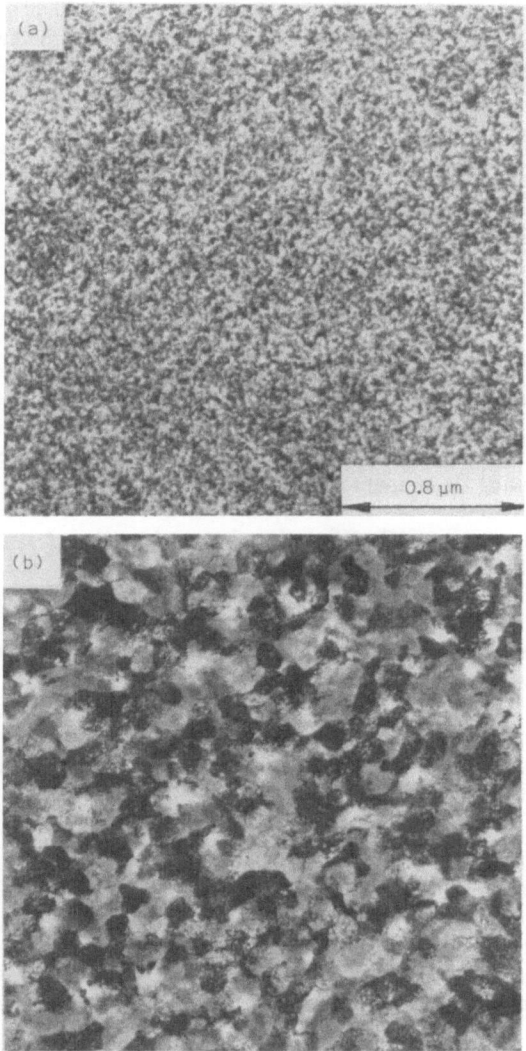

Figure 16: Bright field transmission electron micrographs of $Au_{50}Ni_{50}$ samples (a) as-deposited multilayer, (b) ion-mixed solid solution. A pronounced increase of grain size is observed as a result of ion-beam mixing (from Tsaur and Mäenpää, 1981).

ion-induced segregation discussed Chapter 8 (see also Piller and Marwick, 1978). The thrust of Chapter 8 is that bombardment produced point-defects are responsible for the segregation. This is in agreement with the concepts discussed previously in that the formation (and also the dissociation) of well defined phases with ion-beams is associated with defect migration probably in the form of vacancies.

We believe that ion beam mixing in the thermal-activated regime has many features in common with ion bombardment assisted precipitation and segregation as discussed in

Chapter 8. This latter subject came out of research groups dealing with radiation damage involving high doses of light ions incident on bulk material held at elevated temperatures. The ion beam mixing community has concentrated mainly on ion implantation and sputtering. Only recently (see for example, Averback *et al.*, 1982) have there been attempts to bridge the two communities. The Ni-Si system provides one common area and the eutectics Au-Si and Ag-Si may provide another.

At this point we note also that surface segregation (Gibbsian segregation) due to ion-bombardment probably plays a large role in preferential sputtering; that is, preferential sputtering may be dominated more by thermodynamic driving forces than by preferential energy transfers within the collision cascade.

VII. CONCLUSIONS AND SPECULATIONS

The penetration of heavy ions through the interface between a deposited layer and substrate leads to intermixing between layer and substrate and serves as an attractive alternative to direct, high-dose ion implantation as a method for material modification. Both thermodynamic forces and energy transfers within the collision cascade around the ion track play a role in determining the amount of intermixing.

Thermodynamic Forces:

1) Immiscible systems (Cu-W) tend not to mix whereas systems (silicides) which have equilibrium phases mix efficiently.

2) The amount of intermixing in similar systems is correlated with thermal diffusion properties (silicides) and grain boundary diffusion (Au-Ag).

3) The existence of well-defined phases of uniform thickness indicates that vacancies are at work.

4) Above a certain temperature that depends on the system, the amount of intermixing is temperature dependent indicating a thermally activated process.

5) The formation of metastable phases and beam-induced de-mixing and segregation indicates thermodynamic forces.

Recoil mixing and ballistic effects in collision cascades:

1) The amount of intermixing depends upon the energy dissipation (eV/Å) within the collision cascade.

2) The nearly temperature independent regime of ion-beam mixing has many of the characteristics of recoil mixing effects calculated for thin markers.

3) Efficient mixing occurs primarily over the dimensions of the collision cascade.

At present, we cannot predict in advance the amount of mixing for any given new system. We can guess reasonably well from thermal reaction data but important exceptions have been found. Metastable phases have been formed but we lack a systematic framework for searching for these phases. This is a clear correlation with sputtering phenomena and ion-induced precipitation and segregation. Altogether, an interesting field.

REFERENCES

Andersen, H. H. (1979), Appl. Phys. *18*, 131.

Averback, R. S., Rehn, L. E., Okamoto, P. R. and Cook, R. E. (1981), Nucl. Instr. and Meth. *182/183*, 79.

Averback, R. S., Thompson, L. J., Moyle, J. and Schalit, M. (1982), J. Appl. Phys. *53*, 1342.

Blank, P. and Wittmaack, K. (1979), Radiat. Effects Lett. *43*, 105.

Chiang, S. W., Chow, T. P., Riehl, R. F. and Wang, K. L. (1981), J. Appl. Phys. *52*, 4027.

Chapman, G. E., Lau, S. S., Matteson, S. and Mayer, J. W. (1979), J. Appl. Phys. *50*, 6321.

d'Heurle, F. M., Petersson, C. S. and Tsai, M. Y. (1980), J. Appl. Phys. *51*, 5976.

Kanayama, T., Tanoue, N. and Tsurushima, T. (1979), Appl. Phys. Lett. *35*, 222.

Lau, S. S, Tsaur, B. Y., von Allmen, M., Mayer, J. W., Stritzker, B., White, C. W. and Appleton, B. (1981), Nucl. Instr. Meth. *182/183*, 97.

Liau, Z. L., Brown, W. L., Homer, R. and Poate, J. M. (1977), Appl. Phys. Lett. *30*, 626.

Liau, Z. L., Mayer, J. W., Brown, W. L. and Poate, J. M. (1978), J. Appl. Phys. *49*, 5295.

Liau, Z. L., Tsaur, B. Y. and Mayer, J. W. (1979), J. Vac. Sci. Technol. *16*, 121.

Liau, Z. L. and Mayer, J. W. (1980), in "Ion Implantation" (J. K. Hirvonen, ed.) (Volume 18 in "Treatise on Materials Science and Technology (H. Herman, ed.)), Ch. 2, Academic Press, New York.

Littmark, U. and Hofer, W. O. (1980a), Nucl. Instr. and Meth. *168*, 329.

Littmark, U. and Hofer, W. O. (1980b), Nucl. Instr. and Meth. *170*, 177.

Liu, B. X., Wielunski, L., Nicolet, M. A. and Lau, S. S. (1982), (MRS).

Matteson, S., Roth, J. and Nicolet, M. A. (1979), Rad. Effects *42*, 217.

Matteson, S., Paine, B. M. Grimaldi, M. G., Mezevy, G. and Nicolet, M. A. (1981a), Nucl. Instr. and Meth. *182/183*, 43.

Matteson, S., Paine, B. M. and Nicolet, M. A. (1981b), Nucl. Instr. and Meth. *182/183*, 53.

Mayer, J. W., Eriksson, L. and Davies, J. A. (1970), "Ion Implantation in Semiconductors," Academic Press, New York.

Mayer, J. W., Lau, S. S., Tsaur, B. Y., Poate, J. M. and Hirvonen, J. K. (1980), in "Ion Implantation Metallurgy" (C. M. Preece and J. K. Hirvonen, eds.), p. 37, The Metallurgical Society of AIME, New York.

Mayer, J. W., Tsaur, B. Y., Lau, S. S. and Hung, L. S. (1981), Nucl. Instr. and Meth. *182/183*, 1.

Mayer, J. W., Fastow, R., Galvin, G., Hung, L. S., Nastasi, M., Thompson, M. O. and Zheng, L. R. (1982), in "Metastable Material Formation by Ion Implantation" (S.T. Picraux and W. J. Choyke, Eds.), p. 125, North-Holland.

Newcombe, R., Christodoulides, C. E., Carter, G. and Tognetti, P. (1980), Rad. Effects Lett. *50*, 51.

Ottaviani, G. and Mayer, J. W. (1981), in "Reliability and Degradation" (M. J. Howes and D. V. Morgan, eds.), Chapter 2, Wiley-Interscience, New York.

Pickering, H. W. (1976), J. Vac. Sci. Technol. *13*, 618.

Piller, R. C. and Marwick, A. D. (1978), J. Nucl. Mater. *71*, 309.

Schulson, E. M. (1979), J. Nucl. Mater. *83*, 239.

Sigmund, P. and Gras-Marti, A. (1981), Nucl. Instr. and Meth. *182/183*, 25.

Thompson, D. A. (1981), Rad. Effects *56*, 105.

Tsai, M. Y., Petersson, C. S., d'Heurle, F. M. and Maniscalco, V. (1980), Appl. Phys. Lett. *37*, 295.

Tsaur, B. Y., Liau, Z. L., Lau, S. S. and Mayer, J. W. (1979a), Thin Solid Films *63*, 31.

Tsaur, B. Y., Liau, Z. L. and Mayer, J. W. (1979b), Appl. Phys. Lett. *34*, 168.

Tsaur, B. Y., Matteson, S., Chapman, G., Liau, Z. L. and Nicolet, M. A. (1979c), Appl. Phys. Lett. *35*, 825.

Tsaur, B. Y., Lau, S. S. and Mayer, J. W. (1979d), Appl. Phys. Lett. *35*, 225.

Tsaur, B. Y., Mayer, J. W. and Tu, K. N. (1980a), J. Appl. Phys. *51*, 5326.

Tsaur, B. Y., Mayer, J. W., Graczyk, J. F. and Tu, K. N. (1980b), J. Appl. Phys. *51*, 5334.

Tsaur, B. Y. and Hung, L. S. (1980), Appl. Phys. Lett. *37*, 922.

Tsaur, B. Y. (1980), in "Thin Film Interfaces and Interactions" (J. E. E. Baglin and J. M. Poate, eds.), p. 205, Electrochemical Soc. Princeton, Proceedings, Vol. 80-2.

Tsaur, B. Y., Lau, S. S., Hung, L. S. and Mayer, J. W. (1981a), Nucl. Instr. and Meth. *182/183*, 67.

Tsaur, B. Y. and Mayer, J. W. (1981), Phil. Mag. *A43*, 345.

Tsaur, B. Y., Lau, S. S. and Mayer, J. W. (1981b), Phil. Mag. *B44*, 95.

Tsaur, B. Y. and Mäenpää, M. (1981), J. Appl. Phys. *54*, 728.

Tu, K. N. and Mayer, J. W. (1978), in "Thin Films--Interdiffusion and Reactions," (J. M. Poate, K. N. Tu and J. W. Mayer, eds.), Chapter 10, Wiley Interscience, New York.

van der Weg, W. F., Sigurd, D. and Mayer, J. W. (1976), in "Applications of Ion Beams to Metals," (S. T. Picraux, E. P. EerNisse and F. W. Vook, eds.), p. 209, Plenum, New York.

Wach, W. and Wittmaack, K. (1981), J. Appl. Phys. *52*, 3341.

Wang, K. L., Bacon, F. and Riehl, R. F. (1979), J. Vac. Sci. Technol. *16*, 1909.

Wang, K. L., Chiang, S. W., Bacon, F. and Riehl, R. F. (1980), Thin Solid Films, *74*, 239.

Wang, Z. L., Westendorp, H. and Saris, F. W. (1982), (MRS).

Wielunski, L., Lien, C. D., Liu, B. X. and Nicolet, M. A. (1982), (MRS).

CHAPTER 10

SPUTTERING AND COMPOSITIONAL CHANGES

H. WIEDERSICH

Materials Science Division, Argonne National Laboratory, Argonne, Illinois

CONTRIBUTORS: H. H. Andersen, N. Q. Lam, L. E. Rehn and
H. W. Pickering

I. INTRODUCTION

Sputtering of atoms from a solid surface as a consequence of the impact of energetic particles was first observed more than a century ago (Grove, 1853). However, most of the present experimental knowledge and theoretical understanding of the sputtering phenomenon has been gained during the past 30 years. Extensive, recent reviews of many aspects of the sputtering phenomenon can be found in volumes edited by Varga *et al.* (1980), by Cobic (1980), and by Behrisch (1981, 1982). That the composition of an alloy near the surface can be affected by sputtering was shown experimentally more than twenty years ago by Gillam (1959). An increasing interest in sputtering of multicomponent materials has developed during the past decade, and much of the experimental work and understanding in this area has been reviewed recently by Andersen (1981).

As recently as five years ago the most sophisticated model of sputter-induced compositional changes in the near-surface region of an alloy was that of Ho *et al.* (1976), which described the development of a layer of altered composition during sputtering of a binary alloy essentially as follows: Preferential sputtering removes atoms of one component of the alloy with a greater probability than that given by the composition of the near-surface region. Displacement or cascade mixing maintains an approximately uniform composition within the so-called "altered layer," which has a thickness about equal to the range of bombarding ions. As the composition of the altered layer becomes depleted in the preferentially sputtered species, a steady state is gradually set up in which the composition of the sputtered flux is equal to that of the initially uniform alloy target. This simple model described some observations satisfactorily. However, it soon became apparent that experimentally observed transient times were frequently longer (Ho, 1978) and that the depths of altered layers were substantially larger, especially during sputtering at elevated temperatures (Rehn *et al.*, 1979), than could be accounted for by this simple model. In fact, it now seems that at least five or six distinct processes may contribute to compositional changes of the near-surface region of alloys during sputtering: (a) preferential sputtering, (b) recoil implantation, and displacement mixing, (c) radiation-enhanced diffusion, (d) Gibbsian adsorption, and (e) radiation-induced segregation. The relative importance of these processes in the evolution of the sputtered flux composition and of the compositional profile of the target will depend on the type of alloy, the temperature, and the species, energy and flux density of the sputter ions.

A good qualitative and partially quantitative understanding exists for most of the individual processes mentioned above. However, systematic investigations of alloy sputtering that consider all of these processes and their potential interactions have not been attempted. The purpose of the following discussion is to characterize the processes in simple physical terms and to point out under what conditions they may become important for compositional changes in near-surface regions of sputtered alloys. In Section III, results of recent model calculations are presented which indicate the effects to be expected from the five major processes.

The discussion will be limited to metallic alloys where ionization processes are unimportant for defect production and sputtering. The emphasis is placed here on compositional changes of the target, which are related to the composition of the integrated sputtered flux. Hence, the promising recent work on the angular dependence of the sputtered flux composition during alloy sputtering, such as that by Andersen *et al.* (1980, 1981a), will not be reviewed. Also, we will not deal with the interesting phenomenon of the development of non-planar surface topologies, such as sputter cone formation; this phenomenon is discussed by several authors in the Symposium volume edited by Varga *et al.* (1980), and reviewed by Carter *et al.* (1982).

II. PROCESSES INVOLVED IN SURFACE COMPOSITIONAL CHANGES

In this section we will characterize the individual processes that can contribute to compositional changes in near-surface regions during sputtering. It should be emphasized from the outset that the existence of each of these processes has been demonstrated experimentally, and each has been treated theoretically to some extent. However, the importance of some of the processes in a typical sputter situation remains to be established. We will proceed from those processes which are predominantly athermal, i.e., which do not require thermally activated motion, namely preferential sputtering, recoil implantation and displacement mixing, to those in which thermally activated atomic migration is an essential part of the process, i.e., radiation-enhanced diffusion, Gibbsian adsorption and radiation-induced segregation.

A. Preferential Sputtering

A simple concept and definition of preferential sputtering can be arrived at as follows. The yield or number of A-atoms of an alloy per incident ion in the flux of sputtered atoms can be written as

$$Y_A = \int_0^\infty \sigma_A(x) \cdot (C_A(x)/\Omega) \, dx \tag{1}$$

where $\sigma_A(x)$ is the cross section for A-atoms at a depth $x \geq 0$ to be ejected from the surface, $x = 0$, into the region $x < 0$ per incoming ion; $C_A(x)$ is the atomic fraction of A in the alloy at depth x; and Ω is the mean atomic volume. For the present purpose we will define $p_A(x) \equiv \sigma_A(x)/\Omega$, where $p_A(x)$ is the probability per unit depth that an A-atom present at depth x is ejected by an incoming ion. The yield of A-atoms then takes the form

$$Y_A = \int_0^\infty p_A(x) \cdot C_A(x) \, dx \simeq \bar{p}_A \cdot C_A^s \tag{2}$$

In this form, the distinction, introduced by Sigmund (1981), between the primary and secondary effects in alloy sputtering is made explicit. The primary effects are those related to the individual sputtering events and the physical variables contributing are all contained in the sputter probability, p_A, which depends on the type and energy of the incoming ion, the type of the sputtered atom and its surface binding energy, etc. Generally, the values of the sputter probabilities, p_i, will not be equal for different atomic species and, consequently, preferential sputtering will occur.

The secondary effects in alloy sputtering enter into Eq. (2) via the atomic concentration, C_A, which gives the probability that a site is occupied by an A-atom. All processes mentioned above that affect compositional changes during sputtering, except preferential sputtering, enter primarily through their effects on the near-surface concentration. For example, Gibbsian adsorption of the element A to the surface will increase the yield of A above that of the same alloy having a uniform composition up to and including the outermost atomic layer. Similarly, preferential recoil implantation as discussed by Sigmund (1979), Sigmund and Gras-Marti (1980, 1981), and Gras-Marti and Sigmund (1981) changes the composition of an initially uniform alloy in the near surface region. As a consequence, the sputter yields of the alloying components will be affected through the factor $C_A(x)$ in Eq. (2).

From a practical point of view, the sputtered atoms come predominantly from a rather shallow layer, as shown by Sigmund (1969) and in more detail by Falcone and Sigmund (1981). For low-energy sputtering the contributions fall off approximately exponentially with depth, with a decay length on the order of two atom layers. Therefore, the integral in Eq. (2) can be replaced to a good approximation by $\bar{p}_A \cdot C_A^s$ where $\bar{p}_A$ is the average total probability for an A-atom present in the surface layer to be sputtered off per incident ion and C_A^s is the average atomic concentration of A in the layer. The thickness of this layer is not well defined but should be taken as one or two atomic layers for determining C_A^s since the origin of sputtered ions is heavily weighted toward the first atomic layer.

The differences in the ejection probabilities $\bar{p}$ for the component atoms in an alloy are caused primarily by differences in (a) the amounts of energy and momentum transferred to atoms of different masses from the same projectile in similar collisions, and (b) the energies required by the alloy components to overcome their different surface binding energies. Differences in (a) exist not only for different chemical species but also for isotopes of the same element with different masses; Russel *et al.* (1980) have observed preferential sputtering of the lighter isotope, ^{40}Ca, relative to ^{44}Ca from CaF_2 and other Ca-bearing minerals. Andersen (1981) and Sigmund *et al.* (1981a) find that the results of Russel *et al.* are in good quantitative agreement with sputtering theory.

Preferential sputtering in alloys is generally caused by both the mass and the surface binding effects. An extensive recent compilation of experimental observations on preferential sputtering can be found in the review on sputtering of multicomponent metals by Andersen (1981). As Andersen points out in the discussion of the compilation, many of the observations reported (most frequently in the form of a sputter-induced change in the near-surface concentration) may have resulted not solely or even primarily from preferential sputtering, but from a combination of preferential sputtering and secondary effects to be discussed in the remainder of this section.

Before we leave this subsection, some general comments regarding the steady state or stationary state are in order. It has long been recognized, e.g., by Ho *et al.* (1976), that continuous sputtering of a semi-infinite alloy target of uniform bulk composition must eventually result in a steady state in which the composition of the flux of sputtered atoms leaving the surface equals the composition of the bulk alloy. We consider a slab of the target extending from the surface, $x = 0$, to a fixed depth, x_0, into the bulk. The coordinate origin and x_0 move with the receding surface. The depth, x_0, is chosen sufficiently large so that the slab extends beyond the ultimate thickness of the compositionally altered layer. The surface of an alloy with n components recedes with the velocity

$$v = j \sum_{k=0}^{n} \omega_k \ Y_k - j\omega_0 \equiv j \cdot \alpha \tag{3}$$

where j is the sputter ion current density, ω_k is the partial atomic volume of the kth component of the alloy, and ω_0 and Y_0 are the partial atomic volume and yield of the sputter ions, respectively. The second term in Eq. (3) is the contribution to the velocity from the implanted sputter ions, assuming complete relaxation normal to the surface to accommodate the implanted ions. The net accumulation rate of species i in the slab $0 \leqslant x \leqslant x_0$ is given by

$$\frac{dN_i}{dt} = j\left\{ C_i^b \ (\alpha/\Omega^b) - Y_i \right\} \tag{4}$$

C_i^b and Ω^b are the atomic concentration of component i and the average atomic volume of the bulk alloy, respectively. The first term represents the gain of i-atoms at $x = x_0$ due to the shift of the slab into the bulk at a velocity v, and the second term the loss by sputtering. The net rate of accumulation of the implanted ions is correspondingly

$$\frac{dN_0}{dt} = j\left\{ 1 - Y_0 \right\} \tag{5}$$

where we have neglected back-reflection of ions. Eventually, when steady state is achieved, the term in the curly brackets of eqs. (4) and (5) must vanish. Thus, at steady state,

$$Y_1 : Y_2 : Y_3... = C_1^b : C_2^b : C_3^b... \tag{6}$$

and

$$Y_0 = 1 \tag{7}$$

i.e, the ratios of the alloy components in the sputtered flux are the same as those of the bulk alloy and their concentrations are just uniformly "diluted" by the re-emitted sputter ions. This result is essentially exact, and it is important to realize that the steady-state flux of sputtered atoms does not contain any information on preferential sputtering, or on the secondary processes that are involved in the sputter-induced compositional changes of the target.

Information about the preferential sputtering, however, can be obtained from the steady-state concentrations in the near surface region via the approximation given in Eq. (2). Combining this approximation with Eq. (6) we obtain for the total sputter probabilities

$$\bar{p}_1 : \bar{p}_2 : ... = \left(C_1^b/C_1^s \right) : \left(C_2^b/C_2^s \right) : ... \tag{8}$$

i.e., after steady state is attained, the sputter probabilities are proportional to the ratio of the bulk and surface concentrations of the element in question. It should be noted that C_i^s is the concentration of i properly averaged over the depth of origin of sputtered atoms and, therefore, heavily weighted toward the first and second atom layers (Falcone and Sigmund, 1981). Thus, surface concentration measurements by low-energy Auger electron spectroscopy (AES) with escape depths comparable to those of the sputtered atoms appear to be best suited for obtaining values for preferential sputtering from measurements of steady-state surface concentrations. Low energy ion scattering spectroscopy (ISS), with its high weighting of the top atom layer, appears to be another good choice.

Provided the sputter ions are sufficiently bound once they are incorporated in the alloy, their total sputter probability can be obtained from their steady-state surface concentration, C_0^s,

$$p_0 \simeq 1/C_0^s \tag{9}$$

This equation is not applicable to sputter ions, such as noble gases, that may form bubbles and leave the solid spontaneously when the free surface approaches the location of implanted sputter ions.

B. Recoil Implantation and Displacement Mixing

Recoil implantation and displacement mixing, frequently also termed cascade or recoil mixing, are both consequences of the strong interactions of the incoming ion beam with atoms of the solid leading to significant ($\geqslant$ nearest-neighbor atom distance) displacements of atoms. In recent years, there have been a significant number of theoretical treatments of these processes that deal with the relocation of atoms caused directly by displacements; see, e.g., Andersen (1979), Hofer and Littmark (1979), Sigmund (1979), Littmark and Hofer (1980), Winterbon (1980), Sigmund and Gras-Marti (1980, 1981) and Gras-Marti and Sigmund (1980). The process of displacement mixing is discussed in chapters 7, 8 and 9 of this volume as it relates to collision cascades, precipitate stability and ion beam mixing, respectively. For the present purpose the distinction between recoil implantation and displacement mixing will be made on the following basis: The incoming ion beam imparts its momentum to the atoms (and electrons) of the solid. Hence, the momentum distribution of atoms during the displacement process is not isotropic (Littmark and Sigmund 1975), and atoms will be relocated preferentially in the beam direction. This in itself does not lead to any significant net atom transport in the beam direction because the solid will relax to approximately its normal density, i.e., the flux of recoiling atoms in the beam direction is compensated by a uniform flux of atoms due to relaxation in the opposite direction. In alloys, however, the relocation cross section and the range of the recoiling atoms depend on the charge and mass of the nucleus (see Sigmund 1979), in such a way that, generally, the lighter component atoms will be transported relative to the heavier components in the beam direction. This situation can be described as a flux of atoms of some of the alloy components toward deeper regions in the target, compensated by an opposite flux of the remainder of the components to maintain the atomic density of the solid at a proper value, i.e., the net flux of atoms is approximately zero across any plane parallel to the surface inside the target. The expression "recoil implantation" is used here for this net transport parallel to the beam direction of some types of atoms relative to other types. The rapid transfer of energy from recoiling atoms of the solid to other atoms leads to an efficient randomization of recoil directions within cascades. As a consequence, most of the relocation events of atoms in energetic cascades lead to isotropic mixing rather than to recoil implantation, according to current theory (Littmark and Hofer, 1980; Sigmund and Gras-Marti, 1981). A predominance of isotropic mixing may also extend to rather low primary recoil energies. King and Benedek (1981) found from molecular dynamics calculations that a dominant fraction of mixing occurs by closed replacement chains for primary recoil energies between about 100 and 450 eV (largest energy used in the calculations). Closed replacement chains are loops in which predominantly nearest-neighbor atoms have been replaced in a ring-like exchange fashion during the collisional and subsequent "cooling" phases of a displacement event without leaving a vacancy and interstitial behind. In contrast, open replacement chains are bounded by a vacancy at one end and an interstitial atom at the other. Closed chains lead to approximately isotropic mixing. The molecular dynamics calculations also indicate that the number of replacements (or of nearest or near-neighbor "diffusion jumps") in a displacement event far exceeds the number of surviving defect pairs (by a factor $\simeq 30$). Thus, we can expect a mean square diffusion distance of atoms, $<x^2> \simeq 30b^2$, per defect pair produced, with b the nearest-neighbor distance in the solid. This estimate is about twice that given by Andersen (1979). Expressed in a different way, an atomically sharp interface or feature is broadened by $<x^2>^{1/2}$ or by about five atom planes for a dose of one displacement per atom. The broadening increases proportionally to the square root of dose. This rough estimate is consistent with broadening of sharp Ge-Si interfaces during sputter-profiling as observed by Etzkorn and Kirschner (1980).

Whereas the approximately isotropic, random walk displacement mixing causes a symmetric broadening of a thin marker layer embedded in a target, recoil implantation will shift the center of the distribution of the marker atoms. The broadening by displacement mixing of thin markers of a number of elements in Si has been experimentally confirmed by Tsaur *et al.* (1979) and Matteson *et al.* (1981). These authors also showed that the half width of the marker distribution increases proportionally to the square root of the ion dose, as expected from a diffusion-like process. No experimental measurements on the shift of marker distributions by recoil implantation, or of the dose dependence, which should be linear, seem to be available. The theoretical treatments by Littmark and Hofer (1980) and Sigmund and

Gras-Marti (1981) show that the anisotropy of atom relocation produced by recoil implantation is generally small compared to the isotropic relocation caused by displacement mixing. However, a low concentration "tail" resulting from high-energy primary recoils is to be expected.

Both recoil implantation and displacement mixing should be temperature independent to a good approximation, since the energies of the atoms involved in the displacement events are rather large compared to thermal energies. Experimentally, however, Matteson *et al.* (1979) and Averback *et al.* (1982) have found that displacement mixing by ion beams at Nb-Si and Ni-Si interfaces is temperature independent only below about 500 and 200K, respectively. Figure 1 shows an Arrhenius plot of the mixed-layer thickness at Ni-Si interfaces. The total mixing seems to result from the superposition of a temperature-independent process (dashed line) and a thermally activated process with an apparent activation energy of ~0.1 eV. Averback *et al.* (1982) have shown that the thermal process is dose-rate independent within experimental error. Therefore, ordinary radiation-enhanced diffusion cannot explain the temperature dependence of the mixing process.

C. Radiation-enhanced Diffusion

Radiation-enhanced diffusion has been reviewed recently by Sizmann (1978) and Rothman (1982) in detail. Here we will discuss the enhancement of diffusion with emphasis on its effects on subsurface compositional changes during sputtering. A short discussion from a somewhat different point of view is contained in Chapter 8.

At sufficiently high temperatures, single point defects and small clusters of vacancies and interstitials undergo thermally activated motions. Defect motion involves, of course, motion of atoms. It has been well established that thermal diffusion in crystalline materials proceeds by defect mechanisms. In metals, thermal diffusion occurs predominantly by the exchange of positions between atoms and vacancies, and the diffusion coefficient of A-atoms in an alloy can be expressed in the simple form

$$D_A^v = \left(b^2/6\right) z_v P_v \nu_{Av} \tag{10}$$

where b is the nearest-neighbor distance, z_v is the coordination number, P_v is the probability that a specific nearest-neighbor site of an A-atoms is occupied by a vacancy, and $\nu_{Av} = \nu_{Av}^0 \, exp\,(-E_A^v/kT)$ is the exchange frequency of the A-atom with the vacancy which can be expressed by an attempt frequency, ν_{Av}^0, and a Boltzmann factor containing the activation energy of motion for the exchange, E_A^v. Provided no binding or repulsion between vacancies and A-atoms exists, P_v equals the atomic fraction of vacancies, C_v, present in the alloy. At thermal equilibrium,

$$C_v = \exp(-G^v/kT) \tag{11}$$

where G^v is the Gibbs free energy of formation of a vacancy. During ion bombardment, the vacancy concentration is increased above that given by Eq. (11) by an amount determined by the balance between the rate of formation of vacancies by displacements and the losses due to recombination with interstitials and to vacancy sinks such as surfaces and dislocations.

Diffusion is enhanced not only by the excess vacancy population present during irradiation, but also by the presence of mobile excess interstitial defects. The motions of interstitials and of vacancies move atoms in similar ways, and the diffusion coefficient of A-atoms moved by interstitials, D_A^i, can be written in an analogous way to D_A^v in Eq. (10).

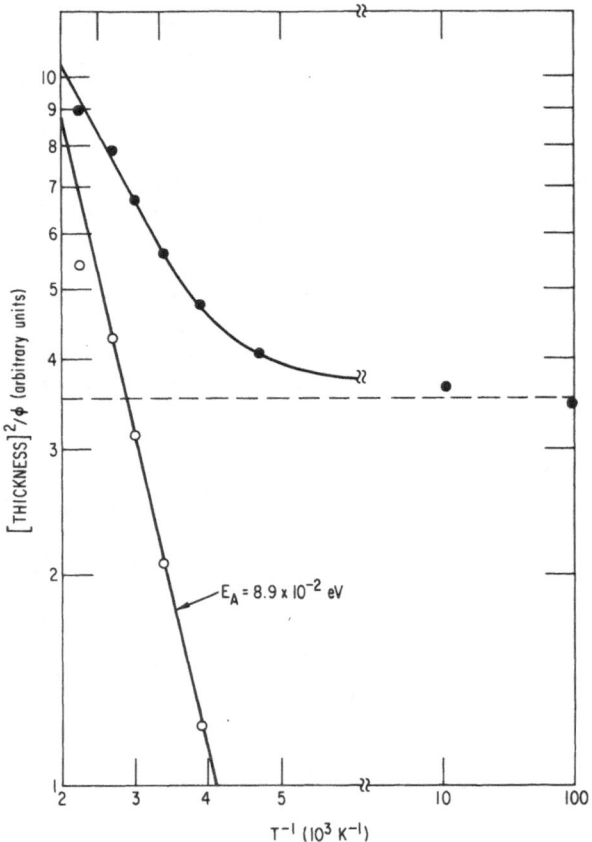

Figure 1   Arrhenius plot of the rate constant for ion beam mixing (solid circles) at a Ni-Si interface for 250-keV Ar⁺ irradiation. The rate constant is proportional to the interdiffusion coefficient. The values shown by open circles represent a thermal contribution to the mixing and are obtained by subtracting the temperature-independent displacement mixing component, obtained from the low-temperature measurements, from the rate-constant values. (Averback *et al.*, 1982)

However, the contribution of interstitials to thermal diffusion is generally negligible in metals because the formation energy of interstitial atoms is large compared to that of vacancies and, therefore, the equilibrium interstitial concentration is many orders of magnitude smaller than that given for vacancies by Eq. (11).

Neglecting contributions of mobile defect clusters and other complications such as correlation effects, as discussed by Manning (1981) and others, we will write the diffusion coefficient of A-atoms due to thermal motion of defects in the simple form

$$D_A = d_{Av} C_v + d_{Ai} C_i \tag{12}$$

where d_{Av} and d_{Ai} are diffusivity coefficients containing jump distances and other geometrical factors, jump frequencies, and factors related to preferential association of A-atoms with vacancies and interstitials, respectively (see Eq. (10)). The diffusivity coefficients are discussed in more detail by Wiedersich *et al.* (1979). Equation (12) relates the diffusion coefficient in an alloy to the concentration of defects; this concentration is determined primarily by the radiation conditions, via the coefficients d_{Av} and d_{Ai}, which are characteristic of the alloy and the defects involved in the diffusion process.

The radiation enhancement of the defect concentration has been discussed by Damask and Dienes (1963), and in more detail (and to various degrees of sophistication) in many papers concerned with the void swelling problem, such as those of Wiedersich (1972), Brailsford and Bullough (1978), and Mansur (1979). Here we make only a few qualitative, general remarks. Under most sputtering and ion-bombardment conditions of interest here, a quasi steady state with regard to point-defect concentrations is rapidly set up in which the production of defects by the incoming ions is balanced by the loss of defects by mutual recombination, clustering, and annihilation at sinks.

At low temperatures, high defect concentrations result and losses occur predominantly by spontaneous recombination of newly created defects with nearby defects left over from previous displacement events. This process is approximately temperature independent, adds to the "diffusion" caused by the displacement mixing discussed in the previous section, and is difficult to distinguish experimentally from the latter process. The average distance over which the defects move between creation and annihilation is approximately equal to, but somewhat shorter than, the average recombination radius. The spontaneous recombination process puts a lower limit on "radiation-enhanced" diffusion.

At somewhat higher temperatures, defects with low activation energies will undergo a number of thermally activated jumps before being annihilated by recombination with pre-existing or newly formed defects. As a consequence of the increased number of jumps of the fast-moving defect species between creation and annihilation, diffusion is increased. Diffusion increases even more quickly as the temperature is increased to levels where the defect species with the higher activation energy, usually vacancies, becomes mobile and contributes to diffusion. As the thermal diffusivity of the defects increases, their lifetime between creation and annihilation decreases and, hence, their concentration decreases. Nevertheless, the contribution to the diffusion of atoms increases because the average number of jumps during the defect lifetime is larger at lower defect concentration as a consequence of the reduced recombination probability.

The steady-state defect concentration may become low enough at intermediate temperatures that most defects annihilate at sinks. If the sink density is constant, the average number of defect jumps between creation and annihilation is constant and, hence, the contribution to diffusion is temperature independent in this regime.

Finally, at high temperatures the thermal vacancy concentration, Eq. (11), becomes large and limits the number of jumps during the lifetime of excess defects. This reduces the radiation-enhancement contribution to diffusion, and thermal diffusion becomes dominant.

Before we leave this subsection some comments should be made on the spatial distribution and extent of radiation-enhanced diffusion, and on the consequences of the finite rate of sputtering on diffusional redistribution in the near-surface region.

The defect production rate during ion bombardment is a function of depth. At high incident ion energies (larger than a few tens of keV), the defect production rate increases with depth as the velocity of the ions decreases, goes through a maximum, and drops quickly to zero as the ions approach the end of the range. At low ion energies the maximum production rate is near the surface. In the spontaneous recombination regime at low temperature, "radiation-enhanced" diffusion is limited in the same way as displacement mixing to the depth of the region in which defects are produced. At higher temperatures where defect migration becomes important, radiation-enhanced diffusion not only varies in accordance with the local defect production rates within the defect production zone, but also extends into the region beyond this zone as a consequence of the escape of mobile defects into the undamaged substrate. These defects will contribute to enhanced diffusion until they are annihilated at sinks or by recombination. The lifetime of defects in the undamaged region should be long compared to that in the damage zone and, therefore, significant diffusion can be expected well beyond the ion range at elevated temperatures. Rehn *et al.* (1979, 1980) have observed copper depletion in Ni-Cu alloys after elevated-temperature sputtering to depths several orders of magnitudes larger than the range of the sputter ions. The observed diffusion distances were significantly larger than could be accounted for by thermal diffusion without enhancement.

As discussed in Section IIA, when sputtering of a uniform, thick alloy target proceeds, a steady state is eventually approached where the composition of the sputtered flux is equal to the composition of the base alloy, and the compositionally altered layer moves with the receding surface into the alloy target. The depth of, and compositional distribution within, the altered layer are determined by diffusion processes within the near-surface region of the target. In the following we will discuss in a highly simplified way the extent of the altered layer and its relation to the diffusion coefficient during sputtering.

Preferential removal of certain alloying elements from the surface of an alloy and the development of a compositionally altered layer are not unique to sputtering but also occur, e.g., during the preferential evaporation or electrolytic dissolution of Zn from Cu-Zn alloys (Pickering 1970). The development of an altered layer during electrolytic dissolution as a consequence of preferential removal of an element from the surface and diffusion in the near-surface region of the alloy has been treated theoretically by Pickering and Wagner (1967) and by Holliday and Pickering (1973). It was applied to sputtering of alloys by Pickering (1976), and in a somewhat more sophisticated way by Ho (1978).

Here we will consider only the steady-state situation in a highly simplified form in order to obtain some rough, order-of-magnitude estimates of the interrelation between the diffusion coefficient, the altered layer thickness, and the surface recession rate during sputtering of a binary alloy. The surface of the target at time t is assumed to be normal to the z-axis of a laboratory coordinate system at the location $z_s(t) = z_s(0) + v \cdot t$, where v is the surface recession velocity due to sputtering. $C_A(z,t)$ is the concentration of A in the semi-infinite target at depth z and time t. Fick's second law relates the rate of change of the local concentration to the spatial variation of the concentration and the diffusion coefficient D as follows:

$$\frac{\partial C_A}{\partial t} = \frac{\partial}{\partial z}\left[D \, \frac{\partial C_A}{\partial z}\right] \tag{13}$$

for $z \geqslant z_s(t)$. If we make the simplifying assumption that the diffusion coefficient does not depend on z, Eq. (13) can be solved analytically for the appropriate starting and boundary conditions as shown by Holliday and Pickering (1973). The solution of Eq. (13) takes a very simple form at steady state,

$$C_A(z,t) = C_A^0 + \left[C_A^s - C_A^0\right] \exp[-(v/d)(z - vt)] \tag{14}$$

for $z \geqslant vt$. In Eq. (14), $C_A^0 = C_A(\infty,t)$ is the concentration of A in the uniform alloy and C_A^s is the steady-state concentration of A at the target surface, which is related to preferential sputtering as discussed in Section IIA. Equation (14) represents the steady-state concentration profile that recedes with the target surface into the alloy. The profile decays exponentially from the surface concentration, C_A^s, to the bulk concentration, C_A^0, with a decay length

$$\delta \simeq D/v \tag{15}$$

Although the assumption of a constant diffusion coefficient in the altered layer is a rather gross simplification for the sputtering process, we will proceed to obtain some order-of-magnitude estimates from Eq. (15).

The thickness of the altered layer, characterized by δ, is given by the ratio of the radiation-enhanced diffusion coefficient, D, and the sputtering velocity, v, both of which depend on the near-surface displacement rate, K. The close connection between sputter yield and displacement rate has been demonstrated by Sigmund (1969) and applied to the problem of depth resolution in sputter profiling by Andersen (1979). Here we will use a very simple approximation, suggested by Rehn (1982), for the sputter velocity:

$$v \simeq K_0 (E_{d,eff}/U_s) \cdot \beta \cdot \gamma \cdot \bar{b} \tag{16}$$

where K_0 is the bulk displacement rate (in displacements per atom per unit time, dpa/sec) near the surface; $E_{d,eff}$ is the effective threshold displacement energy as used in the modified Kinchin and Pease formulation or in several of the commonly used computer codes to calculate displacement rates (see e.g., Averback et al., 1978; Manning and Mueller, 1973); U_s is the average surface binding energy for the atoms in the alloy; $\beta(\leq 1/2)$ is a geometrical factor that takes into account the fact that only those atoms at the surface which receive their recoil momentum in an outward direction can be sputtered; $\gamma(\approx 1/2)$ is a factor to correct for the fact that only cascades originating within the target contribute to sputtering whereas the bulk displacement rate is calculated with cascades originating in all directions from the displaced atom; and $\bar{b}$ ($\simeq \Omega^{1/3} \simeq b$, the nearest-neighbor distance) is the effective layer thickness from which the sputtered atoms are ejected. Andersen (1979) has recently compiled values for the surface binding energy and the threshold displacement energy for a large number of metals and finds that ($E_{d,eff}/U_s$) for most of them, ranges between 7 and 11. Taking 9 for this ratio, $\beta = 1/2$, and $\gamma = 1/2$ as representative values, we estimate a surface recession rate of

$$v \simeq 2.25 \, \bar{b} \cdot K_0 \tag{17}$$

or about two to three monolayers per displacement per atom in the subsurface region.

By combining Eq. (17) with Eq. (15) and taking $\bar{b} = 2.5 \times 10^{-8}$ cm, we obtain an estimate of the thickness of the compositionally modified layer,

$$\delta \simeq D/\left[5.6 \times 10^{-8} \text{cm} \cdot K_0\right] \tag{18}$$

For a displacement rate of 10^{-3} dpa/sec (see Figure 1 in Chapter 8), one estimates a diffusion coefficient of $\simeq 3 \times 10^{-17}$ cm^2/sec in the low-temperature ($\leq 200°$C) displacement mixing regime, and of $\simeq 5 \times 10^{-15}$ cm^2/sec in the sink-dominated regime around 600°C. The corresponding steady-state altered-layer thicknesses are $\simeq 50$Å ($\leq 200°$C) and $\simeq 1$ μm (at $\simeq 600°$).

Several comments should be made about these order-of-magnitude estimates. We have made the assumption that D is constant within the diffusion zone for the derivation of Eq. (15) and, hence, Eq. (18). For the low-temperature displacement mixing regime, this assumption is a reasonable approximation if the damage range of the sputter ions exceeds the estimated altered-layer thickness. Since the diffusion coefficient arising from displacement mixing is proportional to the displacement rate K_0, the altered-layer thickness is expected to be independent of the sputter rate as well as approximately independent of the mass and energy of the sputter ions, provided the damage range exceeds the altered-layer thickness. However, the altered-layer thickness cannot exceed the damage range in the low-temperature athermal displacement mixing regime.

The situation is somewhat different at elevated temperatures where thermal migration of the radiation-produced defects is important, because the radiation-enhanced diffusion is not confined to the damage depth but may exceed it considerably. In fact, the altered-layer depths measured by Rehn and Wiedersich (1980) in a Cu-40 at.% Ni alloy were ~50, 500 and several thousand angstroms after 2 h of sputtering at 300, 350 and 450°C, respectively, as shown in Figure 2. These depths substantially exceed the projected range (~30Å) of the 5-keV Ar$^+$ sputter ions, even before the ultimate steady-state condition is reached. It appears that radiation-enhanced diffusion extends considerably beyond the damage range during elevated-temperature sputtering. Utilizing the measured thicknesses and the sputtering rate of 0.5Å/sec, average diffusion coefficients of 2×10^{-15}, 2×10^{-14}, and several times 10^{-13} cm^2/sec are estimated for 300, 350 and 450°C sputtering, respectively, from Eq. (15). It should be noted that a large radiation enhancement of diffusion must extend to the depths of the altered layers and perhaps beyond. The magnitudes of the estimated diffusion coefficients are larger than, but not inconsistent with, those shown in Figure 1 of Chapter 8 at similar temperatures; the subsurface displacement rate under the sputter conditions used for the data shown in Figure 2 was approximately two orders of magnitude higher than that used in Chapter 8.

The time-dependent solution for the preferential removal and diffusion problem given by Holliday and Pickering (1973) shows that steady state is only approached after times corresponding to the removal of several altered-layer thicknesses. This is entirely consistent with the observations by Rehn *et al.* (1979) and Rehn and Wiedersich (1980).

D. Gibbsian Adsorption

The free energy of an alloy is frequently lowered by a readjustment of the surface composition to a composition that differs from that of the bulk of the alloy. Such readjustment occurs spontaneously at temperatures sufficiently high for diffusion to proceed at reasonable speed. The phenomenon, called Gibbsian adsorption or thermal surface segregation, has been reviewed recently by Wynblatt and Ku (1979). The change in composition in metallic systems can be very substantial and is confined to one or, at most, two atom layers at the surface (Ng *et al.*, 1979). However, the bulk composition is essentially unaffected at thermal equilibrium because of the generally large bulk- to surface-volume ratio.

This type of segregation can be described by considering a surface layer (of composition C_A^s and C_B^s) which differs from the bulk composition, C_A^b and C_B^b. At equilibrium, the composition of the surface is related to that of the bulk by

$$\left[C_A^s / C_B^s \right] = \left[C_A^b / C_B^b \right] \exp \left[- \Delta H_A / kT \right] \tag{19}$$

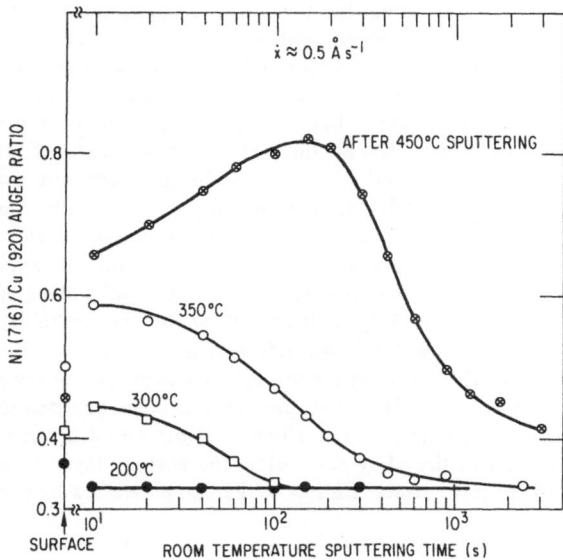

Figure 2   Ratios of the 716-eV Ni and the 920-eV Cu Auger electron transitions (escape depths ≃15Å) measured after sputtering for various periods of time at room temperature. The specimens had been sputtered previously for 2 h at the following temperatures: 200°C; 300°C; 350°C; and 450°C. The sputter rates were approximately 5Å/sec. (Rehn and Wiedersich, 1980)

where ΔH_A is the heat of adsorption, defined as the enthalpy change associated with the exchange of an A-atom in the bulk with a B-atom at the surface.

At elevated temperatures the equilibrium, Eq. (19), is approached by a net flux, J_A, of A-atoms which can described by

$$J_A \Omega = \left[\nu_A^{bs} C_A^b C_B^s - \nu_A^{sb} C_A^s C_B^b \right] b^s \tag{20}$$

where b^s is the surface layer thickness ($\simeq b$, the nearest-neighbor atom distance). The effective surface-to-bulk jump frequency, ν_A^{sb}, is related to the bulk-to-surface jump frequency, ν_A^{bs}, by the equilibrium condition, $J_A = 0$; in conjunction with Eq. (19), this yields

$$\nu_A^{bs} = \nu_A^{sb} exp (\Delta H_A/kT) \tag{21}$$

A simple physical interpretation of Eq. (21) is that the activation enthalpy of a surface segregating element ($\Delta H_A < 0$) for the jump back into the bulk is effectively increased by the heat of adsorption relative to the migration enthalpy in the bulk. The difference in concentration between the surface layer and the bulk is established and maintained by the reduced probability of thermally activated jumps of A-atoms from the surface into the bulk.

The changed surface concentration due to Gibbsian adsorption will result in an increased loss of the surface-active or segregated elements even in the absence of preferential sputtering as defined in Section IIA. Continued preferential loss of an element A as a consequence of Gibbsian adsorption requires, however, that the rate of diffusion of atoms to the surface adsorption layer from the atom layers below is sufficiently rapid compared to the rate of surface recession by sputtering. Using Eq. (15) with $\delta \simeq 5\text{Å}$ (2 atom layers) and a sputtering rate of 1Å/sec, one finds that the diffusion coefficient must be larger than $\simeq 5 \times 10^{-16} \text{ cm}^2/\text{sec}$ if Gibbsian adsorption is to play a role in the sputtering process. In metals, thermal bulk diffusion coefficients typically start to exceed this value at about one half of the absolute melting temperature, T_m. Because sputtering enhances thermal diffusion, the limit above which Gibbsian adsorption can affect the development of an altered layer is shifted to lower temperatures ($\simeq 0.3\ T_m$). It should be emphasized that the athermal process of displacement mixing opposes Gibbsian adsorption and, hence, no effects of surface segregation are expected in the low temperature displacement mixing regime.

It has been shown in Section IIA that the surface layer composition is determined by preferential sputtering at steady state. The effect of Gibbsian adsorption at steady state is to suppress the concentration in the alloy just below the surface layer to an appropriate value that maintains the surface layer concentration at the value dictated by preferential sputtering.

E. Radiation-Induced Segregation

As already pointed out, the excess point defects produced by the sputter ions migrate, provided the temperature is not too low, until they are eliminated by recombination or annihilation at sinks such as the surface, dislocations, and defect clusters. The spatial separation between the production and annihilation of defects frequently leads to persistent defect fluxes, e.g., toward the surface, or from the peak damage region toward the regions of lower defect production in front of and behind the peak damage region. Motion of defects requires motion of atoms, e.g., a vacancy exchanges sites with a neighboring atom, an interstitial atom jumps into a neighboring interstice, or interstitialcy motion forces a substitutional atom into an interstitial position while returning a different atom to a substitutional site. In alloys, generally, defects will migrate preferentially via atoms of some of the alloy components. Therefore, a preferential coupling exists between defect fluxes and fluxes of certain alloying elements. The combination of defect fluxes, which persist for some time, and the preferential coupling of these fluxes to fluxes of certain alloy components leads to a non-uniform distribution of the elements within the microstructure of an initially uniform alloy phase. This phenomenon, called radiation-induced segregation, has been studied intensely during recent years. The proceedings of a Workshop on Solute Segregation and Phase Stability during Irradiation, edited by Stiegler (1979), and those of a Symposium on the same subject edited by Holland et al. (1981) provide much recent information. Experimental and theoretical aspects of radiation-induced segregation have been reviewed by Rehn and Okamoto (1982) and Wiedersich and Lam (1982), respectively. Chapter 8 provides an overview of the phenomenon with the emphasis on its effects on alloy surface modifications.

Here, we will formulate radiation-induced segregation in a binary alloy, $A-B$, with the simple concept of partitioning the defect fluxes into fluxes of A- and B-atoms, and atom fluxes into fluxes via vacancies, v, and interstitials, i, as proposed by Wiedersich et al. (1979). The fluxes of atoms, J_A and J_B, and those defects, J_v and J_i, can then be expressed in terms of the concentration gradients of all species present as

$$\Omega J_A = - \left[d_{Av}C_v + d_{Ai}C_i \right] \nabla C_A + d_{Av}C_A \nabla C_v - d_{Ai}C_A \nabla C_i$$

$$\Omega J_B = - \left[d_{Bv}C_v + d_{Bi}C_i \right] \nabla C_B + d_{Bv}C_B \nabla C_v - d_{Bi}C_B \nabla C_i$$

$$\Omega J_v = - \left[d_{Av}C_A + d_{Bv}C_B \right] \nabla C_v + d_{Av}C_v \nabla C_A + d_{Bv}C_v \nabla C_B$$

and

$$\Omega J_i = - \left[d_{Ai}C_A + d_{Bi}C_B \right] \nabla C_i - d_{Ai}C_i \nabla C_A - d_{Bi}C_i \nabla C_B \tag{22}$$

The factor Ω, the average atomic volume, is required in Eq. (22) because all concentrations are expressed in atomic fractions. The diffusivity coefficients d_{Av}, d_{Ai}, d_{Bv} and d_{Bi} are defined as described in Section IIC in conjunction with eqs. (10) and (12) so that, e.g., $d_{Av}C_v$ and $d_{Ai}C_i$ are the partial diffusion coefficients of A-atoms migrating via vacancies and interstitials, respectively, and $d_{Ai}C_A$ and $d_{Bi}C_B$ are the partial diffusion coefficients of interstitials migrating via A- and B-atoms, respectively. The first right-hand-side terms in Eq. (22) are those portions of the flux of the species arising from its own concentration gradient; the second and third terms are those portions of the flux induced by fluxes of other species, e.g., $d_{Av}C_A \nabla C_v$ and $-d_{Ai}C_A \nabla C_i$ are A-atom fluxes caused by those portions of the vacancy and interstitial flux, respectively, attributable to migration via A- atoms. We note that the cross terms between vacancies and atoms have positive signs because the motion of a vacancy induces a motion of an atom in the opposite direction and vice versa.

It can be shown from Eq. (22) (Wiedersich et al., 1979) that enrichment of the component A at a defect sink, such as the surface, occurs when $d_{Ai}/d_{Bi} > d_{Av}/d_{Bv}$, i.e., when preferential migration of interstitials via A-atoms outweighs preferential migration of vacancies via A-atoms. Conversely, depletion of the A-element occurs at sinks when $d_{Ai}/d_{Bi} < d_{Av}/d_{Bv}$. The surface is an important defect sink during sputtering, and radiation-induced segregation of an element to the surface will increase its loss by sputtering even in the absence of preferential sputtering. As in the case of Gibbsian adsorption, radiation-induced segregation ultimately depletes the subsurface region of the target of the segregating element to an appropriate level so that the surface concentration, dictated by preferential sputtering at steady state, can be maintained despite segregation.

Strong radiation-induced segregation of Si in solution in Ni has been firmly established, e.g., by Okamoto and Rehn (1979). Rehn and Boccio (1982) sputtered Ni-Si alloy targets at temperatures between 30 and 600°C and monitored low- and high-energy Auger transitions during sputtering. Their results show that the composition of the first few monolayers remains close to the bulk alloy composition at all temperatures, indicating that preferential sputtering is insignificant in this alloy. However, considerable Si depletion in a subsurface layer is observed during elevated-temperature sputtering, consistent with expectations of preferential loss of silicon arising from radiation-induced segregation and perhaps also from Gibbsian adsorption of Si.

III. MODEL CALCULATIONS

The number of experimental investigations on sputter-induced compositional changes of alloys aimed at the elucidation of the roles of the diverse physical processes discussed here is rather limited. Rehn *et al.* (1979, 1980), Shikata and Shimuzu (1980), and Okutani *et al.* (1980) investigated Cu-Ni alloys and came to the conclusion that surface segregation of Cu is important at elevated temperatures. Nakamura *et al.* (1981) sputtered thin films of Ni-Au alloys at elevated temperatures and interpreted the results by a two-step segregation model. Thus far, experimental studies have led to few definite conclusions. Therefore, we will discuss in this section some results of model calculations of Lam and co-workers (1980, 1981, 1982a, 1982b) which give some insight into the effects that may be expected.

The calculational model is based on a set of coupled, partial differential equations describing the spatial and temporal evolution of the alloy composition and defect concentrations during sputtering of a binary alloy. The model takes into account preferential sputtering, displacement mixing, radiation-enhanced diffusion, Gibbsian adsorption, and radiation-induced segregation, but does not include recoil implantation. The set of diffusion and reaction rate equations, i.e., Fick's second law with source and sink terms, describing the time-rate of change of the alloy composition and defect concentrations can be written as

$$\frac{\partial C_A}{\partial t} = - \nabla \cdot \left[\left(\Omega J_A \right) - D_A^{\text{disp}} \nabla C_A \right] \qquad (23a)$$

$$\frac{\partial C_v}{\partial t} = - \nabla \cdot \left(\Omega J_v \right) + K_0 - R \qquad (23b)$$

$$\frac{\partial C_i}{\partial t} = - \nabla \cdot \left(\Omega J_i \right) + K_0 - R \qquad (23c)$$

where K_0 and R are the local, spatially dependent rates of production and recombination of vacancies and interstitials (in number of events per atom per second). The factor Ω is retained in Eq. (23) in order to use atomic units for the C's, K_0, and R. The fluxes J_A, J_v, and J_i, as given by Eq. (22), include the contributions from thermal and radiation-enhanced diffusion and from radiation-induced segregation, but not that from displacement mixing in the presence of a composition gradient. This term has been added as the second term in Eq. (23a), where D_A^{disp} is the diffusion coefficient caused solely by the displacement process as discussed in Section IIB. The equation for B-atoms does not need to be included in Eq. (23) because it is not independent; i.e., $C_B = 1 - C_A$, neglecting the small defect concentrations.

The time-dependent atom and defect concentration distributions can be determined by solving Eq. (23) numerically for a semi-infinite target and appropriate starting and boundary conditions as described by Lam *et al.* (1980, 1982a). The effects of preferential sputtering and Gibbsian adsorption are accommodated in the model by treating the surface layer as a separate phase.

The results shown here were obtained with physical parameters believed to be representative for a Cu-40 at.% Ni alloy (Lam and Wiedersich, 1981). Cu atoms were assumed to exchange preferentially with vacancies. The energies of vacancy migration via Cu and Ni atoms were taken to be 0.77 and 0.82 eV, respectively. The former value is the vacancy migration energy for pure Cu, and the latter is assumed to be appropriate for vacancy motion via Ni atoms in Cu-Ni alloys. The interstitial migration energy was taken to be 0.12 eV, the same for both alloying elements. This set of parameters leads to Cu depletion at the surface and Cu enrichment at the peak damage depth, consistent with radiation-induced segregation in Cu-Ni alloys observed experimentally by Rehn *et al.* (1981). To include Gibbsian adsorption, a heat of adsorption of −0.25 eV was used for Cu atoms. Sputtering and damage production rates employed in the calculations were appropriate to

5 keV Ar^+ ions. The sputtering probabilities for Cu and Ni were 5.5 and 2.75 atoms/ion per unit concentration, respectively. The damage distribution for 5 keV Ar^+ ions was estimated using the Edgeworth expansion. The peak damage occurs at 7Å and the total range is ~30Å. An ion flux of 2.5×10^{14} ions/cm^2sec (i.e., 40 μA/cm^2) was assumed, yielding a peak damage rate of $K_0 = 1$ dpa/sec. The diffusion coefficient resulting from displacement mixing was taken as $D_A^{disp} = 30(b^2/6)K_0$, i.e., 30 nearest-neighbor replacements per displacement.

The evolution of the spatial redistribution of Cu atoms during sputtering at 400°C is shown in Figure 3 as a function of time (Lam and Wiedersich, 1982b). The spatially dependent damage rate, K_0, used in the calculations is shown by the dashed curve in the top illustration ($t = 0$). Since a logarithmic scale is used for the distance from the sputtered surface to effectively illustrate rapid compositional variations near the surface and moderate ones at greater depths, the compositions of the first two atomic layers, each 2.06Å thick, are shown in step-like form. The thickness of the surface layer removed by sputtering, the rate of which is time-dependent because of the changing surface composition, is also indicated for various times. The Cu concentration at the surface, C_{Cu}^s, increases at short sputtering times owing to dominant-enhanced Gibbsian adsorption and then decreases toward the steady-state value given by $\left[C_{Cu}^s/C_{Ni}^s \right]_{s \cdot s} = \left[\bar{p}_{Ni}/\bar{p}_{Cu} \right] \left[C_{Cu}^b/C_{Ni}^b \right]$. At steady state, the surface alloy composition is determined by the sputtering probabilities of the alloy components, $\bar{p}_{Cu}$ and $\bar{p}_{Ni}$, and the bulk composition, C_{Cu}^b and C_{Ni}^b, in such a way that the composition of the sputtered flux equals that of the bulk alloy, as discussed in Section IIA.

The effects of the different processes on the time evolution of the surface concentration and on the steady-state profiles are illustrated in Figure. 4 and 5, respectively. The calculations were performed with various combinations of preferential sputtering, displacement mixing, radiation-enhanced diffusion, Gibbsian adsorption and radiation-induced segregation included.

Figure 4 shows the time dependence of the Cu concentration at the alloy surface, calculated at 400°C (Lam and Wiedersich, 1981). Without irradiation, Gibbsian adsorption leads to a strong Cu enrichment (up to 99.1 at.%) in the first surface atom layer (curve 1). With irradiation, if only preferential sputtering and radiation-enhanced diffusion are taken into account (curve 2), the surface concentration of copper, C_{Cu}^s, decreases monotonically to the steady-state value defined by the sputtering probability ratio and bulk composition. A true steady-state is attained only when the rate of loss of Cu and Ni atoms from the surface altered layer is balanced by a corresponding gain from the bulk as the altered-layer recedes into the alloy. At steady-state, $C_{Cu}^s \simeq 43$ at.%. Now, if Gibbsian adsorption is also included (curve 3), C_{Cu}^s increases rapidly at short bombardment times (<1 sec) owing to radiation-enhanced adsorption, and then decreases slowly to the steady-state value. The inclusion of displacement mixing reduces the effect of Gibbsian adsorption (curve 4). On the other hand, if one considers only preferential sputtering, radiation-enhanced diffusion, and radiation-induced segregation (curve 5), C_{Cu}^s decreases rapidly to the steady-state value, owing to the dominant effect of segregation. If Gibbsian adsorption is added (curve 6), then the effect of radiation-induced segregation is masked; dominant radiation-enhanced Gibbsian adsorption, causing an increase in C_{Cu}^s at short times, is followed by dominant preferential sputtering. Finally, with the addition of displacement mixing (curve 7), i.e., when all the processes are included, C_{Cu}^s increases initially and then decreases toward the steady-state value. Compared to curve 4, curve 7 is lowered by radiation-induced segregation during the transient state.

The effect of different combinations of processes on the steady-state Cu concentration profile is illustrated in Figure 5 for $T = 400$°C (Lam and Wiedersich, 1982a). As in Figure 3, a logarithmic scale has been used for the distance axis to show rapid compositional

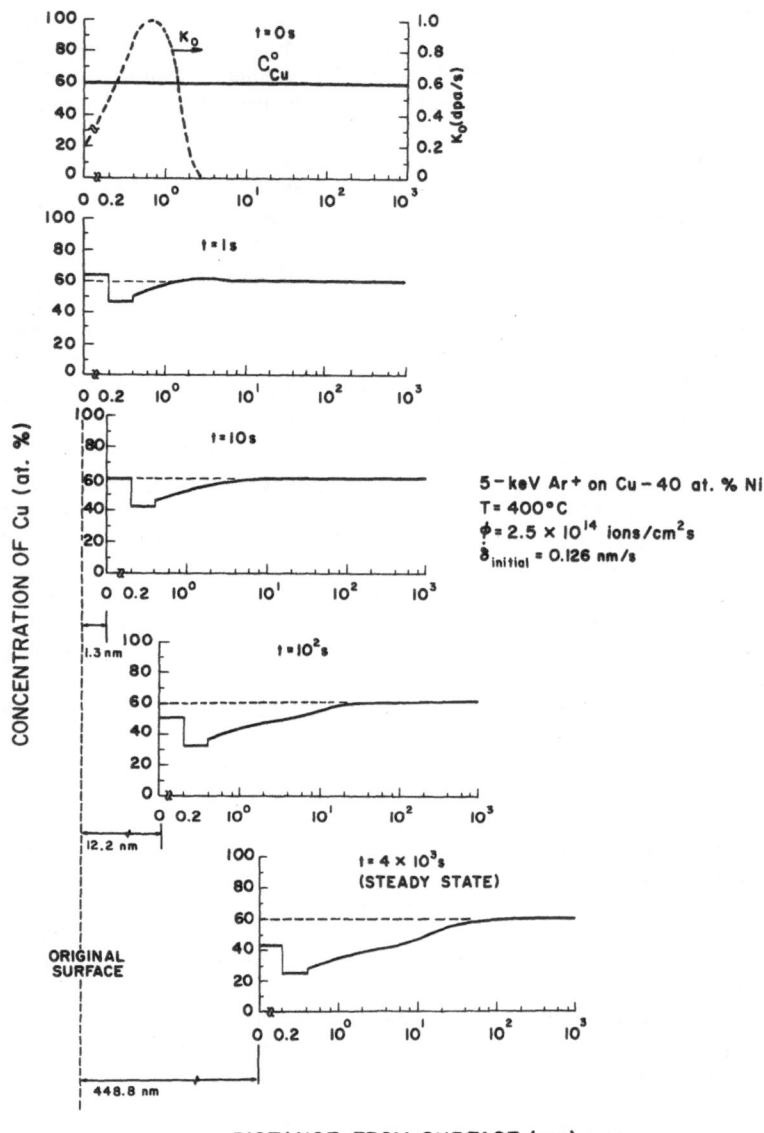

Figure 3   Time evolution of Cu concentration profiles in a Cu-40 at.% Ni alloy sputtered at 400°C with 5 keV Ar⁺ ions. The sputtering rate for the bulk composition is 1.26Å/sec. The profile of the damage rate, K_0, and the thickness of the surface layer sputtered off are indicated. (Lam and Wiedersich, 1982b)

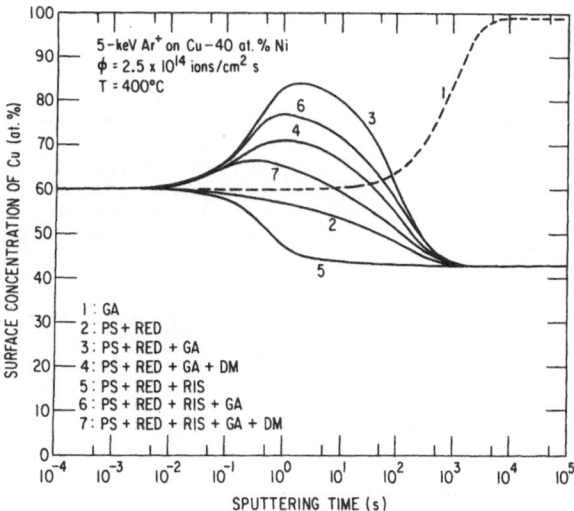

Figure 4   Effects of various combinations of Gibbsian adsorption (GA), preferential sputtering (PS), radiation-enhanced diffusion (RED), displacement mixing (DM), and radiation-induced segregation (RIS) on the time evolution of the surface concentration of Cu during sputtering at 400°C. (Lam and Wiedersich, 1981)

changes near the surface, and the composition of the first two atomic layers is shown in step-like form. Without irradiation, Gibbsian adsorption causes a strong Cu enrichment in the surface atom plane. When only preferential segregation and radiation-enhanced diffusion are included in the calculations, the subsurface region becomes moderately depleted of Cu during sputtering. The surface concentration of Cu attains the steady-state value defined by the sputtering coefficient ratio and the bulk composition regardless of which transport mechanisms are included. The addition of Gibbsian adsorption induces a larger Cu depletion in the near-surface region. This depletion at intermediate depth is reduced if radiation-induced segregation is also taken into account, because of a net influx of Cu atoms into regions near the peak damage, in the direction opposite to that of the vacancy flow. The inclusion of displacement mixing tends to reduce the concentration gradient and the concentration difference between the surface "phase" and the bulk; thus lessening Cu depletion. The profile obtained when all the processes are included shows more Cu depletion in the subsurface region than that calculated for only preferential segregation and radiation-enhanced diffusion. The depth of depletion is determined by radiation-enhanced diffusion. The decay length $\delta \simeq 300\text{Å}$, with the sputter velocity of $v \simeq 1.2\text{Å/sec}$, would yield an enhanced diffusion coefficient of $D \simeq 4 \times 10^{-14}$ cm^2/sec from Eq. (15), and it should take $\simeq 250$ sec to sputter off one altered-layer thickness, consistent with the time to approach steady-state (see Figure 4). Thus the rough estimates given at the end of Section IIC are borne out by the model calculations.

   The temperature dependence of the sputter-induced compositional changes is illustrated with the aid of the time evolution of C_{Cu}^s plotted in Figure 6 (Lam and Wiedersich, 1981). Near room temperature, Gibbsian adsorption is negligible and C_{Cu}^s changes monotonically to the steady-state value. With increasing temperatures, the effect of radiation-enhanced Gibbsian adsorption becomes stronger; C_{Cu}^s increases more rapidly at short sputtering times,

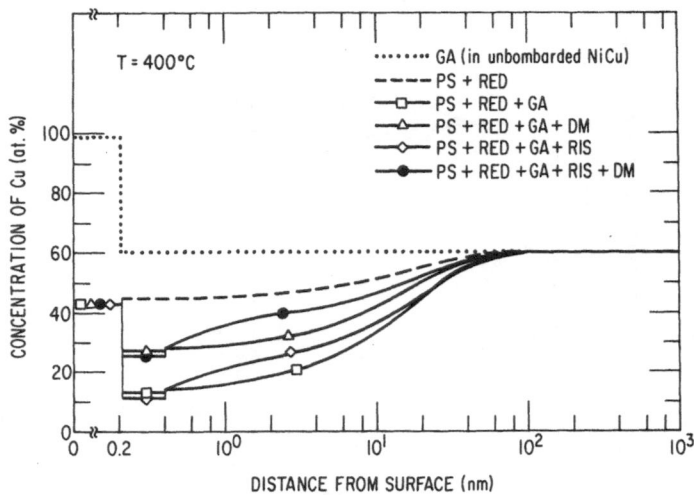

Figure 5   Effects of various combinations of the five basic processes defined in Figure 4 on steady-state Cu concentration profiles during sputtering at 400°C. (Lam and Wiedersich, 1982a)

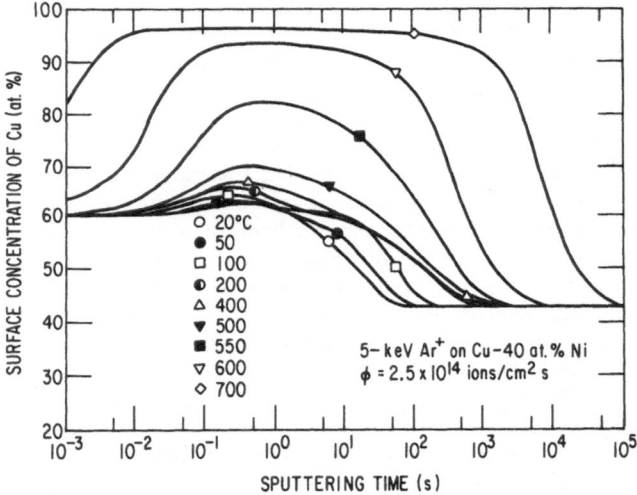

Figure 6   Time dependence of Cu concentration at the alloy surface during sputtering at various temperatures. (Lam and Wiedersich, 1981)

and takes longer to decay to the steady-state value. The time required to attain steady state increases from 80 to 10^5 sec as the temperature increases from 20 to 700°C. This increase is a consequence of the increase in thickness of the steady-state altered-layer with increasing temperature, as shown in Figure 8 of Chapter 8. Below 100°C, where point-defect mobility is limited, preferential sputtering and displacement mixing govern the development of the alloy composition in the altered-layer, which extends to a depth approximately equal to that of the damage range, and short times are required to remove a thickness corresponding to a few times the altered-layer thickness. Above 100°C, Gibbsian adsorption, radiation-enhanced diffusion, and radiation-induced segregation become increasingly significant, and the altered-layer thickness and, therefore, the time required to reach steady-state increase rapidly with temperature.

The increase in transient time and in severity of Cu depletion (or Ni enrichment) in the near-subsurface region with increased temperature is illustrated by the calculated time dependence of the Auger ratio I_{Ni}(716 eV)/I_{Cu}(920 eV) plotted in Figure 7 for sputtering at various temperatures (Lam and Wiedersich, 1982a). Below ~500°C, steady-state is reached in relatively short times. At higher temperatures, the buildup time to steady state is much longer; e.g., ~10^5 sec at 700°C. The rate of Cu depletion increases to a maximum value at ~600°C and then decreases with increasing temperature. The Auger ratios calculated for steady-state do not vary strongly with temperature below 500°C, because displacement mixing strongly suppresses Gibbsian adsorption, thus leading to a small temperature dependence of the Cu depletion in the near-surface region. Above 500°C, Gibbsian adsorption prevails, the subsurface Cu depletion becomes strongly temperature dependent, and the Auger ratios increase rapidly with temperature, reaching maximum values at ~700°C. At higher temperatures, because of decreased Gibbsian adsorption at thermal equilibrium, the Cu depletion near the surface is smaller than at 700°C, but extends to much larger depths. As a result, the Auger ratio at steady-state is expected to decrease with temperature above 700°C. Both the calculated time evolution and temperature dependence of the Auger ratios are in good qualitative agreement with the AES measurements by Rehn et al. (1979, 1980).

IV. SUMMARY AND CONCLUSION

The present state of knowledge of sputter-induced compositional modifications of alloy surfaces has been reviewed. During the past few years it has become recognized that at least six distinct processes can contribute to the compositional changes: preferential sputtering, recoil implantation, displacement mixing, radiation-enhanced diffusion, Gibbsian adsorption, and radiation-induced segregation. As a consequence of the interactions between these processes, sputter-induced surface modifications can be rather complex. Each of these processes has been investigated experimentally and theoretically with various degrees of sophistication in the past, often in conjunction with phenomena, such as radiation effects in solids or thermal segregation, where sputtering is unimportant or nonexistent. Even though only a limited number of studies have specifically addressed the phenomenon of sputter-induced surface modifications, a good physical understanding of all component processes exists.

The phenomenon is simplest at low temperatures where only the athermal processes, i.e., preferential sputtering, recoil implantation, and displacement mixing, play a role. In an initially uniform alloy target, preferential loss of components by sputtering, redistribution of components by recoil implantation, and implantation of the sputter ions initiate the formation of a compositionally altered-layer that approaches steady-state characteristics after the

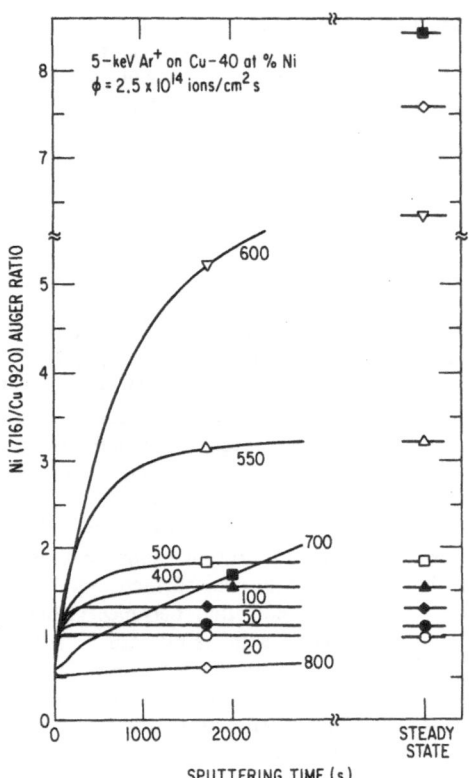

Figure 7   Ratio of the higher-energy Ni (716 eV) and Cu (920 eV) Auger transitions as a function of time during sputtering at various temperatures. The escape depths are approximately 15Å. (Lam and Wiedersich, 1982a)

surface has receded by a few times the altered-layer thickness. The thickness of this layer is approximately equal to the range of the sputter ions. The composition profile is affected by displacement mixing.

As temperature is increased, the radiation-produced defects become increasingly mobile, and radiation-enhanced diffusion, Gibbsian adsorption, and radiation-induced segregation affect the development and extent of, and the composition distribution within, the altered layer. These processes, which require thermally activated defect motions, can produce transient surface compositional changes in the opposite direction from those in the ultimate steady-state. At steady-state, the surface composition is always determined by preferential sputtering and the bulk composition. Most dramatically, the depth of the altered-layer can become orders of magnitude larger at elevated temperatures than the range of the sputter ions, because radiation-enhanced diffusion and radiation-induced segregation can extend into the undamaged microstructure well beyond the defect production region. The increased depth of the altered-layer causes a corresponding increase in transient time to steady state.

Although we have a good grasp of the mechanisms that produce and affect the near-surface compositional modifications of alloys during sputtering, we are a long way from making reliable quantitative predictions. The two major reasons for this inability are a lack of knowledge of (1) many of the physical quantities, such as defect motion, formation, and binding energies in alloys, that are required for reliable model calculations, and (2) defect clustering processes that lead to temperature-dependent sink structures within and beyond the defect production region. The extent and magnitude of radiation-enhanced diffusion and radiation-induced segregation are controlled by the latter processes. This problem has not been addressed in the present review because quantitative knowledge in this area is sparse.

The complexity and multitude of processes that affect surface layer composition and structure during sputtering, however, provide great flexibility and opportunity to modify alloy surfaces. The energy and mass of the incoming ion influence the relative magnitude of the sputter rate and total defect production as well as the extent of the damage range. Temperature affects the lifetime of defects and, hence, the extent of radiation-enhanced diffusion and radiation-induced segregation. An interesting manifestation of the effect of temperature has been observed by Cohen *et al.* (1982) in the form of a significantly different distribution of implanted atoms at low and at room temperature. Phosphorus ions (125 keV) implanted into Ni at 90K remain located in the projected range region and form an amorphous phase when the local P concentration exceeds $\approx$20 at.%. In contrast, 125 keV P^+ ions implanted at 300K migrate toward the surface during implantation and the amorphous phase $Ni_{0.8}P_{0.2}$ grows from the surface into the target as additional P arrives at the interface, presumably by radiation-induced segregation. The implant energy is sufficiently high to permit long-range defect transport processes to dominate over sputter recession of the surface.

The flexibility in the choice of energy, type, mass, and dose rate of the incoming ion as well as implantation temperature and post-implantation annealing will provide a wide latitude in the optimization of desired alloy surface properties. This flexibility is, however, equally important for future experimentation to further our understanding of the complex phenomenon of sputter-induced surface modifications.

ACKNOWLEDGEMENTS

The author has greatly benefited from discussions on the subject matter of this paper with many of the participants of the NATO Advanced Research Institute in Trevi as well as with his co-workers at Argonne National Laboratory. Special thanks are extended to L. E. Rehn and N. C Lam for valuable suggestions for clarification throughout the manuscript. Ms. S. L. Ruffatto's contributions in competently formatting and typing the manuscript are also gratefully acknowledged. The author thanks Ms. E. M. Stefanski for editing the final manuscript.

Work performed at Argonne and Penn State was supported by the U.S. Department of Energy and National Science Foundation, respectively.

REFERENCES

Andersen, H. H. (1979), Appl. Phys. *18*, 131.

Andersen, H. H. (1980) in Physics of Ionized Gases (SPIG 1980), B. Cobić, ed., Boris Kidric Institute of Nuclear Science, Belgrade, Yugoslavia, in press.

Andersen, H. H., Besenbacher, F. and Goddiksen, P. (1980), in Symposium on Sputtering, P. Varga, G. Betz and F. P. Viekböck, eds., Inst. Allgem. Physik, Vienna, 446.

Andersen, H. H., Chernysh, V., Stenum, B., Soǿrenson, T. and Whitlow, H. J. (1981a) in Proceedings of the Conference on Interaction of the Atomic Particles with Solids, Minsk, USSR, Vol. III, in press.

Averback, R. S., Benedek, R. and Merkle, K. L. (1978) Phys. Rev. B *18*, 4156.

Averback, R. S., Thompson, L. J., Jr., Moyle, J. and Schalit, M. (1982), J. Appl. Phys. *53*, 1342.

Behrisch, R., ed. (1981, 1982) Sputtering by Particle Bombardment I (1981); II, III (in press); Topics in Applied Physics, Springer, Berlin.

Brailsford, A. D. and Bullough, R. (1978), J. Nucl. Mater. 69/70, 434.

Carter, G., Navinsek, R. and Whitton, J. L. (1982) in Sputtering by Particle Bombardment, R. Behrisch, ed., Topics in Applied Physics, Springer, Berlin, in press.

Cobić, B., ed. (1980) Physics of Ionized Gases (SPIG 1980), Boris Kidric Institute of Nuclear Science, Belgrade, Yugoslavia, in press.

Cohen, C., Drigo, A. V., Barnas, H., Chaumont, J., Krolas, K. and Thomë, I. (1982) to be published. See also Abstracts, Materials Research Society, Annual Meeting, Nov. 16-19, 1981, p. 199.

Etzkorn, H. W. and Kirscher, J. (1980), Nucl. Instrum. Methods, 168, 395.

Falcone, G. and Sigmund, P. (1981), Appl. Phys. 25, 307.

Gillam, E. (1959) J. Chem. Phys. Solids 11, 55.

Gras-Marti, A. and Sigmund, P. (1981) Nucl. Instrum. Methods 180, 211.

Grove, W. R. (1853) Phil. Mag. 5, 203.

Ho, P. S. (1978), Surf. Sci. 72, 253.

Ho, P. S., Lewis, J. E., Wildman, H. S. and Howard, J. K. (1976) Surf. Sci. 57, 393.

Hofer, H. O. and Littmark, U. (1979) Phys. Lett. 71A, 457.

Holland, J. R., Mansur, L. K. and Potter, D. I., eds. (1981) Phase Stability During Irradiation, AIME, New York.

Holliday, J. E. and Pickering, H. W. (1973) J. Electrochem. Soc. 120, 470.

King, W. E. and Benedek, R. (1981) in Proceedings of the Yamada Conference V, Point Defects and Defect Interactions in Metals, in press.

Lam, N. Q., Leaf, G. K. and Wiedersich, H. (1980) J. Nucl. Mater. 88, 289.

Lam, N. Q. and Wiedersich, H. (1981) in Second Topical Meeting on Fusion Reactor Materials, Aug. 9-12, 1981, Seattle, Wash., J. Nucl. Meter. 103 & 104, 433.

Lam, N. Q. and Wiedersich, H. (1982a) in Proc. of the Materials Research Society Symposium on Metastable Materials Formation by Ion Implantation, Nov. 16-18, 1981, Boston, Mass., in press.

Lam, N. Q. and Wiedersich, H. (1982b) Radiat. Effects Lett., in press.

Littmark, U. and Hofer, H. O. (1980) Nucl. Instrum. Methods 168, 329.

Littmark, U. and Sigmund, P. (1975) J. Phys. D. 8, 241.

Manning, J. R. (1981) in Phase Stability During Irradiation, J. R. Holland, L. K. Mansur and D. J. Potter, eds., AIME, New York, p. 3.

Manning, J. and Mueller, G. P. (1974) Comp. Phys. Comm. 7, 85.

Mansur, L. K. (1979) J. Nucl. Mater. 83, 109.

Matteson, S., Paine, B. M., Grimaldi, M. G., Mezey, G. and Nicolet, M. A. (1981) Nucl. Instrum. Methods 182/183, 43.

Matteson, S., Roth, J. and Nicolet, M. A. (1979) Radiat. Eff. 42, 217.

Nakamura, H., Morita, K. and Itoh, N. (1981) Nucl. Instrum. Methods 191, 119.

Okamoto, P. R. and Rehn, L. E. (1979) J. Nucl. Mater. 83, 2.

Okutani, T., Shikata, M. and Shimuzu, R. (1980) Sur. Sci. 99, L410.

Pickering, H. W. (1970) J. Electrochem. Soc. 177, 8.

Pickering, H. W. (1976) J. Vac. Sci. Technol. 13, 618.

Pickering, H. W. and Wagner, C. (1967) J. Electrochem. Soc. 114, 698.

Rehn, L. E. (1982) private communication.

Rehn, L. E. and Boccio, V. T. (1982) private communication.

Rehn, L. E., Danyluk, S. and Weidersich, H. (1979) Phys. Rev. Lett. 43 1764.

Rehn, L. E. and Okamoto, P. R. (1982) in Phase Transformations and Solute Redistribution in Alloys during Irradiation, F. V. Nolfi, Jr., ed., Appl. Sci. Publ., Ltd., Barking, Essex, England, in press.

Rehn, L. E., Wagner, W. and Wiedersich, H. (1981) Scripta Met. 15, 683.

Rehn, L. E. and Wiedersich, H. (1980) Thin Solid Films 73, 139.

Rothman, S. J. (1982) in Phase Transform and Solute Redistribution in Alloys During Irradiation, F. V. Nolfi, Jr., ed., Appl. Sci. Publ., Ltd., Barking, Essex, England, in press.

Russel, W. A. Papanastassiou, D. A. and Tombrello, T. A. (1980) Radiat. Eff. 52, 41.

Shikata, M. and Shimuzu, R. (1980) Surf. Sci. 97, L363.

Sigmund, P. (1979) J. Appl. Phys. 50, 7261.

Sigmund, P. (1969) Phys. Rev. 184, 383 and 187, 768.

Sigmund, P. (1981) in Sputtering by Particle Bombardment I, R. Behrisch, ed., Topics in Applied Physics, Springer, Berlin, pp. 11-71.

Sigmund, P. and Gras-Marti, A. (1980) Nucl. Instrum. Methods 168, 389.

Sigmund, P. and Gras-Marti, A. (1981) Nucl. Instrum. Methods *182/183*, 25.

Sigmund, P., Oliva, A. and Falcone, G. (1981a) Ninth International Conference on Atomic Collisions in Solids, Lyon, France, 1981, Nucl. Instrum. Methods, in press.

Sizmann, R. (1978) J. Nucl. Mater. *69/70*, 386.

Stiegler, J. O., ed. (1979) Proceedings of the Workshop on Solute Segregation and Phase Stability During Irradiation, J. Nucl. Mater. *83*, No. 1.

Tsaur, B. Y., Matteson, S., Chapman, G., Liau, Z. L. and Nicolet, M. A. (1979) Appl. Phys. Lett. *35*, 825.

Varga, P., Betz, G. and Viekböck, F. P., eds. (1980) Symposium on Sputtering, Inst. Allgem. Physik, Vienna.

Wiedersich, H. (1972) Radiat. Eff. *12*, 111.

Wiedersich, H. and Lam, N. Q. (1982) in Phase Transformations and Solute Redistribution in Alloys During Irradiation, F. V. Nolfi, Jr., ed., Appl. Sci. Publ., Ltd., Barking, Essex, England, in press.

Wiedersich, H., Okamoto, P. R. and Lam, N. Q. (1979) J. Nucl. Mater. *83*, 98.

Winterbon, K. B. (1980) Radiat. Eff. *48*, 97.

Wynblatt, P. and Ku, R. C. (1979) in Interfacial Segregation, W. C. Johnson and J. M. Blakely, eds., American Society for Metals, Metals Park, OH, pp. 115-136.

CHAPTER 11

SURFACE MODIFICATION AND ALLOYING: ALUMINUM

S. T. PICRAUX and D. M. FOLLSTAEDT

Sandia National Laboratories, Albuquerque, New Mexico

CONTRIBUTORS: G. Della Mea, L. F. Donà dalle Rose, T. Hussain,
J. A. Knapp, G. Linker, P. Mazoldi, P. S. Peercy and
W. R. Wampler

I. INTRODUCTION

Ion implantation in combination with electron or laser beam pulsed heating provides a way to adjust the surface composition and microstructure of alloys independently of the usual equilibrium constraints. A variety of such surface modification studies have recently been carried out in aluminum by several groups (Battaglin *et al.*, 1981; Hussain *et al.*, 1980; Peercy *et al.*, 1982; and Picraux *et al.*, 1981). Selected results from these studies are summarized in the present chapter, with the objective of better clarifying the fundamental mechanisms involved for the case of metals. Thus for a single host system, we illustrate the various aspects of directed energy modification of materials discussed individually in previous chapters.

Surface modification and alloying require two considerations. First, it must be possible to tailor the composition of the near surface region. Second, a means is required to alter the microscopic state (e.g., solutionized, precipitated, amorphous,...) while maintaining the desired composition. Ion, electron, and laser beams may be viewed as general tools for achieving these requirements. The composition may be altered directly by ion implantation, with adjustments in composition from the starting alloy values typically limited to ≤ 30 at.% for metallic species, due to the sputter erosion of the surface. Alloying to higher concentrations can be achieved by vapor deposition of films, followed by the atomic mixing of these films with the substrate. Atomic mixing may be induced by either ion bombardment or by liquid state diffusion during pulsed melting to achieve a desired composition. The metallurgical states which can be achieved by ion implantation alone have been discussed in some detail previously both for equilibrium and nonequilibrium final states (Borders, 1979; Kaufmann and Buene, 1981; Myers, 1980; Picraux, 1980; Poate and Cullis, 1980) and will not be dealt with here. The use of ion beam mixing will be illustrated briefly at the end of this chapter. At this time the primary mechanisms which control ion beam mixing are much less well understood than the ion implantation process.

Ion implantation and pulsed melting followed by rapid solidification can alter the microstructure of the surface layer (Follstaedt, 1982) in ways which at present cannot be easily predicted. One way to better understand what surface structures can be achieved is to examine for a single system the various mechanisms which operate during such surface processing and the microscopic states which result. This method is adopted here for aluminum with emphasis given to both the resulting state of the quenched system and the mechanisms which produced it. Many studies have been carried out for pulsed heating of silicon (Appleton and Celler, 1982), but there have been far fewer investigations of metallic systems.

Earlier studies of metallic systems, sometimes referred to as laser glazing, have emphasized laser melting of conventional bulk alloys or coated alloys (Breinan *et al.*, 1976; Copley *et al.*, 1979). The present approach differs from laser glazing studies in that we concentrate on implantation alloyed surfaces and on uniform area pulsed heating.

In this chapter we first summarize in Section II the similarities and differences between nanosecond timescale electron, laser, and ion beam pulses used for the rapid heating of aluminum. Then in Section III we consider the combined use of ion implantation and pulsed heating for forming surface alloys. Section III is divided into two areas of emphasis. The first half concentrates on the microscopic state of the implanted species, considering in turn segregation, diffusion, metastable solid solution formation, and precipitation. The second half is concerned primarily with the structural order of the resolidified alloy and considers disorder annealing, introduction of vacancies, dislocation formation, slip, and amorphous phase formation. In Section IV the various microstructures which can be formed by the above approach are summarized. Then in Section V ion beam mixing and pulsed melting of layers are briefly discussed to illustrate further surface alloying possibilities in aluminum.

II. ELECTRON, LASER AND ION BEAM HEATING

Electron, laser, and ion beams have been used for the pulsed heating of metals (Chapter 2; Appleton and Celler, 1982). For electron and ion beams the energy deposition profiles are qualitatively similar; typically, the energy is deposited within the first few microns of the surface in an approximately Gaussian distribution. Also, electrons and ions deposit their energy into both electronic and lattice excitations, with a greater fraction of the energy going into lattice excitations for ion beams. In contrast, laser beams initially deposit their energy into the electronic system, with maximum intensity at the surface in an exponentially decreasing function with depth. The characteristic absorption length over which the energy is deposited typically is ≤ 0.1 μm. For each source, the electronic energy is rapidly converted to thermal lattice energy; typically the electron-phonon relaxation time is ~ 1 picosecond (Von Allmen, 1980). The examples discussed in the present chapter all use electron beams or lasers for pulsed heating.

For pulsed heating of metals the three most important differences between electron and laser beams are illustrated in Figure 1. For the present studies of Al and its implanted alloys the major difference between ruby laser ($\lambda = 0.69$ μm) and 10-50 keV electron beam irradiation is in the quench rates achieved (Figure 1c). This difference derives directly from the different depths of the energy deposition (Figure 1b). For electron beams with typical pulse lengths of 50 nsec and ranges of ~ 1 μm for the primary energy deposition (e.g., ~ 20 keV electrons in Al), the quench rate is controlled primarily by the depth of energy deposition. This depth determines the resulting thermal gradients within the solid and thus the time required for the heat to flow away from this region. The cooling rates in the liquid phase just after the pulse ends are 10^8-10^9K/sec (Wampler et al., 1981); cooling rates in the solid phase just after resolidification are similar. Shorter pulse lengths do not result in substantially faster quench rates for a given material unless the depth of energy deposition is made shallower. The deposition depth also controls the minimum melt depth for pulsed electron beams. Heat flow calculations show that for a 50 nsec, 1.5 J/cm^2 pulse, melt depths of ~ 2.6 μm and times of ~ 500 nsec are obtained, with a resolidification interface velocity of ~ 8 m/sec.

In contrast, for ruby lasers of pulse lengths of 10 to 50 nsec the absorption depths are ~ 0.02 μm in Al and the quench rates are typically $\sim 10^9$-10^{10}K/sec (Figure 1). The quench rates are determined primarily by the pulse length and energy deposited, rather than the depth of energy deposition. This is because the distance that the absorbed heat energy can diffuse (~ 1 μm) during a 10 nsec pulse exceeds the absorption depth. Thus by utilizing both electron beams and lasers for the pulsed heating studies, it is possible to vary the effective quench rate upon solidification from the liquid phase over a wide range. Also, varying the substrate temperature during the pulse changes the thermal gradient and provides additional control on the quench rate.

The energy deposition dependence for an electron beam can be calculated from a knowledge of the incident energy and the target atomic number (Spencer, 1959). Typically only ~ 5 to 10% of the incident energy is reflected, and the total absorbed energy (Figure 1a) is measured directly from the current to the sample during the pulse. For lasers, the energy deposition depends on both the material absorption length and the surface conditions. For example, for ruby laser light with wavelength 0.69 μm incident on Al, about 90% of the incident energy is reflected. The reflectivity (and correspondingly the absorbed energy) changes upon melting, and also is altered by ion implantation and other surface treatments. However, accurate determination of the energy deposition can be obtained by the temperature rise of small samples (Peercy et al., 1982). For pure Al the absorbed fraction of the incident laser energy was found to increase from 0.104 in the solid phase for an

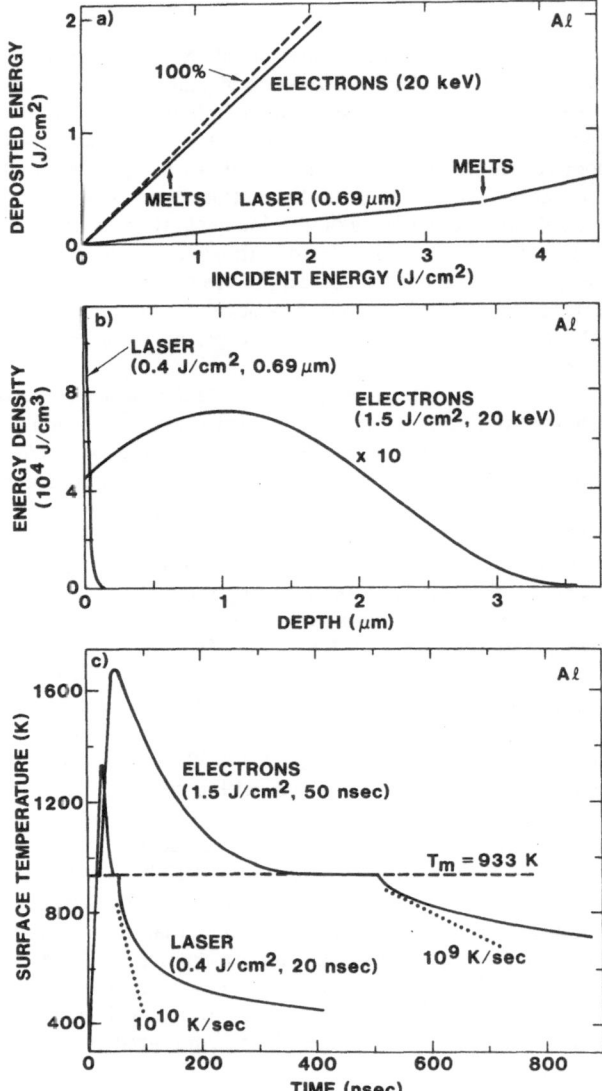

Figure 1   Comparison of pulsed heating of Al by electron beam (20 keV) and by ruby laser ($\lambda = 0.69\ \mu m$)
            illustrating the differences in: a) fraction of incident energy deposited, where threshold energy for
            melting is indicated by arrows for 50 nsec electron pulse and 25 nsec FWHM Gaussian laser pulse;
            b) depth profile of deposited energy density where the absorption depth for the laser light has been taken
            to be 0.02 μm and the curve for the electrons is shown multiplied x10; and c) calculated surface
            temperature (1.5 J/cm² deposited for e-beam and 0.4 J/cm² deposited for laser). The finite difference
            heat flow calculations include the temperature dependent values for thermal conductivity and specific
            heat, and are for an electron beam with a square intensity vs. time pulse of 50 nsec duration and a laser
            beam with triangular intensity pulse of 20 nsec FWHM. After Wampler *et al.*, 1981; Knapp, 1982;
            Peercy *et al.*, 1982; Spencer, 1959.

electropolished (110) surface to 0.13 upon melting. For a typical laser pulse of 20 nsec (FWHM) and 3.7 J/cm^2; heat flow calculations indicate that this gives a melt depth of ~0.5 μm and a melt time of ~40 nsec, with a resolidification interface velocity of ~30 m/sec (Figure 1).

III. ION IMPLANTATION AND PULSED HEATING

A. Segregation

In both silicon and aluminum, resolidification after pulsed melting occurs by epitaxial growth from the maximum melt depth back out to the surface. Impurities which have an equilibrium distribution coefficient ($k_0 \equiv$ solid solubility/liquid solubility) of less than unity will preferentially segregate into the liquid phase and be swept to the surface during equilibrium solidification with slow cooling rates. However for the high interface velocities (1-3 m/s) found in pulse melting, equilibrium conditions cannot be established, and the degree of segregation decreases. This interesting crystal growth process has been studied extensively for silicon and it has been found that the observed segregation is well described by an effective, velocity-dependent distribution coefficient, k' (see, for example, Chapter 4, White et al., 1980; Wood, 1980). Here we illustrate the situation for metals by comparing aluminum to silicon for similar pulse heating and equilibrium segregation conditions.

The equilibrium phase diagrams exhibit retrograde solubility behavior for Sn in both Al and Si, with equilibrium distribution coefficients $\approx 4 \times 10^{-3}$ and $\approx 1.6 \times 10^{-2}$, respectively. Thus, to compare segregation behavior in Si and Al after pulse melting, Sn was implanted in both crystals to a peak concentrations of 1 at.% (Picraux et al., 1981). Pulsed electron beam heating proved to be particularly convenient for this comparison since essentially identical energy deposition profiles can be achieved under similar pulse conditions. This would be difficult to achieve with lasers due to the strong differences in absorption depths and in reflectivities between semiconductors and metals.

As seen in Figure 2 appreciable surface segregation of Sn is found for Si, whereas no observable segregation is seen for Al. The slight broadening of the profile in Al is expected, due to liquid phase diffusion. Even for Si the segregation is less than for equilibrium conditions, with an estimated value of $k' \approx 10k_0 = 0.16$ (Campisano et al., 1980). Transmission electron microscopy of the Si(Sn) shows no evidence of cell formation or of the presence of either phase of pure Sn. The Sn segregation occurs for Si, but not Al, in spite of the fact that the equilibrium distribution coefficient is even smaller for Al. This nonequilibrium incorporation of the implanted impurity into the solid phase upon resolidification is referred to as interface trapping. The rather complete interface trapping of Sn into Al suggests that a greater degree of nonequilibrium may be obtained by this melt quenching technique for Al than for Si.

The difference in segregation in Si and Al can be qualitatively understood to result from two primary factors (Picraux et al., 1981). First, the liquid-solid interface velocity is appreciably larger for Al than for Si. Figure 3 shows the melt depths as a function of time obtained from heat flow calculations for the conditions of Figure 2. The interface velocities are ~8.3 m/sec for Al and ~1.7 m/sec for Si, a difference of a factor of 5. The higher interface velocity for Al is due to its greater thermal conductivity, which is generally true for metals in comparison to semiconductors. The second factor is that the liquid phase diffusivity of Sn is greater in Si than Al, thus allowing faster equilibration of the solute at the liquid solid interface in Si. The combined influence of these two factors can be seen in Table I. By estimating the Sn velocity in the liquid state (v_{Sn}) as the liquid phase diffusivity divided by the distance the interface moves in solidifying one atomic layer, we can obtain from the

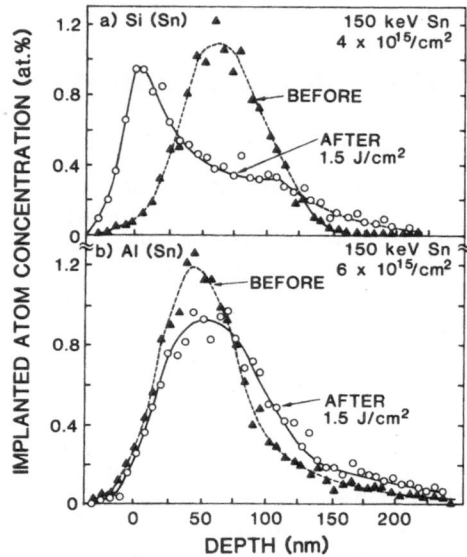

Figure 2   Implanted Sn depth profiles in a) Si and b) Al before and after pulsed electron beam melting under identical conditions. From Picraux *et al.*, 1981.

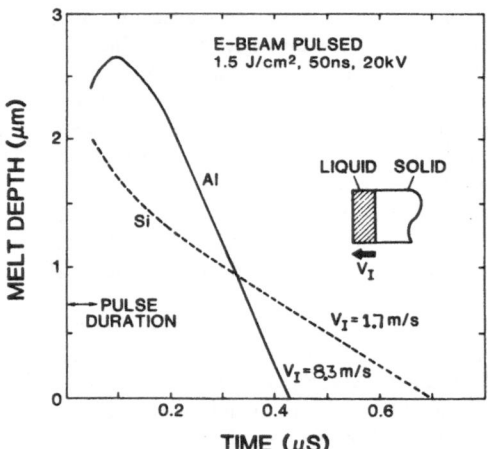

Figure 3   Melt depth vs. time in Si and Al for the same pulsed electron beam conditions of Figure 2 obtained by finite difference numerical heat flow calculation. After Picraux *et al.*, 1981.

velocity ratio a rough value for the number of jumps a Sn atom can make away from and to the interface before a new atomic layer is solidified. This is seen to be a factor 50 larger for Si than Al. For Si the value is $\gg 1$, and some degree of segregation can be achieved. For Al the number is ~ 1, indicative of the regime where nonequilibrium behavior and efficient interface trapping would be expected. These estimates are thus consistent with the observations of Figure 2. This comparison illustrates that the degree of metastability which can be achieved for implanted and pulse melted metals, such as Al, should be as great as, or greater, than that achieved in Si.

TABLE I

Parameters which lead to the observed Sn segregation in Si but not in Al under identical pulsed electron beam heating treatments of 1.5 J/cm², 50 nsec (After Picraux et al., 1980). The k_0 is the equilibrium distribution coefficient, $D_{liq.}$ the liquid phase diffusivity, v_{Sn} the Sn velocity in the liquid and $v_{int.}$ the interface velocity.

	Segregation Observed	$k_0^{(a)}$	$D_{liq.}^{(b)}$	v_{Sn} $(\sim D/a)$	$v_{int.}$	N_{jump} $(\sim v_{Sn}/v_{int.})$
			(cm²/sec)	(m/sec)	(m/sec)	
Si(Sn)	Yes	1.6×10^{-2}	$\sim 2 \times 10^{-4}$	80	1.7	~ 50
Al(Sn)	No	$\sim 4 \times 10^{-3}$	$\sim 2 \times 10^{-5}$	8	8.3	~ 1

(a) Trumbore, F. A. (1960), Bell Sys. Tech. J. *39*, 205; Hanson, M. (1958), "Constitution of Binary Alloys," McGraw-Hill, New York.

(b) Morehead, F. (1980), in "Laser and Electron Beam Processing of Materials," (C. W. White and P. S. Peercy, eds.), p. 143, Academic Press, New York; Keita, M., Steinemann, S., Kunzi, H. U. and Guntherodt, H. J., (1977), in "Liquid Metals," Inst Phys. Conf. Ser. No. 30, p. 655.

B. Diffusion

We next consider the diffusion of implanted impurities during Al pulsed melting (Wampler et al., 1980). We choose the Zn-implanted Al system for our example here, since Zn is soluble in Al at concentrations of several at.%. The upper part of Figure 4 shows the depth profiles for Zn as determined by ion backscattering both before and after 50 nsec pulsed electron beam heating with an energy (1.6 J/cm²) calculated to easily melt Al. A broadening of the Zn profile is observed, and the calculated profile (solid line),which assumes a constant diffusivity with depth and a reflecting surface boundary, is seen to accurately fit the observed profile for a Dt product of 2×10^{-11} cm². From the calculated melt time of 0.5×10^{-6} sec one obtains an effective diffusivity of 4×10^{-5} cm²/sec, which is close to the reported liquid phase value for Zn in Al of 6×10^{-5} cm²/sec. The assumption of a constant Dt value with depth is a good one for the pulsed electron beam case due to the deep melting. As seen from Figure 3 the time ($\sim 0.04 \times 10^{-6}$ sec) required for the liquid-solid interface to pass through the ~ 0.3 μm deep Zn profile is a small fraction of the total melt time. The above diffusivities are more than 2 orders of magnitude greater than the maximum solid phase diffusivity of Zn in Al, 4.7×10^{-8} cm²/sec. Thus significant atomic motion occurs primarily during the time that the surface remains liquid for this submicrosecond pulsed heating.

Solid phase (furnace) diffusion of Zn in Al at 196°C is shown in the lower part of Figure 4. Even though 9 orders of magnitude longer times are needed to achieve the same Dt value, the shape of the diffusion profile is seen to agree closely with the liquid diffusion case. This implies that other effects during electron beam pulsed melting, for example due to the thermal gradient (Soret effect), do not have a large influence on the diffusion profiles. Some evidence for this effect in the case of the steeper thermal gradients resulting from pulsed laser melting of Al has been discussed (Miotello and Donà dalle Rose, 1981). These authors include a nonzero Soret coefficient, S_T, in the diffusion equation which gives a contribution to the atomic transport due to the thermal gradient, $\frac{\partial T}{\partial x}$ (see Chapter 2). The Zn depth profile after laser pulsed melting is shown in Figure 5, with fits for $S_T = 0$ (solid line, Wampler et al., 1981) and $S_T = + 10^{-2} \, K^{-1}$ (dashed line, Miotello and Donà dalle Rose, 1981). The use of a positive S_T value here results in a lower concentration at the surface and the profile being shifted ~100Å down the temperature gradient (10^7 K/cm), and is more consistent with the observed profile.

The diffusional broadening of implanted profiles can provide a convenient marker for pulsed heating studies. It can be used to confirm that melting has occurred and to obtain a relative measure of the time spent in the liquid phase for miscible systems. Furthermore, a rough estimate of the liquid phase diffusivity can be obtained when the melt time is known through calculations or measurements. Finally, the above example demonstrates that for typical 10-50 nsec pulsed melting studies, the heating times are sufficiently short that significant solute-solute interactions, such as precipitation, would be expected to occur primarily in the liquid phase.

C. Metastable Solid Solutions

Since resolidification is usually sufficiently rapid after pulsed melting to allow efficient interface trapping of the implanted impurities into the Al lattice (Section III.A), one might expect to obtain metastable solid solutions by this approach. Miscibility within the liquid phase is a sufficient condition to maintain a homogeneous liquid and thus allow a metastable solution to be quenched in from the melt. Other requirements will become apparent in later subsections and are summarized in Section IV. Indeed, metastable solutions are found to be readily formed by implantation and pulsed melting, and we show several examples here of substitutional solutions with significant enhancements in solubility over equilibrium values. Furthermore, greater enhancements in the solubility have been achieved by implantation and pulsed melting than by implantation alone. We have hypothesized that the greater degree of enhancement upon pulsed melting for Al is due to the fact that there is a significantly lower availability of point defects upon resolidification than during implantation (Picraux et al., 1981).

The formation of metastable substitutional solutions by pulsed laser heating is nicely demonstrated by the case of Cr in Al (Battaglin et al., 1982). Chromium has a maximum solid solubility of only 0.4 at.%, whereas peak substitutional concentrations of 7 at.% have been achieved by the combination of ion implantation and pulsed ruby laser heating. This result is demonstrated by the channeling angular distribution data of Figure 6. The Cr implantation was done at room temperature and pulsed melting was inferred from surface morphology changes for both the 1.4 and 2.0 J/cm² laser conditions here. The lack of significant reduction in the as-implanted Cr yield (open circles) relative to the Al yield (filled

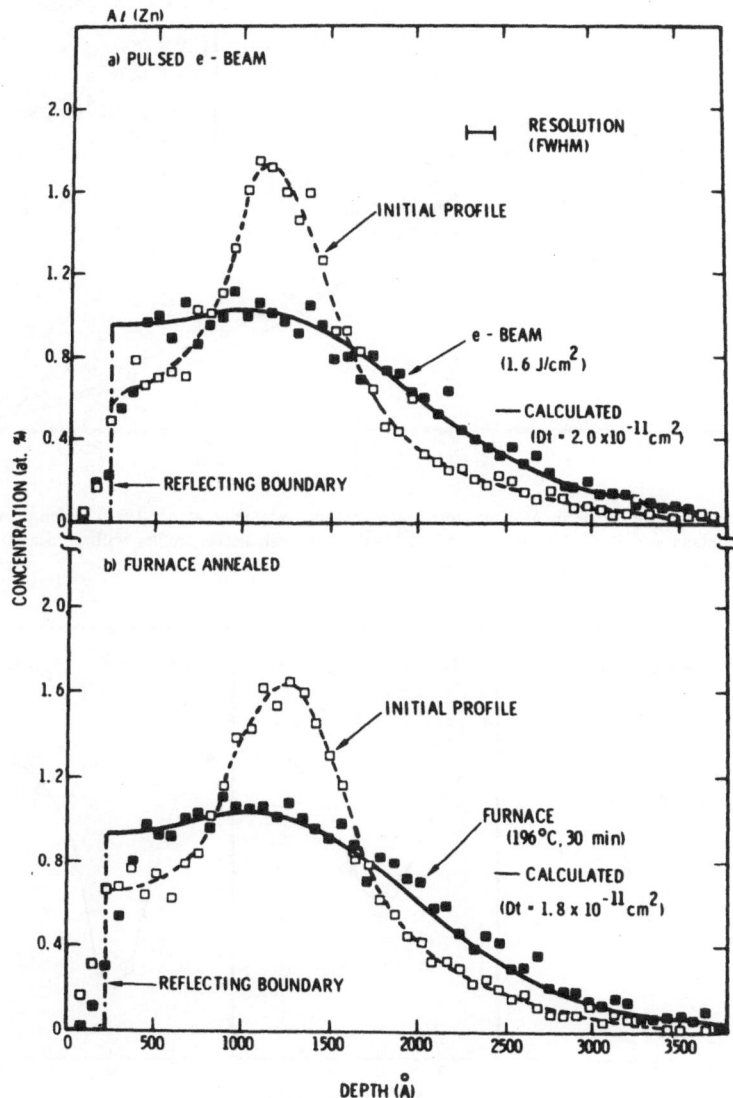

Figure 4   Implanted Zn depth profiles in Al a) before and after pulsed electron beam melting and b) before and after furnace heating. From Wampler *et al.*, 1980.

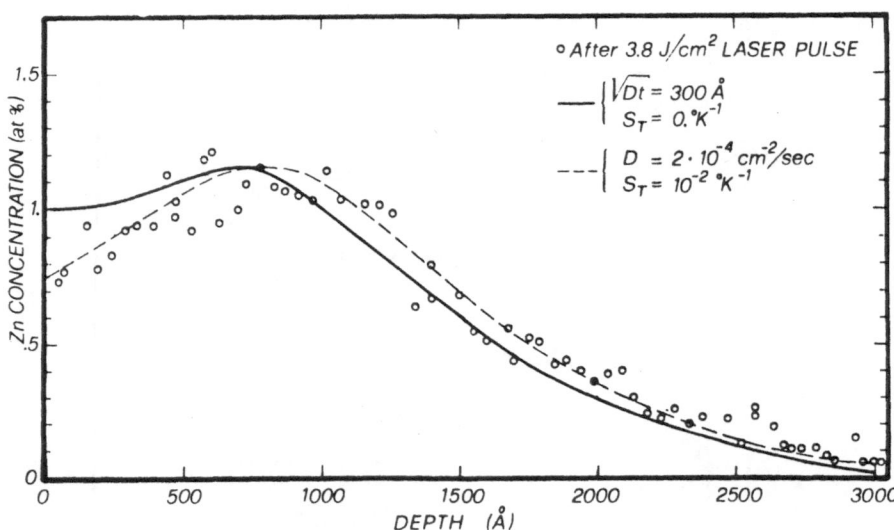

Figure 5   The Zn depth profile in Al after pulsed laser melting (Wampler *et al.*, 1981) which is being compared (Miotello and Donà dalle Rose, 1981) to best fits of calculated profiles with zero and nonzero Soret coefficients. From Miotello and Donà dalle Rose, 1981.

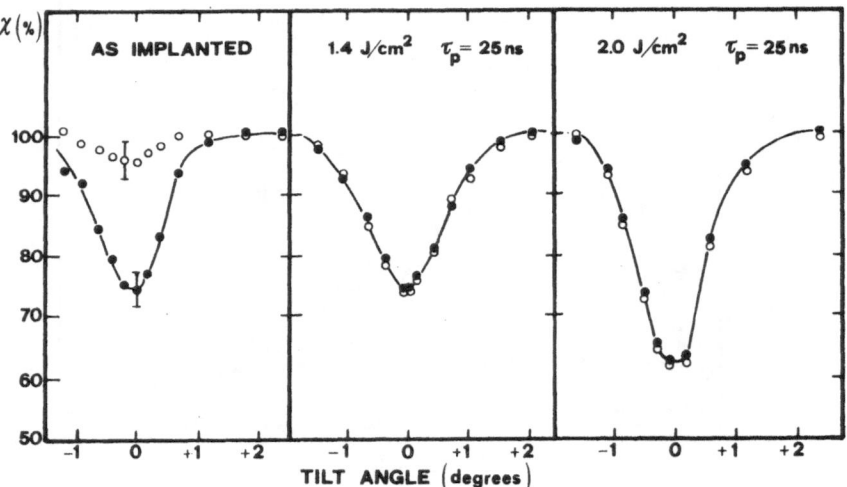

Figure 6   Angular scans about the <100> channeling direction for backscattering from the Al host (filled circles) and Cr (open circles) implanted to 1.2×10^{17} Cr/cm^2 for as-implanted and after pulsed ruby laser heating. From Battaglin *et al.*, 1982.

circles) illustrates that ion implantation alone is not always sufficient to form metastable substitutional solutions. However, after pulsed laser heating with 1.4 J/cm^2, the angular scan for the Al host and Cr solute are identical, indicating essentially 100% substitutionality for this peak Cr concentration of 7 at.%. Further reduction in the yield for both Cr and Al at 2.0 J/cm^2 indicates a reduction in the disorder present while maintaining complete substitutionality.

In Figure 7 the Cr depth profiles (filled circles) and reduced channeling yield (open circles) are shown after 2.0 J/cm^2 pulsed laser treatment as well as after subsequent furnace annealing. The furnace annealing results confirm that the melt quenched Al(Cr) is a metastable solution, since the Cr reverts to nonsubstitutional sites upon thermal annealing (indicated by the open and filled circles coinciding at 600°C). Furnace annealing of samples which were only Cr-implanted did not result in a measurable substitutionality.

Pulsed electron beam melting can also result in metastable substitutional solutions in Al, as illustrated in Figure 8 for the weakly soluble species Sn and Ni in Al (Picraux et al., 1981). Here the channeled and random yields give aligned fractions of 0.67 for Sn and 0.41 for Ni for peak concentrations of 1 at.% after pulsed melting; the corresponding aligned fractions after implantation were ≤0.04 and 0.18, respectively. The substitutional concentration is enhanced over the maximum solid solubility by a factor of 35 and 17 for Sn and Ni, respectively (see Table II). As for Cr-implanted Al, significantly greater enhancements in solubility can be achieved by implantation and pulsed melting than by ion implantation alone. For example, the Sn substitutional concentration increases from <0.05 at.% after implantation to 0.7 at.% upon pulsed melting.

A further example of pulsed melting enhancing the solubility over that achieved by implantation alone is given by the pulsed electron beam results for Cu in Al of Figure 9 (Hussain and Linker, 1982). Here as for Cr and Sn, the pulsed melting significantly enhances the substitutionality. In this case a substitutional solubility of 2.1 at.% is obtained. While this does not exceed the maximum solid solubility (2.5 at.%) for Cu in Al, it is, for example, significantly above the equilibrium solubility of 0.19 at.% at 300°C. Only smaller enhancements in the substitutional concentrations were achieved by furnace annealing at 300°C alone. Improvements in substitutionality over furnace annealing have also been obtained for Ga-implanted Al by pulsed electron beam heating for the case where melting did not occur (Hussain and Linker, 1980). Solid phase pulsed annealing might be expected to require longer heating times, and for the Ga in Al case four pulses of 300 nsec duration were used, in contrast to the typical 10 to 50 nsec single pulse for liquid quenching.

Solutes in aluminum for which metastable substitutional quenching from the liquid phase are listed in Table II. While the number and detail of investigations in Al have not been nearly so extensive as for Si, it is nevertheless clear that metastable substitutional solutions can be readily formed in Al with a wide variety of elements. Furthermore, ion implantation followed by melt quenching appears to be appreciably more effective in bringing about this metastable state in Al than implantation alone or than implantation in combination with furnace annealing.

Pulsed quenching from the liquid phase may not necessarily be more effective in forming metastable solutions with implantation than furnace annealing for other metal hosts, even though it clearly is for the case of Al. We suggest that the difficulty in achieving high metastable substitutional concentrations in Al by implantation and furnace annealing alone is due to the presence of defects. Many displacements per atom are typically created during implantation to alloy concentrations ≥1 at.% and these defects can interact with the implanted solute to form impurity-defect centers. Oversized atoms in the Al lattice such as Sn are known to attract vacancies, whereas similar or undersized atoms in Al tend to trap Al interstitials, with the latter centers usually being less stable (Swanson et al., 1978 and

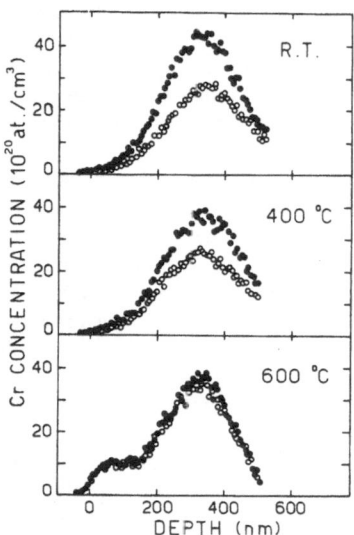

Figure 7   Depth profiles from random yields (filled circles) and the <100> channeling yields (open circles) for the Cr-implanted Al case of Figure 6 after 2.0 J/cm², 25 nsec pulsed laser heating at room temperature and after subsequent 15 min. isochronal annealing to the indicated temperatures. From Battaglin *et al.*, 1982a.

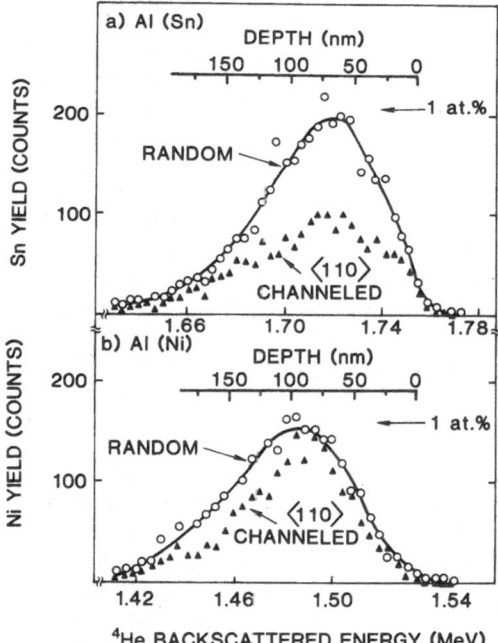

Figure 8   The <110> channeled and random orientation backscattering yields for a) Sn and b) Ni implanted Al (7 × 10¹⁵/cm², 150 keV) after 1.1 J/cm², 50 nsec pulsed electron beam melting. From Picraux *et al.*, 1981.

TABLE II

Comparison of maximum equilibrium solid solubilities (C_0) in Al to those achieved by ion implantation and pulsed melting (C_p) as determined by ion channeling measurements. Experiments reported have not necessarily attempted to systematically vary parameters to determine if the C_p values are the maximum achievable.

Element	$C_0^{(a)}$	$C_p^{(b)}$	C_p/C_0	Reference
	(at.%)	(at.%)		
Cr	0.4	7.0 (ℓ)	18	Battaglin *et al.*, 1982
Cu	2.5	2.1 (e)	1	Hussain and Linker, 1982
Ni	0.023	0.4 (e)	17	Picraux *et al.*, 1981
Mo	0.07	1.5 (ℓ)	20	Battaglin *et al.*, 1982a
Sn	0.02	0.7 (e)	35	Picraux *et al.*, 1981
Sb	$\leq$0.03	0.9 (ℓ)	30	Peercy *et al.*, 1982

(a) Maximum equilibrium solid solubility reported in M. Hanson, "Constitution of Binary Alloys," McGraw Hill, New York, (1958) or subsequent supplements.

(b) Pulsed heating method of ruby laser (ℓ) or electron beam (e) indicated in parentheses.

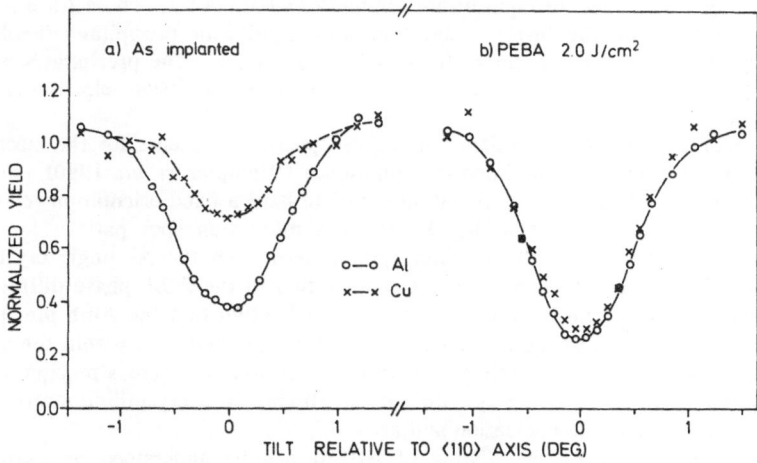

Figure 9   Angular scans about the <110> channeling direction for backscattering from the Al host and the implanted Cu (2.9×10^{16}/cm²) a) as implanted and b) after pulsed electron beam melting with 2.0 J/cm², 300 nsec. From Hussain and Linker, 1982.

1980). For example, vacancies are mobile in Al at room temperature and have been shown to cluster to Sn in Al when introduced by ion beam irradiation (Swanson *et al.*, 1980). These clusters distort the Sn from a purely substitutional site. In addition, insoluble impurities may interact with dislocations or other extended defects and thus lower their free energy by coming out of substitutional lattice sites. In contrast to the many defects created per implanted atom during implantation, only ~0.01 at.% vacancies are found to be retained in Al after quenching from the melt (see Section III.F below). Thus for impurity concentrations ~1 at.%, almost all the atoms are expected to be in regions free of point defects.

D. Precipitation

Precipitation or clustering of implanted impurities in the solid phase is greatly restricted during pulsed heating due to the short times spent at temperatures where the impurities are mobile and due to the lower diffusivities in the solid phase. For example, substitutional impurities with typical high temperature diffusivities $\leq 10^{-8}$ cm^2/sec would have diffusion lengths of less than 10Å for times ~1 μsec at elevated temperature. In contrast, liquid phase diffusivities are many orders of magnitude higher (typically ~10^{-4} cm^2/sec), giving rise to sufficient atomic mobilities for precipitates to form within the liquid. Therefore, the phase equilibria within the liquid state and the melt time can be important factors in determining the resulting alloy after pulsed heating.

Liquid phase precipitation in pulse melted Al is illustrated by the case of Sb-implanted Al. The broadened Sb profile due to liquid phase diffusion is shown in Figure 10 after pulsed melting with an electron beam (Wampler *et al.*, 1980). Although the liquid phase diffusivity for Sb in Al is not known, it can be estimated from these measurements to be ~3×10^{-5} cm^2/sec. In contrast, furnace annealing at a temperature and time sufficient to produce about the same Dt diffusional broadening, based on the known solid phase diffusivity of Sb in Al, results in no measurable change in the Sb profile. This lack of diffusional broadening is due to local precipitation of Sb into the AlSb phase, since the Sb concentration is orders of magnitude above its equilibrium solubility (≤ 0.03 at.%). These AlSb precipitates immobilize the Sb from the further migration since significant precipitate dissolution and solute transport do not occur at these times and temperatures. The precipitates are readily observed by diffraction and dark field imaging with transmission electron microscopy (Figure 11c and d).

As seen in Figure 11, precipitation of the AlSb phase also occurs for the electron beam pulsed melting case, but with one important difference (Wampler *et al.*, 1980). For furnace annealing the AlSb precipitates are always observed to have a fixed orientation relative to the crystalline Al matrix, as indicated by the electron diffraction spot pattern for the AlSb precipitates which aligns with the brighter spot pattern from the Al single crystal matrix (Figure 11c). In contrast, the predominant contribution to the AlSb phase diffraction after pulsed melting is in the form of rings (Figure 11a), indicating that the AlSb precipitates are randomly oriented relative to the Al matrix. Thus AlSb precipitation within the solid phase gives diffraction spots, due to the orientation with the Al matrix, whereas precipitation within the liquid phase gives diffraction rings, due to the absence of a crystalline matrix to fix the precipitate orientation during nucleation and growth.

The above results for pulsed melting of Al(Sb) can be understood by examining the equilibrium phase diagram shown in Figure 12. Upon pulsed heating the implanted Al(Sb) alloy first reaches sufficiently high temperatures (up to ~2000°C in the 1.6 J/cm^2 pulse of Figure. 10 and 11) so that the system is in the miscible liquid regime, and the liquid phase diffusion observed in Figure 10a occurs. While cooling, the system passes through the two-phase regime which in equilibrium consists of liquid Al and solid AlSb. For the final Sb

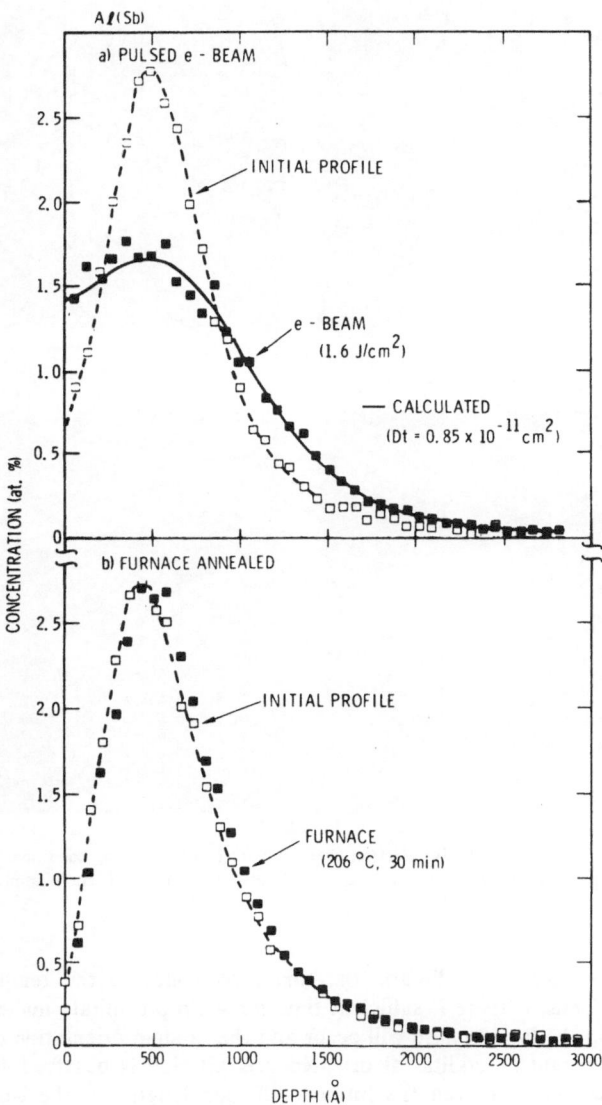

Figure 10   Implanted Sb depth profiles in Al a) before and after pulsed electron beam melting and b) before and after furnace heating. From Wampler *et al.*, 1980.

PULSED ANNEALING

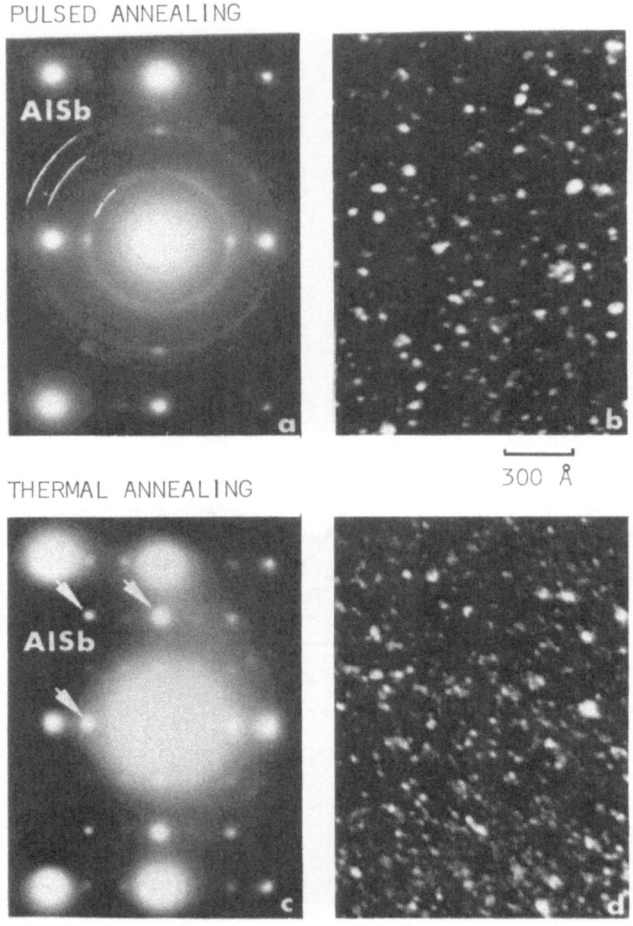

THERMAL ANNEALING

300 Å

Figure 11   Electron diffraction for the (a) electron beam and (c) furnace heating conditions of Figure 10 and with
the corresponding dark field images (b and d) of AlSb precipitates. Form Wampler *et al.*, 1980.

concentration of ∼1.5 at.% in Figure 10a, this corresponds to the temperature region of
∼730 to 657°C. Thus, if there is sufficient time for AlSb precipitate nucleation and growth,
precipitation within the liquid phase will occur and the random orientation of precipitates will
result. A small amount of additional oriented precipitation is observed (see faint spots in
Figure 11a), indicating that even the limited diffusion lengths in the solid phase are just
sufficient to produce detectable precipitation.

   Probably the most interesting aspect of the ability to observe precipitation within pulse
melted alloys is the possibility it raises of measuring submicrosecond nucleation times for
precipitation within the liquid state. This has been demonstrated for the Al(Sb) system by
going to still shorter quench times with pulsed laser melting (Peercy *et al.*, 1982). The
electron diffraction pattern in Figure 13 shows that with the Al substrate at 20°C
precipitation within the liquid phase does not occur for pulsed laser melting. Finer

adjustments in the time spent in the two phase regime are obtained by changing the substrate temperature for a given pulsed laser energy deposition. As shown in Figure 13, liquid phase precipitation (AlSb diffraction rings) is observed for 100°C but not for 20°C substrate temperatures. The energy deposited by the laser into the sample can be measured by a thermocouple attached to the sample and this, together with heat flow calculations, allows the time in the two phase regime to be accurately determined (Table III). The time to grow the observed 50Å precipitates is calculated to be 10 nsec using the previously determined liquid phase diffusivity, which indicates that the AlSb precipitation is nucleation limited. Accounting for the growth time gives a measured nucleation time of 15(±10) nsec for AlSb in liquid Al (Peercy et al., 1982). This is believed to be the first measurement of precipitate nucleation time in liquid metals in the nanosecond time regime.

TABLE III

Calculated time spent in two phase (solid AlSb + liquid Al) regime for Sb-implanted Al after pulsed laser melting as a function of the temperature of the Al substrate (after Peercy et al., 1982).

Temperature of Substrate (°C)	Time in Two-Phase Regime (nsec)	AlSb Precipitation in Liquid Al
20	19	No
100	30	Yes
200	55	Yes

E. Disorder Annealing

In the remaining parts of Section III we discuss the disorder remaining in Al after pulsed heating. We will consider in turn point defects, extended defects, and amorphous phase formation. However, before embarking on a detailed description of the disorder in Al after pulsed heating, we point out the contrast to Si. For Si good epitaxial regrowth with few defects can be achieved in most cases (see Chapters 3-5). The major requirements are that the melt depth extends all the way through and below the disorder of the implanted layer to provide a good epitaxial seed, and that the liquid-solid interface velocity not be too great (i.e., ≤5 m/sec). For appropriate pulsed annealing conditions, then, ion channeling and electron microscopy measurements indicate good crystalline quality similar to that for virgin Si. Only at high implanted concentrations of certain species in Si is the epitaxial regrowth process impeded such that it is difficult to find any pulsed melting conditions which do not leave appreciable disorder.

In Al, epitaxial regrowth after pulsed melting. also occurs. This is perhaps the only close analogy with Si. Pulsed heating of virgin Al invariably results in increased disorder for good single crystals. For ion implanted Al the results are complex. Ion channeling measurements of the disorder may show either an increase or decrease in the net disorder after pulsed melting. Examples of this are illustrated in Figure 14 for Mo and Cd implanted Al after pulsed laser melting (Mazzoldi et al., 1981). The channeling disorder just after

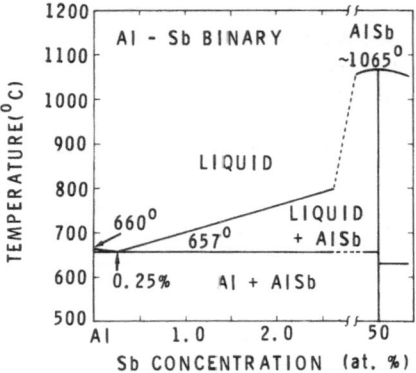

Figure 12 Aluminum-rich portion of the equilibrium phase diagram for the Al-Sb binary.

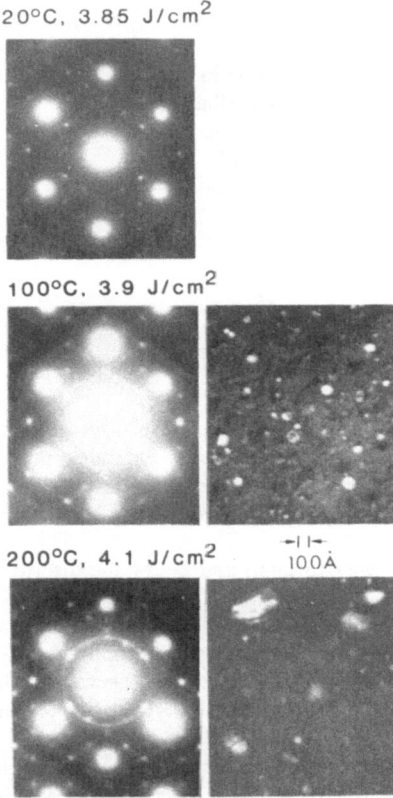

Figure 13 Electron diffraction and dark field images of AlSb precipitates in Sb-implanted Al after pulsed laser melting (25 nsec) as a function of Al substrate temperature. From Peercy *et al.*, 1982.

implantation is about the same for similar fluences of Mo and Cd. However, after identical laser treatments, the disorder for Al(Mo) decreases slightly while that for Al(Cd) approximately doubles. Thus the term "pulsed annealing" is somewhat confusing in the context of the metals and therefore we avoid this term in the present discussion. To understand this more complex situation, generally observed for metals after pulsed melting, it is useful to first examine and characterize the nature of the disorder which can be introduced into virgin (unimplanted) Al.

F. Vacancy and Dislocation Formation

We first consider the possibility of point defect introduction due to pulsed heating of Al. Studies of solid phase quenching of Al into liquid baths from temperatures near the melting point have shown that large quantities of vacancies can be retained (see, for example, Gauster and Wampler, 1980). This results from a significant fraction of the vacancies present in thermal equilibrium concentrations at the high temperatures being retained within the lattice for the quench rates of $\sim 10^4$ K/sec. At room temperature the vacancies are mobile and they coalesce to form dislocation loops, which are readily observable by transmission electron microscopy. Recently it has been demonstrated that pulsed melting of Al (solid phase quench rates $\sim 10^8$ K/sec) also quenches in appreciable numbers of point defects which result in dislocation loops (Follstaedt and Wampler, 1981). This is illustrated in Figure 15, which shows that, after pulsed melting, a high density of loops is observed whose images are similar to those observed after conventional furnace quenching. Weak beam electron microscopy (Figure 16) reveals that these are hexagonal dislocation loops lying on {111} planes with edges along [110] directions and with diameters of ~ 150Å. The loops have a structure consistent with those observed after furnace quenching and are thus ascribed to vacancy coalescence.

For pulsed heating of metals it is anticipated that the vacancies will be quenched in at appreciable concentrations only for the case where melting has occurred. This is because the times are insufficient for appreciable vacancy introduction by solid phase diffusion from available sources such as the surface. For example, examination of Al pulse heated to just below the melt threshold showed far fewer loops (Picraux et al., 1982a). However, upon melting vacancies may be directly introduced during resolidification. The observed loop densities in Al correspond to vacancy concentrations (~ 100 ppm) a factor of ~ 10 lower than the thermal equilibrium values just below the melt. Whether this is related to nonequilibrium effects analogous to the lack of segregation at high interface velocities, or other effects (see Follstaedt and Wampler, 1981) is not known. Thus, in Al, and most likely in other metals as well, appreciable concentrations of vacancies may be quenched in by pulsed melting. Secondary defects such as dislocation loops may result, for example, for substrate temperatures where the vacancies are mobile, as is the case for Al at room temperature.

In addition to the loops, dislocation lines are seen in both unimplanted and implanted Al after pulsed melting. For example, in Zn-implanted Al (see Figure 1 in Follstaedt et al., 1980) extended dislocations are observed; however, dislocation loops are not. The latter are believed absent due to interactions between the vacancies and the Zn impurities, similar to Al(Cu) (Gauster and Wampler, 1980). It may be that the dislocation lines are introduced at the liquid-solid interface during the rapid resolidification.

For the case when the melt depth does not extend all the way through the region of original implantation disorder, such as in lower power pulsed laser melting, the residual disorder may be particularly extensive. This can even result in the blocking of epitaxy in extreme cases (as for Al(Sb), Peercy et al., 1982), and is one mechanism by which polycrystalline surface layers can be formed (see also Section III.H).

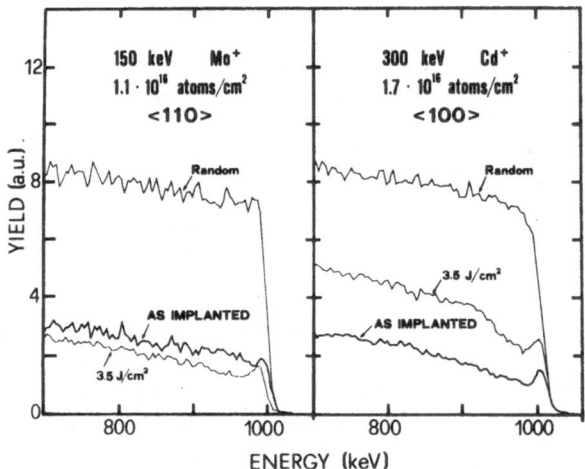

Figure 14   Backscattering spectra for channeled and random orientations before and after pulsed laser heating (25 nsec) of Al single crystals implanted with Mo (left side) and Cd (right side). From Mazzoldi *et al.*, 1981.

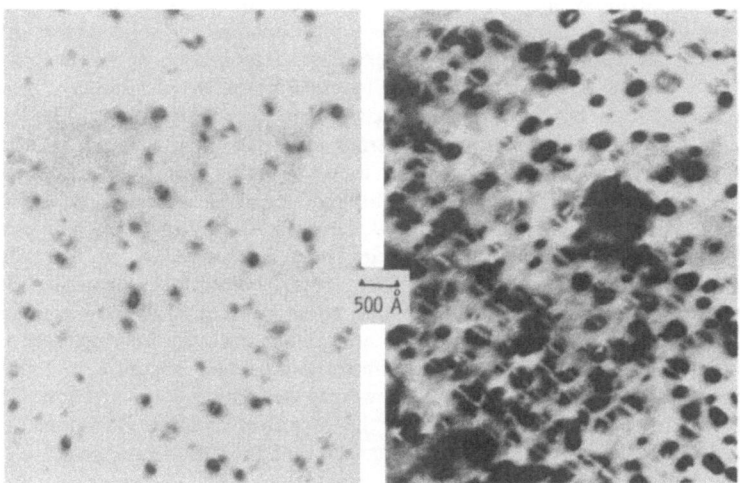

Figure 15   Bright field transmission electron micrographs showing dislocation loops in Al after quenching from a furnace at 550°C (left side) and after pulsed electron beam melting with 1.6 J/cm², 50 nsec. From Follstaedt and Wampler, 1981.

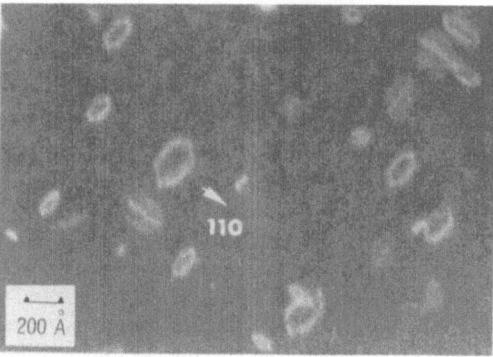

Figure 16   Weak beam image of dislocation loops for g = (200) and viewing in the <100> direction in Al after pulsed electron beam melting with 1.6 J/cm², 50 nsec. From Follstaedt and Wampler, 1981.

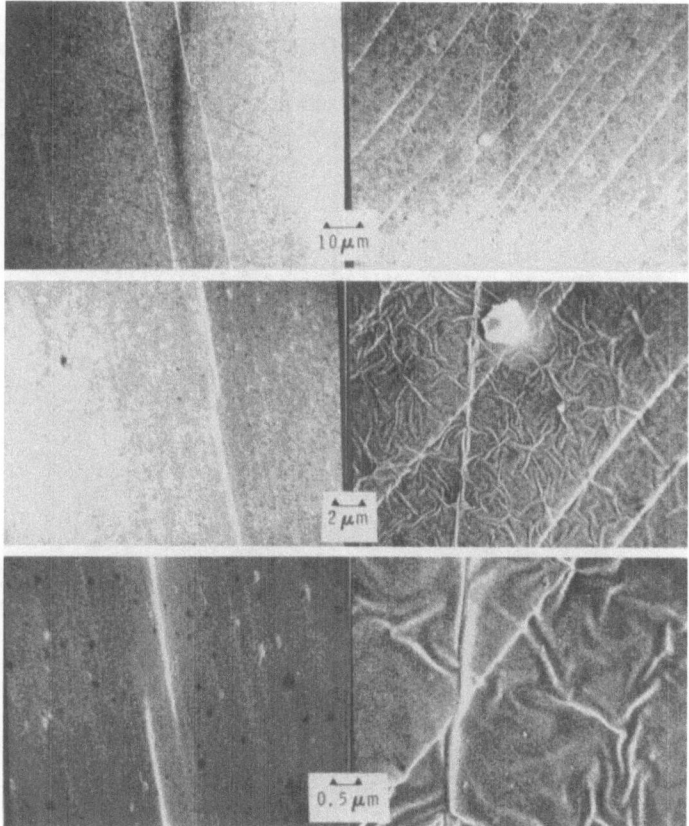

Figure 17   Scanning electron micrographs at three different magnifications of laser pulse (20 nsec) heated (110) Al for 2.9 J/cm² (left side; corresponding to solid phase heating) and for 3.5 J/cm² (right side; corresponding to surface melting). After Follstaedt et al., 1981.

G. Slip

In addition to defects which may be introduced during epitaxial regrowth at the liquid-solid interface, there is also the possibility of defect introduction due to the thermal stresses present during pulsed heating. Lateral thermal stress results from the difference in lattice expansion along the large temperature gradient between the surface pulse heated region and the substrate. Such stress can lead to slip deformation when material yield stresses are exceeded. Slip proceeds by the mechanisms of dislocation motion and slip traces generally appear as planes of heavily damaged material with associated dislocations present. The degree of defect introduction by thermal stresses and the associated mechanical deformation of the surface layer directly depends on the thermomechanical properties of the material and may occur for conditions of pulsed heating into either the liquid or solid phase. Thus, slip formation will depend on the material, crystal orientation, alloying elements, substrate temperature and pulsed heating parameters. For example, the thermal gradients differ significantly between laser and electron beam heating.

Slip deformation can readily be observed in Al upon pulsed laser heating both below and above the threshold for melting (Follstaedt *et al.*, 1981). In Figure 17 the parallel lines observed for energies below (left) and above (right) the threshold for melting are due to surface steps where slip planes have intercepted the crystal surface. The rippled surface appearance after 3.5 J/cm², 20 nsec treatment (right-hand side of Figure 17) is indicative of melting. Surface melting will erase surface steps due to slip which occurred in the solid during the heating phase and only the slip subsequent to the resolidification due to cooling of the recrystallized layer would be expected to be observed in the scanning electron

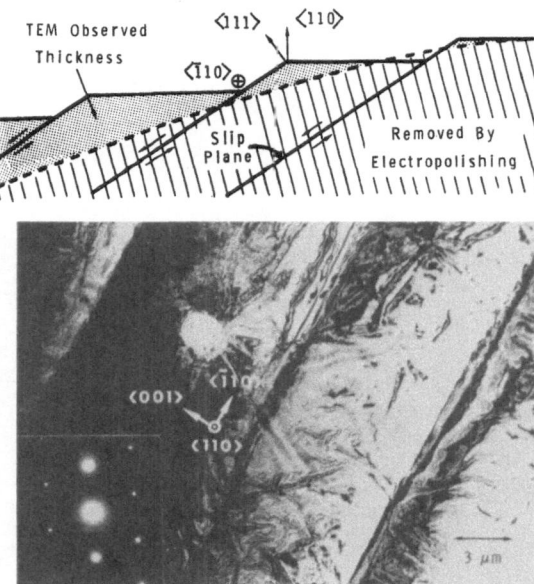

Figure 18   Transmission electron micrograph taken normal to surface and schematic cross section of surface showing slip bands in (110) single crystal Al after 3.5 J/cm², 20 nsec pulsed laser melting. From Follstaedt *et al.*, 1981.

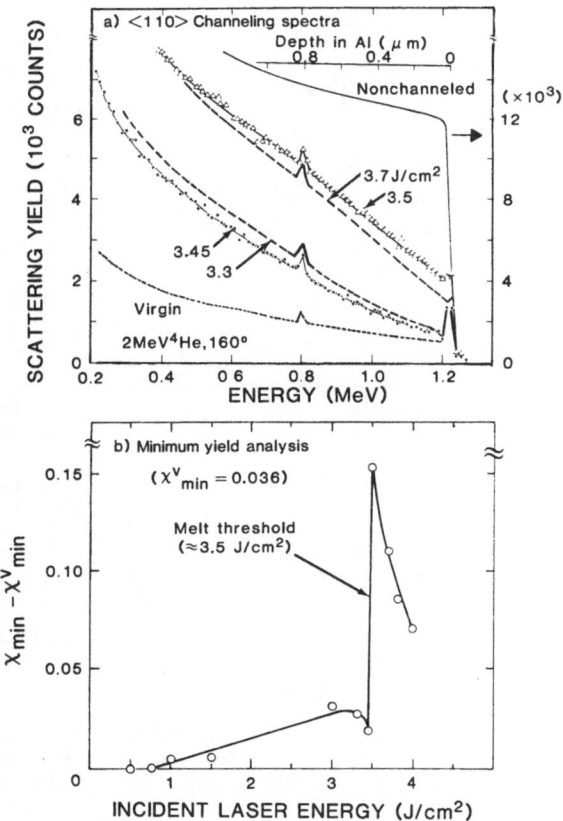

Figure 19   a) Ion backscattering random and channeling spectra for virgin Al and as a function of 20 nsec pulsed laser energy. b) Increase in minimum yield over virgin value obtained from the channeling spectra as a function of laser energy. From Follstaedt *et al.*, 1981.

micrographs. Transmission electron microscopy on the same 3.5 J/cm^2 sample (Figure 18) shows that these slip traces are heavily deformed regions within the single crystal Al, and the observed thickness changes are due to slip along inclined {111} planes as illustrated schematically above the micrograph. The {111} planes are the expected slip planes for the fcc lattice. Dislocations are observed within the heavily deformed regions, and dislocation loops (not shown) are found in the slip bands between the traces.

Slip deformation can contribute significantly to the greater channeling yields for pulse heated Al crystals. This is shown for virgin Al in Figure 19a where the curves for laser energies of 3.3 and 3.45 J/cm^2 correspond to below the melt threshold and the 3.5 and 3.7 J/cm^2 curves correspond to surface melting. The higher dechanneling rate (slope) throughout the first $\sim$2 μm probed by the channeled beam is consistent with the slip bands extending throughout this region.

The increase in the channeling minimum yield observed in Figure 19a at the sample surface can arise from misorientations of the crystal lattices in the slip bands (Follstaedt *et al.*, 1981). The 1 mm^2 channeled beam probes many of these areas and if a given slip band is not raised or lowered exactly parallel to adjacent regions due to more slip at one end of the band than the other, then there will be angular misorientations between slip

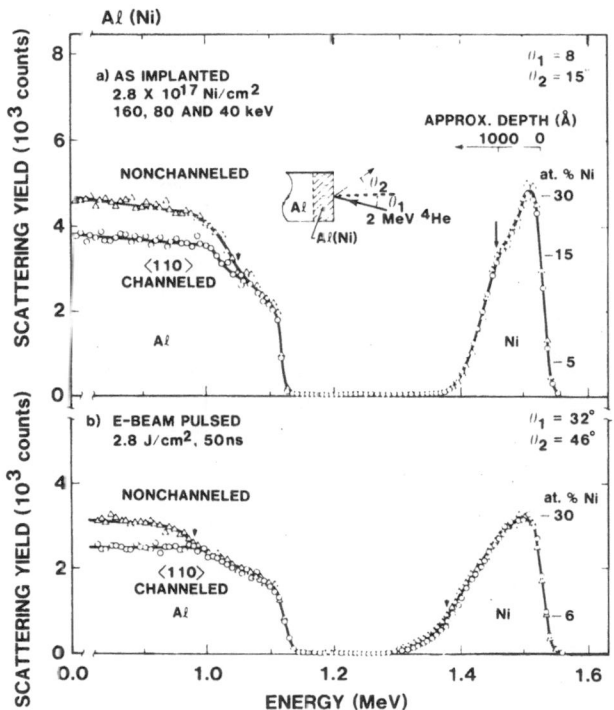

Figure 20   Ion channeling spectra for Al implanted with 2.8×10^{17} Ni/cm² a) as-implanted and b) after 50 nsec pulsed electron beam melting with 2.8 J/cm². The arrows indicate the positions on Al and Ni backscattering yields corresponding to the amorphous surface layer/crystalline substrate interface. From Follstaedt and Picraux, 1982.

bands. This angular misorientation is much like mosaic spread, which is known to give a steep increase in the channeling yield over that for a perfect crystal. In Figure 20b this increase in minimum yield over that for a virgin crystal is plotted vs. laser energy. From the increase the mean misorientation between slip bands can be determined. For a Gaussian variation in the angular misorientation with characteristic angular deviation sigma, the peak at 3.5 J/cm² in Figure 19b corresponds to a sigma value of 0.28 degrees. This value is in good agreement with the misorientation observed by electron diffraction (~0.2°).

A rapid increase in the relative misorientations is observed just above the threshold for melting (Figure 20b). This effect is attributed to the fact that the slip, which moves in one direction upon heating and the other upon cooling, might partially cancel the misorientations below the threshold for melting. However, above this threshold the heating phase of the slip is erased and the subsequent slip after resolidification is in one direction only, maximizing the possible angular misorientation. As the energy is further increased the thermal gradients are reduced after resolidification, resulting in a lesser degree of slip and misorientation (Figure 19b).

The above example of slip damage introduction upon pulsed heating has been for a virgin crystal. Implanted impurities, precipitates or defects would be expected to influence the extent of any slip damage just as these entities influence mechanical properties of a bulk solid (for an example of this in Sn-implanted vs. unimplanted Fe see Knapp and Follstaedt, 1982).

H. Amorphous Phase Formation

The previous two sections on disorder have emphasized lattice defects which can be introduced by pulsed heating of an undoped crystal. For most metallic elements amorphous phase formation cannot be achieved without alloying additions. Therefore in this section we examine the combined influence of ion implantation alloying and pulsed melt quenching on amorphous phase formation.

We illustrate amorphous phase formation in Al after pulsed heating to the liquid state using the case of Ni-implanted Al. For this system it has been recently found that Al can be converted to an amorphous phase by implantation of Ni to concentrations ~15-30 at.%, whereas after electron beam pulse melting the amorphous phase can be formed at Ni concentrations as low as ~6 at.% (Follstaedt and Picraux, 1982). This compositional dependence of amorphous phase formation, as measured by the ion channeling technique, is illustrated in Figure 20. The upper panel corresponds to the as-implanted alloy whose composition extends from a maximum of greater than 30 at.% near the surface to lower concentrations with increasing depth. This tailored composition was achieved by multiple energy implants (160, 80 and 40 keV). In the lower panel the spectra are shown for the same alloy after pulsed melting, which is indicated by the broadening of the Ni profile. The point at which the channeled and nonchanneled spectra for the Al matrix coincide determines the depth of the amorphous layer, and the corresponding point on the Ni profile is indicated by the arrow for the as-implanted and the pulse melted conditions. The amorphous layer is seen to extend appreciably deeper and to lower Ni concentrations after electron beam pulsed melting. This technique of backscattering and channeling provides a convenient way to correlate the thickness of the amorphous layer to the alloy composition at which the amorphous phase is formed.

To confirm that the surface region which shows no channeling is amorphous, transmission electron microscopy was carried out on the same samples. A schematic of the thinned sample is shown in Figure 21a, and the corresponding electron diffraction pattern and bright field micrograph are shown in Figure 21b and c after Ni implantation and pulsed melting. The white arrows indicate the diffuse ring corresponding to the amorphous phase. Similar diffuse scattering from amorphous Al(Ni) is observed in the as implanted alloy. In addition, after pulsed melting the AlNi crystalline phase is present at the higher concentrations near the surface. These AlNi precipitates are indicated by the diffraction rings at black arrows in Figure 21b and by the dark regions in the bright field image of Figure 21c. This compound was not observed in similarly treated alloys with 14 at.% Ni (Picraux et al., 1981); evidently the minimum concentration needed for AlNi formation is between 14 and 30 at.% Ni.

These observations have several interesting implications. The first is that the range of compositions for which an amorphous phase is formed by pulsed melting will not necessarily be the same as by ion implantation alone. It is significant that in this system a lower Ni concentration is sufficient to stabilize the amorphous phase for melt quenching than for ion irradiation. Irradiation-induced collision cascades are sometimes referred to as "thermal spikes", i.e., a nonequilibrium localized zone of highly energetic atoms, which are considered to quench to the lattice temperature in ~1 picosecond. Since this "quench time" is more rapid than that of pulsed melting, it might be expected that a lower Ni concentration would be sufficient to retain an amorphous structure after implantation. However, it must be realized that local configurations characteristic of a liquid cannot be achieved during the short cascade lifetimes. This suggests that the local short range order in amorphous alloys formed by quenching from the liquid state may differ in important ways from those formed directly at low temperatures by ion bombardment. The local atom configurations over a few nearest neighbor distances which are quenched in may be closely related to equilibrium configurations in the liquid state and might not be sustained under the randomizing influence

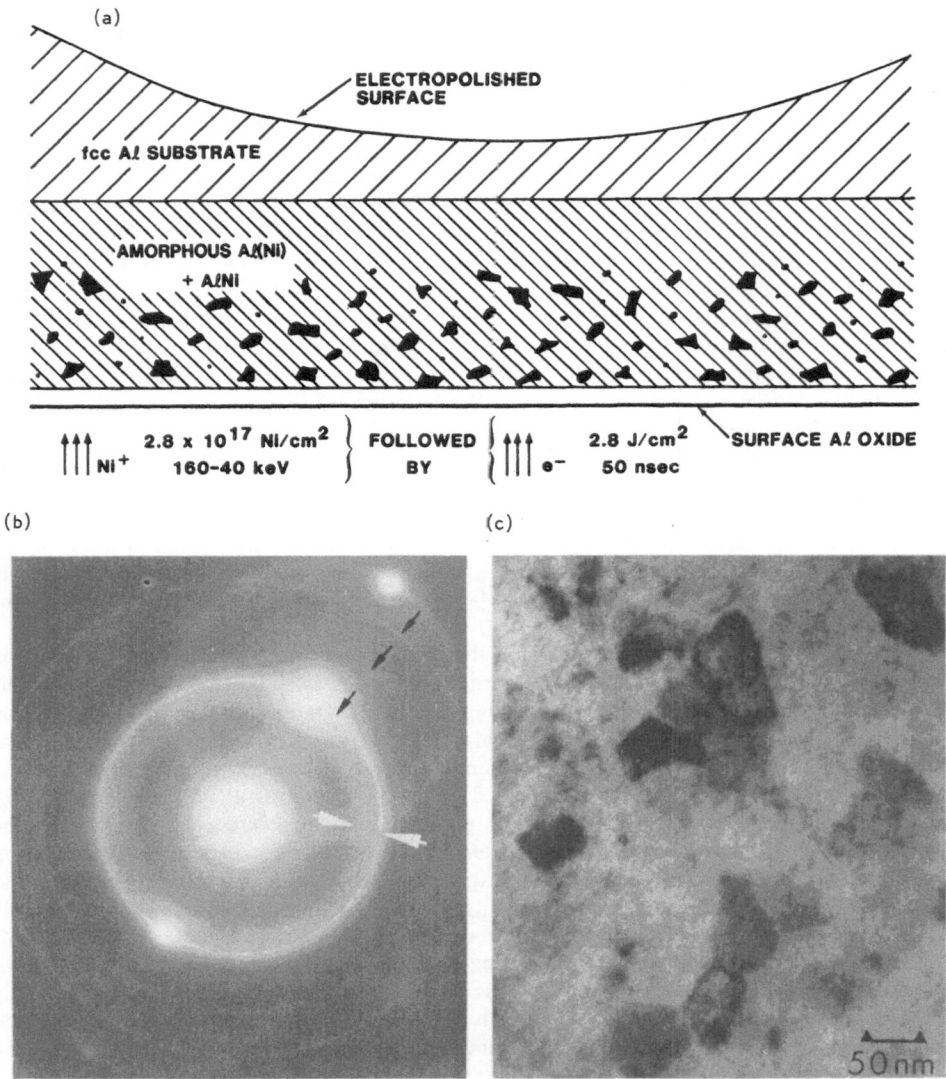

Figure 21   Transmission electron microscopy analysis of Ni-implanted and pulsed electron beam melted sample of
Figure 20b showing: a) schematic cross section of thinned sample; b) diffraction pattern with rings at
black arrows corresponding to AlNi precipitates, diffuse ring between white arrows corresponding to
amorphous Al(Ni) and spots corresponding to Al substrate; and c) bright field and micrograph showing
AlNi precipitates. From Follstaedt and Picraux, 1982.

of ion bombardment. In addition, the many more defects available for implanted alloys (e.g., 10 to 30 displacements per atom in the above Al(Ni) case) may be significant. This area would appear to be a fruitful one for study, particularly with techniques which are sensitive to short range order.

A second implication involves the formation of unusual layered microstructures. The above Al(Ni) alloy was formed by multiple energy implants designed to produce the maximum Ni concentration at the surface and have it continuously decrease with depth in the Al crystal. For a single monoenergetic implant the nickel concentration will follow an approximately Gaussian profile with the peak concentration buried below the surface. This occurs even for relatively high Ni fluences of $\sim 10^{17}/cm^2$ because there is little modification of the profile due to sputtering of the Al host. As illustrated in Figure 22 this can lead to a condition where the Ni concentration is sufficient to form the amorphous phase within a buried layer but the crystalline phase is maintained both below and above this layer. Under this condition the crystal seed for epitaxial regrowth during resolidification is lost after the liquid-solid interface reaches the amorphous layer. Thus the resulting surface layer should nucleate randomly, resulting in a polycrystalline surface layer. Furthermore, if the above interpretation is correct, only 6 at.% Ni will be sufficient to block epitaxy, even though this does not result in an amorphous layer by implantation alone. This prediction is confirmed in Figure 23 where the electron diffraction patterns and electron micrographs show that at 6 and 14 at.% peak Ni concentrations the surface Al layer is polycrystalline and randomly oriented, whereas at lower concentrations good single crystal epitaxial regrowth occurs (Picraux et al., 1981). Thus, in this pulse melted alloy a rather unique microstructure is formed, consisting of a thin buried amorphous layer coated by a thin polycrystalline layer, both upon a single crystal substrate.

A third interesting implication involves the above observation of only the amorphous and the AlNi phases within the 6 to 30 at.% Ni composition range after pulsed melting. From the equilibrium phase diagram one would also anticipate precipitation of the intermediate crystalline phases Al_3Ni and Al_3Ni_2, in addition to AlNi. The quenching here is from the miscible liquid region into two phase regions, where precipitation is expected within liquid Al if sufficient time is available for nucleation. From the pulsed electron beam conditions and the heat flow calculation it is possible to determine that the time available for precipitation, for example for Al_3Ni, must be ≥ 1 μsec. This is significantly longer than the 15 nsec nucleation times measured for AlSb precipitation in liquid Al (Section III.D) and provides an interesting contrast to that study.

Although numerous earlier studies of rapid solidification of alloys have been carried out by scanned laser and liquid splat cooling techniques, the approach of large-area, uniform pulsed heating on nanosecond timescales has been explored very little. Particularly in the area of amorphous alloy formation, the combination of ion implantation and pulsed melt quenching would appear to offer a powerful set of new possibilities, and should provide useful comparisons to previous splat quenching and related approaches. This results primarily because the concentrations that can be achieved within a surface layer by ion implantation are independent of solid (or liquid) equilibrium boundary constraints, and the heating times are sufficiently short that the concentrations within the alloyed layer do not fall appreciably due to diffusion into the bulk. Additional insight into amorphous alloy formation should be gained by comparing other implanted alloys before and after pulsed melting.

IV. SUMMARY OF PULSED MELTING MICROSTRUCTURES

In the previous section we examined the various processes and the wide variety of resulting microstructures which have been observed in pulse heated Al. Here we summarize

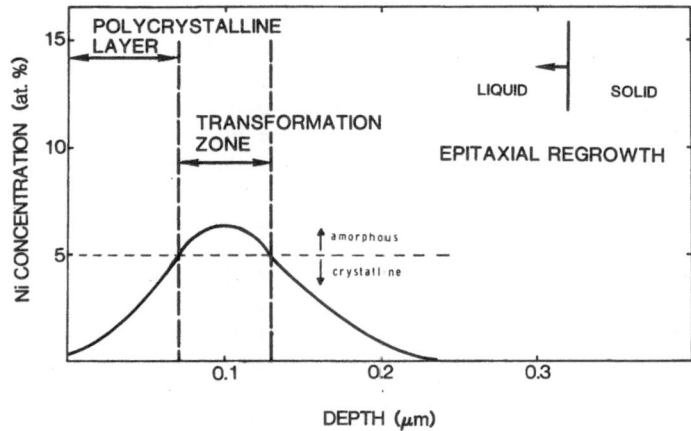

Figure 22   Schematic for buried Ni depth profile in Al with peak concentration sufficient to be melt quenched to the amorphous phase. After Picraux *et al.*, 1981.

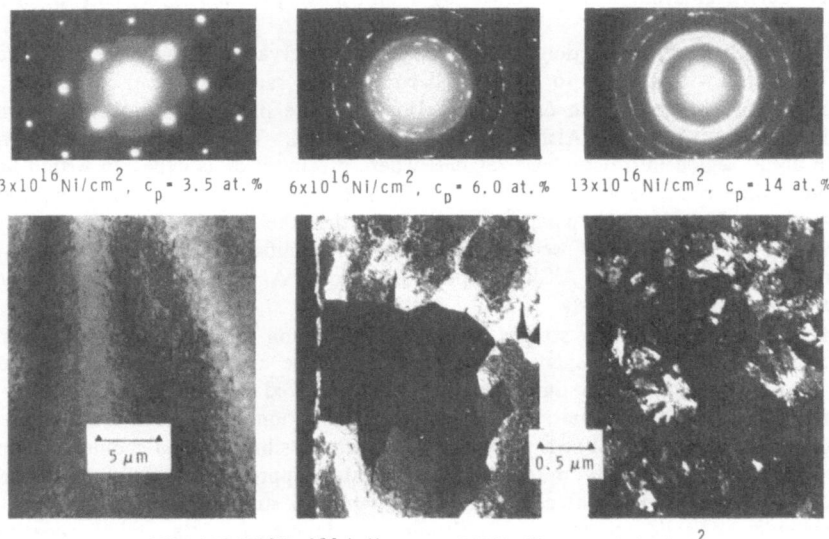

Figure 23   Transmission electron diffraction and bright field micrographs of Ni implanted and pulsed electron beam melted (1.6 J/cm², 50 nsec) Al for three peak buried Ni concentrations of 3.5, 6.0 and 14 at.% showing the transition from single crystal to polycrystalline surface layer. From Picraux *et al.*, 1981.

the lattice defects as well as the single and multiple phase alloy microstructures that can occur with solute additions (Table IV).

The number of microstructures and their possible multiplicity point to the need for transmission electron microscopy to characterize the alloy precisely. In addition, ion backscattering and channeling provide a valuable probe of the near surface composition profiles and general structural state. In Table IV we give a summary of the microstructure details observed by electron microscopy and by ion backscattering/channeling. We do this with the warning: with multiple microstructures and multiple phases, ion channeling alone cannot conclusively delineate the microstructure except in special cases. However once the microstructure has been identified, ion channeling can give additional quantitative information which is difficult or impossible to get with microscopy, such as substitutional concentration and solute site location. Thus, both experimental techniques are required to fully characterize the microstructure.

We first discuss lattice defects and then the solute microstructures resulting from alloying additions. The simplest alloy microstructure and the one often desired is a solid solution in the host phase. Thus, we end this section by giving the requirements on alloy systems and pulsed melting conditions needed to achieve a metastable solid solution.

A. Lattice Defects

Pulse melted Al, both pure and alloyed, always contains dislocations. The density of dislocation lines in an areal projection of the thickness examined by transmission electron microscopy (~ 0.3 μm for Al) may depend upon the solute species, but appears to be less than observed after a nominal fluence implant $\geq 10^{16}$at./cm^2 (Follstaedt et al., 1980). Small (≤ 200Å) dislocation loops are also observed after pulsed melting of pure Al. They yield a vacancy concentration of ~ 100 ppm; however the exact concentration may yet be found to depend upon the melt time, cooling rate and substrate orientation. Vacancy loops have not been observed after pulsed melting with solutes present, presumably because of solute-vacancy interactions (Gauster and Wampler, 1980). However, the observed number of vacancies is $\leq 1/100$ of the solute concentration of typical implanted alloys (≥ 1 at.%) and thus good substitutionality of even strongly interacting solutes is achievable.

Slip deformation has been observed both below and above melt threshold when the lateral thermal stress exceeds the mechanical yield stress. Slip deformation is largest just above the melt threshold where the thermal gradients are largest. Because of the orientation dependence of the yield strength and angle of {111} slip planes to the sample surface, the slip deformation can be expected to be orientation dependent. Substrate temperatures during pulsed melting will influence the degree of slip through the temperature gradient. For example, a significant reduction in the extent of slip is observed at 200°C, relative to 100 and 23°C, for laser pulse melted Al implanted with Sb (Peercy et al., 1982). There has not been a detailed characterization of the effect of solute additions on slip deformation, although solute additions would be expected to influence slip deformation by affecting the material yield stress.

B. Solute Microstructures

No liquid segregation of solutes to the surface has been reported in Al for these fast pulsed melting experiments, due to the high interface velocities and efficient interface trapping in Al. For sufficiently slow liquid-solid interface velocities, solute segregation and constitutional supercooling with the solute segregating laterally to form cell walls would be expected, similar to the case for Si (Chapter 4; Narayan et al., 1982).

TABLE IV

Summary of pulsed melting microstructures in Al

Microstructural Feature	TEM Observations	Backscattering/ Channeling	General Comments
Lattice Defects			
1. Extended Dislocations	Always observed after pulsed melting.	Increased dechanneling slope.	Generally less than produced by implantation.
2. Dislocation Loops	Small ($\leqslant$ 200Å)	Increased dechanneling slope.	~100 ppm; seen only in pure Al.
3. Slip Deformation	{111} planes; best contrast on inclined planes.	Increased x_{min} at surface; increased dechanneling slope.	Slip due to lateral thermal stress; greatest just above melt threshold; may be influenced by solutes.
Solute Microstructures			
1. Substitutional Solid Solutions	Single (host) phase observed.	Reduced impurity channeling yield; angular yield tracks that of the host.	Single phase liquid and other requirements.
2. Segregation	Possible call formation	Solute at surface; channeling absent or different from host.	Not yet observed in Al.
3. Precipitation a) In liquid	Ring diffraction pattern.	Direct scattering from elements in the precipitate.	Second phase may be a liquid or a solid; nucleation times determine whether precipitation occurs.
b) In Solid	Weak diffraction.	Complicated; specific orientations and sizes must be considered.	Minimal due to rapid cooling.
4. Amorphous Phase Formation	Diffuse ring diffraction; possible polycrystalline surface layer.	Direct scattering; no channeling at the amorphous and polycrystalline layer depths.	Needs minimum impurity concentration to stabilize.

Alloying additions which are insoluble in the host liquid can lead to second phase formation within the host-rich liquid. For immiscible liquids, melt times are expected to usually be sufficiently long for segregation of the liquids, which then would solidify into two phases. A two phase liquid plus solid region in the equilibrium phase diagram at the temperatures reached also provides a necessary condition for such phase separation to occur. However, as demonstrated by Al(Sb) this condition is not sufficient to predict precipitation of solids within the liquid phase. The additional requirement is that sufficient time must be spent in the two phase temperature interval for precipitate nucleation to occur. Once nucleation occurs, liquid phase diffusivities are sufficiently rapid to produce observable precipitates during typical melt times (≥ 10 nsec). The absence of the compounds Al_3Ni and Al_3Ni_2 in electron beam pulse melted Al(Ni) demonstrates that the faster cooling rates obtainable with the laser are not necessarily needed to bypass predicted nucleation.

Supersaturated solid solutions are often formed in melt quenched Al, since liquid solubility limits generally exceed solid solubilities. It is often assumed that cooling after resolidification will be sufficiently rapid to prevent precipitation within the solid. However, solid state phase separation can occur for both laser and electron beam pulse melted Al. For example, in Figure 13 the faint spot reflections of AlSb indicate there is some solid phase precipitation. Thus, although diffusion in the solid state is minimal, precipitation may still occur in some instances, and this can result in precipitate distributions of extremely high density and small size.

Amorphous phases may be quenched in after pulsed melting of implanted Al. For Al(Ni) it was demonstrated that a minimum concentration of 6 at.% Ni was necessary to stabilize the amorphous phase. In general, the amorphous phase may be considered an alternative metastable state to a super-saturated solution when the solubility limits are exceeded and phase separation prevented. However if precipitation occurs, the local concentration may be reduced to such an extent that crystallization is no longer inhibited. In the case of AlNi precipitation, such large Ni concentrations are required before AlNi precipitation can occur that a mixed system of amorphous Al(Ni) and crystalline AlNi precipitates results.

C. Metastable Solid Solution Criteria

After the above discussion, we now ask what are the minimum requirements for solid solution formation. This question has also been addressed by Sood, who examined the substitutional nature of solutes in pulse melted Al, Ni and Cu (Sood, 1981). He states that the rapid interface velocity provides complete interface trapping and that cooling rates in the solid phase are too rapid to permit precipitation. However, the numerical arguments given to support these hypothesis are made for very rapid heating and cooling by laser pulses. The following additional requirements were then deduced: (a) good liquid phase epitaxy, (b) liquid phase miscibility, and (c) cooling rapid enough to suppress precipitate nucleation and growth.

To cover a broader range of possible conditions and alloys, we here state four more general criteria for metastable solid solution formation:

1) The alloyed layer must remain a single phase liquid solution prior to resolidification.

2) The liquid-solid interface velocity must be sufficiently rapid to provide complete solute trapping and thus prevent segregation.

3) The solute concentration must be less than that needed to stabilize an amorphous phase upon resolidification.

4) Solute diffusion and cooling times must be insufficient to permit precipitation.

Criteria 1) is a more general statement of Sood's b) and c). Precipitation must not remove the solute from solution. Second phase separation (solid or liquid) must either not be expected during cooling or its nucleation must be suppressed (Wampler *et al.*, 1981). Criterion 2) appears to be met for most Al studies, including an example with the slower cooling from pulsed electron beam melting (Picraux *et al.*, 1981). Nonetheless, for slightly slower interface velocity or faster solute diffusion, segregation could occur. Criterion 3) results from our observation of the amorphous layer in Al(Ni) for concentrations above 6 at.% Ni (Follstaedt and Picraux, 1982). We require the solid solution to be crystalline. Criterion 4) is generally met by most studies, but precipitation observations in Al(Sb) show that this requirement cannot be overlooked. Sood's criterion a) seems overly restrictive. Crystallization is required, but the quality of the epitaxy does not limit solid solution formation. For example, a polycrystalline layer may also be a metastable solid solution.

Finally, we note that intermediate situations exist: Precipitation may remove only part of the solute out of solution; part of the solute may segregate to the surface, leaving the rest in solution; part of the alloyed layer may exceed the minimum concentration for amorphization while the rest does not. In addition, an isolated solute in the host lattice may move slightly off the perfect lattice position, for example due to defect interactions. One or more of these effects may explain why alloys like Al(Ni) and Al(Sn) (see Table II) are not 100% substitutional.

V. ION BEAM MIXING AND PULSED MELTING OF LAYERS

An alternative approach to ion implantation for forming a given surface alloy composition is to deposit thin layers by conventional deposition and then induce atomic scale mixing between the surface layers and the substrate. The mixing may be achieved by ion bombardment or by pulsed melting of the surface (see, for example, Chapters 9 and 13). Furnace alloying of such layers can be greatly limited by such effects as oxide barriers, unwanted intermediate phases, low solubilities and disparate melting points. In this section we give a brief example of surface alloying of Al by ion beam mixing to illustrate this technique.

Figure 24 schematically illustrates 400 keV Xe ion beam mixing of a thin Sb film with a single crystal Al substrate (Picraux *et al.*, 1982). Ion scattering indicates that the collision cascades result in Sb motion into the Al substrate and Al motion into the Sb film. In the actual experiments the Sb layer is covered by a thin Al surface layer to reduce loss of the surface alloy by sputtering. The presence of an aluminum oxide layer at a surface results in a very low sputtering coefficient. At room temperature and below, the Al migrates rapidly throughout the Sb film and uniformly increases in concentration with Xe bombardment. Simultaneously a much lower concentration tail of Sb is mixed into the Al substrate. In Figure 25 this evolution with Xe fluence is shown by backscattering spectra for 400 keV Xe bombardment of a 260Å Al surface layer over a 460Å Sb film on an Al <110> crystal substrate. The rate of mixing of Al atoms into Sb proceeds at ~15 Al/Xe, whereas the rate of mixing of Sb into Al is only ~0.5 Sb/Xe at 300K.

The lower Sb spectrum of Figure 25 corresponds to an AlSb alloy of approximately 50-50 composition, and transmission electron microscopy on multiple layer films showed that only the intermetallic phase, AlSb, can be readily formed by Xe ion beam mixing (Picraux *et al.*, 1982). In contrast, 50 keV Sb implantation into Al (Kant *et al.*, 1979) results in Sb peak alloy concentrations of only ≤30 at.% at the sputtering limit. Thus one advantage in the use of ion beam mixing of surface layers is in the formation of alloys with high concentrations. In addition, when the rate of mixing in terms of atoms per incident beam particle is appreciably greater than one, as in the case of 15 Sb/Xe above, significant reduction in the

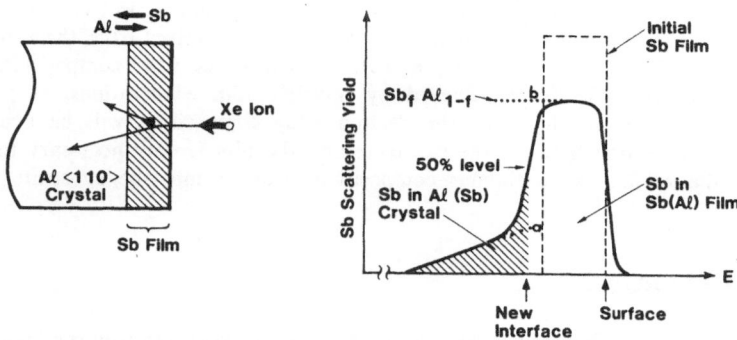

Figure 24   Schematic He ion scattering spectra from Sb film on Al before and after ion beam mixing. From Picraux *et al.*, 1982.

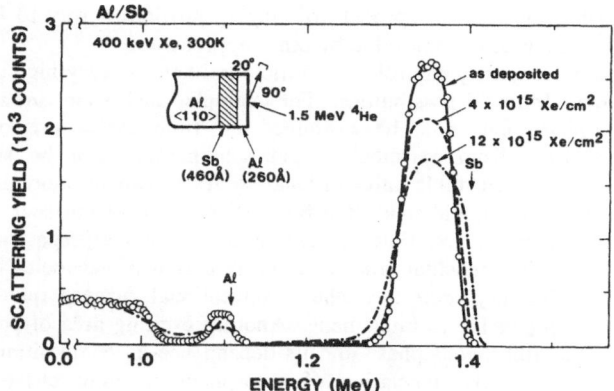

Figure 25   Ion scattering spectra for Al/Sb/Al <110> substrate system before and after 400 keV Xe ion beam bombardment to 4 and 12 × 10^{15} Xe/cm² at 300K. From Picraux *et al.*, 1982.

number of implanted ions (and thus the implantation time) can result. This may be a particular advantage in the case of the high fluences often required for metallurgical applications.

The primary driving force for ion beam mixing is the atomic collision cascade created by the bombarding ions. Higher atomic number ions, which create denser cascades, are usually more efficient at inducing mixing. Structural and chemical aspects of the system being mixed have also been found to be important in determining the final mixing rate (Picraux *et al.*, 1982). For example, the soluble system Ag in Al exhibits a much faster mixing of Ag into Al (~2.6 Ag/Xe) than found under identical conditions for the highly insoluble system Sb in Al (0.5 Sb/Xe). Another such example is more than an order of magnitude difference observed in the mixing rate between Sb into Al and Al into Sb. Also the mixing has been found to be faster for the motion of Sb into polycrystalline Al than into single crystal Al. More work is required to fully understand all the processes involved in ion beam mixing.

The number of atoms mixed across the interface between film and substrate may be considered a kind of efficiency factor or gain achieved over implantation alone. For ion beams the mixing results from the motion of energetic atoms in the solid state (see Chapter 9), whereas for pulsed melting, the mixing process derives from the rapid diffusion coefficients in the liquid state (see Chapter 13). In both cases, fixed compositions appear to be most rapidly and effectively achieved by multiple film evaporations, such that upon complete mixing between the films the desired alloy composition will be achieved. For multiple films, migration between the substrate and the film is only necessary to the extent that good adhesion and a continuous composition gradient into the substrate need to be assured.

VI. CONCLUSIONS

The purpose of this chapter has been to illustrate in an organized way the many aspects of surface alloying of metals by ion implantation and pulsed heating techniques. We have shown for a single system, aluminum, the wide variety of microscopic alloy states which can be achieved. In so doing we have illustrated some of the new, and as yet little explored, aspects of these techniques. This approach to alloying provides a way to form thin surface alloys, some of which may be unachievable by other approaches.

The ability to simultaneously control composition and achieve very high quench rates (10^8 to 10^{10} K/sec) offers additional possibilities. For example, nucleation times for precipitation within the liquid phase can now be examined on time scales previously unexplored. Microstructures resulting from extremely rapid solidification can be studied and such nonequilibrium states as metastable substitutional solutions can be more fully explored. A key parameter is that the liquid-solid interface velocities (~ 10 m/sec) are so fast that efficient interface trapping occurs, thus preventing the normal segregation of solutes upon resolidification. The rapid quenching and observed retention of vacancies can be applied to high melting point refractory materials where conventional furnace quenching techniques have been difficult or impossible to implement. Another exciting area of possible application is the formation of the amorphous phase and its dependence on composition and short range order. The differences in the formation of amorphous metallic phases between liquid quenching at $\sim 10^9$ K/sec and ion bombardment can be directly explored by the combined approach of ion implantation and pulsed melting.

REFERENCES

Appleton, B. R. and Celler, G. K., eds. (1982), "Laser and Electron Beam Interactions with Solids," North Holland, New York (and previous MRS proceedings of this Symposium listed therein).

Battaglin, G., Carnera, A., Della Mea, G., Mazzoldi, P., Jain, A. K., Kulkarni, V. N. and Sood, D. K. (1981), in "Laser and Electron-Beam Solid Interactions and Materials Processing," (J. F. Gibbons, L. D. Hess and T. W. Sigmon, eds.) p. 615, North-Holland, New York.

Battaglin, G., Carnera, A., Della Mea, G., Mazzoldi, P., Jannitti, E., Jain, A. K. and Sood, D. K. (1982), J. Appl. Phys. (submitted).

Battaglin, G., Carnera, A., Della Mea, G., Mazzoldi, P., Jain, A. K., Kulkarni, V. N. and Sood, D. K. (1982a), to be published.

Borders, J. A. (1979), Ann. Rev. Mat. Sci. 9, 313.

Breinan, E. M., Kear, B. H., Banas, C. M. and Greenwald, L. E. (1976), in "Proc. 3rd Intl. Symp. Superalloys: Metallurgy and Manufacture," p. 435, Seven Springs.

Campisano, S. U., Baeri, P., Grimaldi, M. G., Foti, G. and Rimini, E. (1980), J. Appl. Phys. 51, 3968.

Copley, S. M., Beck, D., Esquivel, O. and Bass, M. (1979), in "Laser-Solid Interactions and Laser Processing - 1978," (S. D. Ferris, H. J. Leamy and J. M. Poate, eds.) p. 161, Amer. Inst. of Phys., New York, (AIP Conf. Proc. No. 50).

Follstaedt, D. M., Picraux, S. T. and Wampler, W. R. (1980), in "Laser and Electron Beam Processing of Materials," (C. W. White and P. S. Peercy, eds.) p. 708, Academic Press, New York.

Follstaedt, D. M., Picraux, S. T., Peercy, P. S. and Wampler, W. R. (1981), Appl. Phys. Lett. *39*, 327.

Follstaedt, D. M. and Wampler, W. R. (1981), Appl. Phys. Lett. *38*, 140.

Follstaedt, D. M. (1982), in "Laser and Electron Beam Interactions With Solids," (B. R. Appleton and G. K. Celler, eds.), North Holland, New York, p. 377.

Follstaedt, D. M. and Picraux, S. T. (1982), to be published in J. Appl. Phys.

Gauster, W. B. and Wampler, W. R. (1980), Phil. Mag. *41*, 145.

Hussain, T., Geerk, J., Ratzel, F. and Linker, G. (1980), Appl. Phys. Letters *37*, 298.

Hussain, T. and Linker, G. (1982), Solid State Comm. (submitted).

Kant, R. A., Myers, S. M. and Picraux, S. T. (1979), J. Appl. Phys. *50*, 214 and unpublished.

Kaufmann, E. N. and Buene, L. (1981), Nucl. Instrum. Methods *182/183*, 327.

Knapp, J. A. and Follstaedt, D. M. (1982), in "Laser and Electron Beam Interactions with Solids," (B. R. Appleton and G. K. Celler, eds.) North Holland, New York, p. 407.

Knapp, J. A. (1982), private communication.

Mazzoldi, P., Battaglin, G., Carnera, A., Della Mea, G., Jannitti, E., Jain, A. K. and Sood, D. K. (1981), private communication.

Miotello, A. and Donà dalle Rose, L. F. (1981), submitted to Phys. Lett. A.

Myers, S. M. (1980), in "Treatise on Materials Science and Technology," Vol. 18 (J. K. Hirvonen, ed.), p. 51, Academic Press, New York.

Narayan, J., Naramoto, H. and White, C. W. (1982), J. Appl. Phys. *53*, 912.

Peercy, P. S., Follstaedt, D. M., Picraux, S. T. and Wampler, W. R. (1982), in "Laser and Electron Beam Interactions with Solids," (B. R. Appleton and G. K. Celler, eds.), North Holland, New York, p. 401.

Picraux, S. T. (1980), in "Site Characterization and Aggregation of Implanted Atoms in Materials," (A. Perez and R. Coussement, eds.), p. 307 and p. 325, Plenum Press, New York.

Picraux, S. T., Follstaedt, D. M., Knapp, J. A., Wampler, W. R. and Rimini, E. (1981), in "Laser and Electron Beam Solid Interactions and Materials Processing," (J. F. Gibbons, L. D. Hess and T. W. Sigmon, eds.), p. 575, North Holland, New York.

Picraux, S. T., Follstaedt, D. M. and Delafond, J. (1982), in "Metastable Materials Formation by Ion Implantation," (S. T. Picraux and W. J. Choyke, eds.), North Holland, New York, (in press).

Picraux, S. T., Peercy, P. S., Follstaedt, D. M. and Wampler, W. R. (1982a), to be published.

Poate, J. M. and Cullis, A. G. (1980), in "Treatise on Materials Science and Technology," Vol. 18 (J. K. Hirvonen, ed.), p. 85, Academic Press, New York.

Sood, D. K. (1981), Radiation Effects Lett. *67*, 13.

Spencer, L. V. (1959), U.S. National Bureau of Standards Monograph No. 1.

Swanson, M. L., Howe, L. M. and Quenneville, A. F. (1978), J. Nucl. Mater. *69/70*, 372.

Swanson, M. L., Howe, L. M. and Quenneville, A. F. (1980), Phys. Rev. *B22*, 2213.

Von Allmen, M. (1980), in "Laser and Electron Beam Processing of Materials," (C. W. White and P. S. Peercy, eds.), p. 6, Academic Press, New York.

Wampler, W. R., Follstaedt, D. M. and Picraux, S. T. (1980), Appl. Phys. Lett. *36*, 366.

Wampler, W. R., Follstaedt, D. M. and Peercy, P. S. (1981), in "Laser and Electron Beam Interactions and Materials Processing," (J. F. Gibbons, L. D. Hess and T. W. Sigmon, eds.), p. 567, North-Holland, New York.

White, C. W., Wilson, S. R., Appleton, B. R. and Young, F. W., Jr., (1980), J. Appl. Phys. *51*, 738.

Wood, R. F. (1980), Appl. Phys. Lett. *37*, 302.

CHAPTER 12

MATERIALS MODIFICATION BY ION IMPLANTATION

J. K. HIRVONEN

Naval Research Laboratory, Washington, D.C.

C. R. CLAYTON

State University of New York, Stony Brook, New York

CONTRIBUTORS: H. Herman, P. Mazzoldi, O. Meyer, S. M. Myers, D. A. Rigney, P. Heilmann and G. K. Wolf

I. INTRODUCTION

The use of ion implantation for the controlled modification of surface sensitive properties has been a rapidly growing research field in the last few years, evidenced by several topical symposia and conference proceedings (see for example Benenson *et al.*, (1981)). The two principal thrusts of current research involve the use of implantation as (i) a metallurgical tool for studying basic mechanisms in areas such as aqueous corrosion, high temperature oxidation and impurity trapping as well as (ii) a means of beneficially modifying the mechanical or chemical properties of materials. This chapter includes examples of both usages and will review the present status of some of the most active research fields outside of the semiconductor area with an attempt to indicate the required direction and potential payoff of further work. The application of lasers to improve surface sensitive properties will be referred to in the next chapter. Table I shows a compilation of material properties influenced by ion implantation from Dearnaley (1978).

II. ION IMPLANTATION - ADVANTAGES AND LIMITATIONS OF THE TECHNIQUE

Some of the advantages and limitations of ion implantation in comparison to other surface treatments (such as coatings) are listed in Table II. An intrinsic basic limitation of ion implantation is that it is a line-of-sight process; it will not be feasible to apply it to samples

TABLE I

Material Properties Influenced by the Surface Composition

Friction	Corrosion resistance	Bonding
Wear	Electrochemistry	Lubrication
Hardening	Catalysis	Adhesion
Fatigue	Decorative Finish	Reflectance

TABLE II

Advantages and Limitations of Ion Implantation as a Surface Modification Technique

Advantages	Limitations
(1) Solid solubility limit can be exceeded	(1) Line-of-sight process
(2) Alloy preparation independent of diffusion	(2) Shallow penetration
(3) Allows fast screening of the effects of changes in alloy composition	(3) Relatively expensive equipment and processing costs
(4) No sacrifice of bulk properties	
(5) Low temperature process	
(6) No significant dimensional changes	
(7) No adhesion problems since there is no sharp interface	
(8) Controllable depth concentrations	
(9) Clean vacuum process	
(10) Highly controllable and reproducible	

having complicated re-entrant surfaces, for example. The shallow penetration would in itself make it appear useless as a technique for engineering applications; however, there are several situations involving both physical and chemical properties in which the effect of the implanted ions persists to depths far greater than the initial implantation range. Specific examples involve wear and oxidation.

As for the advantages, the fact that ion implantation is a nonequilibrium technique permits the formation of surface alloys whose formation is highly independent of solubility limits and diffusivities governing conventional alloy formation. There are examples in which the solid solubility limits of implanted alloys can exceed equilibrium values by several orders of magnitude.

Ion implantation also allows the convenient production and subsequent study of surface alloys with well-defined compositions. In this manner the technique can be used as a powerful research tool to examine the physical state of alloys as a function of varying alloy composition. Using implantation alloying one can sometimes avoid changing other parameters (e.g., grain size) that may affect the property of interest, such as oxidation behavior.

Ion implantation has other potential advantages for treating limited-area critical parts. The surface properties can be optimized independently of the bulk properties, and implantation can be carried out at low temperatures without producing any significant dimensional changes. In addition the surface alloy produced by implantation should not suffer from adhesion problems since there is no sharp interface.

III. TRIBOLOGICAL AND MECHANICAL PROPERTY CHANGES

A. Background

Tribology ("the science and technology of interacting surfaces in relative motion") is a term encompassing an old, important, and often complicated set of phenomena relating to friction and wear. The application of ion implantation to these areas has not only yielded surfaces with improved properties but has been important for the study of basic mechanisms. The first studies in this field were initiated by Hartley and colleagues at Harwell in the early 1970's (Hartley et al., 1973; Hartley, 1975).

Since that time there has been a steadily increasing number of researchers publishing in the area. The increased interest in implantation into metals and insulators is also reflected in a growing number of conferences in this area, as well as special symposia. These meetings include Application of Ion Beams to Metals held at Albuquerque, New Mexico, in 1973 (Picraux et al., 1974) and Applications of Ion Beams to Materials held at Warwick, England, in 1975 (Carter et al., 1976). Implantation effects in non-semiconductors have also constituted a major part of topical symposia including Ion Implantation: New Prospects for Materials Modification, held at Yorktown Heights in 1978 (Brown, 1978) and Surface Modification of Materials by Ion Implantation held in Cambridge, Massachusetts, in November 1979 (Preece and Hirvonen, 1980). More recent meetings include Ion Beam Modification of Materials held in Albany, New York, in 1980 (Benenson et al., 1981) and Modification of the Surface Properties of Metals by Ion Implantation held in Manchester, England, in 1981 (Ashworth et al., 1982).

Several excellent reviews are contained in the proceedings of the aforementioned meetings including the following subject areas: aqueous corrosion (Clayton, 1981), electrochemistry and catalysis (Wolf, 1981a), surface mechanical properties (Herman, 1981), and oxidation related properties (Dearnaley, 1981a). Readers who are particularly interested in the use of ion implantation in oxidation studies, a topic not covered in this chapter, are also urged to see the following sources (Ashworth et al., 1982; Grabowski and Rehn, 1982). Finally, a compilation of comprehensive review chapters (Hirvonen, 1980) covers the use of implantation for both basic studies and improving surface sensitive properties.

B. Experimental Results

Hartley (1976) has reviewed the work investigating the effects of ion implantation on friction which was done primarily at Harwell. They measured the friction force between a heavily loaded (up to 10^5 g/cm^2) tungsten carbide ball and case-hardened steel (EN 352) implanted with Mo, Sn, Pb, In and Kr at fluences between 10^{16} and 10^{17} ions/cm^2. They observed no effect for Kr but noted an increase for Pb and a decrease for Sn of up to 50%. It is interesting to note that the reduction in friction following overlapping implanted distributions of Mo ions and twice as many S ions is greater than for either element alone. However, the presence of MoS_2 (a widely used solid lubricant) was not determined. More work needs to be done to test the feasibility of providing low-friction, low-wear surfaces by this technique. However, in principle, ion implantation provides a novel method of producing precipitates (such as MoS_2) in the near surface region of hard, wear-resistant materials while maintaining good thermal contact and adhesion with the substrate. These two factors are problems frequently encountered with coatings. More recently, Carosella et al. (1980) have found decreased friction in 52100 bearing alloy implanted with high dose Ti. The implanted Ti getters C and eventually forms an amorphous Ti-C-Fe surface region with reduced friction and wear. Overlapping implants of Ti plus C have been found to produce the same effects (Yost et al., 1982; Jeffries et al., 1982).

In general, the wear properties of materials are more important than either their hardness or friction properties. Implantation has thus been mostly studied for improving wear resistance. As for friction, the first experiments involving the use of ion implantation for improving the wear resistance of metals were conducted by Hartley (1975) and colleagues at Harwell using a pin-on-disc wear tester as depicted in Figure 1. A weighted pin bears against an ion implanted disc and the wear rate of the pin (or disc) is determined via optical microscopy. Figure 1 also shows how high dose nitrogen implantation can significantly reduce the wear rate. These experiments also ascertained that the implantation of inert species (such as Ne^+ or Ar^+) to create compressive stresses in the near surface region was not effective in reducing the wear rate. However implantations of interstitial species (i.e., B^+, C^+ and N^+) at fluences exceeding 10^{17} ions/cm^2 were effective and have been reported to be persistent in reducing wear for times greatly in excess of that required to wear away a layer of thickness corresponding to that of the range of implanted ions. This persistence effect has been postulated by the Harwell group (Dearnaley and Charter, 1981) as being due to the following phenomena.[†]

"In the case of steels, there is now a reasonably good understanding of how the implantation of light interstitial atoms such as nitrogen, carbon or boron brings about resistance to wear and fatigue. These species have the property of segregating, even under room temperature conditions, at dislocations. In turn, this impedes the movement of the dislocations, which will obviously render the surface harder and more resistant to wear. The initiation of fatigue cracks is also inhibited by the restriction of dislocation movement. This property of interstitial impurities has long been recognized as the cause of strain aging in steel."

"Under conditions of sliding wear, two other processes take place, which may well account for the persistent effect of ion implantation, observable well beyond the stage at which the original shallow implanted layer has been worn away (Hartley, 1979). The first is the production and propagation of fresh dislocations by the heavy load exerted at contacting asperities. A dense dislocation network is constantly generated ahead of the worn surface, and this can drag forward associated impurity atoms (solute drag). The second effect is that of localized frictional heating, which occurs at the same points on the surface. The work of Rowson and Quinn (1980) has demonstrated that, under typical pin-on-disc test conditions, the asperity temperature may rise to 600° - 700°C. Migration of mobile impurities will take place under the strong temperature gradients so created. Pipe diffusion along forest dislocations is the most likely transport process, and we and other groups have used nuclear reaction analysis to detect significant quantities of implanted nitrogen at the base of wear tracks which are deeper than the original depth of implantation."

[†]   Quoted with permission of the authors

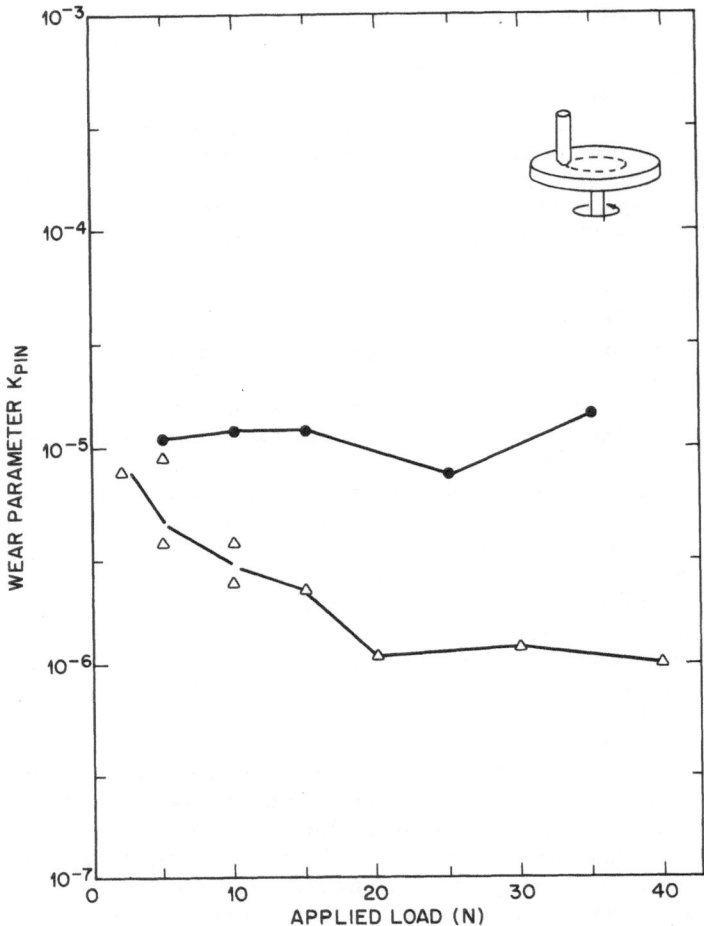

Figure 1   The Archard wear parameter measured on a 440C steel pin of Vickers hardness 230 as a function of applied load. Intense N implantation produces a decrease in wear. $10^{18}N^{+}ions/cm^{2}$, 35 keV, white spirits lubricant (Hartley *et al.*, 1975). Triangles refer to the implanted surface.

"The coefficient of friction is also reduced by ion implantation into steel and this may be due to two effects: (i) junctions formed between the two surfaces will be more brittle, due to the lessened dislocation movement, and (ii) an oxide film is more likely to be retained under these circumstances, and the presence of oxide will itself reduce adhesion. Nuclear reaction analysis has again been used to show that more oxide is present in the wear track on nitrogen implanted steel than on unimplanted steel, worn under the same conditions."

"When we consider the composite material, cobalt-cemented tungsten carbide, the situation is less simple and our understanding is not as great. Wear at high temperatures is believed to involve a diffusion mechanism, cobalt migrating outwards and species such as iron (from worked metal) diffusing inwards to weaken the structure and undermine carbide grains. At low temperature (where, as we shall see, ion implantation has proved effective) it is more likely that adhesion and tearing of cobalt occurs, and this process may be accentuated if the transverse shear forces are sufficient to cause an extrusion of softer cobalt between the grains of carbide."

"Ion implanted nitrogen or carbon may segregate to dislocations in cobalt, as it does in ferrous alloys, and since cobalt, unlike iron, forms no stable nitrides or carbides the implanted atoms must remain in solid solution. Mazey (private communication) has observed, in the electron microscope, martensitic transformations that have occurred in the cobalt binder phase in nitrogen-implanted cemented tungsten carbide. This is evidence of lattice distortion, and probable hardening, brought about by the nitrogen in solid solution."

"Alternatively, the nitrogen may segregate to the boundaries between the carbide grains and the cobalt binder and, by the formation of strong chemical bonds, may so strengthen the composite.†"

"The beneficial effects of ion implantation into Co-cemented tungsten carbide persist to depths which far exceed the dimensions of WC grains. This is strong evidence that transport of implanted nitrogen or carbon takes place through the interstices of the cobalt binder, and once again this is likely to be driven thermally by the strong thermal gradient generated at load-bearing asperities. For this reason, if the conditions of wear are particularly mild, and plenty of coolant is available, the implanted atoms may not move, and it is to be expected that the process would be less effective: this is in accordance with the tests that have been carried out using an extremely fine abrasive slurry of alumina in water."

In response to these initial and somewhat unexpected results a series of wear tests have been performed involving both adhesive and abrasive wear (e.g., Hirvonen et al., 1979; Carosella et al., 1980; Singer et al., 1981) in an attempt to ascertain the pertinent parameters and mechanisms involved. Hirvonen et al., (1979) found a 10×-100× reduction in sliding wear for the AISI type-416 stainless steel (SS) and 304 SS wear couple under lubricated conditions when the sliding 304 SS surfaces were implanted with high dose N or a number of other species (e.g., B,C,Ti+C,Ti+B).

In a later study, Singer and Bolster (1980) have N-implanted a low C alloy steel and AISI type-304 stainless steel, and have examined abrasive wear with micron-sized particles. The analysis is augmented through surface electron spectroscopy, enabling careful determinations of the implanted species location. Since it is known that abrasive wear is related to hardness, the authors assert that this measurement can yield information on implantation-modified hardness. On implanting 10^{17} N^+ ions/cm^2 at 40 keV into the quenched and tempered steel, the relative (abrasive) wear resistance shows significant increases, whereas implantation yields the *opposite* results for the 304 stainless steel. In fact, austenitic stainless steel, which is known to transformation-harden, shows considerable softening, which these authors suggest may be due to the N acting to stabilize the austenite. These thoughts have recently been given further credence by TEM observations that implanted N transforms stress-induced martensite in 304 SS back to the austenite phase (Vardiman et al., 1982). Singer and Bolster also found that the abrasion hardening effect as reflected in the relative abrasive wear resistance observed in C steel was limited to the depth of implantation. For the stainless steel, on the other hand, the softening effect persisted, the N apparently being forced deeper during the wear process.

---

† The possibility of such an interfacial segregation is discussed by the authors in a subsequent section of their paper in order to explain some recent results in friction between carbides.

As an example of some recent wear studies, Lo Russo *et al.* (1979; 1982) have studied the effect of N ion implantation on the unlubricated sliding wear of steel by using a reciprocal motion machine in which a cemented 19 CN5 (Italian standard) steel block slides on a parallelepiped bar with the following chemical composition (wt.%): C/0.35, Ni/0.82, Cr/0.72, Mo/0.20, Mn/0.73, Si/0.20. Nitrogen ions were implanted in the bar at 30 keV energy and with a beam current of about 80 μA. For implantation doses lower than 10^{17} N^+/cm^2 no significant difference in the wear behavior of implanted and unimplanted samples was observed. On increasing the dose, a wear reduction of a factor 3 was measured, with the lowest slope for the wear vs. time curve in Figure 2 corresponding to the $2 \times 10^{17} N^+$/cm^2 implantation dose. This reduction factor is lower than that obtained in other laboratories, probably due to the more severe wear conditions (absence of lubricant and load of 20N).

Nevertheless, the wear reduction also persists after removal of some microns of material, a thickness which is orders of magnitude greater than the range of implanted ions. This persistence has been connected to the presence of about 20% of the implanted N ions in the steel subsurface region, after removal by wear of 5 microns of material. The remaining N was measured by using the N^{14} $(d,\alpha)C^{12}$ nuclear reaction induced by a 1.4 MeV deuteron beam. These results are shown in Figure 3.

The low dose N implantation, however, is not sufficient to affect the wear behavior of the steel. Their tentative explanation for the role of implanted N ions in the wear reduction is the interaction between interstitial N ions and defects in such a way that dislocations are decorated and impeded in their motion. The wear reduction is thus thought to be connected to the interaction of atoms which can be introduced by direct implantation (e.g., N, C and B) or created in the matrix (i.e., C atoms dissolved by irradiation from precipitates in high C steels by implantation of inert ions such as Ne).

In order to clarify this hypothesis, the effect of double implantation has been studied to separate the role of the radiation damage and the interaction between interstitial atoms and defects. Lo Russo *et al.* (1982) first implanted Ar ions at high dose (2×10^{17} Ar^+/cm^2) to produce a damage level comparable to that produced by 2×10^{17} N^+/cm^2 implantation and subsequently N ions at a low dose (5×10^{16} N^+/cm^2) to decorate the produced defects. The results obtained in the wear behavior are shown in Figure 2. They did not observe wear reduction for low dose N implantation (i.e., for less than 10^{17} N/cm^2) or for high dose Ar^+ implantation. However, there is a significant wear reduction for both the high dose nitrogen implantation and also for the double implantation of Ar^+ (at high dose) and N^+ (at low dose). This last result seems to confirm the previous reported hypothesis.

C. Industrial Applications

Hartley (1982, private communication) states that of the several hundred industrial clients approaching Harwell with problems, approximately 80% of their problems have dealt with wear. This is, in large part, the reason Harwell has concentrated on wear applications of implantation. As previously stated, it can be the final processing step carried out at or near room temperature with no change in dimensions. However, since the doses required for these applications are approximately two orders of magnitude higher than for semiconductor applications, the machines required to perform implantations must deliver sufficiently high beam currents (i.e., 1-10 mA) for the processing times to be reasonable (i.e., 30-3 sec/cm^2). At these rates the economics of ion implantation can compare very favorably with other conventional treatments, (e.g., case carburizing) according to a recent review of the economics and machine requirements for industrial applications (Charter *et al.*, 1982). It should be noted that implantation into already gas-nitrided steel has been reported to further improve its wear resistance (Dearnaley, 1978; Kossowsky *et al.*, 1981). The problem areas

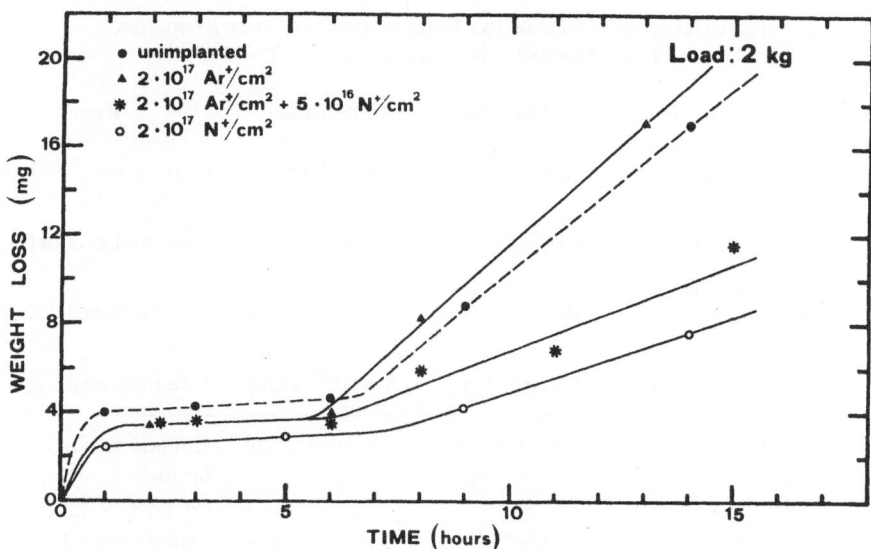

Figure 2   Weight loss following various periods of a reciprocating (unlubricated) wear test for implanted and unimplanted conditions (Lo Russo *et al.*, 1982).

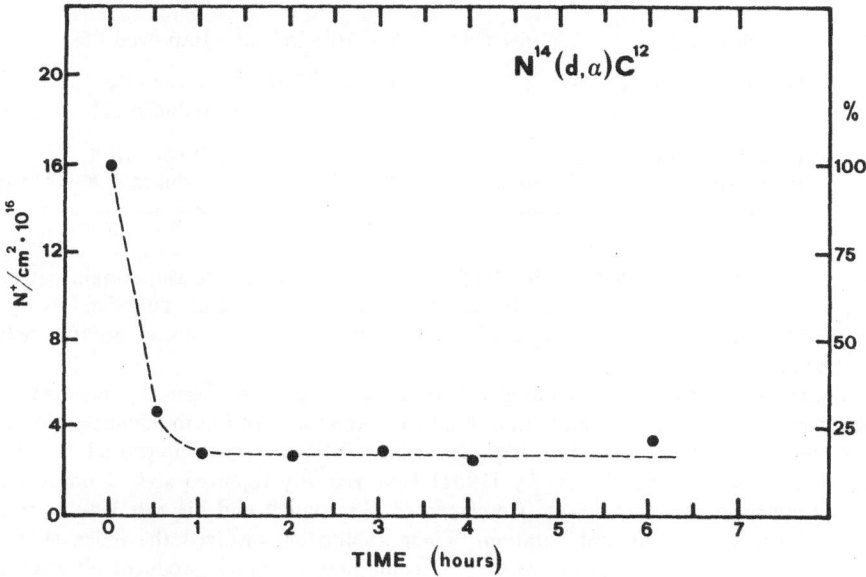

Figure 3   Yield of N^{14} (d,α) C^{12} nuclear reaction indicating retention of nitrogen following wear test (Lo Russo *et al.*, 1982).

TABLE III

Industrial Applications of Ion Implantation for Wear Reduction
(From Hartley, 1980 and Dearnaley, 1980)

Category	Application	Material	Treatment	Result
	Paper slitters	1C1.6 Cr steel	$8 \times 10^{17} N/cm^2$	Cutting life increased 2X
	Acetate punches	Cr-plate	$4 \times 10^{17} N/cm^2$	Improved product
i	Taps for drilling plastic	HSS	$8 \times 10^{17} N/cm^2$	Life increased 5X
	Slitters for synthetic rubber	WC-6% Co	$8 \times 10^{17} N/cm^2$	Life increased 2X
ii	Tool inserts	4 Ni 1 Cr steel	$4 \times 10^{17} Co/cm^2$	Contamination on inserts reduced by 3X
	Forming tools	12 Cr 2 C	$4 \times 10^{17} N/cm^2$	Much reduced adhesive wear
	Dies for copper rod	WC-6% Co	$5 \times 10^{17} C/cm^2$	Throughput increased 5X
iii	Drawing dies	WC-6% Co	$2 \times 10^{17} Co/cm^2$	Improved life
	Dies for steel wire	WC-6% Co	$3 \times 10^{17} Co/cm^2$	Wear rate reduced 3X
	Injection molds for plastic	Cr-plate	$4 \times 10^{17} N/cm^2$	Wear rate reduced $4 \times 10^{17} N/cm^2$

that they have been able to help by implantation can be grouped into three main categories; (i) cutting and slitting operations, (ii) corrosive applications and adhesive wear, and (iii) extrusion operations and applications where large surface forces occur. Selected examples are given in Table III.

Dearnaley (1982, private communication) relates that over seventy companies are presently working with Harwell evaluating the commercial uses of ion implantation in metals. Figure 4 shows a set of wire drawing dies whose lifetimes have improved 5X by C^+ implantation. Fromson and Kossowsky (1981) have recently reported sixfold improvements in the lifetime extension of precision Co-cemented WC punch and die sets similar to those shown in Figure 5 following implantation. Their application involved the index slotting of electric motor rotor laminations in low C steel containing Si which produces abrasive wear. In order to minimize burr formation which would prevent close stacking of the laminations, the tools must be very precise and wear tolerances are small (Charter et al., 1982). It should be noted that N implantation has not been found to enhance the wear resistance of tool steels or Co-cemented WC tools for ferrous metal cutting applications. For tools steels this is ascribed to the high temperatures at the cutting edges and hence instability of

Figure 4   Co-cemented WC wire drawing dies of the type improved by ion implantation (courtesy of G. Dearnaley, Harwell).

produced nitrides. For Co-cemented WC tools the wear mechanism during high temperature metal cutting is thought to be one of Co out-diffusion resulting in the loss of carbide grains.

D. Synthesis of Quasi-refractory Carbides by High Dose Ti Implantation

Carosella *et al.* (1980) observed a very significant reduction of friction for AISI 52100 bearing alloy steel after high dose Ti implants as well as for Ti plus C implantation. They also measured the wear characteristics of these high dose Ti-implanted surfaces using a ball on disc geometry and found an apparent incubation period prior to wear at which time the wear rate was similar to that of the unimplanted (Figure 6). In addition, it was found that the implantation of interstitial species like B and N had an insignificant effect. Hubler (1982) discusses the situations under which N implantation provides wear resistance. Further characterization of these surfaces (Singer *et al.*, 1981) by analytical techniques showed the existence of an amorphous Ti-C-Fe surface region. The excess C has been postulated to be gettered onto the surface by the sputter-exposed Ti and subsequently incorporated into the bulk by the implanted Ti atoms. The wear resistance of these quasi-refractory carbide surface is 6-7 times that of the martensitic base alloy which is itself a low wear material. Unlike implantation of interstitials, these quasi-refractory carbides have imparted wear resistance in metal cutting operations (Smidt *et al.*, 1981). However, their relative utility is expected to be limited to a relatively small class of applications where dimensionality is crucial and where cheaper (e.g., CVD) coatings of refractory carbides or nitrides cannot be applied because of the high temperatures (900°C) required. The co-implantation of Ti and C reduces the doses needed for effective wear protection (Yost *et al.*, 1981; Jeffries *et al.*, 1981). Clearly other elements (e.g., Hf) whose carbides exhibit higher hot-hardness than TiC should be (co)implanted and wear evaluated.

Figure 5   Co-cemented WC precision punch and die set whose lifetime has been improved 6X by ion implantation
(courtesy of G. Dearnaley, Harwell).

An obvious extension of this work would be to produce thin film nitrides by N
implantation. Duckworth and Wilson (1979) have already demonstrated the feasibility of
producing such nitride films. The use of ion-beam activated deposition techniques such as
used by Pranevicius (1979) and Weissmantel (1979) promise control over both composition
and interface bonding without any constraint on film thickness as found with direct ion
implantation. This technique involves simultaneous deposition and ion bombardment. Films
produced in this way often exhibit substantially improved mechanical properties, however
there presently appears to be more interest in their electrical properties. It appears that the
combination of ion beams and conventional coating techniques could offer considerable
potential for the controlled production of future thin film structures for wear or corrosion
protection.

E. Ion-Beam Induced Mixing

The use of ion-beam induced mixing for changing surface sensitive chemical and mechanical
properties has received much less attention than for electronic applications described in
previous chapters. The different classes of ion beam induced diffusion phenomena has been
reviewed recently (Dearnaley, 1981b). Dearnaley (1980) describes the in-diffusion of
deposited Co and Ir films on Au substrates as a result of N_2^+ implantation at elevated
substrate temperatures (620°C). The Ir-Au combination was selected because neither species
has any significant equilibrium solubility in the other, and therefore the achievement of a
metastable solid solution of high concentration is more striking in this case than in the Au-Co
system in which a limited range of equilibrium solid solution exists. It should be noted that
for diffusion to occur both the elevated sample temperature and bombarding N beam are

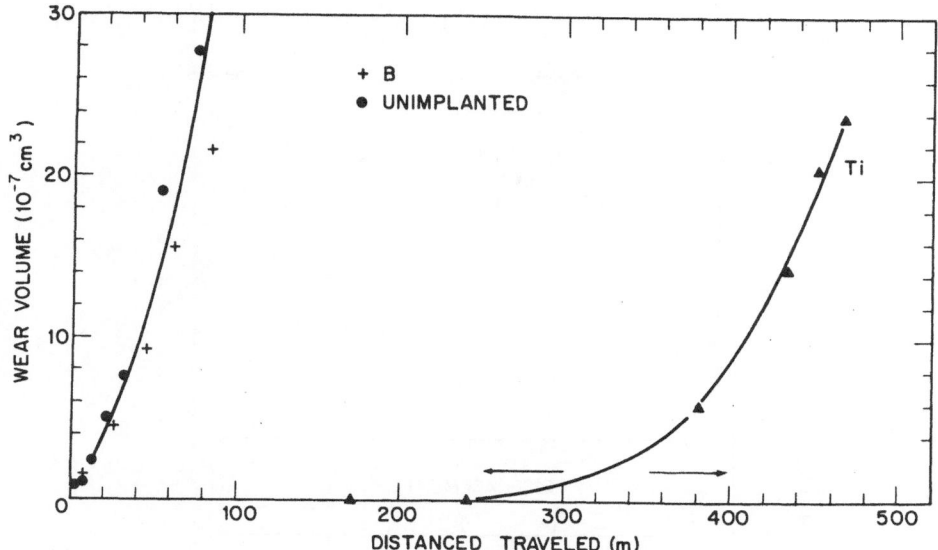

Figure 6   52100 "pin"-on-52100 disk wear experiments. The arrows beside the Ti implantation wear curve indicate that the onset of severe wear is highly variable but always occurs well after severe wear begins for the boron implanted and unimplanted samples (Carosella *et al.*, 1980).

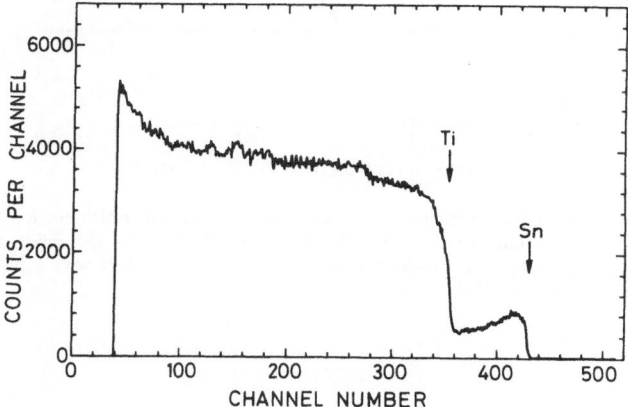

Figure 7   Ion backscattering spectrum of 2 MeV ^{4}He$^+$ ions from a specimen of titanium alloy (Ti-6 Al-4 V) into which tin has been introduced by bombardment diffusion using nitrogen ions at 100 keV energy (Dearnaley and Goode, 1981).

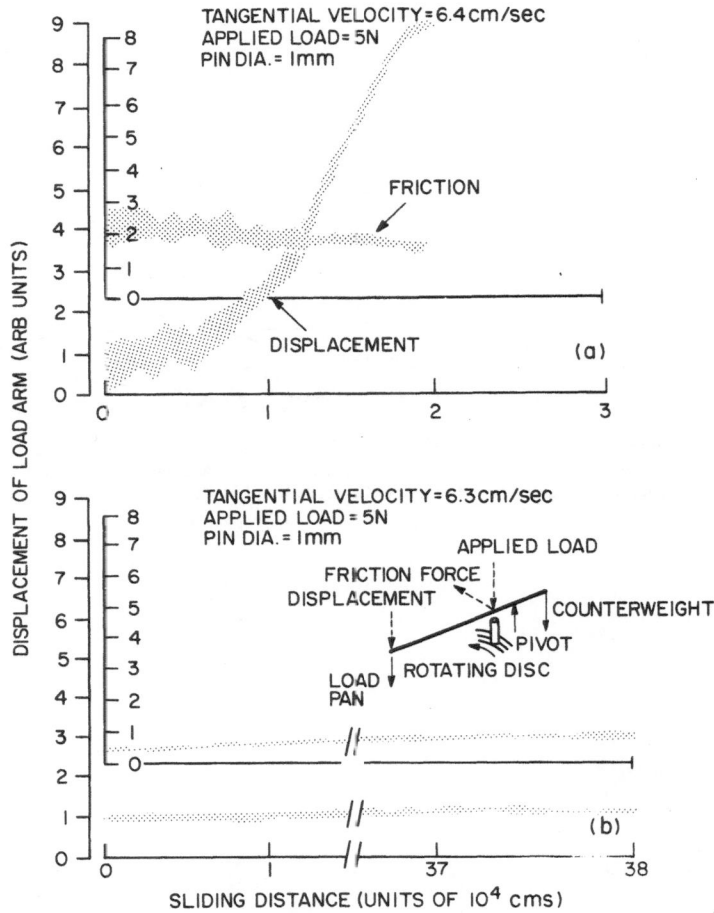

Figure 8   Wear and coefficient of friction in a specimen of titanium into which tin has been introduced by bombardment with nitrogen ions at 150 keV energy, 400°C. In the upper trace is shown the corresponding behavior of untreated titanium. Applied load was 5 N with a pin diameter of 1 mm (Dearnaley and Goode, 1981).

required. He suggests the application of these surface alloys lies in providing wear resistant contacts for microelectronic circuit boards, switches and similar systems. More recently, Dearnaley and Goode (1981) have reported on the use of thermally activated radiation enhanced diffusion mechanisms to diffuse a 700Å thick deposited Sn layer into the Ti/6%Al/4%V (wt.%) alloy. Figure 7 shows a backscattering spectrum of the sample after N_2^+ bombardment performed at an elevated temperature. The Sn is seen to penetrate 3-5 μm into the Ti alloy whereas the N_2^+ range is only a fraction of a μm. Dearnaley (1982) suggests and gives evidence for a pairing between the implanted N atoms and Sn atoms which may be involved in their diffusion in association with radiation-induced vacancies. The friction and wear properties of the resulting surface exhibit superior performance as seen in Figure 8. Interestingly, neither N^+ or Sn^+ implantations by themselves produce any significant changes.

IV. A MECHANISM OF FRICTION AND WEAR

A. Introduction

In this section, some of the results from recent work by Heilmann and Rigney (1981a; 1981b) on mechanisms of sliding friction and wear will be described. The work has concentrated on near-surface deformation and fracture during sliding, effects of microstructure and microstructural changes, and energy dissipation. The results suggest promising directions for future work on friction and wear mechanisms for simple sliding systems as well as for those involving coatings and layers modified by treatments such as ion implantation.

B. Plastic Deformation

A typical wear sample shows clear evidence of extensive plastic deformation at and near the sliding interface. In many cases this deformation is visible on the wear scar even without magnification. By using cross-sections of these samples one can measure the depth of the deformed region and one can observe a sequence of microstructural features associated with different amounts of plastic strain. A longitudinal section of a Cu sample is shown in Figure 9. The cell microstructure provides evidence that plastic strains near the surface are large.

The conditions of loading during sliding encourage large plastic strains even in materials which are not generally considered ductile. For example, in 1095 pearlitic steel the lamellae of cementite (Fe_3C) have been observed to bend over and reverse direction without fracturing. Even materials like alumina, silicon nitride, and diamond are known to deform plastically during sliding.

Plastic deformation is important for friction because it is a principal way in which energy can be stored and dissipated during sliding. Also, by concentrating on plastic deformation it is possible to interpret many of the frictional phenomena which have not been clearly explained by using other approaches to understanding friction mechanisms (Heilmann and Rigney, 1981a).

C. An Energy-based Friction Model

If one assumes that the frictional work during sliding is equal to the work of plastic deformation, it is natural to use this equality to develop an energy-based model of friction Heilmann and Rigney, 1981a, 1981b). A simple work balance yields

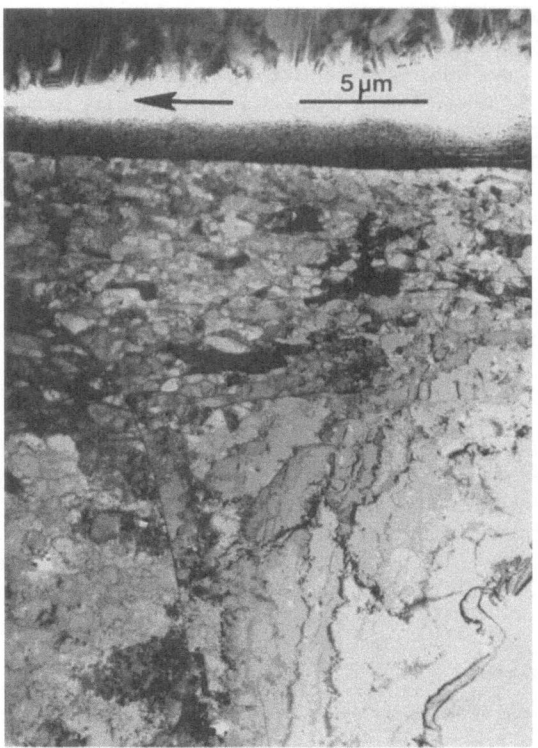

Figure 9   TEM micrograph showing a longitudinal section through a wear sample. Sliding direction is indicated
by the arrow. The microstructure, which results from plastic deformation, varies with depth below the
sliding interface. Sliding speed 1 cm/sec., total sliding distance 12m (100 cycles, sample block on ring),
normal load 6.8 kg (15 lb.), sample material OFHC copper. The thickness of the plastically deformed
region extends beyond the field of view (Heilmann and Rigney, 1981a).

$$F \, \delta x_s = \int \tau \gamma d \, V, \tag{1}$$

where F is the friction force, δx_s is an incremental displacement of a slider, τ is the local
shear stress, γ is the shear strain, and V is the volume deformed. Materials properties can be
incorporated by using appropriate stress-strain relations for each deforming material.

The simplest case would involve a hard non-deforming slider moving on a deforming
sample. The model leads directly to the prediction that the friction coefficient will be
independent of the hard slider material, as found experimentally (Heilmann and Rigney,
1981a). If the slider also deforms, then the total friction will include contributions from each
deforming material. Since the energy terms are additive, the contributions from all
deforming regions will also be additive. The same principle applies for more complex systems
such as lamellar structures and systems with layers or coatings of various kinds. If data are
available for $\tau(\gamma)$ for each region, it should be possible to account for observed friction
coefficients.

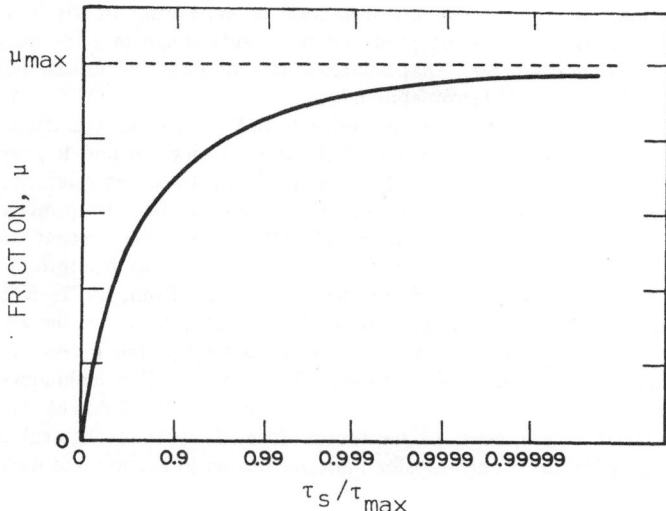

Figure 10   Calculated friction coefficient μ as a function of surface stress τ_s for a simple $\tau(\gamma)$ function which saturates at τ_{max} (Heilmann and Rigney, 1981b).

As an example, one can use a simple analytical function, such as $\tau = \tau_{max}\,(1-\exp(-c\gamma))^{1/2}$, to approximate $\tau(\gamma)$. This function is parabolic at low strains, but saturates at τ_{max} for large strains. This is consistent with experimental results from mechanical tests involving large shear strains (Sevillano *et al.*, 1980). Displacement profiles near the surface can be approximated by $\delta x_s = \delta x_s\,\exp(-az)$, where z is the depth below the surface. The strain profile has a similar depth dependence.

For the simple case of one deforming material the model yields (Figure 10)

$$\mu = \frac{NA}{L}\,\tau_{max}\,F\left[\frac{\tau_s}{\tau_{max}}\right] = \mu_{max}\,F\left[\frac{\tau_s}{\tau_{max}}\right], \tag{2}$$

where NA is the total asperity contact area and L is the normal load. $F(\tau_s/\tau_{max})$ is a function (Heilmann and Rigney, 1981b) which rises smoothly and approaches unity asymptotically as the surface stress τ_s approaches the maximum shear strength τ_{max}. The function F accounts for changes during break-in as the system approaches steady-state conditions. It should be noted that when $F \to 1$, μ takes a form similar to that predicted by simple adhesion models of friction (Bowden and Tabor, 1964).

A similar procedure can be used for coated and layered systems if one uses reasonable boundary conditions at each materials interface. One can then calculate the dependence of friction on coating thickness for various materials combinations. The results include minima (soft coating on a hard substrate) and maxima (hard coating on a soft substrate). Either of these cases could be produced by ion implantation with controlled thickness of implanted material and proper selection of implanted species.

D. Sliding Wear; Transfer

Plastic deformation near the surface can produce a wear scar without forming loose wear debris particles. In such cases, friction and wear are related, since they both depend on plastic deformation. Wear involving the formation of loose wear debris is more difficult to understand. Debris particles can be produced in a wide range of sizes ranging from sub-micron to millimeters in diameter. Shapes include irregular chunks, flakes, cylinders, spheres, and loose clusters (especially if ferromagnetic).

Researchers have devoted particular attention to flake debris; several delamination models have been proposed to explain their origin (Suh, 1977; Heilmann and Rigney, 1981c). The simplest models are based on the delamination of substrate material near the surface. However, recent fracture mechanics calculations do not predict the propagation of cracks which could lead to flake debris (Rosenfield, 1981). In fact, recent work in several laboratories indicates that flake debris are commonly generated by fracture in a surface layer which includes material transferred from the counterface (Suh, 1977; Rice *et al.*, 1981; Norose and Sasada, 1980). A remnant of such a layer appears as the dark fine-grained material in Figure 9. Typical layers are several microns thick, but layers as thick as 20 μm have been studied with the aid of SEM/EDAX, TEM, and STEM techniques. Such studies show that transfer layers consist of very small pieces (50-300Å) of both sample and counterface material. They seem to have been formed by a mechanical mixing process, probably aided by adhesion. The transfer material begins to accumulate during early stages of sliding.

Transfer layers have been found on both lubricated and unlubricated copper samples (Heilmann and Rigney, 1981c). Furthermore, flake debris particles seem to have the same microcomposite structure as the transfer layer, indicating that these flakes were generated by delamination of transfer layer material. It appears that the accumulation of a transfer layer is one of the processes which occur during break-in. Steady-state conditions may be reached when the rate of formation of debris is equal to the rate of formation of the transfer material. Environmental factors could affect both of these processes.

The processes involved in the formation of the component particles of the transfer layer are not at all clear. They probably involve deformation, adhesion and fracture, and possibly segregation. Surface modification treatments such as ion implantation can affect all of these processes, but at this stage in our understanding the net effect on wear resistance is difficult to predict. If implantation inhibits the formation of a transfer layer, it is likely that wear through formation of flake debris will be decreased. If implantation inhibits the formation of what seem to be precursor particles for transfer layer formation, then wear through formation of fine debris particles will also be decreased.

The presence of a transfer layer also affects friction, since energy is dissipated by continuing deformation of this layer. Depending on the mechanical properties of layer and substrate, the accumulation of a layer of transfer material can increase or decrease friction. Implantation will affect friction both directly, by creating a surface region with properties different from the substrate, and indirectly, by changing the tendency to form a transfer layer.

It has been found by many investigators that sliding friction and wear are surprisingly insensitive to differences in initial microstructure. One reason could be that sliding modifies the near-surface microstructure and generates similar highly-strained material for a wide range of initial conditions. Another explanation could be that the accumulation of transfer material dominates the response of the system. Chemical differences rather than microstructural differences seem to be important for sliding behavior. Therefore modifications such as ion implantation can produce significant changes if ion species are carefully selected. Whether or not these are persistent effects will depend on the load and the surface finish.

E. Abrasion

The preceding sections have concentrated on sliding friction and wear. At least a few remarks should also be devoted to abrasion. In some respects, abrasion is better understood than sliding. Effective abrasion involves efficient removal of material by cutting processes (Samuels, 1978). Cutting requires penetration, which depends on hardness. Hardness in turn depends on microstructure. Therefore, different factors must be considered for an understanding of abrasion resistance. In particular, the small depth of penetration for ion implantation suggests that significant changes in abrasion resistance will be limited to situations involving very small abrasive grit sizes and very light loads. Otherwise the implanted layer will be removed very quickly and unprotected material will be exposed.

F. Concluding Remarks

The modification of surface material by ion implantation and by other techniques can affect both friction and wear. It should be possible to understand the observed friction effects with the aid of an energy-based model which emphasizes plastic deformation. Sliding wear is more complicated. One way in which it could be modified is through changes in the transfer layer. Abrasion resistance depends more on structure-sensitive properties such as hardness. Surface treatments for sliding wear resistance and for abrasion resistance will generally by different.

V. FATIGUE

A. Introduction

Fatigue represents a singularly dangerous mode of material failure, in that no obvious prior warning is given of impending fracture. Generally, such failure occurs upon the cyclic loading at some stress below the static fracture stress. High loading amplitudes give rise to short lifetimes (low-cycle fatigue), whereas relatively low loads yield longer lifetimes (high-cycle fatigue). In the case of ferrous alloys there is a stress level, referred to as the endurance limit with lifetimes exceeding 10^8 cycles.

The commonly employed rotating-bend test, involving a specimen of circular cross-section with a reduced central section is seen in Figure 11. In this test, one end of the specimen is gripped in a chuck and loaded at the other end. As the specimen rotates, the stress at a given point on the surface of the reduced section varies sinusoidally, being in tension at the upper region and compression in the lower region. The maximum stress amplitude is always the same; therefore this test is referred to as one of constant stress. Any work-hardening which occurs during cycling will give rise to a decreased strain. In the constant strain test, the strain amplitude is constant, so that the stress amplitude, will generally vary. While constant stress tests are the most common, constant strain amplitude tests are used as well, especially in studies of dislocation phenomena.

Fatigue fracture represents the initiation and propagation of cracks in a sequential manner (Fine, 1980; Kocanda, 1978). It is generally thought that crack initiation occurs at or near the specimen's surface. Crack propagation is a macroscopic phenomenon which gives rise to failure when the fracture toughness of the material ultimately is exceeded. Dislocation mechanisms are thought to give rise to those defects which will finally form the fatigue crack. The means by which dislocations are affected by implantation will thus be the central question on the matter of fatigue failure as well as any surface controlled failure mechanism.

B. Experimental Results

The initial work on fatigue of ion implanted metals was carried out by the Harwell group (Hartley, 1976) showing the influence of N implantation into low C steel. These early experiments showed significant extensions of fatigue life. Hartley (1980) has reported that fatigue lifetimes for N-implanted stainless steel, Ti, and maraging steel are 8-10 times longer than for the unimplanted materials. This work prompted other workers to attempt to quantify this effect. As part of a Stony Brook-NRL program, Hu *et al.* (1980) implanted molecular N at 150 keV into AISI 1018 steel (0.18 wt.%C) to doses of 2×10^{17} ions/cm^2 and observed increases in fatigue life (Herman, 1981). However, major fatigue lifetime enhancement was not observed until the specimen was aged for several months at room temperature ("naturally aged") or for 6h at 100°C ("artificially aged"). These results are reviewed in Figure 12 for three specimen conditions.

A common way of displaying fatigue behavior is with a stress versus number of cycles to failure plot (i.e., a S/N plot) such as seen for these experiments in Figure 13. Each data point represents the lifetime for a given stress amplitude. Of particular interest is the horizontal regime at high cycles, the so-called endurance limit, below which no failure occurs.

It should be noted that the S/N plot represents the division between safe and unsafe regimes of operation, although notches and other surface imperfections will significantly modify the plot for a wide range of metals and alloys. Ion implantation is more likely to affect crack initiation than crack propagation. The S/N plot is actually a reflection of the latter, so that its modification with implantation should be treated with caution.

Clearly, N implantation into a low C steel improves fatigue resistance, especially following a low temperature annealing treatment. For commercial material, it must be noted that the metallurgical considerations are most complex, the alloy being comprised of a relatively soft ferrite phase into which is embedded pearlite, a duplex structure made up of alternate layers of iron carbide (Fe_3C) and ferrite. In addition, prior to implantation, C will be dissolved in the ferrite matrix. Nitrogen gives rise to considerable radiation damage in the implant region ($\sim$1000Å). It is also likely that the pearlitic structure is disrupted on irradiation. Some indication of the extent of changes which occur within the pearlite can be seen in TEM studies. These studies show a very high degree of phase decomposition and defected structure, which would be expected to modify dislocation behavior.

Nitrogen implantation at the energies employed in this study is expected to damage the metal and carbide lattices and to modify structure. Supersaturation of N will accompany implantation, but as a result of the expected high interstitial solute diffusion at the implantation temperature, and damage-induced nucleation of precipitates, phase decomposition is likely to occur. The aging phenomenon indicates that relief of excess free energy has not occurred during implantation. Precipitates and clusters which form during implantation are expected to be redissolved through ion collision with these phases giving, in effect, a steady state precipitate distribution. The steady-state metastable structures which are implantation-induced can, in fact, decay to more stable phases, involving, for example, the decomposition of supersaturated N martensite (formed through implantation) into a metastable nitride. Evidence from TEM studies supports this concept. The N-implanted steels contain 100Å particles of $Fe_{16}N_2$ at the surface, whereas annealing gives rise to an additional size range: a fine ($\sim$ 20Å) precipitate of $Fe_{16}N_2$, plus a mottled background indicating the precipitation of some electron contrast-inducing phase associated with an acicular product (i.e., martensite). It is suggested that the martensite was formed during implantation, but was not observed until decorated by the precipitating nitride (and/or carbide) (Hu *et al.*, 1980).

The central question to resolve is the manner in which the products of N implantation can have so significant an influence on fatigue behavior. A simple working model can be

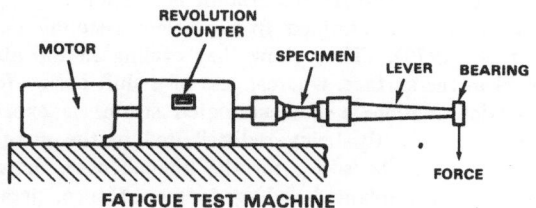

FATIGUE TEST MACHINE

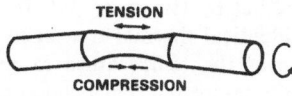

FATIGUE — TEST SAMPLE

Figure 11   Schematic of rotating-bend fatigue machine and standard specimen.

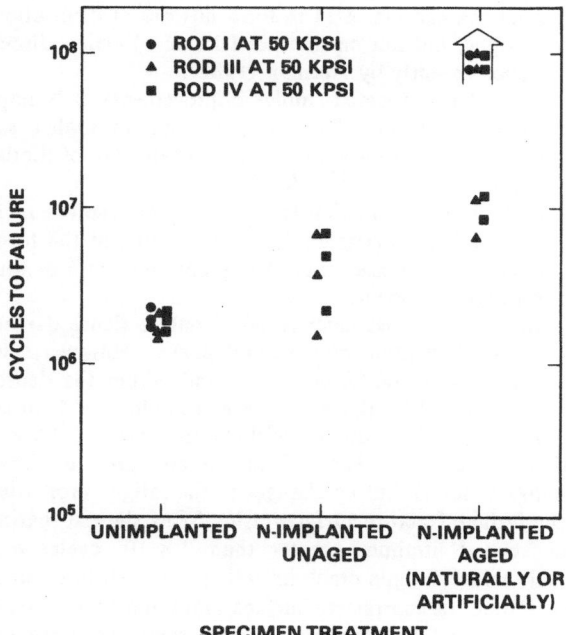

SPECIMEN TREATMENT

Figure 12   Fatigue lifetime in ambient air for three different rods of AISI 1018 steel (0.18 wt.% C) for (i) as-received material (unimplanted), (ii) implanted with molecular nitrogen at 150 keV to a dose of 2×10^{17} ions/cm^2, and (iii) implanted and aged. The rotated bending mode had constant stress amplitude at 5000 rpm (Herman, 1981).

suggested: the fine precipitates of $Fe_{16}N_2$ act to both strengthen the ferrite phase and to make dislocation motion, and, consequently, surface-emerging slip more homogeneous. Major slip inhomogeneities are thus reduced, and fatigue lifetime is extended.

Nitrogen profiles have been determined from nuclear reaction analysis of the fatigue specimens (Herman et al., 1979). They show that cycling in the absence of prior aging gives rise to a shift of N to the surface, whereas less of a shift occurs for the aged specimen. More integrated dislocation motion (i.e., accumulated strain) is expected to occur in the unaged case. More N is apparently being redistributed in the unaged specimen as well. Since the surface is expected to be softer for the unaged case, more dislocations will be present at the surface for the implanted-and-aged case. Hence, greater pipeline diffusion would be expected or alternatively, more interstitial-dislocation association.

In an effort to examine the possible association of the implantation species and dislocations, experiments were carried out by Hu et al. (1980) on the internal friction of the same type of steel, subjected to the identical thermal treatments and implantation conditions. The experiments were carried out over the temperature range from liquid helium to over 100°C. A complex modification of the entire internal friction spectrum is achieved following implantation and implantation-and-aging. The greater than 40% decrease in mechanical damping at room temperature (at a 1 kHz resonant frequency), due to N implantation with 2×10^{17} ions/cm^2 is of particular interest. Furthermore, the damping is linearly dependent upon dose and appears related to an amplitude-dependent dislocation-based mechanism. Aging, while being influential on fatigue lifetime, has less effect on mechanical damping. It may be due to substantial segregation of N to form nitrides at dislocations. Aging therefore has a major effect on fatigue, but not on internal friction. Further discussion on the role of implanted N has been given recently by Herman (1982).

Lo Russo et al. (1980) have reported similar improvements in N-implanted steel fatigue samples without the need for aging. This may be due to higher sample temperatures produced by high beam currents during implantation. Their data of lifetime versus N fluence shows a maximum near a dose of 2×10^{17} N/cm^2.

Vardiman and Kant (1982) have implanted N and C into the commercial Ti-based alloy, Ti-6Al-4V and have studied fatigue resistance. Their results, in the form of S/N plots, are shown in Figure 14, where it can be seen that N implantation, with or without annealing, has only a modest effect on fatigue behavior.

A TEM study of the N-implanted layer revealed only a dense damage structure for the implanted material with no indication of a second phase. However, after C implantation, precipitates of TiC with sizes of 100-200Å were found within the dense damage structure. This difference may be explained by the much lower solubility of C in α-Ti than N in α-Ti and by the greater mobility of C compared with N (a factor of 3.5×10^5) at 100°C, the estimated implantation temperature. The TiC precipitates grew to sizes of 200-500Å after vacuum annealing at 400°C for 1h but no changes in the fatigue properties occurred.

Examination of the fatigue fracture surfaces using SEM showed that most cracks found in samples which demonstrated lifetimes greater than 2×10^5 cycles originated 25-150 μm below the surface. Subsurface fatigue crack initiation is known to occur in this material and does not change the lifetime, in contrast to surface crack initiation. Ion implantation in less than the first micron would not be expected to affect this deep crack initiation. Thus it is inferred that implantation inhibits crack growth to the surface in the alloy Ti-6Al-4V.

A number of experiments have been carried out on implantation-modified fatigue resistance of fcc metals which show significantly smaller effects than ferrous or titanium alloys. Normal fatigue studies of single and polycrystal Al and Cu have been reviewed elsewhere, (Kocanda, 1978). A major goal in many of these studies has been to evaluate the influence of alloying additions on early-stage crack initiation. For example, alloying elements

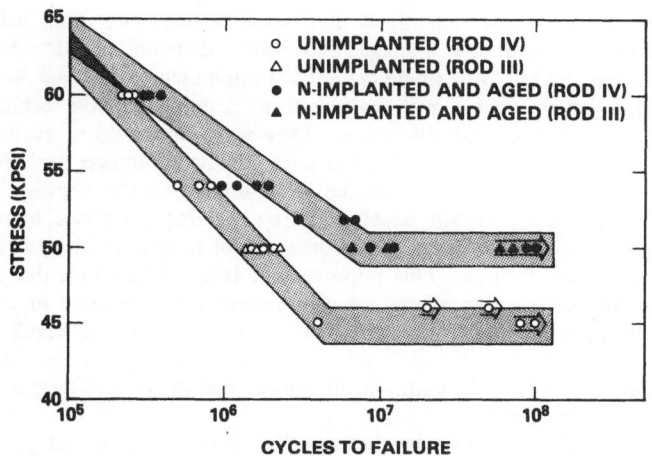

Figure 13   S/N plot for AISI 1018 steel. See Figure 12 for details.

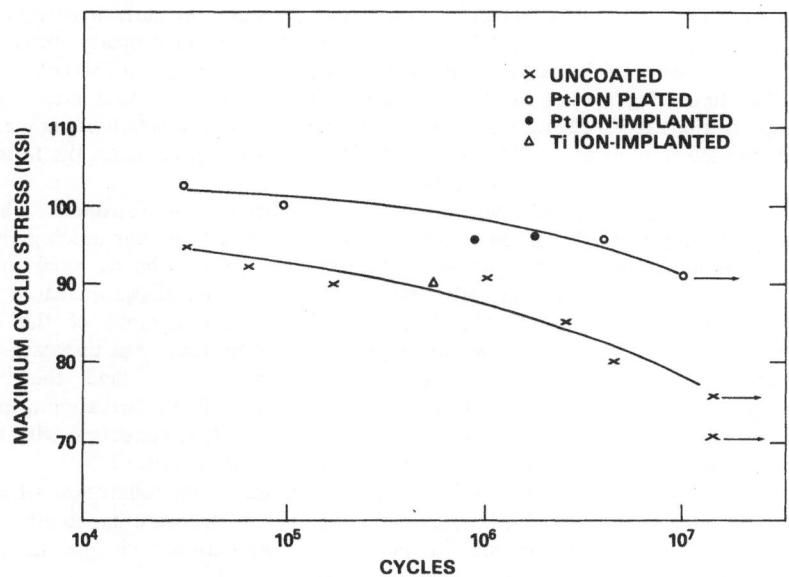

Figure 14   Lifetime versus N-fluence for fatigue samples. (Lo Russo *et al.*, 1980.)

can change the stacking fault energy in Cu, thus modifying cross-slip dynamics and changing work hardening behavior. This behavior is thought to play an important role in cyclic-strain-hardening and, therefore, fatigue. Precipitation of a metastable second phase or a compound can also have a great influence on dynamic mechanical properties.

Kujore *et al.* (1981) have examined ion implantation into Cu, using X-ray and microscopic methods to evaluate damage and solute distribution. Much of their work concentrated on crack initiation in single crystal Cu implanted with Al or Ar. In addition to fatigue studies, they carried out X-ray analyses using a double-crystal technique to examine diffuse scattering and Bragg peak distortion. They were thus able to evaluate the state of stress at the surface and to detect the occurrence of lattice defects and their products of coagulation. A sub-surface area was detected in Al-implanted Cu, showing heavy radiation damage. Spooner and Legg (1980) conclude that annealing gives rise to migration of the implanted Al out of the area of damage. An increase of fatigue lifetime was found for both high cycle and low cycle fatigue. This improvement is attributed to a decrease in stacking fault energy and the introduction of surface compressive stress (Kujore *et al.*, 1980). They attribute the improvement of fatigue behavior to a reduction of crack initiation as a consequence of fewer "persistent slip bands," thought generally, to be the crack initiator (Fine, 1980; Kocanda, 1978). In addition, implanted Cu shows less fatigue hardening than does the unimplanted specimen.

The fatigue properties and microhardness of B-implanted single and polycrystal Cu and Ni have been studied by Preece *et al.* (1980). Although the equilibrium solubility is low for both metals, the B concentration in the implanted surface layers was of the order of 15 at.%. Comprehensive analyses were carried out using the techniques of Rutherford backscattering, channeling and TEM. In these experiments Cu and Ni single crystals were used for the channeling studies, whereas polycrystals were used for mechanical property measurements. Implantation doses were in the range 2.3 to 18.5×10^{15} ions/cm^2 at 25 to 150 keV.

Ion beam channeling studies for B implantation into Cu indicate that many radiation-induced imperfections (e.g., dislocation loops) are formed, but that crystallinity is maintained. Boron implantation into Ni as determined by RBS and channeling, indicates the formation of an amorphous or microcrystalline layer (Preece, 1980).

For B implanted into Cu, there was no significant effect on microhardness. This is not particularly surprising since the hardness indentor will penetrate to depths much greater than that of the implanted layer hence no effect of implantation will be detected unless the hardness is greatly enhanced. On the other hand, fatigue behavior shows increased lifetimes when the amount of B exceeds 60%. Scanning electron micrographs of the fatigued, implanted Cu show no extruded and intruded regions, indicating that gross dislocation motion did not penetrate through the implanted surface. Implantation had the effect of strengthening the surface, therefore limiting dislocation egress and the formation of persistent slip bands and other crack-initiating surface defects. This result is consistent with the work reported above on the implantation of Al into Cu (Kujore *et al.*, 1980).

For the case of B implanted into Ni, there is an increase in microhardness which varies with depth of penetration and load. A highly disordered structure is obtained by the implantation of B into Ni. This is reflected in the observed hardness changes in which the penetrator breaks through a brittle, non-plastic layer (Preece, 1980).

Of special interest is the fatigue behavior of Ni following implantation in which an increase in fatigue lifetime greater than 100% is observed. A hard, amorphous (or microcrystalline) layer is presumed to be limiting the establishment of persistent slip bands.

Burr *et al.* (1980) have used the limitation of formation of persistent slip bands to explain some interesting implantation experiments in Cu. A number of species have been implanted into Cu at MeV energies to doses ranging from 10^{15} to 5×10^7 ions/cm^2. Increases in lifetime were found for B, Cl, He, Ni, N, and Ne. A hardened regime has been established through which glide dislocations cannot penetrate easily, thus limiting the formation of obvious and intense persistent slip bands.

Burr *et al.* (1980) also implanted 150 keV Be ions into Cu to doses from 10^{17} to 1.5×10^{18} ions/cm^2. A modest increase in fatigue lifetime was observed. However, on annealing for one hour at 800°C, there was a major loss of fatigue resistance. This effect may result from redissolution and dispersion of fine coherent precipitates which formed during annealing. Conversely, large, incoherent particles (e.g., Be_2Cu) may have been formed, permitting deformation inhomogeneity in the middle of the grain. This would favor the easy formation of persistent slip bands, leading to cracking. Careful TEM is needed to resolve these questions.

Little consideration has been given to establishment of residual stresses on ion implantation. Residual stresses, especially compressive surface stresses are known to have an important influence on fatigue resistance. Shot peening, for example, is a straightforward way of introducing such stresses, and improving fatigue properties of engineering alloys. In general, little effect on fatigue has been found by implantation with inert ion species which would be expected to give surface residual stresses. Implantation-induced stresses are assumed to be of little consequence in enhancing fatigue resistance, but it is a point which requires further study.

C. Fretting Fatigue

Dearnaley and Goode (1981) have recently given the following report[†] on the use of implantation by Syers and Dearnaley (unpublished work) for reducing fretting fatigue. "Titanium alloys are particularly susceptible to fretting fatigue, in which fatigue crack initiation is brought about at a point on the surface that is subjected to low-amplitude cyclical wear. Wear debris accumulates and enhances the dislocation movement at such sites. They have investigated the effect of implanting those species (e.g., Ba) which are known to inhibit thermal oxidation of titanium. This is aimed towards a reduction of the oxidative wear, and the belief is that these beneficial species act by precipitating as pinning agents at the dislocations that serve as short-circuit diffusion paths for oxygen in titanium. The preliminary results have been successful: the implantation of about 10^{16} Ba$^+$ions/cm^2 into shot-peened Ti-6 Al-4 V was tested at Rolls-Royce Ltd. alongside 45 other methods of surface treatment for titanium in a standardized fretting fatigue test, and ranked third in order of merit. The assessment is made in terms of the fraction of the normal high-cycle fatigue endurance (10^7-10^8 cycles) that is retained when fretting pads are applied to the specimen. Barium implantation gave a value of 55%, compared with a best value of 62% for a tungsten carbide and chromium detonation-gun treatment. However, the ion implantation process is preferred because it entails no change in dimensions of precision components such as the fir-tree roots of compressor blades. Analysis of the surface after the test showed that about 40% of the barium can still be present after 10^7 cycles of fretting. This result is interesting because it throws more light on the mechanisms taking place during fretting fatigue, and emphasizes the importance of oxidation and dislocation movement at the surface."

---

† Quoted with permission of the authors.

D. High Temperature Fatigue

In an investigation of Ti compressor blade alloys, Hubler (1982) discusses measurements of the fatigue life of Pt-implanted specimens which are oxidized in air for 250h at 465°C prior to fatigue testing. The alloy composition is Ti-6Al-2Sn-4Zr-2Mo. The pre-oxidation is done to simulate the hot environment experienced deep in the jet engine. At these temperatures of 400°C to 500°C, strengthening by C implantation is probably not operative due to the decomposition of the TiC precipitates and the high mobility of C. The fatigue test itself is run at 465°C in air. It is thought that the primary cause of fatigue failure in this alloy is the process called "α-casing" where dissolved O promotes transformation of the β phase in the near surface region to the more brittle α phase so that the entire piece becomes encased in the α phase material. Fatigue cracks then easily form at flaws in the brittle surface and fatigue life is sharply diminished. In this application, it is perhaps more important to confer oxidation resistance to the material than to strengthen the surface.

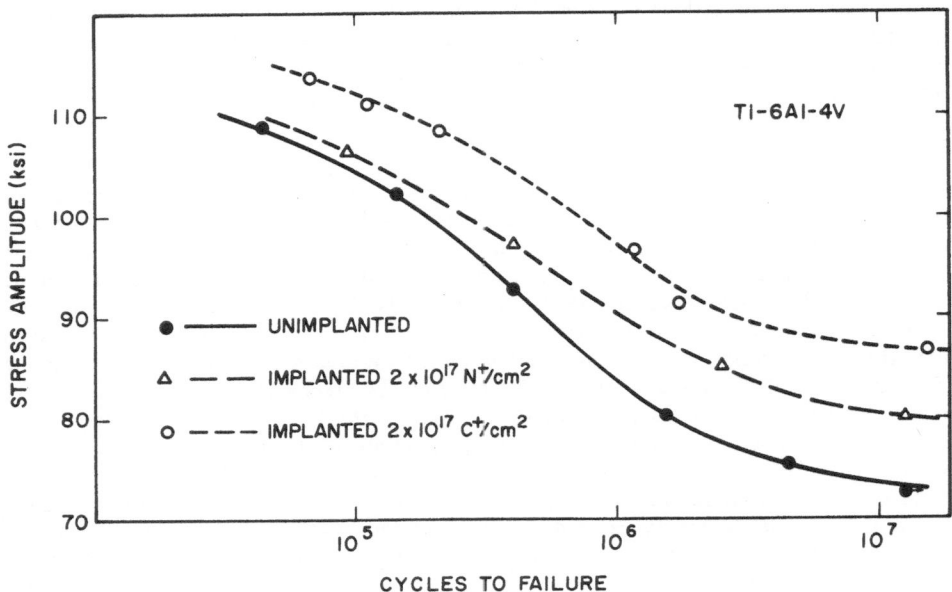

Figure 15   S/N plot for carbon and nitrogen implanted Ti-6Al-4V alloy. (Vardiman and Kant, 1982.)

Figure 15 shows the preliminary fatigue data at two points above the untreated specimen baseline curve. The fatigue life is improved substantially by implantation of Pt to a fluence of 1 to 2 × 10^{16}ions/cm^2 at 150 keV, and the implanted region of the samples appears to be less oxidized (metallic color) than the unimplanted regions (deep blue color). Preliminary results for Ba-implanted samples (2 × 10^{16}/cm^2, 125 keV) show similar fatigue life improvement. It should be noted that Dearnaley et al. (1982) have also recently reported fatigue life improvement in Ti-alloys at 500°C by means of Pt and Ba implantation (Dearnaley, 1982). The smooth curve in Figure 14 is for Pt ion-plated specimens with a coating thickness of 1 μm of Pt (Fujishiro and Eylon, private communication).

VI. APPLICATIONS OF ION IMPLANTATION TO CORROSION SCIENCE AND ENGINEERING

A. Introduction

This section will review the application of ion implantation as a research tool in the study of the corrosion mechanisms of conventional alloys and a corrosion protection treatment for engineering components. The review will begin with a critical discussion of the possible influence of the implantation process upon the corrosion behavior of surface alloys and will emphasize the need for more compositional and structural studies to be carried out on surface alloys prior to electrochemical analysis and corrosion testing. Two groups of studies will then be reviewed: (a) Implantation into pure metals, (b) Implantation into engineering alloys. Studies concerned with the electrochemical behavior of amorphous surface alloys formed on engineering alloys will then be discussed. Finally, studies which have attempted to improve the corrosion fatigue resistance of Fe and 1018 steel will be presented.

Several studies have demonstrated that ion implantation may be used to modify the corrosion behavior of metals and alloys (Ashworth *et al.*, 1980a; Clayton, 1981). In these studies metallic systems have been doped with suitable elements in order to systematically modify the nature and rate of the anodic and/or cathodic half-cell reactions which control the rate of corrosion. These studies fall into two categories: (a) Studies of novel surface alloys and (b) attempts to improve the corrosion resistance of some commonly used engineering alloys.

A common limitation of the corrosion studies of surface alloys formed by ion implantation is the absence of both microstructural and compositional analysis prior to electrochemical analysis or corrosion testing. In their absence, some assumptions have to be made about the nature of the surface alloy formed. However, there is a growing awareness that the ion implantation process is complex, thus making predictions of the final implant distribution and alloy structure difficult.

In addition to the alloying concepts which are considered in the design of the surface alloy, the corrosion behavior of an implanted system will also depend on the aspects of the implantation process to be discussed in the following sections.

1. Surface and Bulk Contamination

A common contaminant found in or on surface alloys formed by ion implantation is C, a by-product of hydroC cracking. It may be incorporated into the near-surface region of an alloy as a result of recoil implantation. Carbon contamination can severely alter the corrosion behavior of the surface alloy by either forming a corrosion inhibiting layer on the metal surface or by stifling passivation and promoting localized corrosion due to the reaction of passivators with bulk C to form inert carbide precipitates. Covino *et al.* (1978b) have reported C surface contamination after implantation of Cr into Fe. PIXE depth profiling also found C to be distributed throughout the layer. No effect of the C on the electrochemical behavior of the surface alloy was reported. Chan *et al.* (1982) have recently reported an AES depth profiling study of AIS 52100 and pure Fe implanted with Cr in which C entrapment led to the formation of carbides near the surface of the alloys. The carbides tended to lower the pitting resistance of the surface alloy.

2. Oxide Films

A common feature of implanted metals is the tendency for the air-formed oxide films to be thicker than those on the virgin metals. Diffusion and oxidation might possibly be enhanced by the high defect concentrations resulting from radiation damage. The sample temperature during oxidation will also determine the oxidation kinetics. Some controversy exists as to whether the oxide film is formed immediately after implantation, while the sample is still inside the vacuum chamber, or on exposure to the atmosphere. Dearnaley and Goode (1981) have reported on the formation of a radiation enhanced oxide film under particular conditions.

Modification of the structure and composition of an air-formed oxide film on exposure to an aqueous environment can facilitate passivation. Ashworth *et al.* (1976a) have considered the electrochemical behavior of the air-formed oxide film covering surface alloys. They have made use of a three sweep polarization technique (Figure 16), which monitors the anodic behavior of the oxide covered alloy, as well as the anodic and cathodic behavior of the oxide stripped surface alloy.

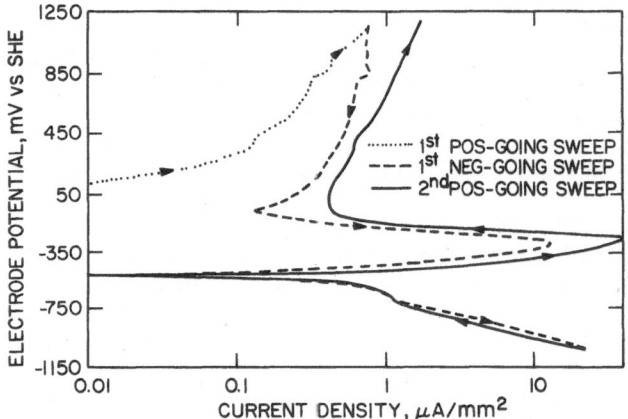

Figure 16   Potentio-kinetic polarization curves of Ar implanted Fe (Ashworth *et al.*, 1976b).

3. Implant Distributions

Systematic variations in doping levels are commonly required in order to facilitate a corrosion study or to obtain the optimum doping level for a surface treatment. In the low dose range, LSS theory (Lindhard *et al.*, 1963) is often used to predict the final implant distribution. However, in the high dose regime the observed implant distribution may significantly deviate from the theoretically predicted distribution. Such effects as sputtering and radiation enhanced diffusion (see Chapters 8 and 10) may contribute to such deviations. Since the majority of corrosion studies involve high dose implantation it is important to monitor any deviations from the predicted implant distribution. Iwaki *et al.* (1977) have compared theoretical profiles with SIMS profiles of Ni and Cr separately implanted into mild steel, at a fluence of 5×10^{16} ions/cm^2 at 150 keV. Nickel showed only a slight deviation from the theoretical profile, exhibiting some deeper penetration probably due to radiation enhanced diffusion. Enhanced diffusion calculations (Gamo *et al.*, 1970) suggested a diffusion coefficient of 10^{-15} cm^2/sec which corresponds approximately to the thermal diffusion coefficient at 600°C. However, in both implantation studies the sample temperature only reached 180°C.

The Cr profile obtained by SIMS exhibited one peak at the surface and another peak corresponding to the position predicted by LSS theory after accounting for sputtering. The surface peaks were attributed to (i) out-diffusion caused by radiation enhanced diffusion, (ii) the sample temperature, and (iii) by the chemical affinity of Cr for O at the metal surface.

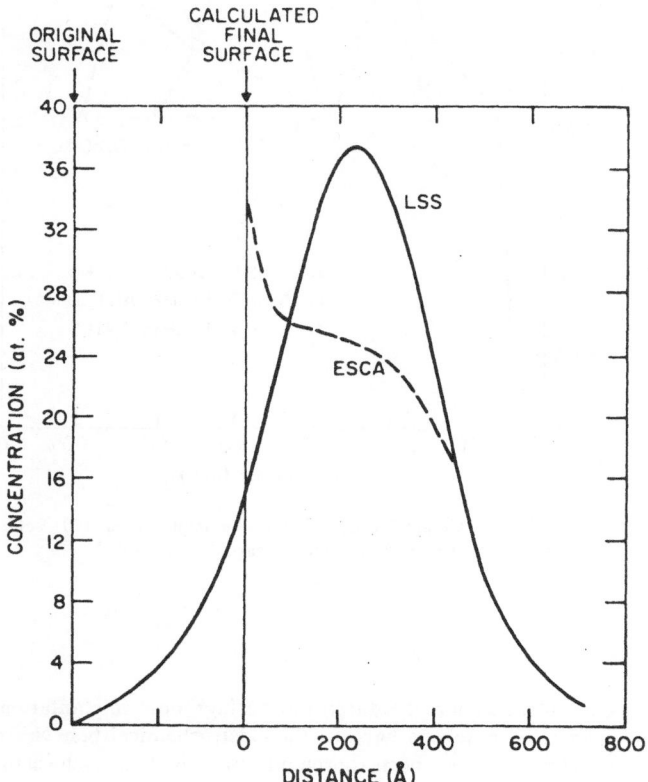

Figure 17   Comparison of calculated (LSS) and observed (ESCA) distributions of Ni-implanted 430 stainless steel (150 keV, $5 \times 10^{17} Ni^+ ions/cm^2$) (Agarwal *et al.*, 1979).

It is also important to point out that the composition of a sample being implanted can also contribute greatly to the final implant distribution. Agarwal *et al.* (1979) studied a range of Ni implanted type-430 stainless steel surface alloys. In this work, XPS depth profiling revealed significant deviations between the derived LSS theoretical profiles (corrected for sputtering) and those observed. Figure 17 indicates that Ni was enriched at the outermost layers of the surface alloy, while Figure 18 shows that Cr was depleted at the surface regions and enriched at deeper levels in the implanted zone. This behavior could not be explained on the basis of selective elemental sputtering, rather it appeared to conform to the misfit model of radiation enhanced diffusion postulated by Okamoto and Wiedersich (1976). Clearly this major redistribution of the two passivators, Cr and Ni, prevented any further systematic study of this alloy to be carried out.

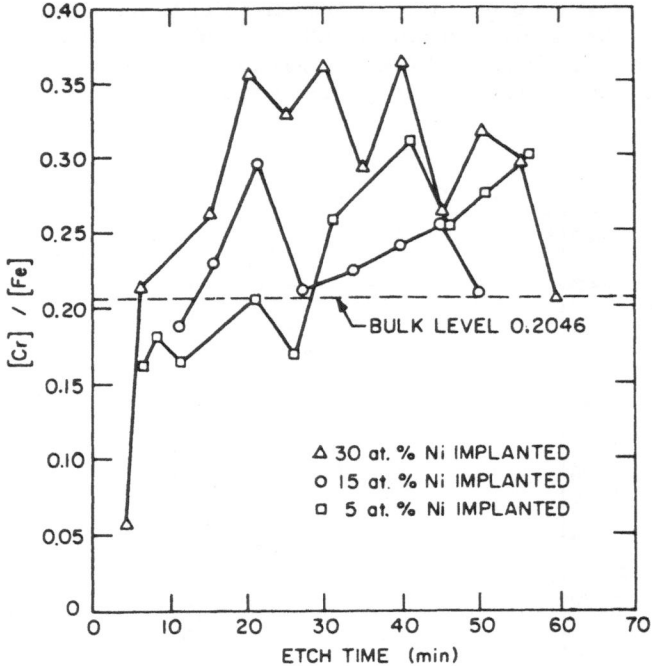

Figure 18   Relative concentrations of Cr and Fe in the region of implantation, (□) 5 at.% Ni implanted, (0) 15 at.% Ni implanted, (Δ) 30 at.% Ni implanted (Agarwal *et al.*, 1979).

4. Defect Concentrations

Since high defect concentrations are produced during high dose implantation, it is important to determine what effect these defects have on the electrochemical behavior of ion implanted surface alloys. The subject has not been thoroughly investigated. Ashworth *et al.* (1976a); Al-Saffar *et al.* (1980) observed that Ar implantation into Fe and Al resulted in thickening of the air formed oxide film, but did not alter the intrinsic electrochemical properties of the underlying metal. Sartwell and co-workers (Covino *et al.*, 1978a) compared the electrochemical behavior of Ni implanted Fe and Cr implanted Fe. Since defect production rates of Ni and Cr ions are known to be approximately equal at the same energy (Winterbon, 1975) the authors used the same implantation conditions in each case, namely 8.5×10^{15}ions/cm^2 at 25 keV. The differences observed between the two surface alloys were attributed only to some unspecified alloy characteristics and not defects.

5. Phase Structure

The nature of the microstructure of a surface alloy can have a significant influence on corrosion behavior. It is well known that multi-phase alloys tend to be susceptible to localized galvanic corrosion between phases of different chemical reactivity. Thus it is always desirable to produce single phase alloys to avoid such effects. Chemical homogeneity in single phase alloys is also desirable. Ion implantation may be used to form single phase solid solutions often far in excess of the equilibrium composition (Poate and Cullis, 1980). From the corrosion scientist's viewpoint this is a major advantage of the use of ion implantation as a surface alloying technique. Efficient heat sinking and the selection of suitable implantation parameters can often prevent second phase precipitation. In the case of the ion implantation of steels and other engineering alloys it is especially important that electron microscopy of the implanted region is carried out in order to determine whether pre-existing second phase particles have undergone dissolution or growth during the implantation process.

Amorphous surface alloys are a special case in which the absence of grain boundaries and other defects can contribute greatly to the overall corrosion resistance of a surface alloy. However, partial micro-crystallinity in such an alloy can be extremely detrimental to corrosion behavior (Clayton et al., 1980). Microstructural analysis of such surface alloys is, therefore, essential.

B. Ion Implantation of Fe

Bearing in mind the important role of Cr in stainless steels, it is perhaps not surprising that several workers have considered the effects of Cr implantation on the passivation behavior of pure Fe. Ashworth et al. (1976b) were the first workers to study this system. They carried out low energy implantations (20 keV) at fluences of 5×10^{16} and 2×10^{17} Cr$^+$ions/cm^2. Polarization studies were carried out in a deaerated acetic acid/sodium acetate buffer solution of pH 7.2. This work showed conclusively that the surface alloys formed by ion implantation behaved in a very similar manner to conventionally formed binary alloys of a similar composition. A later paper by Covino et al. (1978a) reported similar results. These workers considered the implantation of Fe with Cr$^+$ ions using an energy of 25 keV and fluences of: 1.25×10^{16}, 2.2×10^{16}, and 4.0×10^{16}ions/cm^2. Polarization studies were carried out in a deaerated borate buffer solution of pH 8.5 with 2400 ppm Cl$^-$ ions. In this work the general passivation and pitting behavior of the surface alloys was compared with conventional binary alloys. The results obtained from the implanted alloys were comparable to those obtained for conventional alloys. However, upon pitting, the surface alloys reverted to the corrosion behavior of pure Fe.

Takahashi et al. (1981) have carried out multi-sweep voltammorgrams of ion implanted pure Fe in 0.5M acetate buffer solution (pH5). The ions implanted into Fe included: Ar$^+$, N$_2^+$, Cr$^+$, Ni$^+$, Zn$^+$, and Cu$^+$ with doses ranging from 3×10^{16} to 1×10^{17}ions/cm^2 at 150 keV. In order to compare the reactivity or durability of the surface alloys during multi-sweep voltammetry, the peak current density (I_p) was plotted against the number of cycles (N_c). It is clear from Figure 19 that the Cr implant (10^{17}ions/cm^2) was the most inert. After 30 cycles I_p had only reached a value of 4 mA/cm^2. SIMS was used to determine the profiles of the alloys before and after polarization. It was shown that the highest Cr fluence exhibited no significant loss of Cr after 30 cycles.

In a further study by Ashworth et al. (1977), the authors reported the electrochemical behavior of Ta-implanted Fe (5×10^{16} and 2×10^{17} Ta$^+$ ions/cm^2 at 20 keV). The surface alloy had the behavior of a single phase solid solution, which cannot be made by conventional means because of the insolubility of Ta in Fe. This study again demonstrated that the implanted passivator was capable of beneficially modifying the passivation of Fe, presumably by incorporation into the passive film. Glancing angle RBS measurements indicated that selective dissolution of Fe facilitated the retention of Ta in the surface alloy.

Since interstitial elements such as B and N are often implanted in Fe and steel to improve surface strengthening, they are likely candidates for studies of stress corrosion and corrosion fatigue. Therefore, knowledge of the influence of these elements on the corrosion behavior of Fe is of importance. Bonora et al. (1981) carried out a study of the corrosion of Fe implanted with B$^+$ (50 keV) and N$^+$ (30 keV) in the following media: (a) De-aerated 0.5 M H$_2$SO$_4$, (b) De-aerated Borate buffer solution pH = 8.5, (c) De-aerated Borate buffer solution pH = 8.5 plus 2400 ppm Cl$^-$ (d) De-aerated 1 M NaCl (pH4) acidified by HCl.

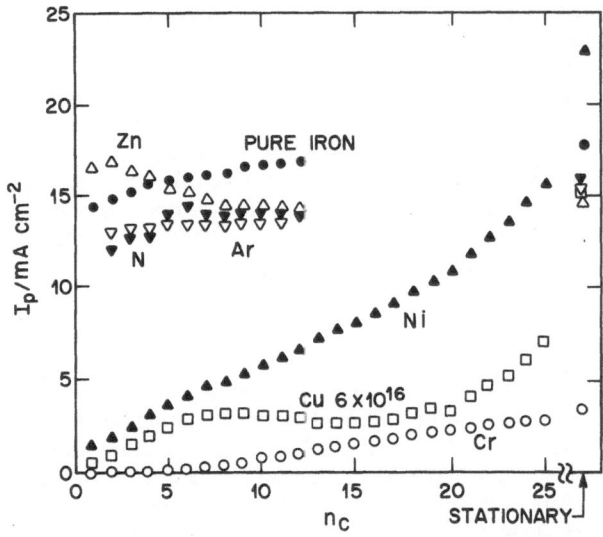

Figure 19   Effect of ion-implantation of various kinds of ions on Ip-N$_c$ characteristics, Zn. ($\triangle$), Ar ($\triangledown$), N ($\triangledown$), Ni ($\triangle$) and Cr (o) implantations with a dose of 1×10^{17}ions/cm^2 and Cu ($\square$) implantation with a dose of 6×10^{16}ions/cm^2; pure Fe (o) as a reference (Takahashi *et al.*, 1981).

TABLE IV

Calculated Corrosion Rate (mm/y)
ARMCO Fe
Doses in ions/cm^2

Solution	nonimpl.	5×10^{14}N$^+$	10^{15}N$^+$	10^{16}N$^+$	10^{15}B$^+$	10^{16}B$^+$
0.5M H$_2$SO$_4$	15.9	6.0	5.6	4.3	7.5	7.4
1M NaCl (pH=4)	0.24	0.16	0.14	0.17	0.06	0.06

Bonora *et al.* (1981) found that B and N implantation generally tends to lower the corrosion rate in acid and acidic chloride media. The calculated corrosion rates supporting this observation are shown below in Table IV.

Corrosion morphology was found to depend on the ion type and dose. The 10^{16}B$^+$/cm^2 implant produced the most uniform corrosion attack in both H$_2$SO$_4$ and NaCl solutions. In borate buffer solutions the best active-passive behavior was produced by 10^{16} N$^+$/cm^2 implantation. This resulted in significant lowering of (i) corrosion potential, (ii) passivation potential, (iii) critical current density, and (iv) passive current density. A 10^{16} B$^+$/cm^2 implantation resulted in an even greater reduction in critical current density and passive current density, little change in passivation potential and an anomalously higher corrosion potential.

Ferber *et al.* (1980) have investigated the possibility of lowering the active corrosion rate of pure Fe in sulphuric acid by lowering the kinetics of the cathodic reaction. In order to investigate the possibility of using ion implantation to introduce an element having a low exchange current density for the cathodic hydrogen reaction in an acidic solution, they implanted a range of elements having different hydrogen exchange current densities, namely: Cu, Pb and Au. Neon and Ar implantations were also carried out in order to determine the effect of defects on the cathodic reaction. It was generally found that Au increased the cathodic kinetic rate while Cu and to a greater extent Pb lowered the cathodic kinetics. Some discrepancies in the weighted exchange current densities were observed. Defects formed during implantation with Ar and Ne resulted in some increase in the cathodic kinetics.

C. Ion Implantation of Engineering Alloys

There is a widespread bearing corrosion problem in military aircraft propulsion systems which is typified by localized pitting along the contact region between the rollers and races. The corrosion pits may act as initiation sites for fatigue spalling which can lead to catastrophic engine failure. Another serious problem is that replacement bearings can have a short shelf-life again due to corrosion. Ion implantation was applied to M50 (Wang *et al.*, 1979; Hubler *et al.*, 1980) and AISI 52100 bearing steels (Hubler *et al.*, 1981). It was found that in addition to maintaining dimensional stability, mechanical integrity in the form of rolling contact fatigue resistance was not altered by the implantation process.

TABLE V

Ion Implantation Treatments for M50 Alloy

Specie(s)		M50 Fluence (ions/cm²)	Energy (keV)
Cr		2×10^{17}	150
Mo		5×10^{16}	100
Ti		2×10^{17}	55
Cr,Mo	Cr	1.5×10^{17}	150
	Mo	5×10^{16}	100

Ion implantation treatments were carried out as shown in Tables V and VI.* Two sets of corrosion tests were carried out on the unimplanted and implanted bearing steel. Survey studies based on electrochemical polarization measurements were performed to determine the active-passive behavior in 1M H_2SO_4, and the pitting resistance in a pH6 buffered NaCl solution (for M50; 0.1 M NaCl solution, and for 52100; 0.01 M NaCl solution). Simulated field service tests, using a geometry simulating an actual bearing, were also carried out over a period of several weeks. The simulated field test has been described fully elsewhere (Valori and Hubler, 1981). The test involved exposure of samples to a seawater contaminated

---

* M50 and AISI 52100 steels are martensitic steels with the following compositions (in wt.%).
  M50:   C(0.8), Mn(0.15-0.35), Si(0.1-0.25), Mo(4.0-4.5), Va(0.9-1.1), Fe(bal)
  52100: C(0.96), Mn(0.36), Si(0.22), Cr(1.36), Fe(bal).

TABLE VI

Ion Implantation Treatments for AISI 52100

Specie(s)		Fluence (ions/cm^2)	Energy (keV)
Cr		2×10^{17}	150
Cr,Mo	Cr	2×10^{17}	150
	Mo	3.5×10^{16}	100
Cr,P	Cr	2×10^{17}	150
	P	5×10^{17}	40
Ta		1×10^{17}	150
Mo		0.5×10^{17}	100

lubricant (neopentyl polyolester) which was thermally cycled (60°C for 8h and 4°C for 16 hrs) to simulate a turbojet engine environment.

The M50 steel tests indicated that the combined Cr plus Mo implantation improved both the active-passive behavior in an acidic environment and the pitting resistance in NaCl solution. Long term field tests also showed remarkable improvements in pitting resistance over a testing period of several weeks. The synergistic behavior of Cr and Mo in the improvement of pitting resistance in stainless steel is well known.

The Ta and Cr plus P implantations were the most beneficial treatments given to 52100 steel (Hubler *et al.*, 1981). For the polarization measurements conducted in the 0.01 M NaCl solution, two conditions were considered (i) anodic polarization following immersion and (ii) anodic polarization after a cathodic pretreatment to remove the oxide film. Simulated field tests also yielded evidence that the implantation treatments had all significantly improved pitting resistance. Figure 20 shows optical micrographs of the Cr plus Mo implanted 52100 surfaces after a 4 week test. The remarkable improvements in the pitting resistance of Ta and Cr + P implanted 52100 is attributed to the formation of passive films in which each of the elements plays an important role. The P implant tends to promote the formation of a phosphate-type passive film (Clayton *et al.*, 1980).

Misra and Kustas (1982) recently attempted to use ion implantation to improve the localized corrosion resistance of precision gears made from 303 SS and 2024 aluminum alloy and 52100 bearing components of a fine positioning micromechanism used in aerospace systems for precision control over long operation periods. Corrosion of these components (chosen on the basis of their machineability) reduces their operating lifetimes. Misra and Kustas reported the ion implantation treatments shown in Table VII.

Samples were exposed to a $NaCl-H_2O_2$ solution at room temperature. After 95.5 hrs. the pitting resistance of 303 SS was significantly improved by Cr implantation. However, the combined Cr plus Mo implant resulted in complete immunity from pitting. After 11 hrs. the 52100 steel was found to have a considerable resistance to pitting in both of the implantation treatments. After 8.5 hrs. of testing the 2024 alloy produced significant pitting with some improvement being attributed to the implantation treatment. The wear resistance of the gears was significantly enhanced by these anticorrosion treatments.

Al-Saffar *et al.* (1980) reported a study of the influence of Ar and Mo implantation on the corrosion behavior of pure Al and a 7075-T6 high strength aluminum alloy. TEM

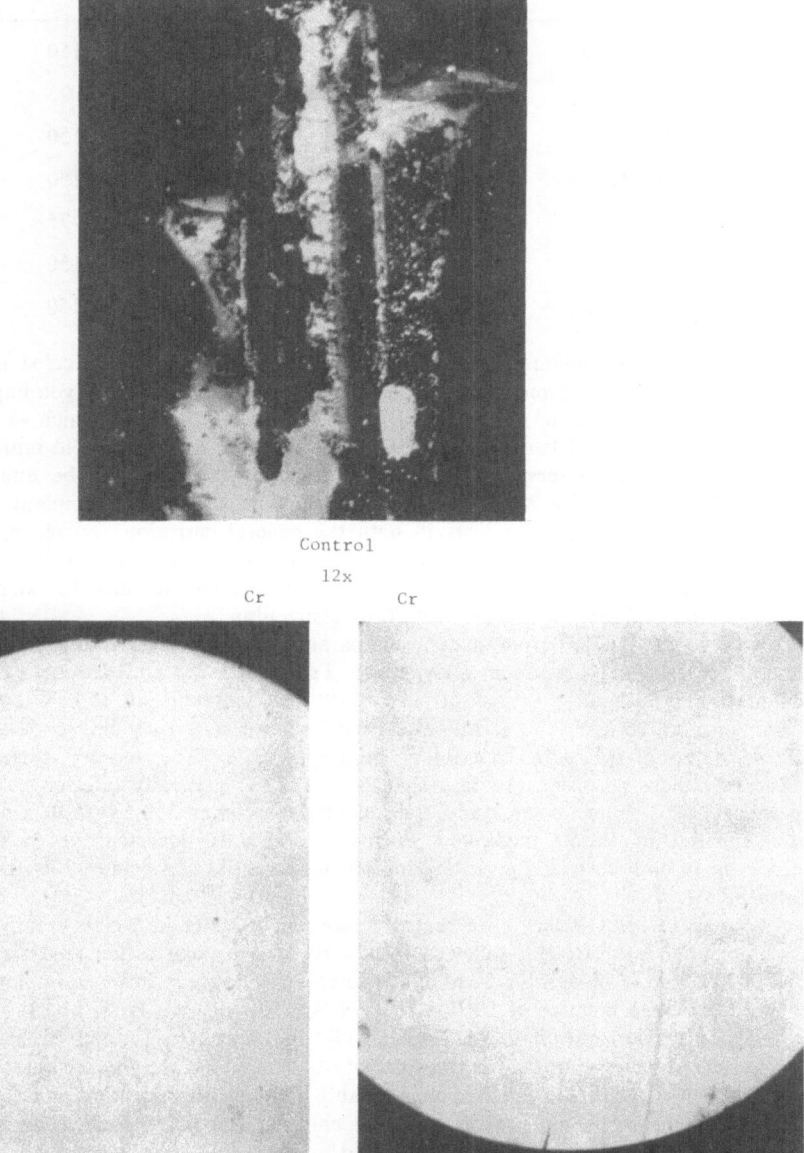

Figure 20 Optical micrographs of the surfaces of 52100 steel polished flats (3/8 inch diameter) after 4 weeks exposure to the corrosion test. The upper sample control is unimplanted and the two lower samples were implanted with 1×10^{17}Cr at 150 keV (Hubler *et al.*, 1981).

TABLE VII

Material	Implant		Fluence (ions/cm^2)	Energy (keV)
303 SS	Cr		1×10^{17}	150
	Cr + Mo	Cr	1×10^{17}	75
		Mo	5×10^{16}	150
52100 steel	Cr		1×10^{17}	150
	Cr + Mo	Cr	1×10^{17}	75
		Mo	5×10^{16}	150
2024-T3 Alloy	Cr		1×10^{17}	150

analysis of pure Al implanted with 10^{17} Mo$^+$ions/cm^2 at 20 keV revealed a single phase crystalline solid solution even though Mo is virtually insoluble in Al. Argon implantation had only a transient influence on the electrochemical behavior of pure Al, which was attributed to air-formed oxide film thickening. In the Cl$^-$ free solution, some persistent influence of Ar on anodic behavior was observed. The authors suggest that this may be due to elemental enrichment of Zn and/or Mg resulting from selective sputtering. Implantation with Mo resulted in significant improvements in both the general corrosion and pitting resistance of pure Al and 7075-T6 alloy. Rutherford backscattering and potential-time measurements suggested that this may be due to either incorporation of Mo into the passive film or to dissolution and replating of Mo ions onto the passive film.

Covino et al. (1978a) have made a comparative study of the electrochemical behavior of ternary Fe/10Cr/15Ni surface alloy, pure Fe, 316L SS and Inconel 625. The dual implantation treatment consisted of 2.05×10^{16} Ni$^+$ions/cm^2 at 25 keV and 2.08×10^{16} Cr$^+$ions/cm^2 at 25 keV. Polarization experiments were carried out in de-aerated borate buffered solution (pH 8.5) containing 2400 ppm Cl$^-$. The ternary surface alloy was significantly more corrosion resistant than pure Fe, but generally inferior to both 316L SS and Inconel 625. In the same study, unimplanted Vascomax 250 (V-250), a maraging steel, was compared to V-250 implanted with Cr (2.08×10^{16}ions/cm^2 at 25 keV). Modest enoblement in both corrosion potential and pitting potential was achieved by the implantation treatment.

Ashworth et al. (1980a) have recently reported a study in which a range of elements were implanted into 304 SS, followed by a three sweep polarization analysis in de-aerated 0.5M H$_2$SO$_4$ and in de-aerated 0.1M NaCl solution. The study involved implantation carried out at 20 keV with fluences of 10^{17}ions/cm^2 of Ar, Ta, Mo, W, Ti, Si and P. The results of the analysis carried out in 0.5M H$_2$SO$_4$ indicated that Ar, Ti, and Si had little effect. Phosphorus reduced the critical current density but increased the passive current density and Ta, W, and Mo facilitated passivation, probably through incorporation into the passive film. Polarization measurements made in the chloride media indicated that the largest improvement in the pitting resistance observed was due to Mo. However, Mo implanted 304 was not as resistant to pitting as 316 SS.

Hubler and McCafferty have studied the influence of Pd implantation (10^6 Pd$^+$ions/cm^2 at 90 keV) on the corrosion behavior of Ti exposed to boiling 1M H$_2$SO$_4$ (McCafferty and Hubler, 1978; Hubler and McCafferty, 1980). The authors reported that the implantation treatment reduced the corrosion rate by a factor of 1000. Backscattering measurements indicated that selective dissolution of Ti resulted in surface retention of Pd, which is a

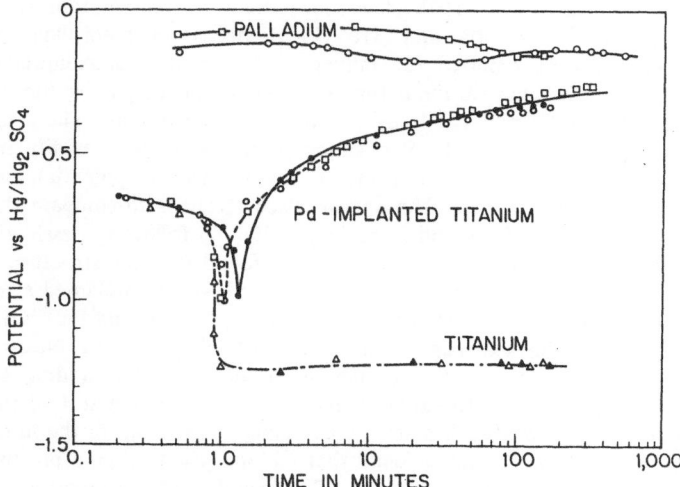

Figure 21   Open circuit corrosion potentials as a function of time in boiling 1M H_2SO_4 for pure titanium, pure palladium and palladium implanted titanium. The different symbols refer to replicable experiments (McCafferty and Hubler, 1978).

catalyst for the cathodic hydrogen evolution reaction. Electrochemical analysis revealed that the steady state open circuit potential in boiling 1M H_2SO_4 for the implanted Ti (Figure 21) was very close to the highly noble value of Pd. The high corrosion resistance of Pd-implanted Ti was attributed to galvanic effects between anodic Ti and cathodic Pd sites which drove the mixed potential into the passive potential region of Ti.

Lichter and co-workers have studied the corrosion behavior of Pt-implanted Ti in de-aerated (100°C, 1M) H_2SO_4 (Thompson et al., 1980; Appleton et al., 1981a). Platinum was implanted into Ti with a range of fluences (2×10^{12} - 3×10^{16}ions/cm^2) and energies (100-270 keV). Platinum was found to accumulate on surface alloys having a Pt concentration greater than 0.12 atomic percent. Platinum was found to act in a similar fashion to Pd, rendering the Ti passive in 1M H_2SO_4. However, in long term open circuit tests, the high corrosion resistance of the implanted Ti was eventually lost, except in the surface alloy containing the highest Pt concentration. Backscattering measurements indicated that the Pt was not lost to the solution, but became "inactive", presumably forming local agglomerates and clusters via a surface diffusion mechanism.

D. Amorphous Surface Alloys

In addition to the conventional approach to designing corrosion resistant alloys, ion implantation offers some scope for the formation of amorphous surface alloys. Hashimoto et al. (1978) and Asami et al. (1976) have shown that amorphous alloys formed by rapid quenching often exhibit superior corrosion resistance provided that the alloy has a sufficient concentration of a strong passivator such as Cr. One advantage of such alloys is that the absence of grain boundaries allows for the formation of a continuous passive film which is not disrupted at the grain boundary region. Furthermore, the splat-quenched alloys exhibit a high degree of compositional uniformity. The majority of highly corrosion resistant alloys studied so far conform to a composition of approximately 80 at.% transition elements and 20 at.% metalloids.

Recent work (Whitton *et al.*, 1978) has shown that P^+ implantation at 40 keV and a fluence of 10^{17}ions/cm^2 into 304 and 316 SS produces an amorphous surface alloy. Clayton *et al.* (1982) have reported the influence of both the contribution of P and the amorphous structure of the alloy on the nature and stability of the passive film formed on 304 SS in de-aerated 1M H_2SO_4 and 1M H_2SO_4 with 2% NaCl solution. The P-implanted steel is seen to be significantly improved. Passivation is achieved more readily in both of the solutions used, as indicated by the lowering of the passivation potential, critical current density and passive current density. The authors used RHEED to compare the structure of the passive film formed on 304 SS and P-implanted 304 SS following passivation for 1 hour in 1M H_2SO_4 at 250 and 550 mV (vs SCE). It was found that the structure of the passive film formed on 304 SS was crystalline, while that formed on the implanted steel appeared to be essentially amorphous with some evidence of crystalline Cr PO$_4$ and Fe (PO$_3$).

The stability of pre-formed passive films in 1M H_2SO_4 + 2% NaCl solution was determined by extending the passivation time by 10 minutes after adding acidic chloride solution to the 1M H_2SO_4. RHEED analysis indicated that the structure of the film formed on the implanted steel was modified, and that no change was found in the film found on the implanted steel. Auger depth profiling found that Cl$^-$ ion penetration, a precursor to pitting, was far greater in the crystalline passive film. The stability of the amorphous passive films formed on the P-implanted 304 SS was considered by the authors to be due to (a) the effectiveness of the amorphous passive film to act as a diffusion barrier, and (b) the corrosion inhibitive properties both of the phosphates contained within the amorphous passive film and of an outer-layer of crystalline phosphates.

Hubler *et al.* (1982) have recently reported the corrosion behavior of Ti implanted 52100 (4×10^{17}ions/cm^2, 190 keV). Singer *et al.* (1981) had previously shown that these implantation conditions produced an amorphous surface alloy. Hubler *et al.* investigated the corrosion behavior of the amorphous surface alloy in 1M H_2SO_4 and 0.1 M NaCl solutions. Polarization studies in 1M H_2SO_4 resulted in some improvements in the general corrosion behavior, but in 0.1 M NaCl solution no improvement in pitting was observed. The authors therefore found surprisingly little improvement over the unimplanted 52100. They gave evidence suggesting that surface carbide or impurity inclusions acted as locally active sites at which pitting initiates. Penetration of the pits, therefore, led eventually to galvanic attack at the base of the amorphous alloy.

It is clear from the above studies that attempts to attain by ion implantation the remarkable corrosion resistance which has been exhibited by amorphous alloys fabricated by rapid quenching techniques, may be frustrated by the inherent compositional heterogeneity of the surface being implanted.

E. Corrosion Fatigue

Few studies have been reported on the effect of ion implantation on corrosion fatigue behavior. De Anna *et al.* (1980) studied the corrosion fatigue behavior of N-implanted Fe in neutral solution containing 0.3% Na$_2$SO$_4$ in distilled water at equilibrium with atmospheric oxygen. The study involved the monitoring of the corrosion potential during the fatiguing of an oscillating cantilever beam sample. The aim of the work was to combine the beneficial effects of N implantation on fatigue resistance and on corrosion behavior in neutral solutions. Implantations were carried out at 30 keV with fluences of 10^{15}, 10^{16} and 10^{17} N$^+$ions/cm^2. Calculated values of the corrosion current during various periods of the fatigue experiments indicated that up to 10^4 cycles, the implanted samples exhibited a lower corrosion current than the unimplanted sample. Beyond 10^4 cycles the implanted samples tended to corrode at a higher rate. In general the corrosion fatigue life was shortened by the ion implantation treatment. The authors attribute this to the tendency for the N-implanted region to be cathodic to the base material. Therefore, upon the opening of surface microcracks, crack development of the implanted samples was accelerated by galvanic corrosion.

Sartwell *et al.* (1982) have recently investigated the use of ion implantation to improve the corrosion fatigue resistance of 1018 steel in air saturated 3% NaCl solution. The implantation treatments were carried out on cylindrical fatigue samples of 1018 steel as shown in Table VIII.

TABLE VIII

Implantation Parameters for Corrosion Fatigue

Specie(s)		Fluence (ions/cm^2)	Energy (keV)
Ti$^+$		2.9×10^{16}	50
		3.7×10^{16}	50
		5.2×10^{16}	50
Mo^{++}Ta$^+$	Mo$^+$	5.3×10^{15}	30
	Ta$^+$	5×10^{15}	60

Using a rotating fatigue machine equipped with an electrochemical cell the authors monitored the variations in corrosion potential with the number of cycles to failure. Fatigue was carried out 4900 cycles per minute at a maximum loading of 318 M.Pa. It was found that Ti implantation was detrimental to the corrosion fatigue resistance of 1018 steel, shortening the cycles to failure by 50%. The combined Ta and Mo implantation was reported not to have altered the corrosion potential or the cycles to failure.

VII. CHEMICAL APPLICATIONS OF ION IMPLANTATION - CATALYSIS AND HYDROGEN MIGRATION

A. Ion Implantation and Catalysis

This field has been reviewed in detail elsewhere (Wolf, 1981a). Therefore the aim of this section is to outline in general the present knowledge and the problems concerning ion implanted catalysts. A brief discussion of the key aspects will be given followed by an outline of continuing difficulties. Ion implantation and sputtering in general are useful methods for preparing catalysts on metals and insulators on substrates. This has been shown for reactions at gas/solid and at liquid/solid interfaces. The main advantage of these methods is the small consumption of precious materials during preparation (typically 10 μg/cm^2). Sputtering should be the choice if only surface coverage is needed with active metals. It is less expensive and easier to perform than implantation. Ion implantation should be chosen in cases where one needs good adhesion of the active metal to the substrate or one wants to produce novel materials with catalytic properties different from either the substrate or the pure active metal. Ion beam mixing promises very interesting prospects for the preparation of catalysts.

The following are seen to be continuing difficulties. Possible corrosion of the substrate and of the implanted or sputtered active layer; this is the main factor in the long-term stability of the catalyst. Ion implanted metals may be buried below the surface layer of the substrate and hence show no activity. Preparation of catalysts with high surface areas presents problems for the ion beam techniques.

1. Reactions at Solid/Gas Interfaces

Cairns (1980) and Rabette *et al.* (1979) both studied the Pt-on-Al$_2$O$_3$ substrate system for the catalytic reaction involving the hydrogenation of unsaturated hydrocarbons. In the former study, sputtering was used for the preparation of the Pt containing surface layer, while in the latter ion implantation was used. For the sputtered catalysts, the ability to produce butane from butadiene was better than for a conventionally prepared catalyst and much less Pt was necessary for the sputter processing. For the implanted catalyst, the initial activity for the hydrogenation of ethylene was very low, but after annealing up to 1300K a large increase in activity took place. This was caused by an initially low surface coverage of Pt which during the heating process diffused from the bulk to the surface and segregated into small precipitates (Matzke, 1981). Consequently, annealing may be of value. Cairns *et al.* (1980) demonstrated the possibility of preparation of high surface area catalysts by sputtering the active metal on small substrate particles kept under motion.

2. Reactions at Solid/Liquid Interfaces

All reactions studied in this field have been concerned with electrocatalysts (electrodes), mainly in systems important for the development of fuel cells or water electrolysis. Wolf (1982) and Kasten and Wolf (1980) made a model study implanting Pt and other metals into Fe electrodes. They demonstrated a three orders of magnitude increase of the H evolution rate in acidic solutions, compared to unimplanted Fe and more than two orders of magnitude in comparison to smooth Pt. A selective dissolution procedure was used. A carefully controlled amount of iron was dissolved from the electrode surface resulting in a porous structured layer, 10^3-10^4Å thick, containing Pt in a very active state. The long-term stability of these electrodes was poor. Platinum was lost by corrosion and agglomeration into an inactive form. Similar observations were made for Pt in Ni by Akano *et al.* (1981) for water electrolysis and by Hayes *et al.* (1978) for Ti electrodes for Ce production.

The difficulty of getting the implanted active element to the surface of the substrate was also demonstrated for the Pt/C system. Voinov *et al.* (1974) and more recently Wolf *et al.* (1981b) found no significant activity for implanted electrodes, for hydrogen evolution or for the oxygen reduction reaction. The latter work also included Rutherford backscattering measurements showing more than one at.% of Pt to be in the near-surface region. But scattering experiments with low energy He ions (Schmiedel and Wolf, 1981) showed the Pt content in the first monolayer to be below 0.1 at.%. These findings were in agreement with unpublished measurements by Wolf and Kasten using sputtered Pt/C electrodes.

Figure 22 shows the current density for the H evolution reaction in acidic solution as a function of the applied potential. In contrast to the nearly inactive implanted case, the activity of the catalyst prepared by sputtering was higher than for the smooth Pt metal. A very interesting additional finding was the further increase of the activity, by nearly one order of magnitude, after intermixing the sputtered Pt layer with the substrate by means of an Ar ion beam.

The following is an example of the creation of a novel catalytically active material by ion implantation. RuO$_2$ is used as an electrode for chlorine production because of its superior corrosion resistance. O'Grady and Wolf (1981) implanted Pt in RuO$_2$ and tested the performance of the catalyst with respect to the oxidation of formic acid and methanol (fuel cell reactions). Figure 23 shows the current density of the formic acid oxidation as a function

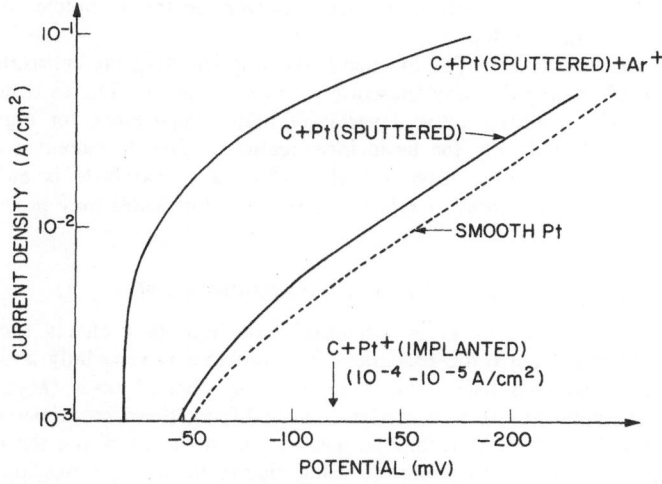

Figure 22 The current densities for the hydrogen evolution reaction as a function of the potential achieved for smooth platinum, platinum sputtered onto carbon substrates and sputtered + Ar⁺-mixed platinum catalysts (Wolf *et al.*, 1981b).

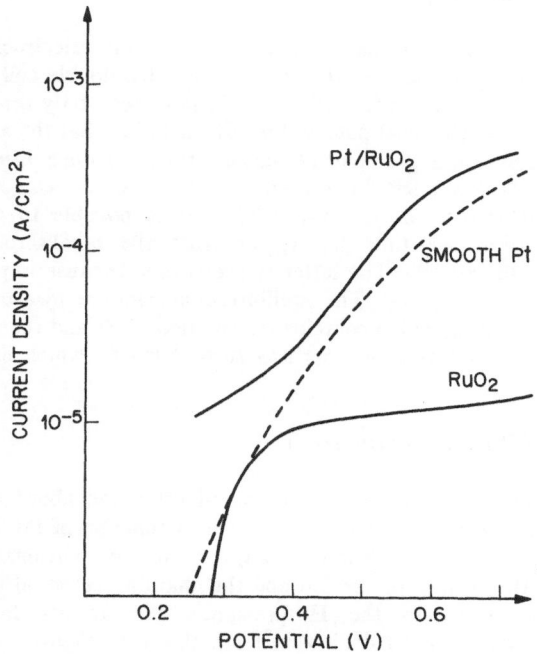

Figure 23 The current densities for the formic acid oxidation as a function of the potentials. RuO₂, smooth platinum and platinum implanted in RuO₂ were used as catalysts (O'Grady and Wolf, 1981).

of the applied potential. The implantation of Pt produces a catalytically active electrode, whose performance is superior to both pure RuO_2 and smooth Pt. It also has a good long-term stability. The most interesting finding, however, is the complete inactivity of the electrode for the methanol oxidation.

For future studies it is essential to choose the implanted specie/substrate combinations carefully, instead of bombarding any substrate with any element. The corrosion resistance of the substrate and the resulting active layer is of major importance for applications. This property may be neglected only for basic investigations. The development of methods for obtaining either implanted or sputtered high surface area catalysts is an essential task. Treatment of powdered substrates or selective dissolution processes may prove to be suitable procedures.

B. Electrochemical Studies of H Migration in Implanted Metals

The migration and trapping of H is responsible for important effects such as hydrogen embrittlement and stress corrosion cracking. The problem is especially serious for ferrous alloys. Ion implantation has been used in a limited number of cases (Myers *et al.*, 1979; Rauch, private communication) as a method for modifying these properties of metals. The nuclear reaction analysis (NRA) technique has been used for analyzing the hydrogen in the metal. Only one study has been published using electrochemical permeation measurements, namely Pt implanted Fe (Zamanzadeh, 1980). The permeation measurement as an equilibrium dynamic technique is an ideal complementary procedure to NRA which is a more static method. Table IX summarizes the information to be obtained from both techniques.

1. The electrochemical method

Figure 24 represents a schematic description of the electrochemical permeation measurements (Boes and Zuchner, 1976). On one side of a double cell H is precipitated at the surface of the metal sample under study. The H migrates partly through the sample and on the other side the electrochemical potential is adjusted such that the arriving H is oxidized to H^+. The H concentration gradient at various time intervals after the beginning of charging is shown in the lower left. The time dependency of the current one needs for the oxidation of the H is shown on the lower right. It is possible to deduce the diffusion coefficient D either from the time lag τ_L or from the breakthrough time τ_b where $\tau_b = S^2/2\pi D$ (sample thickness). The latter is preferable, because impeding surface layers such as oxides affect τ_b less than τ_L. The equilibrium current one measures at large values of τ is the permeation current i_p, which contains information on D and the solubility of H in the metal. I_p is also strongly affected by impeding surface layers, which have to be avoided in these measurements.

2. Selected Results of Permeation Studies

A number of implanted Fe foils were measured using the above technique. Diffusion coefficients for H and permeation current densities as a function of the H charging potential or charging current were obtained. The charging current is proportional to the total amount of H precipitating at the surface of the foil and the charging potential is proportional to the surface coverage by H or to the H pressure. The results for $Fe(10^{17}Ar^+/cm^2)$, $Fe(5 \times 10^{16}Pt^+/cm^2)$ and $Fe(5 \times 10^{16}Pb^+/cm^2)$ are shown in Figure 25 in comparison with pure Fe. The H permeation rate is lowered a little by Pt implantation in spite of the fact that the total amount of H precipitating at a fixed charging potential is more than for 10 times higher than pure Fe. These values are in reasonable agreement with the earlier measurements by Zamanzadeh *et al.* (1980). Lead implantation causes a considerable

TABLE IX

Information on Hydrogen Migration in Metals

Type of Information	Nuclear Reaction Analysis	Electrochemical Measurements
Equilibrium solubility	no	yes
Local concentration	yes	no
Diffusion coeff. -with traps-	yes	yes, + limited temperature range
Binding enthalpies	yes	Yes, + limited temperature range
Permeation data	no	yes

lowering of the permeation rate in comparison with pure Fe. In both cases surface effects are mainly responsible for the action of the implanted element. Platinum catalyzes the desorption of the precipitating H, preventing it from entering into the material. Lead on the other hand, lowers the total amount of H precipitating and acts as a barrier against the H entry.

The effect arising from the Ar implantation is probably a bulk effect. The permeation rate increases somewhat, but the measured diffusion coefficients decrease slightly compared with pure Fe. Consequently, since the permeation rate is proportional to the product of the diffusivity and solubility, the solubility of the implanted layer for H must be much higher than the normal solubility of H in Fe. This was confirmed by measurements using the $^{1}H(^{15}N,\alpha)^{12}C$ reaction (Frech et al., 1981). Figure 26 shows the H profile in Fe implanted with $10^{17} Kr^{+}$ions/cm^2 at room temperature. The damaged region below the surface contains around 2 at.% H, while the normal solubility in Fe is less than 1 ppm. This difference is consistent with the permeation data for Ar. These few examples illustrate the usefulness of electrochemical experiments on H migration in implanted metals especially when combined with nuclear reaction analysis.

VIII. RADIATION DAMAGE AND PULSED ANNEALING OF SUPERCONDUCTORS WITH A15 STRUCTURE

A. Introduction

Implantation effects in superconductors have been extensively reviewed elsewhere (Meyer, 1980). Accordingly, only pulsed laser and electron beam effects will be discussed in this section which reviews some recent work on the influence of radiation damage on the superconducting properties of A15 superconductors. Results dealing with the influence of laser and electron beam irradiation on the nucleation and regrowth of irradiated and amorphous A15 superconductors are summarized.

MEASURING CELL

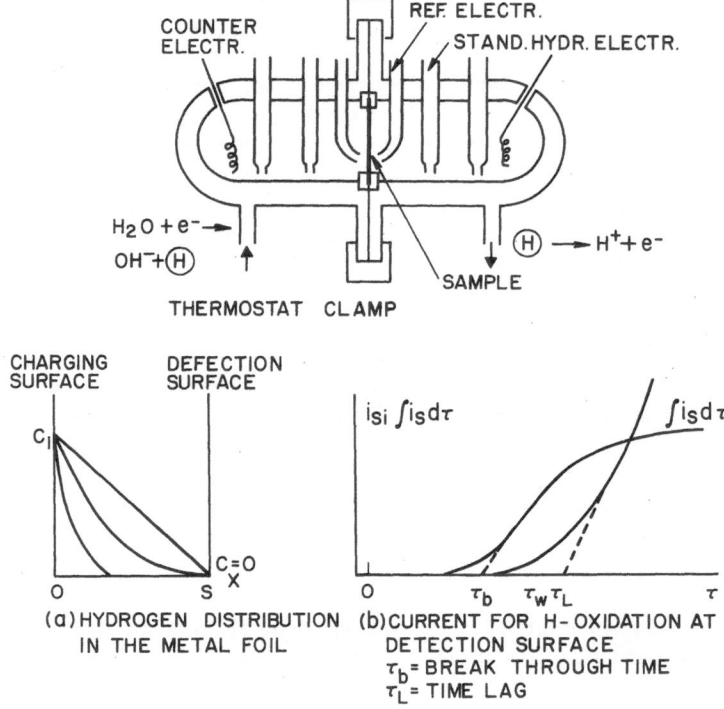

(a) HYDROGEN DISTRIBUTION
    IN THE METAL FOIL

(b) CURRENT FOR H-OXIDATION AT
    DETECTION SURFACE
    τ_b = BREAK THROUGH TIME
    τ_L = TIME LAG

Figure 24   Schematic representation of the electrochemical determination of hydrogen diffusion and permeation in metals. Upper part: Double cell for permeation studies. Lower left: Concentration gradients for hydrogen in a metal foil after electrochemical charging. Lower right: Current signal recorded for the oxidation of hydrogen, arriving at the detection side of the double cell (Boes and Zuchner, 1976).

B.   Radiation Damage

Compounds with the A15 structure reveal the highest transition temperatures, T_c, known up to now (Nb_3Ge, $T_c = 23K$) and show the best performance as high field magnets (Nb_3Sn, $T_c = 18.2K$, $H_{c2} = 20T$ at 4K). In this structure the non-transition metals form a bcc-sublattice and two transition metals are situated in each face thus forming three densely packed orthogonal chains. The high T_c values are thought to be caused by this chain structure. Any disorder in this structure will therefore result in a decrease of T_c. During particle and neutron irradiation large depressions of T_c have been observed together with an increase of the specific resistivity and the lattice parameter.

The nature of these defects has attracted much attention in recent years due to the assumption that the presence of such defects may ultimately limit the maximum T_c values observed for these materials (Brown *et al.*, 1978). Usually the decrease T_c is attributed to a perturbation of the long range order. The Bragg-Williams long range order parameter, S, is found to decrease with a T_c depression caused either by irradiation or by deviations in composition from stoichiometry (Sweedler and Cox, 1975; Blaugher *et al.*, 1969).

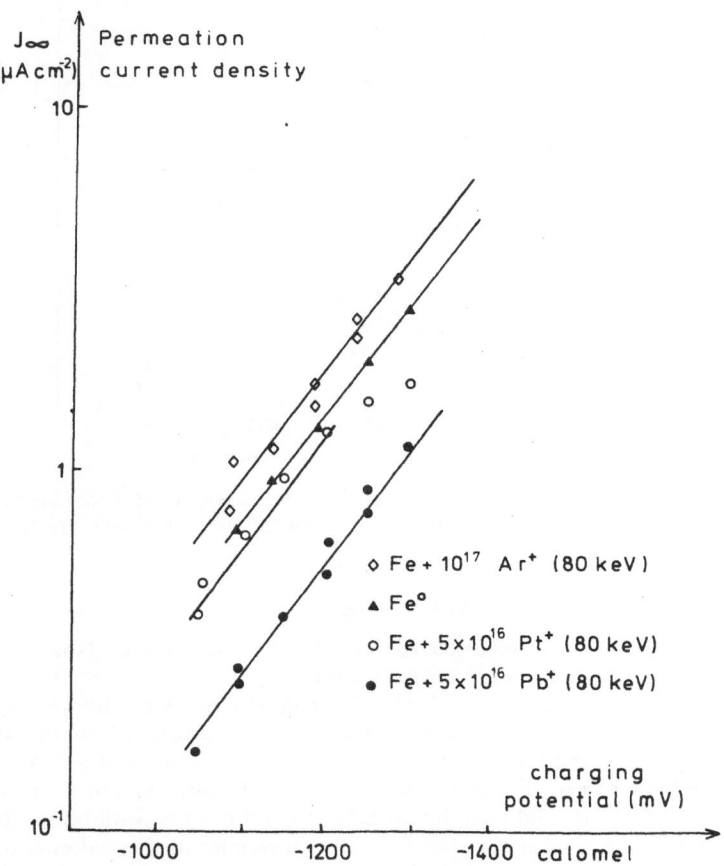

Figure 25   Permeation current density for hydrogen migration through iron foils as a function of the charging potential (proportional to the H-pressure). Pure iron is compared with samples bombarded with $10^{17}Ar^+/cm^2$, $5 \times 10^{16}Pt^+/cm^2$, and $5 \times 10^{16}Pb^+/cm^2$ (Boes and Zuchner, 1976).

In irradiated A15 superconducting thin films and single crystals, small static displacements of the atoms from their lattice sites have been shown to exist with amplitudes between 0.06Å and 0.1Å in the fluence region where T_c decreases. This was found in recent experiments performed with X-rays (Burbank *et al.*, 1979; Pfluger and Meyer, 1979) as well as with ion channeling (Meyer and Seeber, 1977; Testardi *et al.*, 1977a; Meyer *et al.*, 1981). The observation of anti-site defects and displacements in all cases studied leads to the assumption that these defects might be inherently connected (Schneider *et al.*, 1982).

With increasing ion fluence, T_c values reach saturation between 1K and 4K depending on the compound. At very high fluences, phase transformations are observed which depend on the ion mass and irradiation temperature. Whereas some A15 phases such as Nb₃Ge, Nb₃Ir, and V₃Ge become amorphous during irradiation, Nb₃Al transforms completely into an A2 phase (Schneider *et al.*, 1982). From these results one may conclude that similar phase transformations will occur during ion implantation. The regrowth of well-ordered high-T_c A15 phases will depend on the amount of disorder and the techniques used for annealing.

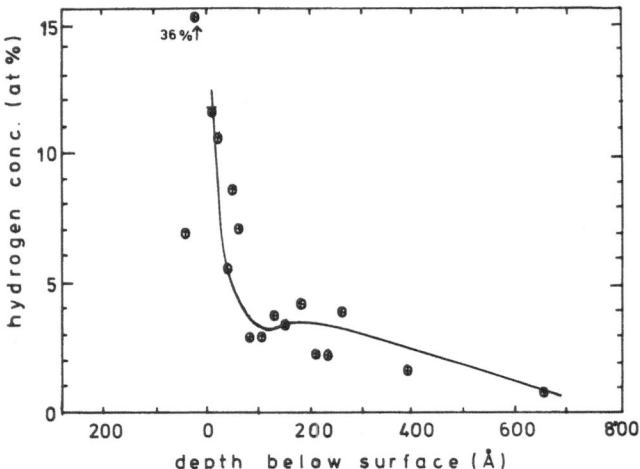

Figure 26 Hydrogen depth profile measured with the $^1H(^{15}N,\alpha)^{12}C$ reaction. Iron foils were used, which have been bombarded with $10^{17}Kr^+/cm^2$ (200 keV) and electrolytically charged with hydrogen (Frech *et al.*, 1981).

C. Pulsed Electron and Laser Beam Annealing

The first experiments on the influence of laser beam irradiation of Nb_3Ge, Nb_3Ga, and $Nb_{0.8}Sn_{0.2}$ bulk samples, prepared in an Ar arc furnace, resulted in a remarkable increase of T_c for all samples (Lekhtvar *et al.*, 1976). The biggest effect was observed for $Nb_{0.8}Sn_{0.2}$, where T_c increased from 9K to 18.2K. This effect is attributed to the formation of stoichiometric Nb_3Sn at the sample surface after laser irradiation. The energy density of the laser pulse was 1 J/cm² and thus one may assume that the formation of a well ordered A15 compound occurred in the solid state by Sn diffusion and reaction with Nb to form the A15 phase, a process, which is also observed for diffusion couples in thermal annealing. Recent results on pulsed laser annealing of V-Si multilayer samples in order to form V_3Si and on irradiated V_3Si showed that although the A15 phase was formed, it was not well-ordered (Appleton *et al.*, 1981b).

In the following we will compare the effects of pulsed laser beam annealing (LABA), pulsed electron beam annealing (PEBA), and thermal annealing on thin films of V_3Si, Nb_3Al, and Nb_3Ge with thicknesses between 0.1 μm and 3 μm. These films appear to be either amorphous, if prepared by cosputtering onto sapphire substrates at room temperature, or crystalline single phase films when predamaged to various levels of disorder by particle irradiation. During irradiation the energetic ions penetrate the films and come to rest in the substrate. Thermal annealing results from high dose ion implanted samples are also included for comparison. The purpose is to study the influence of growth velocity and of various microscopic states on the formation of well-ordered A15-phases.

Figure 27 presents a comparison between LABA on crystalline V_3Si films (Meyer *et al.*, 1982), predamaged with B^+ ion irradiation (T_c decreased from 16K to 2K, but the films are still crystalline) and PEBA on amorphous V_3Si films. With increasing pulse energy density, T_c is found to increase for both films. Differences are noted for the threshold energy densities and for the width of the transition curves. Since PEBA of crystalline films yields results similar to those obtained from LABA (not shown) we conclude that the different threshold energy densities are due to the different microscopic states. The recovery of T_c for the crystalline film starts at 0.5 J/cm², and at energy densities above 1 J/cm², the samples exhibit cracking and peeling from the sapphire substrates. Using a cumulative deposition of laser energy, the recovery could be improved slightly (Figure 27). The annealing of amorphous layers which starts at 1.5 J/cm^2 seems to occur by solidification from the liquid phase. The rather broad transition widths indicate large inhomogeneities in the composition of the compound. The question arises why the T_c recovery is only about 10K whereas in a thermal annealing process a nearly complete recovery is observed (Figure 28). It may be argued that the regrowth velocity is too high to form a well-ordered compound. In order to prove this argument, PEBA was applied to films, heated to about 580 and 680°C within 5 min. The T_c onset values increased to about 13K (Figure 28) and were about 1 to 2K higher than the T_c values of the thermally annealed part of the sample. The difference decreased with increasing substrate temperature during PEBA. From the solution of the heat flow equation we estimated that for this case the regrowth velocity is slowed down by a factor of 3. The velocity seems still to be too high, however, to form a well-ordered V$_3$Si phase.

Well-ordered high T_c Nb$_3$Al is difficult to form in co-evaporation and sputtering processes (Kwo et al., 1980). T_c values are usually about 3K below the maximum value of about 18.5K. Thermal annealing from the amorphous phase results in a rather low maximum T_c value as shown in Figure 29. Crystalline Nb$_3$Al layers with a T_c value of about 15K were heavily damaged by irradiation with N^{++} ions (Schneider et al., 1982). During this irradiation the A15 phase was totally transformed into the A2 phase with a lattice parameter, a, of about 3.24Å, (a(Nb) = 3.3Å) indicating that the Al is dissolved in Nb. Thermal annealing of this A2 phase leads to retransformation into A15 with considerably improved Tc values (Figure 29). It is interesting to note that high dose Al$^+$ implantation into Nb single crystals (25 at.% Al in Nb) at room temperature forms an A2 phase in a similar way. Thermal annealing caused a sharp increase of T_c up to 17K between 800 and 850°C (Figure 29). From this T_c value and from x-ray analysis it could be concluded that a well-ordered Nb$_3$Al A15 phase had formed. Further improvement of T_c could possibly be reached by ordering anneals at 600°C, as is routinely done for as-cast bulk material.

The stable A15 phase of the Nb-Ge system has about 20% of the Nb atoms on Ge lattice sites (Nb$_3$(Ge$_{0.8}$Nb$_{0.2}$)) and a T_c-value of 6.5K. Crystalline layers with nearly stoichiometric composition and T_c values of 21.5K were irradiated with H$^+$ and Ar$^+$ ions to various damage levels and subjected to an isochronal annealing process. From Figure 30 it can be seen that the maximum T_c value, which is reached at about 850°C, is strongly dependent on the amount of disorder present in the sample. Complete recovery is obtained for the H$^+$ irradiated samples, which were only slightly damaged in a sense that there was only a minor reduction in the X-ray line intensity and a slight increase of the lattice parameter. Samples irradiated with high dose Ar ions appeared amorphous in X-ray analysis and revealed similar recovery curves to those obtained by annealing sputtered amorphous Nb$_3$Ge films. It is also seen in Figure 30 that with increasing disorder, the temperatures with the highest annealing rates shift to higher values whereas the annealing rate itself is not strongly affected. It is interesting to note that the annealing of the Ge-implanted layer on bulk Nb$_3$(Ge$_{0.8}$Nb$_{0.2}$) (Geerk, 1980) which is amorphous after Ge-implantation at room temperature does occur about 100°C prior to the annealing of amorphous films. From X-ray measurements it was concluded that the implanted film did regrow epitaxially on the crystalline bulk sample. It may therefore be speculated that epitaxy facilitated the nucleation and growth process in this case. The lattice parameter of Nb$_3$Ge grown from amorphous phases was about 5.127Å, which is smaller than that of the best crystalline films with maximum T_c values and with a = 5.145Å (note that T_c increases and a decreases with x approaching zero in Nb$_3$(Ge$_{1-x}$Nb$_x$)). A possible explanation would be the incorporation of point defects during regrowth from amorphous phases. At annealing temperatures above 900°C, the Nb$_3$Ge A15 phase starts to decompose and the stable tetragonal Nb$_5$Ge$_3$ phase starts to form.

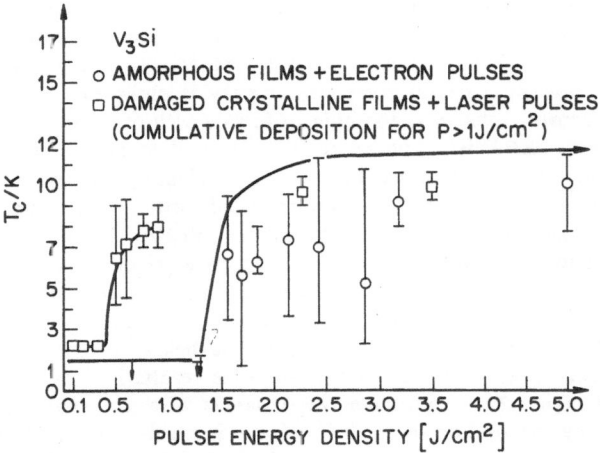

Figure 27   Influence of LABA on the T_c of crystalline V_3Si films predamaged with B^+ irradiation and of PEBA on amorphous V_3Si films.   Error bars indicate the transition width (Meyer *et al.*, 1982).

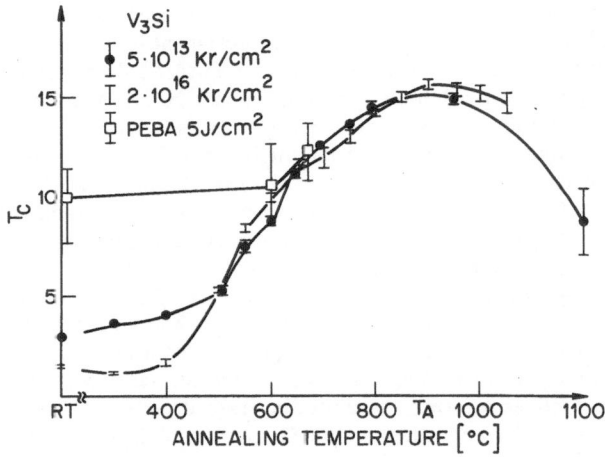

Figure 28 T_c-recovery of V_3Si films, predamaged by Kr^+-irradiation, during an isochronal annealing process. The results of PEBA on preheated films are included for comparison.

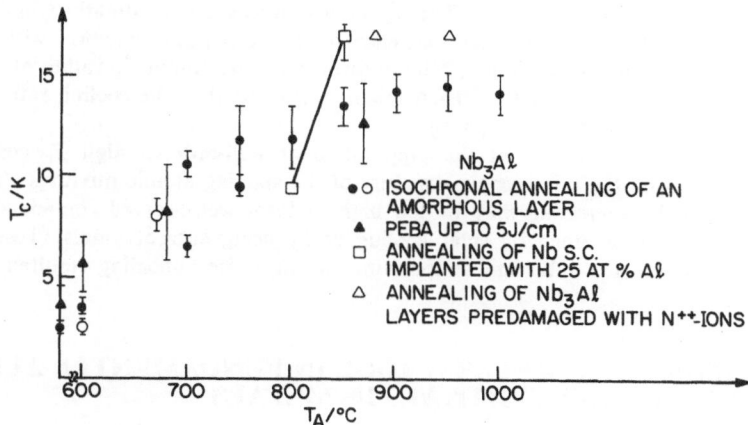

Figure 29   Nucleation and regrowth of Nb₃Al from various metallurgical conditions: (a) thermal annealing and PEBA on amorphous layers. (b) thermal annealing of Nb-single crystals implanted with 25 at.% Ge. (c) thermal annealing of Nb₃Al-layers predamaged with N⁺⁺-irradiation.

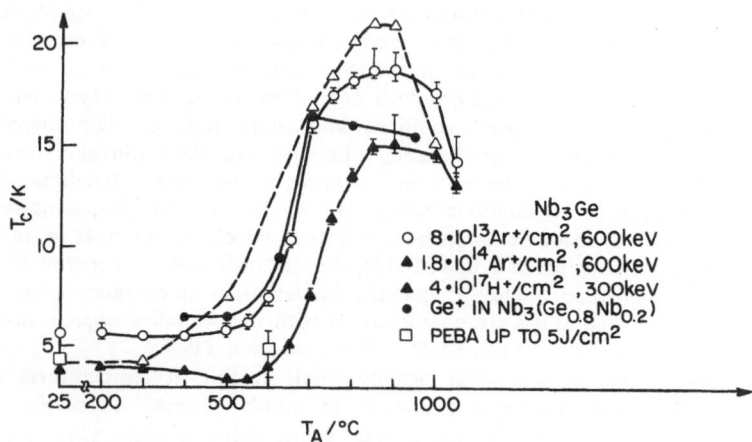

Figure 30 T_c-recovery of Nb₃Ge films, predamaged to different levels of disorder. Thermal annealing of the Ge-implanted Nb₃(Ge₀.₈Nb₀.₂) A15 phase is included for comparison.

PEBA experiments have been performed at RT and 600°C on amorphous Nb$_3$Ge layers with energy densities up to 5 J/cm^2. The T$_c$ values increased only slightly (Figure 30) and thin-film X-ray analysis indicated the existence of the A15 phase together with the stable tetragonal phase. Using splat cooling, Nb$_3$Ge forms with maximum T$_c$ values of about 17K. Thus a possible explanation for the PEBA results would be that the cooling rate is too high for a formation of the high T$_c$ A15 phase.

In summary we will note that the preparation of well-ordered high T$_c$ compounds is strongly dependent on the microscopic structure of the starting atomic mixtures. The growth velocity during pulse annealing may be too high to form well-ordered compounds. This is inferred from results obtained by splat cooling or by using current pulses (Testardi et al. (1977b)). Increasing the substrate temperature during pulse annealing resulted in a slight improvement.

IX. ION IMPLANTATION AS A TOOL IN FUNDAMENTAL ALLOY RESEARCH - INTERSTITIALS IN METALS

A. Introduction and Background

The unique properties of ion implantation have been extensively exploited in fundamental, microscopic studies of alloys. For instance, the capability of producing an atomic mixture of tailored composition, independent of the equilibrium phase diagram, has proved particularly advantageous in work with metastable phases. Furthermore, metastable implanted alloys can be driven to a condition of local thermodynamic equilibrium through annealing. In this way equilibrium entities such as precipitates are introduced at exceptionally high densities, making it possible to observe effects which otherwise would be hidden. Another significant feature is the control over the depth distributions of the implanted species. For example, several layers with different compositions can be introduced into a single matrix, and when the configuration is suitably chosen, the observed flow of solutes between layers elucidates such complicated properties as multiphase equilibria and solute trapping. The microscopic depth scale of these experiments is also beneficial because the short diffusion distances allow thermally activated processes to be examined at lower temperatures. Irradiation damage, an inevitable consequence of implantation except at very low ion energies, is useful in certain cases; it can serve, for example to introduce lattice defects of interest, or to produce an amorphous state. When damage is undesirable, the constraints of the experiment may permit post-implantation annealing or, less frequently, implantation at energies below the damage threshold. The implanted alloys are characterized with various microscopic probes, including ion backscattering/channeling, AES, SIMS, Mossbauer, and TEM.

Experiments on equilibrium alloys usually entail heating at temperatures sufficient to anneal out defects and cause precipitation of stable phases. Properties successfully investigated in the equilibrium regime include solute-diffusion rates (Myers et al., 1974; Myers and Rack, 1978), solid solubilities (Myers and Smugeresky, 1977; Myers and Rack, 1978), ternary phase diagrams (Myers and Smugeresky, 1978), and mechanisms of solute trapping at precipitates (Myers et al., 1980a; Myers and Follstaedt, 1981). The information obtained in these ion-beam experiments would have been difficult to extract using conventional metallurgical procedures.

A substantially larger effort has been devoted to metastable alloys. Here the thermally activated evolution from the initial, implanted condition toward thermodynamic equilibrium does not go to completion. Under these conditions the state of the material depends upon its detailed thermal history, and a particular implantation treatment may ultimately give rise to a number of qualitatively different alloys. Consequently the range of metastable materials which can be produced with implantation and heating is very large, and this is reflected in the wide variety of experimental investigations. As one example, a number of amorphous phases have been discovered (Grant, 1981; Poate, 1978; Poate and Cullis, 1980). Moreover, new crystalline materials have also been produced, including highly supersaturated solid solutions (Borders and Poate, 1976; Kaufmann et al., 1977) and intermetallic compounds which depart from the equilibrium phase diagram in structure as well as composition (Mayer et al., 1981). The implanted solutions, when analyzed by ion channeling, have proved particularly useful in studying the lattice position of solutes (see especially Kaufmann et al., 1977). Another important application is in the study of trapping phenomena. In metastable materials, the trapping processes investigated by implantation are mostly of two types: first, the binding of mobile interstitial solutes to metastable entities such as lattice defects (Anttila et al., 1981; Picraux, 1981) and impurity-defect complexes (Kornelsen and Van Gorkum, 1980; Myers et al., 1981); and second, the interaction of mobile point defects with immobile solutes (Reintsema et al., 1979; Swanson et al., 1978, 1980; Thome et al., 1979). Finally, ion implantation currently provides the only practical means of examining bubble nucleation in metals containing inert gases such as He (Jager and Roth, 1981; Evans et al., 1981; Thomas and Bastasz, 1981).

The use of ion implantation as a research tool will be illustrated with three recent examples, dealing with the behavior of interstitials in metals. ("Interstitial" refers to the lattice site in a solid solution at thermodynamic equilibrium.) The first study was concerned with nucleation of He bubbles in an fcc lattice initially free of defects. In the second example, the binding of H to He bubbles was measured in Fe, Ni and stainless steel. Finally, experiments will be discussed which demonstrate the role of C in stabilizing the amorphous phase in Fe-Ti-C alloys.

B. Nucleation of He Bubbles

Helium has negligible solubility in metals, but in certain materials applications it is introduced athermally. This occurs, for example, through ion implantation from thermonuclear plasmas, decay of tritium, and decay of neutron-excited heavy nuclei. When present the He may have several detrimental effects, including mechanical embrittlement and initiation of void swelling (Farrell, 1980; Ullmaier and Schilling, 1980) as well as trapping of H isotopes (see below). This has prompted a number of mechanistic studies of the behavior of He, all using ion implantation to introduce the solute. Noteworthy success has been achieved in characterizing diffusion (Thomas et al., 1979), trapping of He (Kornelsen and Van Gorkum, 1980), bubble nucleation at defects (Evans et al., 1981; Kornelsen and Van Gorkum, 1980), and bubble nucleation in a perfect lattice. This last investigation, by Thomas and Bastasz (1981), will not be described.

The specific questions of concern to Thomas and Bastasz were whether interstitial He in an fcc metal lattice can collect into a bubble in the absence of defect nucleation sites, and if so, how. Theoretical calculations of lattice energies in Ni as a function of atomic positions (Wilson et al., 1981) lead to the prediction that migrating He will preferentially occupy adjacent interstitial sites, causing clustering. It is further predicted that, when an interstitial cluster grows to five atoms, a host atom will be spontaneously ejected to an interstitial site and the He will collapse into the resulting vacancy, as shown in Figure 31. Thus begins a sequence of events leading ultimately to bubble formation.

The experiments of Thomas and Bastasz were performed on fcc Au, and consisted of He implantation at room temperature or below, followed by room temperature analysis with TEM. A key aspect was the use of He ion energies below the threshold for production of atomic displacements, which is about 0.45 keV. Since the material was well annealed and the concentration of thermal vacancies negligible, this produced a solution of He in an essentially perfect lattice. Figure 32 shows the final microstructure at room temperature following implantation of 0.3 keV He at 100K to a fluence of $1 \times 10^{16}/cm^2$. There is a dense array of small He bubbles with an average diameter of about 10Å, three of which are circled in the figure. The identification of these entities as bubbles is based on the change in their images from light to dark on going from underfocus to overfocus in the electron microscope, a property of cavity-like structures. The other important features in the micrographs are the half-moon images representing dislocation loops, which more detailed analysis showed to be of the interstitial type.

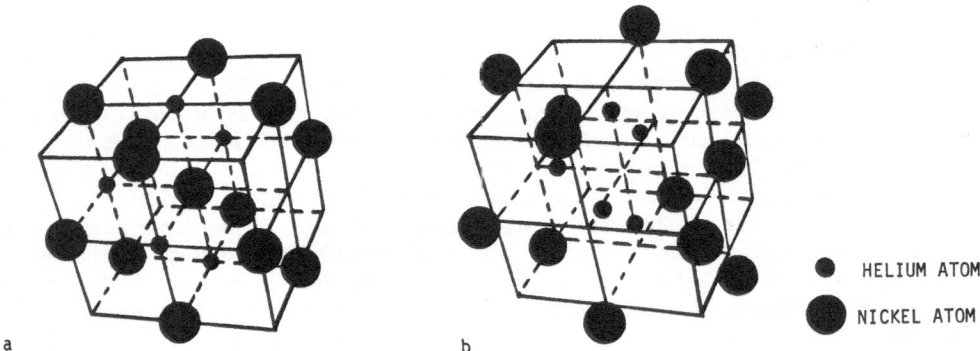

Figure 31   Predicted spontaneous formation of Ni interstitial in He interstitial cluster (Wilson, 1981).

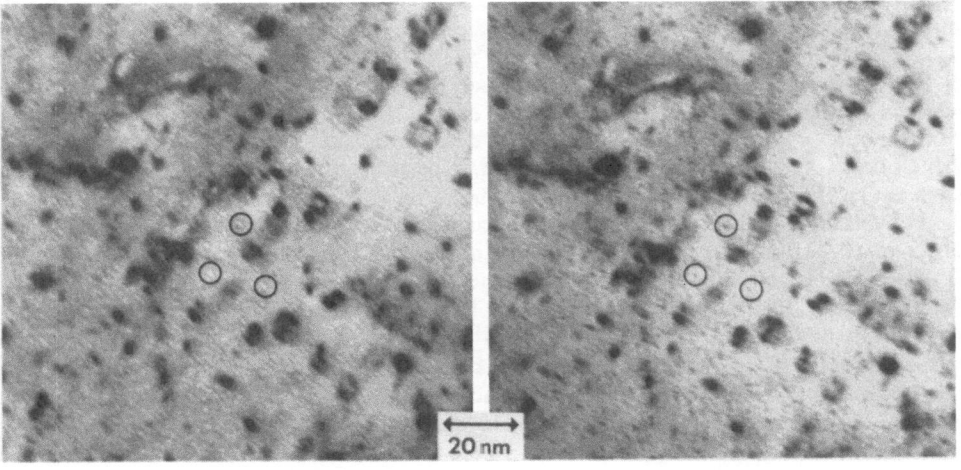

Neg. Defocus   Pos. Defocus

Figure 32   Transmission-electron micrographs of Au implanted with 0.3 keV He at 100K and then warmed to room temperature (Thomas, 1981).

Thomas and Bastasz interpreted the above result as direct evidence for the kind of process depicted in Figure 31, namely, He bubble formation from a perfect fcc lattice through spontaneous formation of host-atom interstitials. The observed dislocations are then presumed to result from the agglomeration of these interstitials. This conclusion was supported by additional experiments. Thus, when the He implantation was performed not at 100K but at room temperature, where interstitial He is believed to be mobile, neither bubbles nor dislocations were observed. Here the He is believed to have diffused to the surface during implantation and escaped so rapidly that the interstitial clusters leading to nucleation could not form. In another room temperature implantation, the He energy was increased to 1 keV, above the threshold for irradiation damage. Then, He bubbles and dislocation loops were both present. In this case irradiation defects are believed to have trapped the He and served as nucleation sites.

The above experiments and other work with implanted He suggest that bubble formation is a rather general phenomenon, occurring both at room and elevated temperatures, either in the presence or absence of lattice defects, and over a broad range of concentrations. The He is expected to remain dispersed only when it is immobilized by low temperatures or by traps whose density is comparable to that of the solute. This conclusion is important in dealing with the influence of He on various alloy properties. One such property is the transport and trapping of H, which has also been investigated extensively by ion implantation.

C. Hydrogen Trapping

F. Besenbacher, J. Bøttiger, S. M. Myers and W. R. Wampler have recently performed a series of experiments to characterize the interaction of H with ion-implanted He in Fe (Myers et al., 1981), Ni (Besenbacher et al., 1982), and austenitic stainless steel (Myers and Wampler, 1981). This interaction is of interest because it is expected to retard the recycling of H isotopes from fusion-reactor materials and, more generally, to influence H embrittlement in the presence of He. The implantation studies are exemplified by the work on Fe. In one typical experiment, a well annealed and electropolished specimen was initially implanted at room temperature with 4×10^{16} He/cm^2 at 15 keV and with 6×10^{16} He/cm^2 at 750 keV. This produced local concentrations of several at.% in two separate layers, centered respectively at depths of 0.07 μm and 1.2 μm. Such a two-layer configuration permitted trapping effects to be distinguished from the influence of the surface permeation barrier. The microstructure of the layer at 0.07 μm was examined by TEM of a similarly implanted foil (Follstaedt and Myers, 1982), and is shown in Figure 33. There is seen to be a dense array of He bubbles with an average diameter of about 10Å. The cavity-like nature of these entities is indicated by the change of their image from light to dark on going from underfocus to overfocus in the electron microscope. Bubbles are presumed to be present also in the deeper He layer at 1.2 μm.

After the introduction of He, the Fe sample was cooled to about 120K, and the deuterium isotope (D) was ion implanted into the near-surface He layer at 0.07 μm. The fluence was 1×10^{16} D/cm^2. Subsequently the temperature was ramped linearly upward at 2 K/min., and periodically the amount of D remaining within the near-surface He layer was measured using the ^{3}He-excited nuclear reaction ^{2}D(^{3}He,p)^{4}He. The resulting retention-versus-temperature data are given by open circles in Figure 34, where loss from the region near 0.07 μm is seen to occur in two distinct stages. Detailed depth profiling at selected temperatures shows that this behavior results from the combined effects of trapping and a permeation barrier at the Fe surface. Thus, at about 360K the D begins to escape from He-associated traps at 0.07 μm, and, prevented from leaving at the nearby surface by the barrier, it redistributes so as to occupy the He traps at 1.2 μm as well as those at 0.07 μm. This

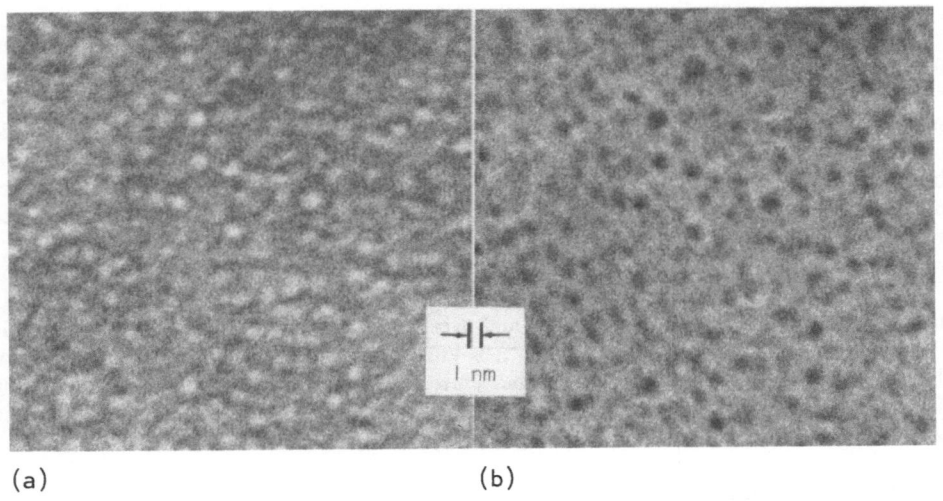

(a)   (b)

Figure 33   Transmission-electron micrographs of Fe implanted with He. (a) Underfocus. (b) Overfocus. (Myers, 1981a.)

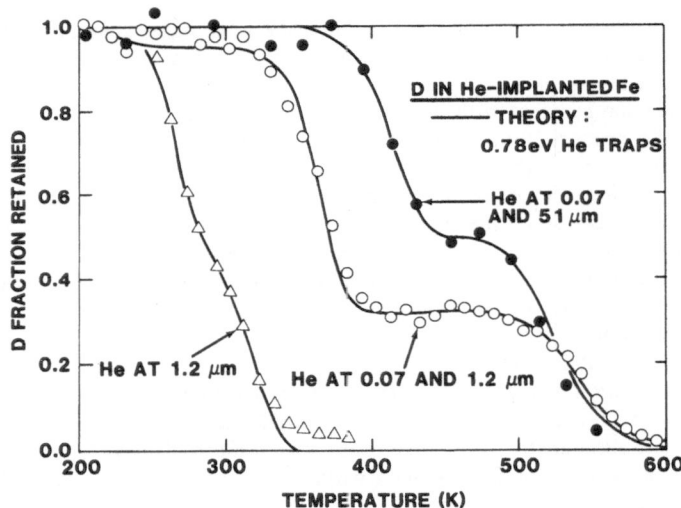

Figure 34   Retention of D within the near-surface implanted layer in Fe during linear ramping of temperature (Myers, 1981a).

constitutes the first of the observed stages. As the temperature rises further, the concentration of untrapped D in solution continually increases, leading eventually to observable transport across the surface permeation barrier. The resulting loss of D from the specimen produces the second stage, at 550K. (Calculations and other experiments indicate that the fraction of implanted D going into the underlying Fe matrix is negligible.)

The temperature of the first of the above stages depends upon the strength of the He-associated D traps, while the second stage is influenced by both trap strength and the effectiveness of the surface permeation barrier. The evolution is described quantitatively by solving the diffusion equation with appropriate trapping terms and with a boundary condition which takes account of the surface barrier (Myers et $al.$, 1981). The solid curve through the open circles in Figure 34 is the result of such a calculation, where the trap binding enthalpy Q_T and the surface recombination coefficient K_L are adjusted to produce agreement with experiment. (The parameter K_L has been discussed by Baskes (1980)). This procedure gives $Q_T \approx 0.78 \pm 0.08$ eV, referenced to a D solution site, and $K_L \approx 0.5 \times 10^{-18}$ cm^4/sec to within a factor of 5.

Two additional sets of data are shown in Figure 34. The open triangles are from an experiment where He was again implanted at 1.2 μm, but the near-surface layer contained only D and ion irradiation damage. The lower-temperature release exhibited by these data reflects migration from weaker, defect traps to the stronger He traps at 1.2 μm. The interpretation is confirmed by detailed depth profiling. Such behavior demonstrates the association of relatively deeper traps with the presence of He itself. Finally, the interpretation and analysis were sensitively tested by increasing the separation of the two He layers by a factor of about 50. This configuration was produced by implanting He into both sides of a 51 μm Fe foil. The results are given by the solid circles. As expected, the stage of redistribution between layers is shifted upward in temperature. The solid line through these data is from the model calculation, all parameters being determined by the earlier fit.

The D is believed to be bound to walls of He bubbles in Fe by a mechanism akin to chemisorption. Indeed, the measured trap strength of 0.78 eV is close to the enthalpy difference between solution and chemisorption sites derived from thermodynamic data, 0.73 eV. Similar mechanisms are proposed for Ni and austenitic (fcc) stainless steel, where the measured strengths of the He-associated traps are 0.55 and 0.42 eV respectively.

D. Amorphous Fe-Ti-C Alloys

The third and final example of fundamental alloy research was concerned with amorphous Fe-Ti-C mixtures, particularly the role of C in stabilizing the disordered state. This investigation, carried out by D. M. Follstaedt, J. A. Knapp, S. T. Picraux and I. L. Singer (Follstaedt et $al.$, 1980; Singer et $al.$, 1981), was motivated in part by observed improvements in the mechanical properties of several steels following implantation with Ti and C (Singer et $al.$, 1981; Yost et $al.$, 1981). The experiments entailed implantation of Ti and C into single-crystal Fe at room temperature, followed by analysis with TEM, ion backscattering/channeling, and AES.

The existence of an amorphous layer within the implanted Fe was demonstrated most directly by transmission-electron diffraction patterns such as that in Figure 35, where the diffuse rings characteristic of structural disorder are seen. In this particular instance, Ti was implanted at a sequence of energies ranging from 90 to 190 keV, and the total fluence of 2×10^{17}/cm^2 produced a concentration of approximately 20 at.% extending from the surface to a depth of about 700Å. Carbon was then implanted into the same region, the fluence again being 2×10^{17}/cm^2.

Ion-channeling analysis as a function of implantation treatment indicates that Ti and C must both be present to produce the amorphous phase at concentrations below about 30 at.%.

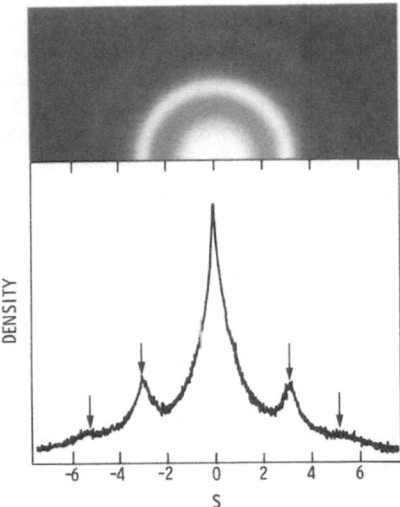

Figure 35   Transmission-electron diffraction pattern and corresponding plot of optical density for Fe implanted with Ti and C (Follstaedt, 1980).

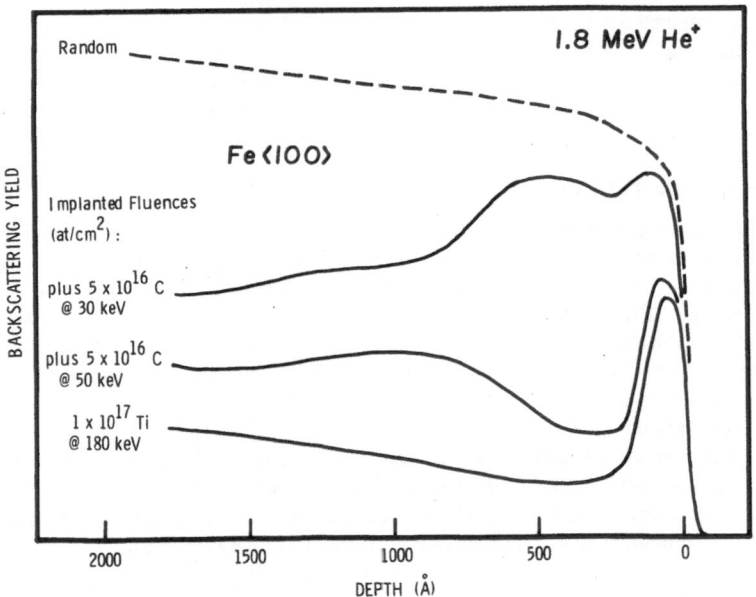

Figure 36   Channeled backscattering spectra for Fe implanted with Ti and then C (Knapp, 1981).

Representative channeling spectra during a sequence of implantations are given in Figure 36, where the horizontal energy scale has been converted to depth (Knapp *et al.*, 1981). Here the Fe crystal was implanted first with 1×10^{17}Ti/cm^2 at 180 keV, producing a concentration of approximately 10 at.% to about 700Å. At this point the implanted region remained mostly crystalline, as is evident from the large channeling dip along the <100> axis. The specimen was then implanted with 5×10^{16}C/cm^2 at 50 keV, an energy which placed the C deeper than the Ti. Once again crystallinity was maintained. The amorphous condition was finally produced by an additional implantation of 5×10^{16}C/cm^2 at 30 keV, so that the C and Ti depth profiles overlapped. The resulting disordered layer is seen, in Figure 36, to extend to about 700Å, conforming to the depth profile of the Ti.

Channeling studies such as these, in conjunction with depth profiling of the Ti and C by AES/sputtering (Singer, 1981), have established approximate composition bounds for the amorphous phase. Thus, with Ti concentrations up to 28 at.%, the concentration of C must be at least 7 at.%. Conversely, at high C densities the concentration of Ti can be reduced by 10 at.%, and perhaps as low as 3 at.%.

ACKNOWLEDGEMENTS

D. A. Rigney and P. Heilmann would like to acknowledge project support from the U.S. National Science Foundation, the Office of Naval Research, and the Army Research Office.

REFERENCES

Agarwal, S. B., Wang, Y-F., Clayton, C. R., Herman, H. and Hirvonen, J. K. (1979), Thin Solid Films *63*, 19.

Akano, U., Davies, J. A., Smeltzer, W. W., Tashlykov, I. S. and Thompson, D. A. (1981), Nucl. Instrum. Meth. *182/183*, 985.

Al-Saffar, A. H., Ashworth, V., Bairamov, A. K. O., Chivers, D. J., Grant, W. A. and Procter, R. P. M. (1980), Corros. Sci. *20*, 127.

Anttila, A., Hirvonen, J. and Hautala, M. (1981), Phys. Rev. B *23*, 1802.

Appleton, B. R., Kelly, E. J., White, C. W., Thompson, N. G. and Lichter, B. D. (1981a), Nucl. Instrum. Methods *182/183*, 991.

Appleton, B. R., Stritzker, B., White, C. W., Narajan, J., Fletcher, J., Meyer, O. and Lau, S. S. (1981b), in "Laser and Electron-Beam Solid Interactions and Materials Processing," (Gibbons *et al.*, eds.), Elsevier, North Holland.

Asami, K., Hashimoto, K., Masumoto, T. and Shimodaira, S. (1976), Corros. Sci. *16*, 909.

Ashworth, V., Grant, W. A., Procter, R. P. M. and Wellington, T. C. (1976a), Corros. Sci. *16*, 393.

Ashworth, V., Baxter, D., Grant, W. A. and Procter, R. P. M. (1976b), Corros. Sci. *16*, 775.

Ashworth, V., Baxter, D., Grant, W. A. and Procter, R. P. M. (1977), Corros. Sci. *17*, 947.

Ashworth, V., Procter, R. P. M. and Grant, W. A. (1980a), in "Ion Implantation" Vol. 18, Treatise on Materials Science and Technology (J. K. Hirvonen, ed.), Acad. Press, New York.

Ashworth, V., Grant, W. A., Mohammed, A. R. and Procter, R. P. M. (1980b), in "Ion Implantation Metallurgy," (C. M. Preece and J. K. Hirvonen, eds.), TMS-AIME, New York.

Ashworth, V., Grant, W. A. and Procter, R. P. M. (eds.) (1982), "Ion Implantation into Metals," to be published by Pergamon Press.

Baskes, M. I. (1980), J. Nucl. Mater. *92*, 318.

Benenson, R. E., Kaufmann, E. N., Miller, G. L. and Scholz, W. W. (eds.) (1981), "Ion Beam Modification of Materials," North-Holland, Amsterdam.

Besenbacher, F., Böttiger, J. and Myers, S. M. (1982), J. Appl. Phys. (in press).

Blaugher, R. D., Hein, R. E., Cox, J. E. and Waterstrat, R. M. (1969), J. Low Temp. Phys. *1*, 539.

Boes, N. and Zuchner, H. (1976), J. Less-Common Metals *49*, 223.

Bonora, P. L., Bassoli, M., Cerisola, G. and De Anna, P. L. (1981), Nucl. Instrum. Methods *182/183*, 1001.

Borders, J. A. and Poate, J. M. (1976), Phys. Rev. B *13*, 969.

Bowden, F. P. and Tabor, D. (1964), "The Friction and Lubrication of Solids II," Oxford.

Brown, W. L., (ed.) (1978), AVS Symp. on Ion Implantation - New Prospects for Mater. Modificat., J. Vac. Sci. Technol. *15*, 1629-1684.

Brown, B. S., Freyhardt, H. C. and Blewitt, T. H., (eds.) (1978), Proceedings of the International Discussion Meeting on Radiation Effects on Superconductivity, J. Nucl. Mat. *72*, 1-300.

Burbank, R. D., Dynes, R. C. and Poate, J. M. (1979), J. Low Temp. Phys. *36*, 573.

Burr, C. R., Bakhur, H. and Gibson, W. (1980), in "Ion Implantation Metallurgy," (C. M. Preece and J. K. Hirvonen, eds.), TMS-AIME, New York.

Cairns, J. A. (1980), Conf. Applications of Low Energy Accelerators, Denton, Texas, 1980.

Carosella, C. A., Singer, I. L., Bowers, R. C. and Gossett, C. R. (1980), in "Ion Implantation Metallurgy," (C. M. Preece and J. K. Hirvonen, eds.), TMS-AIME, New York.

Carter, G., Colligon, J. J. and Grant, W. A. (eds.) (1976), "Applications of Ion Beams to Materials, 1975," Conference Series Number 28, Institute of Physics, London.

Chan, K. W., Clayton, C. R. and Hirvonen, J. K. (1982), in "Corrosion of Metals Processed by Directed Energy Beams," (C. R. Clayton and C. M. Preece, eds.), TMS-AIME, New York (in press).

Charter, S. J. B., Thompson, L. R. and Dearnaley, G. (1981), Thin Solid Films *84*, 355-360.

Clayton, C. R., Doss, K. G. K., Herman, H., Prasad, S. and Wang, Y-F. (1980), in "Ion Implantation Metallurgy," (C. M. Preece and J. K. Hirvonen, eds.), TMS-AIME, New York.

Clayton, C. R. (1981), Nucl. Instrum. Methods *182/183*, 865-873.

Clayton, C. R., Doss, K. G. K., Hubler, G. K., Wang, Y-F. and Warren, J. B. (1982), in "Ion Implantation into Metals," (V. Ashworth, W. A. Grant and R. P. M. Procter, eds.), to be published by Pergamon Press.

Covino, B. S., Sartwell, B. D. and Needham, P. B. (1978a), J. Electrochem. Soc. *125*, 366.

Covino, B. S., Needham, P. B. and Conner, G. R. (1978b), J. Electrochem. Soc. *125*, 370.

De Anna, P. L., Cerisola, G., Bonora, P. L. *et al.* (1980), Werkstoffe Und Korrosion *31*, 783.

Dearnaley, G. (1978), Materials in Engineering Applications *1*, 28-41.

Dearnaley, G. (1980), in "Ion Implantation Metallurgy," (C. M. Preece and J. K. Hirvonen, eds.), TMS-AIME, NY, pp. 1-20.

Dearnaley, G. (1981a), Nucl. Instrum. Methods *182/183*, 899-914.

Dearnaley, G. (1981b), "Bombardment-Diffused Coatings and Ion Beam Mixing," Harwell (AERE) Report-R 10180.

Dearnaley, G. (1982), in "Ion Implantation into Metals," (V. Ashworth, W. A. Grant and R. P. M. Procter, eds.), to be published by Pergamon Press.

Dearnaley, G. and Charter, S. J. B. (1981), "Ion Implantation of Cemented Carbide Cutting and Forming Tools," International Conference on New Tool Materials-Metal Cutting and Forming (London, March 1981).

Dearnaley, G. and Goode, P. D. (1981), Nucl. Instrum. Methods *189*, 117-132.

Duckworth, R. G. and Wilson, I. L. (1979), Thin Solid Films *63*, 289-297.

Evans, J. H., Van Veen, A. and Caspers, L. M. (1981), Nature *291*, 310.

Farrell, K. (1980), Radiat. Eff. *53*, 175.

Ferber, H., Kasten, H., Wolf, G. K., Lorenz, W. J., Schweikert, H. and Folger, H. (1980), Corros. Sci. *20*, 117.

Fine, M. E. (1980), Metall. Trans. A, *11A*, 365.

Follstaedt, D. M., Knapp, J. A. and Picraux, S. T. (1980), Appl. Phys. Lett. *37*, 330.

Follstaedt, D. M. and Myers, S. M. (1982), unpublished.

Frech, G., Wolf, G. K., Kalbitzer, S., Damjantschitsch, H., unpublished.

Fromson, R. E. and Kossowsky, R. (1981), in Proceedings of the 1981 Materials Research Society Meeting, Boston, MA, North Holland (in press).

Gamo, K., Masuda, K., Namba, S., Ishihara, S. and Kimura, I. (1970), Appl. Phys. Lett. *17*, 391.

Geerk, J. (1980), Solid State Commun. *33*, 761.

Grabowski, K. S. and Rehn, L. E. (1982), in "Corrosion of Metals Processed by Directed Energy Beams," (C. R. Clayton and C. M. Preece, eds.), TMS-AIME, New York (in press).

Grant, W. A. (1981), Nucl. Instrum. Methods *182/183*, 809.

Hartley, N. E. W., Swindlehurst, W. E., Dearnaley, G. and Turner, J. F. (1973), J. Mater. Sci. *8*, 900-904.

Hartley, N. E. W. (1975), Tribology *8*, 65-72.

Hartley, N. E. W. (1976), Inst. Phys. Conf. Ser. *28*, 210-223.

Hartley, N. E. W. (1979), Thin Solid Films *64*, 177-190.

Hartley, N. E. W. (1980), in "Ion Implantation" (J. K. Hirvonen, ed.), Academic, New York, 321-371.

Hashimoto, K., Naka, M., Noguchi, J., Asami, K., Masumoto, T. (1978), in "Passivity of Metals" (R. P. Frankenthal and J. Kruger, eds.), J. Electrochem. Soc., Princeton, N.J.

Hayes, M., Kuhn, A. T. and Grant, W. (1978), J. Catal. *53*, 88.

Heilmann, P. and Rigney, D. A. (1981a), "Sliding Friction of Metals," Proceedings, Fundamentals of Friction, Leeds/Lyon Symposium, Sept. 9-12, 1980.

Heilmann, P. and Rigney, D. A. (1981b), Wear *72*, 195-217.

Heilmann, P. and Rigney, D. A. (1981c), "Running-In Processes Affecting Friction and Wear," Leeds/Lyon Symposium, Sept., 1981.

Herman, H., Hu, W. W., Clayton, C. R., Hirvonen, J. K., Kant, R. and MacCrone, R. K. (1979), IPAT-79 (London, 1978), p. 255.

Herman, H. (1981), Nucl. Instrum. Methods 182/183, 887-898.

Herman, H. (1982), in "Ion Implantation into Metals," (V. Ashworth, W. A. Grant and R. P. M. Procter, eds.), to be published by Pergamon Press.

Hirvonen, J. K., Carosella, C. A., Kant, R. A., Singer, I., Vardiman, R. and Rath, B. B. (1979), Thin Solid Films 63, 5-10.

Hirvonen, J. K. (ed.) (1980), "Ion Implantation," Academic Press, N.Y.

Hu, W. W., Clayton, C. R., Herman, H. and Hirvonen, J. K. (1978), Scripta Met. 12, 697.

Hu, W. W., Herman, H., Clayton, C. R., Kozubowski, J., Kant, R., Hirvonen, J. K. and MacCrone, R. K. (1980), in "Ion Implantation Metallurgy," (C. M. Preece and J. K. Hirvonen, eds.), TMS-AIME, New York.

Hubler, G. K. and McCafferty, E. (1980), Corros. Sci. 20, 103.

Hubler, G. K., Hirvonen, J. K., Clayton, C. R., Wang, Y. F., Budnick, J. and Hayden, H., (1981). In Naval Research Laboratory Memorandum Report 4527 (F. A. Smidt, ed.), June 24, 1981.

Hubler, G. K. (1982), in Proceedings of the 1981 Materials Research Society Meeting, Boston, MA, North Holland (in press).

Hubler, G. K., Trzaskoma, P. P., McCafferty, E. and Singer, I. L. (1982), in "Ion Implantation into Metals," (V. Ashworth, W. A. Grant and R. P. M. Procter, eds.), to be published by Pergamon Press.

Iwaki, M., Namba, S., Yoshida, K., Soda, N. et al. (1977), Japan J. Appl. Phys. 16, 1475.

Jager, W. and Roth, J. (1981), Nucl. Instrum. Methods 182/183, 975.

Jeffries, R., Bolster, R. and Singer, I. L. (1982), unpublished.

Kasten, H. and Wolf, G. K. (1980), Electrochem. Act. 25, 1581.

Kaufmann, E. N., Vianden, R., Chelikowsky, J. R. and Phillips, J. C. (1977), Phys. Rev. Lett. 39, 1671.

Knapp, J. A., Picraux, S. T. and Follstaedt, D. M. (1981), unpublished.

Kocanda, S. (1978), "Fatigue Failure of Metals," (Sijthoff and Noordhoff, Leiden, 1978).

Kornelsen, E. V. and Van Gorkum, A. A. (1980), J. Nucl. Mater. 92, 79.

Kossowsky, R., Fromson, R. E. and Ecer, G. M. (1981), as reported at the Modification of the Surface Properties of Metals by Ion Implantation, Manchester, England, June 1981.

Kujore, A., Chakrabortty, S. B., Starke, E. A., Jr. and Legg, K. O. (1980), in "Ion Implantation Metallurgy," (C. M. Preece and J. K. Hirvonen, eds.), TMS-AIME, New York.

Kujore, A., Chakrabortty and Starke, E. A. (1981), Nucl. Instrum. Methods 182/183, 949-958.

Kwo, J., Hammond, R. H., and Geballe, T. H. (1980), J. Appl. Phys. 51, 1726.

Lekhtvar, I.Ya., Ivanov, L. I., Karlov, N. V., KuzMin, G. P., Nishchenko, M. M., Prokhorov, A. M., Rykalin, N. N. and Yanushkevich, V. A. (1976), Sov. J. Quant. Electron. 6, 460.

Lindhard, J., Scharff, M. and Schiott, H. E. (1963), K. Dan. Vidensk. Selsk. Mat. Fys. Medd. 33, 14.

Longworth, G. and Hartley, N. E. W. (1977), Thin Solid Films 48, 95-104.

LoRusso, S., Mazzoldi, P., Scotoni, I., Tosello, C. and Tosto, S. (1979), Appl. Phys. Lett. 34, 627.

LoRusso, S., Mazzoldi, P., Scotoni, I., Tosello, C. and Tosto, S. (1980), Appl. Phys. Lett. 36, 822.

LoRusso, S., Mazzoldi, P., Scotoni, I. and Zhang, G. L. (1982), in publication.

Matzke, H., Turnos, A. and Rabette, P. (1981), Proc. Intern. Conf. Radiation Effects in Insulators, Arco, Italy, July 1981.

Mayer, J. W., Tsaur, B. Y., Lau, S. S. and Hung, L-S (1981), Nucl. Instrum. Methods 182/183, 1.

McCafferty, E. and Hubler, G. K. (1978), J. Electrochem. Soc. 128, 1892.

McLean, M. (1980), Conference on Interface Phenomena, N.P.L. London (unpublished).

Meyer, O. and Seeber, B. (1977), Solid State Commun. 22, 603.

Meyer, O. (1980), in "Ion Implantation," Vol. 18 - Treatise on Materials Science and Technology (J. K. Hirvonen, ed.) Acad. Press., New York.

Meyer, O., Kaufmann, R., Appleton, B. R. and Chang, Y. K. (1981), Solid State Commun. 39, 825.

Meyer, O., Thompson, J., Appleton, B. R. and White, C. W. (1982), unpublished work.

Misra, M. S. and Kustas, F. M., in "Corrosion of Metals Processed by Directed Energy Beams," (C. R. Clayton and C. M. Preece, eds.), TMS-AIME, New York (in press).

Myers, S. M. and Follstaedt, D. M. (1981), J. Appl. Phys. 52, 4007.

Myers, S. M. and Rack, H. J. (1978), J. Appl. Phys. 49, 3246.

Myers, S. M. and Smugeresky, J. E. (1977), Metall. Trans. A 8, 609.

Myers, S. M. and Smugeresky, J. E. (1978), Metall. Trans. A 9, 1789.

Myers, S. M. and Wampler, W. R. (1981), unpublished.

Myers, S. M., Picraux, S. T. and Prevender, T. S. (1974), Phys. Rev. B 9, 3953.

Myers, S. M., Picraux, S. T. and Stoltz, R. E. (1979), J. Appl. Phys. *50*, 5710.

Myers, S. M., Follstaedt, D. M. and Rack, H. J. (1980a), Metall. Trans. A *11*, 1465.

Myers, S. M., Picraux, S. T. and Stoltz, R. E. (1980b), Appl. Phys. Lett. *37*, 168.

Myers, S. M., Besenbacher, F. and Bøttiger, J. (1981), Appl. Phys. Lett. *39*, 450, unpublished.

Norose, S. and Sasada, T. (1980), J. Japan Soc. Lubr. Engins., Int'l. Ed. *1*, 5-9.

O'Grady, W. E. and Wolf, G. K. (1981), Proc. Electrochem. Soc., Minneapolis, May 1981.

Okamoto, P. R. and Wiedersich, H. (1976), J. Nucl. Mater. *53*, 336.

Pfluger, J. and Meyer, O. (1979), Solid State Commun. *32*, 1143.

Picraux, S. T., EerNisse, E. P. and Vook, F. L. (eds.) (1974), "Application of Ion Beams to Metals," Plenum Press, New York.

Picraux, S. T. (1981), Nucl. Instrum. Methods *182/183*, 413.

Poate, J. M. (1978), J. Vac. Sci. Technol. *15*, 1636.

Poate, J. M. and Cullis, A. G. (1980), in "Ion Implantation" Vol. 18 - Treatise on Materials Science and Technology (J. K. Hirvonen, ed.), Acad. Press, New York.

Pranevicius, L. (1979), Thin Solid Films *63*, 77-85.

Preece, C. M. and Hirvonen, J. K. (eds.) (1980), "Ion Implantation Metallurgy," TMS-AIME, New York, NY.

Preece, C. M., Kaufmann, E. N., Staudinger, A., and Buene, L. (1980), "Ion Implantation Metallurgy," (C. M. Preece and J. K. Hirvonen, eds.), TMS-AIME, New York.

Rabette, P., Deane, A. M., Tench, A. J., Che, M. (1979), Chem. Phys. Letters *60*, S. 348.

Reintsema, S. R., Verbiest, E., Odeurs, J. and Pattyn, H. (1979), J. Phys. F. *9*, 1511.

Rice, S. L., Nowotny, H. and Wayne, S. F. (1981), Trans. ASLE *24* (2), 264-268.

Rosenfield, A. (1981), in "Fundamentals of Friction and Wear of Materials," (D. A. Rigney, ed.), ASM 1980 Materials Science Seminar, publ. ASM, 1981.

Rowson, D. M. and Quinn, T. F. J. (1980), J. Phys. D. *13*, 209.

Sartwell, B. D., Walter, R. P., Wheeler, N. S. and Brown, C. R. (1982), in "Corrosion of Metals Processed by Directed Energy Beams," (C. R. Clayton and C. M. Preece, eds.), TMS-AIME, New York (in press).

Samuels, L. E. (1978), Sci. Am., Nov., 1978, 132-152.

Schmiedel, H. and Wolf, G. K. (1981), unpublished results.

Schneider, U., Linker, G. and Meyer, O. (1982), J. Low Temp. Phys.

Sevillano, Van Houtte, P. and Aernoudt, E. (1980), Progress in Materials Science *25*, Pergamon.

Singer, I. L. and Bolster, R. A. (1980), in "Ion Implantation Metallurgy," (J. K. Hirvonen and C. M. Preece, eds.), TMS-AIME, New York.

Singer, I. L. (1981), unpublished.

Singer, I. L., Carosella, C. A. and Reed, J. R. (1981), Nucl. Instrum. Methods *182/183*, 923-932.

Smidt, F. A., Hirvonen, J. K. and Ramalingam, S. (1981), Naval Research Laboratory Memorandum Report 4616, 25 Sept. 1981.

Suh, N. P. (1977), Wear *44*, 1-16.

Spooner, S. and Legg, K. (1980), in "Ion Implantation Metallurgy," (C. M. Preece and J. K. Hirvonen, eds.), TMS-AIME, New York.

Swanson, M. L., Howe, L. M. and Quenneville, A. F. (1978), J. Nucl. Mater. *68/70*, 372.

Swanson, M. L., Howe, L. M. and Quenneville, A. F. (1980), Phys. Rev. B *22*, 2213.

Sweedler, A. R. and Cox, D. E. (1975), Phys. Rev. B *12*, 147.

Takahashi, K., Okabe, Y. and Iwaki, M. (1981), Nucl. Instrum. Methods *182/183*, 1009.

Testardi, L. R., Poate, J. M., Weber, W., Augustyniak, W. M., Barrett, J. H. (1977a), Phys. Rev. Lett. *39*, 716.

Testardi, L. R., Wakiyama, T. and Royer, W. A. (1977b), J. of Appl. Phys. *48*, 2055.

Thome, L., Bernas, H. and Cohen, C. (1979), Phys. Rev. B *20*, 1789.

Thomas, G. J. and Bastasz, R. (1981), J. Appl. Phys. *52*, 6426.

Thomas, G. J., Swansiger, W. A. and Baskes, M. I. (1979), J. Appl. Phys., 6942.

Thompson, N. G., Lichter, B. D., Appleton, B. R., Kelly, E. J. and White, C. W. (1980), in "Ion Implantation Metallurgy," (C. M. Preece and J. K. Hirvonen, eds.), TMS-AIME, New York.

Ullmaier, H. and Schilling, W. (1980), in Physics of Modern Materials, International Atomic Energy Agency, Vienna, pp. 301-397.

Valori, R. and Hubler, G. K. (1981), in Naval Research Laboratory Memorandum Report 4527 (F. Smidt, ed.), June 24, 1981.

Vardiman, R. G., Bolster, R. N. and Singer, I. L. (1981), in Proceedings of the 1981 Materials Research Society Meeting, Boston, MA, North Holland (in press).

Vardiman, R. G. and Kant, R. (1982), J. Appl. Phys. *53*, 690.

Voinov, M., Buhler, D., Tannenberger, H. (1974), Proc. Electrochem. Soc., San Francisco.

Wang, Y. F., Clayton, C. R., Hubler, G. K., Lucke, W. H. and Hirvonen, J. K. (1979), Thin Solid Films *63*, 11.

Weissmantel, C., Reisse, G., Erler, H-J., Henny, F., Bewilogua, K., Ebersbach, U. and Schurer, C. (1979), Thin Solid Films *63*, 315-325.

Whitton, J. L., Grant, W. A. and Williams, J. L. (1978), Proc. Int. Conf. on Ion Beam Modification of Materials, Budapest, Hungary, 1981.

Wilson, W. D., Bisson, C. L. and Baskes, M. I. (1981), Phys. Rev. B *24*, 5616.

Winterbon, K. Bruce (1975), "Ion Implantation Range and Energy Distribution," Vol. 2, Plenum Press, New York.

Wolf, G. K. (1981a), Nucl. Instr. Meth. *182/183*, 875-885.

Wolf, G. K. (1981b), Proc. Intern. Conf. Radiation Effects in Insulators, Arco, Italy, July 1981.

Wolf, G. K. (1982), Chemie Ingenieur Technik (in press).

Wolf, G. K. and Kasten, H., unpublished results.

Yost, F. G., Pope, L. E., Follstaedt, D. M., Knapp, J. A. and Picraux, S. T. (1982), in Proceedings of the 1981 Materials Research Society Meeting, Boston, MA. North Holland (in press).

Zamanzadeh, M., Allam, A., Pickering, H. W. and Hubler, G. K. (1980), J. Electrochem. Soc. *127*, No. 8, 1688-1693.

CHAPTER 13

LASER SURFACE ALLOYING

C. W. DRAPER

Western Electric Engineering Research Center, Princeton, New Jersey

J. M. POATE

Bell Laboratories, Murray Hill, New Jersey

CONTRIBUTORS: L. Buene and D. C. Jacobson

I. INTRODUCTION

In this chapter we examine the alloying of films on metallic or semiconductor substrates by laser heating. Following a brief introduction we will show through the use of selected examples how extended solid solutions and amorphous phases can be made and point out both the thermodynamic and thermophysical limitations that constrain this new surface metallurgy. Selected examples from the ferrous, nonferrous and silicide work will be reviewed.

Figure 1 is a schematic representation of a film and substrate combination and an isolated laser event. The symbols denote the various thermophysical and laser parameters which influence the surface alloying. The symbols D, T and P^0 stand for the thermal diffusivity, melting point and vapor pressure of either the substrate (subscript B) or thin film (subscript A). The role some of them play will be discussed in greater detail below. The focussed laser radiation is characterized by the pulse duration, t_p, the incident energy density, I, and the optical spot size, d_{opt}. As pointed out in Chapter 2, III.A the reflectance (R_0) of most metals at many laser wavelengths is rather high and the absorbed energy density may be very much less than the incident flux. The absorption depths are very small, generally <100Å, and so for most film and substrate combinations it is the optical properties of the film which predominate. The collisional de-excitation of the conduction electrons with the lattice atoms on the time scale of 10^{-12} sec means that the heat flow problem may be treated from the viewpoint of an instantaneous surface source. The time associated with the inward and outward movement of the liquid-solid interface will be comparable with the pulse duration. It should be noted that the melt depth and effective spot size, d_{eff}, have not been drawn to scale in Figure 1. For laser-alloying with pulse lengths <200 nsec the melt depth will be <10,000Å and the melt spot width will be of the order of 10 μm to 1 cm depending on the exact laser source. In Figure 1 the film A has been predeposited (evaporated, sputtered, plated, or painted) which is the most common practice, although codeposition-irradiation (particle or wire injection) is also used. The goal of surface alloying is to controllably incorporate the film element(s) into the substrate B by melting A and some portion of B. Mixing in the liquid state will be followed by a rapid recrystallization. The alloying element(s) will thus be incorporated in the near surface region of the substrate.

Two approaches to material development are being pursued in this area. The first approach is to make use of the rapid quench rates to fabricate novel metastable alloys. The second is the making of "equivalent" surface alloys such as Cr, Co and Ni in Fe and Pd in Ti where the alloying elements are costly and of strategic importance. It makes sense in conservation and financial terms to consider surface alloying because in many of those applications the alloying elements are added to the bulk to affect the surface properties. The dependence (Swager, 1980) of the U.S. and the European community on such metal imports is very extensive as can be seen in Figure 2.

II. METALLURGY

A. Thermodynamic Constraints

In this section we want to examine qualitatively the thermodynamic and thermophysical constraints which control the surface alloying. Table I is a compilation from the published literature of all the binary systems which have been laser surface alloyed. They have been grouped into three categories according to loosely applied thermodynamic considerations. These groupings are i. systems exhibiting both liquid and solid state solubility, ii. systems with liquid state solubility but limited or nonexistent solid state solubility and iii. systems with liquid state immiscibility. Complete solid solubility is displayed, for example, by the Pd-Ni system. Figure 3 shows Rutherford backscattering and channeling spectra of the surface alloying of 1150Å Pd films on single crystal Ni (Buene *et al.*, 1980a). Melt times are on the order of 100 μ sec and the films were melted using a scanning CW CO_2 laser. The Pd has been alloyed from the surface to depths greater than one μm with surface concentrations of 12 at.%. Comparison of the random and <110> spectra indicates that the Pd is substitutional. However, the magnitude of the dechanneling shows that considerable strain or damage has been introduced during the laser processing. It should be emphasized that such concentrations of metal could not be introduced to such depths by ion implantation and ion beam mixing using current technology.

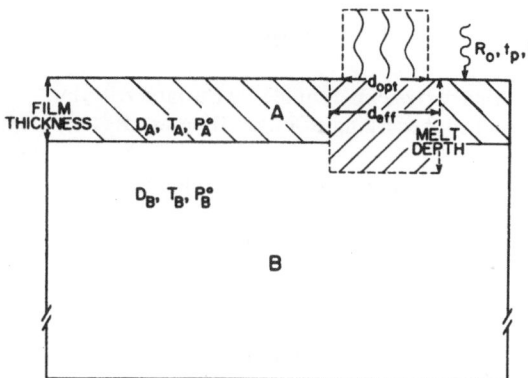

Figure 1   Surface alloying schematic. A thin film (A) is predeposited on substrate (B). The thermophysical properties of importance are annotated (thermal diffusivity, D; melting point, T; and vapor pressure, P^0). An isolated laser event is characterized by the incident the pulse length, t_p, and the normal spectral reflectance, R_0. Controlled surface melting (without surface vaporization) results in a molten film and a partially melted substrate.

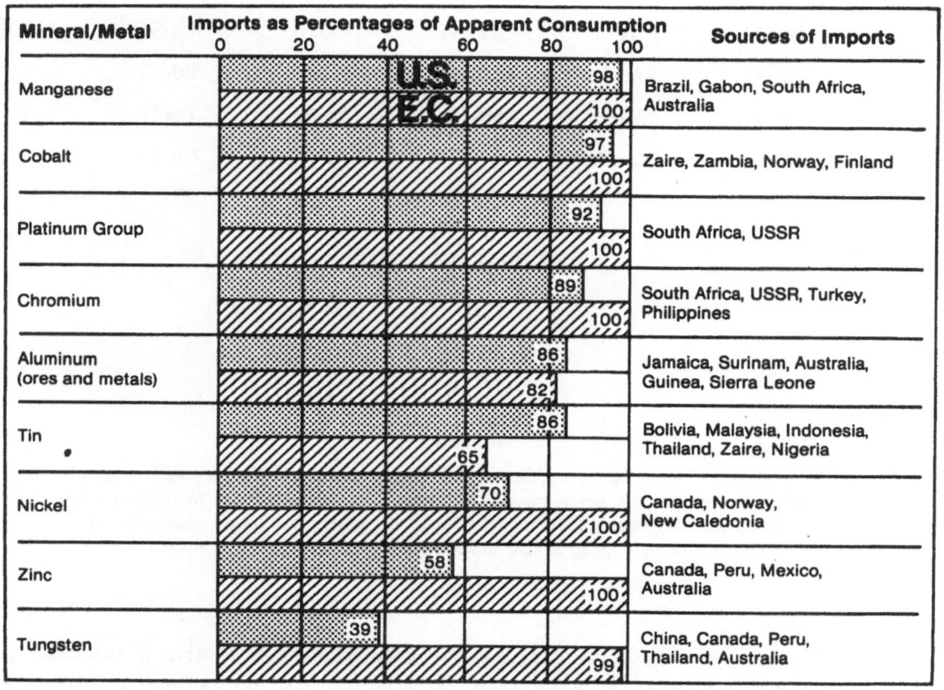

Mineral/Metal	Imports as Percentages of Apparent Consumption	Sources of Imports
Manganese	U.S. 98 / E.C. 100	Brazil, Gabon, South Africa, Australia
Cobalt	U.S. 97 / E.C. 100	Zaire, Zambia, Norway, Finland
Platinum Group	U.S. 92 / E.C. 100	South Africa, USSR
Chromium	U.S. 89 / E.C. 100	South Africa, USSR, Turkey, Philippines
Aluminum (ores and metals)	U.S. 86 / E.C. 82	Jamaica, Surinam, Australia, Guinea, Sierra Leone
Tin	U.S. 86 / E.C. 65	Bolivia, Malaysia, Indonesia, Thailand, Zaire, Nigeria
Nickel	U.S. 70 / E.C. 100	Canada, Norway, New Caledonia
Zinc	U.S. 58 / E.C. 100	Canada, Peru, Mexico, Australia
Tungsten	U.S. 39 / E.C. 99	China, Canada, Peru, Thailand, Australia

Figure 2   Bar graph representation of United States and European Community (EC) dependence on foreign imports for commercially important element groups. The numbers represent percent imported relative to amount consumed.

TABLE I

Binary Systems Studied by Surface Alloying*
Grouped According to Phase Diagram Restrictions

Mutual Solid State Solubility

Cr-Fe

V-Fe

Pd-Ni

Au-Pd

Zr-Ti

Limited or Nonexistent Solid State Solubility

Cu-Ag	Si-Al	C-Fe	Ni-Nb	Zr-Ni	Rh-Si	Co-W
Cr-Al	Sn-Al	Mo-Fe	Au-Ni	Co-Si	Au-Sn	Cu-Zr
Cu-Al	Zn-Al	Nb-Fe	Eu-Ni	Nb-Si	Pd-Ti	
Mo-Al	Zr-Al	W-Fe	Hf-Ni	Ni-Si	Pt-Ti	
Ni-Al	Ni-Be	Zr-Fe	Sn-Ni	Pd-Si	Sn-Ti	
Sb-Al	Co-Cu	Al-Nb	Ta-Ni	Pt-Si	Zr-V	

Liquid State Immiscibility

Cd-Al	Cu-Mo
Pb-Al	Ag-Ni
Cr-Cu	Au-Ru
Pb-Cu	

* References for most of the metal systems can be found in the recent laser surface alloying bibliography (Draper, 1981b). The silicides have been extensively covered in the recent proceedings of the Materials Research Society (Ferris *et al.*, 1979, White and Peercy, 1980 and Gibbons *et al.*, 1981). The binary systems have been arranged within phase diagram restriction groupings alphabetically by substrate.

The equilibrium phase diagram of the Au-Ni system (Figure 4) exhibits a miscibility gap in the solid phase but complete liquid phase miscibility. We have examined this system because it should be a good test case for surface alloying for quenching metastable solid solutions from the melt. Figure 5 shows Rutherford backscattering and channeling spectra of a 250Å Au film melted on single crystal Ni using a Q-switched frequency-doubled Nd-YAG laser. The use of very short 100 nsec laser pulses results in melt times of about 500 nsec. A

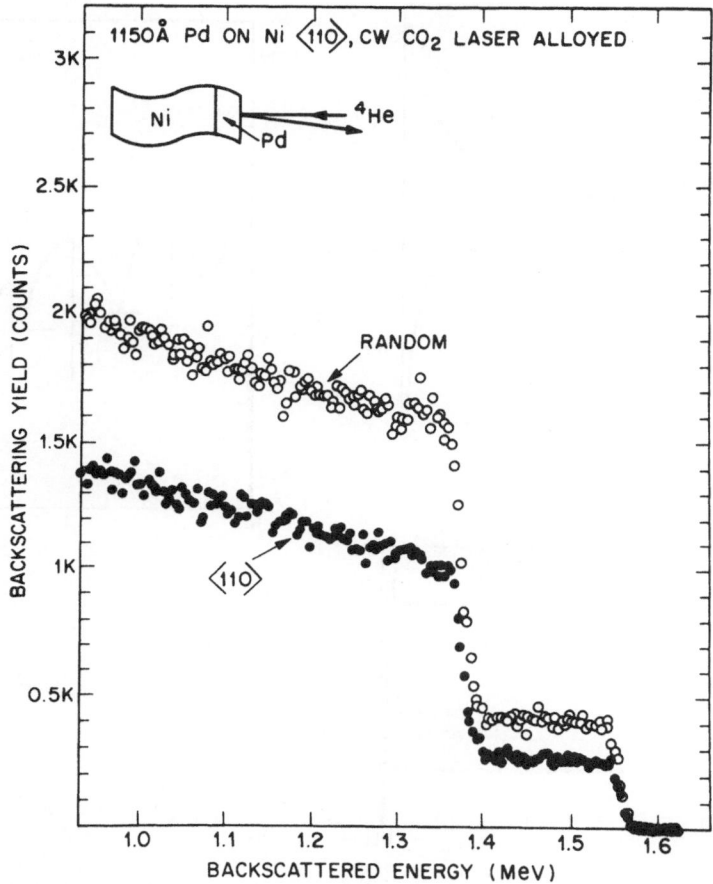

Figure 3   Rutherford backscattering and <110> channeling spectra for 1150Å of Pd on single crystal <110> Ni
which has been laser alloyed with a continuous wave CO_2 laser. The average laser power was 500W and
the dwell time was of the order of 10 μsec. The Pd has been alloyed to depths over 10,000Å and the
surface concentration is 12 at.% Pd. The channeling results show that the Pd is substitutional.

substitutional solid solution has been formed with a maximum surface concentration of
6 at.%. These spectra are similar to those obtained from very high dose implantation in
metals (Poate and Cullis, 1980).

A film-substrate combination which offers one of the worst thermodynamic obstacles is
that of Ag-Ni whose equilibrium phase diagram is shown in Figure 4. There is liquid
immiscibility and very limited solid solubility. Figure 6 shows Rutherford backscattering
spectra of the mixing of a Ag film on Ni using a CW CO_2 laser. The incident power density
was 5 MW/cm^2 and the dwell time was 10 μ sec. The temperature of the melt has,
presumably, been sufficiently high to permit liquid intermixing. However, on quenching
virtually all the Ag has phase separated or, very possibly, never gone into solution. It is
possible to alloy very limited amounts of Ag into Ni by pulse melting thin layers as shown in
Figure 7. Thin films (10Å) of Ag were deposited on single crystal Ni and irradiated with a

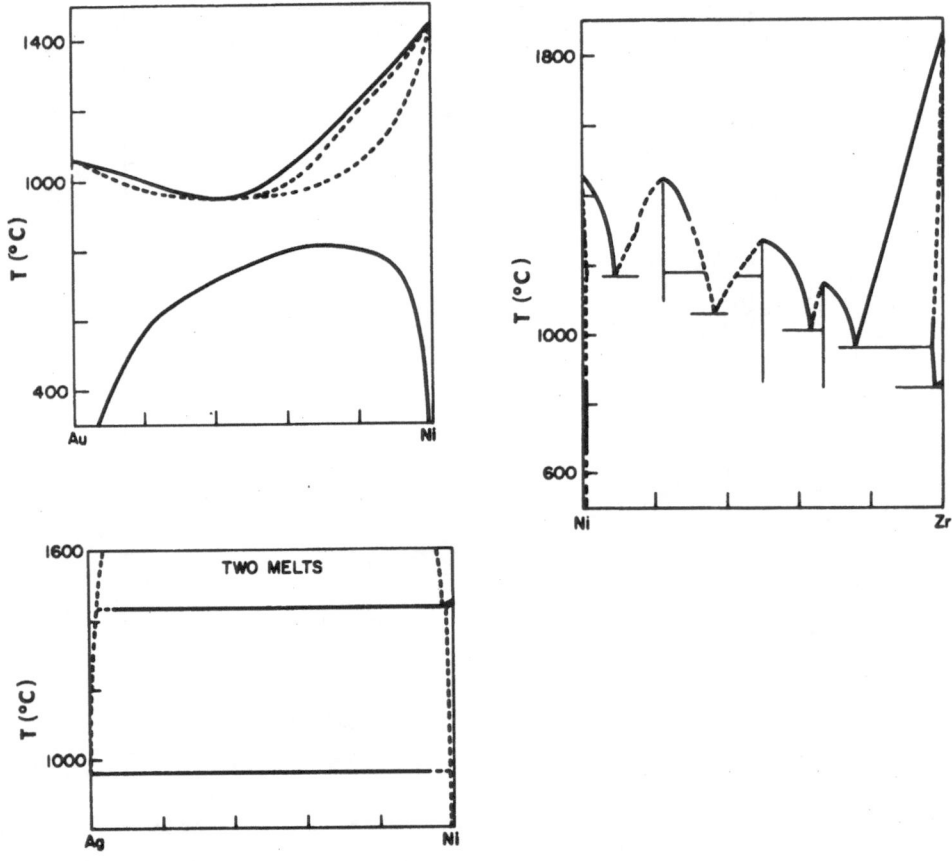

Figure 4   Phase diagrams for Au-Ni (miscibility gap in the solid), Ag-Ni (miscibility gap in the liquid) and Zr-Ni (numerous deep eutectics and intermetallic compounds).

frequency-doubled-Q-switched Nd-YAG laser. The irradiation has damaged the host lattice but the Ag has been incorporated on lattice sites at a concentration of 2 at.%. The melt time for the incorporation of Ag on lattice sites is estimated to be ~200 nsec. From these experiments we can make the following observation. If the film and substrate will not homogenize in the liquid state there is little chance for producing a single phase alloyed region. If (i) the film is thin enough (ii) the melt temperature high enough, and (iii) the time in the melt (at high temperatures) long enough a single phase liquid may result. The extent and shape of the liquid-miscibility gap will dictate whether or not a single phase liquid region is transgressed during the temperature excursion. Even if a homogeneous liquid is produced solute precipitation in the liquid ahead of the resolidification interface may occur. Both the melt time and solute concentration in the liquid are important factors in determining whether second phase nucleation or solute transport will interfere with the solute trapping required to make a metastable solid solution. This has been detailed in chapter eleven for the case of Sb implanted Al and then irradiated on various time scales. For a given concentration of Sb implanted into Al only the shorter melt times resulted in solid solutions.

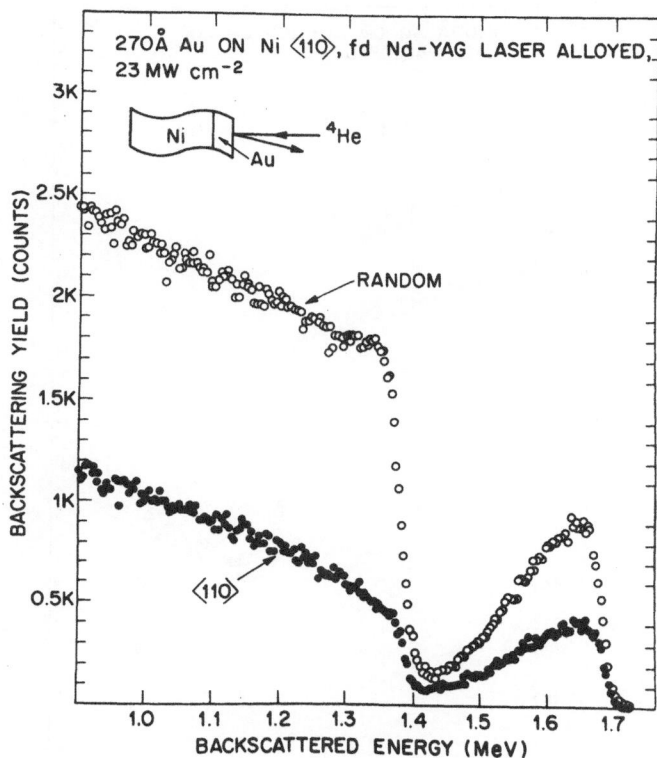

Figure 5   Rutherford backscattering and channeling spectra for 270Å of Au on single crystal <110> Ni which
has been laser alloyed with a Q-switched frequency doubled Nd-YAG laser. The short melt times
(hundreds of nsec) result in diffusion limited Au redistribution. In this case the peak Au concentration is
6 at.% and alloying occurs to a depth of 2500Å. The channeling results show that the Au is highly
substitutional.

In at least one example liquid state immiscibility has been used advantageously. This
system is Au-Ru. In addition to the thermodynamic restriction imposed by the liquid state
miscibility gap there are thermophysical obstacles associated with the much higher melting
point and heat of fusion for the Ru substrate. When the liquid Au interface reaches the solid
Ru it is very effectively quenched (Draper *et al.*, 1981a). There is insufficient energy
available in the molten Au to overcome the large temperature increase and heat of fusion
required for Ru phase transformation. The result is a small quantity of intermixing at the
interface with negligible dilution of the Au throughout most of the quenched film. From the
viewpoint of thin film technology one now has a substrate plus film structure where the film
has been quenched from the liquid as opposed to vapor or sputter deposited and one has
increased the adhesive nature of the film to the substrate. Both of these factors make laser
quenched Au-Ru an attractive material system for development as metal mirrors for high
power laser applications.

The binary systems of the intermediate group, those with a single phase above the liquidus
but two or more phase regimes below it, are the systems receiving considerable attention. It
is this group where laser or electron beam surface alloying offers the greatest opportunity for

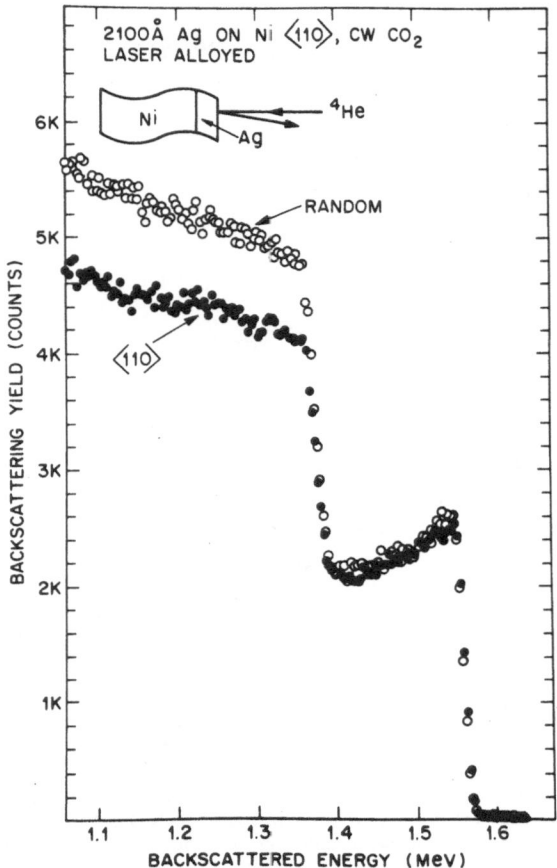

Figure 6   Rutherford backscattering and channeling spectra for 2100Å of Ag on <110> Ni which has been laser alloyed with a CW CO_2 laser. As in the case of Pd on Ni in Figure 3, the longer continuous laser dwell times have resulted in deep intermixing; however, in contrast to the Pd case, the Ag is nonsubstitutional.

development of new metal alloys. Conventional metal forming techniques simply cannot provide these quench rates from the liquid. While the more established rapid solidification techniques, particularly those developed for producing glassy metals, can achieve very high cooling rates, they produce samples of severely limited physical dimensions. Laser or electron beam scanning may offer the best way to produce these surfaces over a relatively large area. Specific examples from this grouping will be discussed below in greater detail.

B. Thermophysical Constraints

The ratio between deposited film thickness and induced melt depth will most strongly influence the final alloy concentration and profile for systems with liquid state miscibility. The pulse duration or dwell time and the incident power density of the laser source will be important processing variables as we have seen in previous Ni examples.

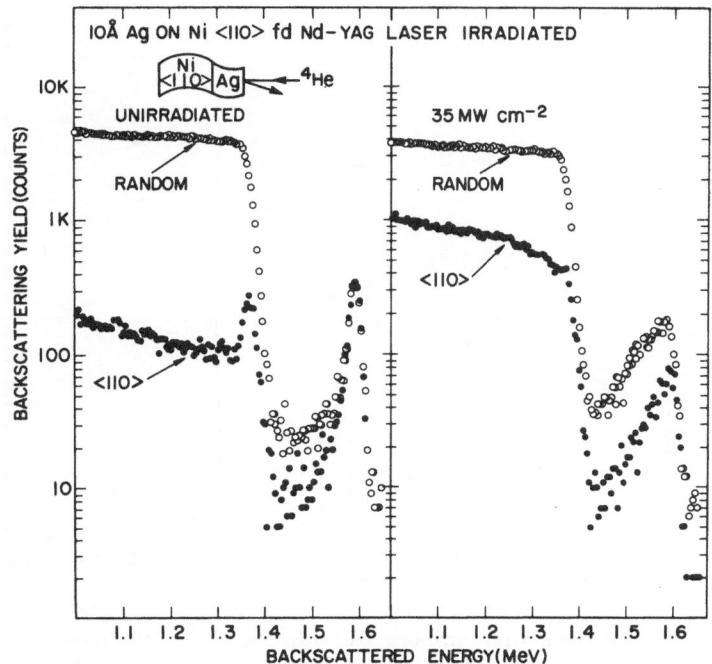

Figure 7   Rutherford backscattering and channeling spectra for ~10Å of Ag on <110> Ni which has been laser alloyed with a Q-switched frequency doubled Nd-YAG laser. Note that the use of very thin Ag films and Q-switched laser pulses results in surface alloying of Ag in Ni which is similar to that of Au on Ni (Figure 5) and contrasts that of thick Ag (Figure 6).

Calculations by Hsu *et al.* (1978) of melt depths vs. total time for several absorbed heat fluxes are shown in Figure 8. The arrows show times at which the surface reaches its vaporization temperature. Clearly at higher power densities (i.e., shorter pulse lengths) the melt depths are constrained by surface vaporization. Some of the other thermophysical properties included in the schematic of Figure 1 may also constrain the surface alloying. Care needs to be taken to choose A-B combinations without large differences in melting points and vapor pressures. It would be difficult, for example, to surface alloy a thin film of low melting point-high vapor pressure Zn onto high melting point Ta or W. The thermal diffusivity can also play a role, influencing the thermal diffusion length. This can be particularly important if the film thickness approaches the dimensions of the heat flow. For example, in a series of laser alloying experiments described in Chapter 2, III.C on Ni(110) + Au(x) + 200Å Ni, where x — 500, 1000, 2000 and 5000Å, it has been found that for a constant laser fluence the melt depth increases significantly with increasing Au thickness. This behavior can be rationalized on the basis of an effective thermal diffusivity which for thin Au is approximately that of Ni, while for thick Au is more closely that of Au, and 5-10 times the value for Ni. Another example, which has been presented in Figure 19 of Chapter 2, comes from experiments on a copper alloy strip material with a bimetallic inlay of Ni and a Pd_xAg_{1-x} alloy. The inlay was subsequently vapor deposited with 1600Å Au and scanned with a Q-switched frequency-doubled Nd-YAG laser source. With increasing Ag content in the Pd-Ag layer the effective thermal diffusivity is increasing, thus producing a greater dilution of the Au.

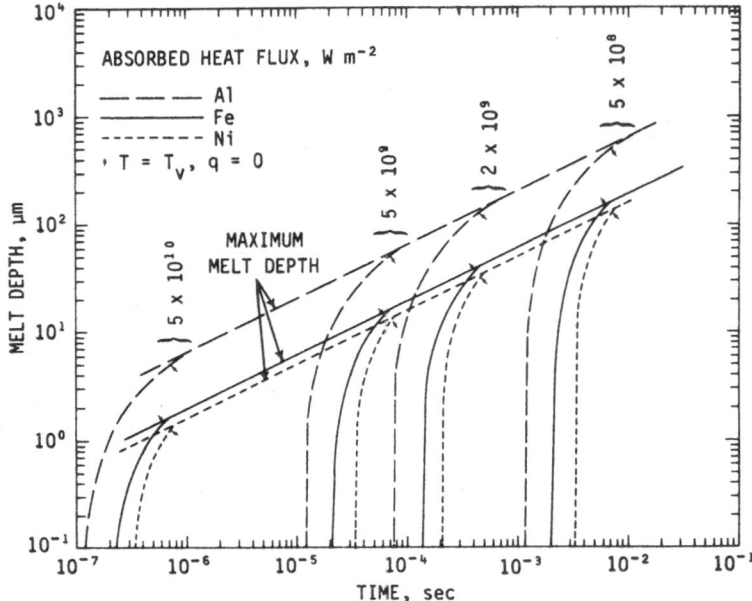

Figure 8   Calculated melt depths versus irradiation time from Hsu *et al.* (1978) based on a one-dimensional computer heat flow model. Results are shown for three common substrate elements - Al, Fe and Ni. The maximum melt depth corresponds to melting without surface vaporization. Note that for 10^{-7} sec (100 nsec Q-switched pulses) typical melt depths are $<5000\text{Å}$, while for $>10^{-4}$ sec (dwell times for continuous lasers) melt depths are $>100,000\text{Å}$.

The normal spectral reflectance of the film at the laser wavelength of interest will determine the size of the processing window - that range of incident power densities over which melting without evaporation can be reproducibly achieved. The processing windows can be very narrow when very high spectral reflectances lead to order of magnitude differences between incident and absorbed laser fluences. When operating with incident fluences high above melting threshold requirements, very small perturbations in local reflectance values can have catastrophic results (Draper, 1981a).

III. FERROUS BASED SYSTEMS

This section and the two that follow will present specific surface alloying examples from the ferrous, non-ferrous and silicide areas. The driving forces behind current interest in the surface alloying of transition metal elements such as Cr, Ni and Mo into Fe and its alloys are both economic and political. In many instances the Cr, Ni and Mo are added in high concentration for protection against the environment, not for bulk structural properties. Transition metal additions account for roughly 1/3 of the cost in highly alloyed steels and surface alloying could reduce that consumption by 90-99%. The political implications are shown in Figure 2.

Figure 9 presents the potentiodynamic polarization scans for surface alloyed carbon steel and bulk AISI 304 stainless steel (Lumsden and Gnanamuthu, 1982). They show nearly

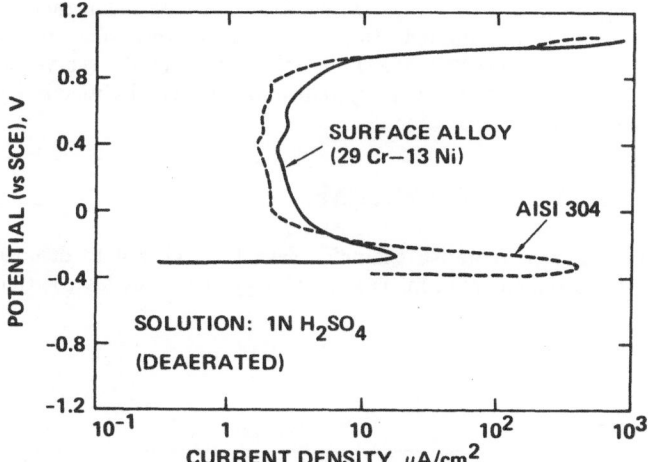

Figure 9   Electrochemical polarization results (Lumsden and Gnanamuthu, 1982) comparing a 29 Cr-13 Ni CO₂ laser prepared surface alloy with conventional bulk 304 stainless steel. Note the lower critical anodic current density at the primary passive potential and the essentially identical behavior throughout the passive region of the bulk and "surface equivalent" alloy.

identical behavior in the passive region. The surface alloy was produced using a continuous CO₂ laser scanning over the substrate steel which was coated with a metal powder (70% Cr-30% Ni)/cellulose alcohol slurry. 12 kW was focussed to a 6.4 × 19 mm rectangular spot and scanned at a rate of 4.7 mm/sec. Similar results have been obtained by Moore *et al.* (1981). The question of diffusion losses in the bulk at elevated temperatures has yet to be addressed. The problems of scaling up to production rate surface area requirements may also prove to be prohibitive. For example, the processing parameters quoted above and used to produce the surface alloy tested in Figure 4 equate to a production rate of roughly 0.4 m²/hr. Clearly much higher rates would be required if large sheet stock applications are proposed.

Another interesting case is that of Mo-Fe. For equilibrium conditions the maximum solubility of Mo in α-Fe at 650°C is 4 at.%. This binary system has been laser surface alloyed by Belotskii *et al.* (1977) utilizing a pulsed Nd-glass laser source and is typical of laser surface alloying work reported in the Soviet literature. Thin Mo films were rolled directly on the Fe substrates. Resulting alloyed regions were 450 to 500 μm deep, with a nonuniform profile and clear evidence for convective fluid flow. The average Mo concentration as determined by changes in the α-Fe lattice spacings was 21 at.%. No other Fe or Mo phases were detected. Thus, even on the recrystallization time scales associated with msec pulsed laser sources metastable solid solutions far removed from equilibrium can be prepared. It is perhaps worth noting that nearly all of the reported Soviet work in this area deals with Fe based alloys and the use of pulsed Nd-glass lasers. There is no published Soviet surface alloying work utilizing Q-switched solid state lasers and very little using continuous CO₂ laser sources.

Lasers are also used in making surface metallic glasses based on the Fe-metal-metalloid family. Important questions regarding glass stability in relation to the substrate crystalline seed and annealing by subsequent laser scans (required to produce area coverage) will have to be addressed. This last question is rather important, for if recrystallization occurs in glassy regions when subsequent laser scans or pulses partially irradiate them, then the usefulness of the processing would be severely limited. Reported results on this point are not in agreement. Bergmann and Mordike (1980, 1981) report no lateral recrystallization, while Borodina *et al.* (1981) report some recrystallization in all Fe-metal-metalloid systems produced using the continuous CO_2 laser.

IV. NONFERROUS BASED SYSTEMS

In this section four representative examples have been chosen to demonstrate both the diversity of interest and the potential for new metallurgy offered by laser surface alloying.

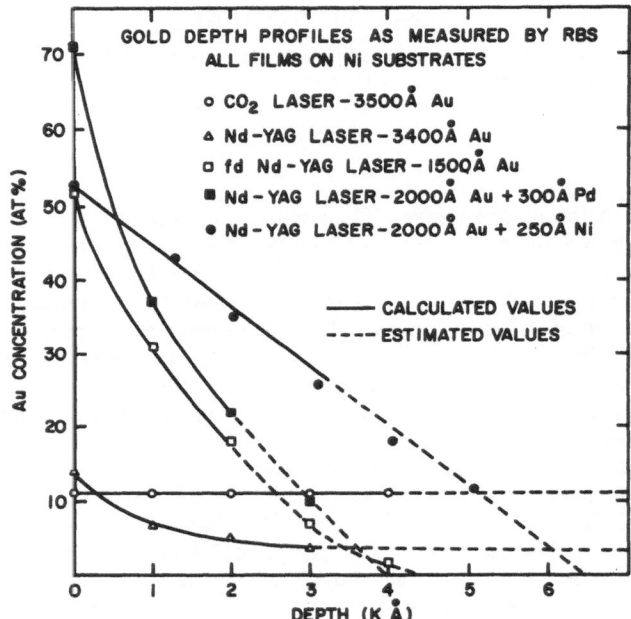

Figure 10   Au-Ni depth profiles derived from RBS spectra like those of Figures 3, 5 and 6. See text for details.

Au-Ni

As the phase diagram in Figure 4 shows, the Au-Ni system is characterized by a large solid state miscibility gap. Since homogeneous liquids and solids at elevated temperatures are easily produced it is not surprising that even bulk single phase solutions of any concentration are easily produced by the use of conventional quenching methods. Au-Ni surface alloys have been prepared using a number of different laser sources. Figure 10 summarizes Au concentration profiles in Ni resulting from various laser treatments of vapor deposited Au thin films on Ni (Draper *et al.*, 1981b). Surface alloys of low concentration and flat deep profiles result (open circles in Figure 10) from the longer times in the liquid state associated with the use of scanning CW lasers. The very high peak power densities associated with the short Q-switched Nd-YAG laser pulses, and the need to operate well above the required absorbed energy densities (because of the high reflectance of Au at 1.06 μm) lead to significant Au loss. These power levels also result in low surface concentrations and sharply sloping concentration gradients that drop to zero in less than 5000Å (open triangles). Efficient coupling to the surface can be achieved with either the frequency doubled Nd-YAG laser or with the addition of a thin reflectance cap of Ni on the Au for the fundamental laser wavelength (open squares and solid circles in Figure 10). Higher total precious metal concentrations can be achieved by the use of Pd caps instead of Ni (solid squares). The problem of a film thickness dependent thermal diffusivity and therefore melt depth has already been addressed above. In summarizing the Au-Ni work, the point to be made is that a wide range of concentrations and profiles have been shown to be readily accessible.

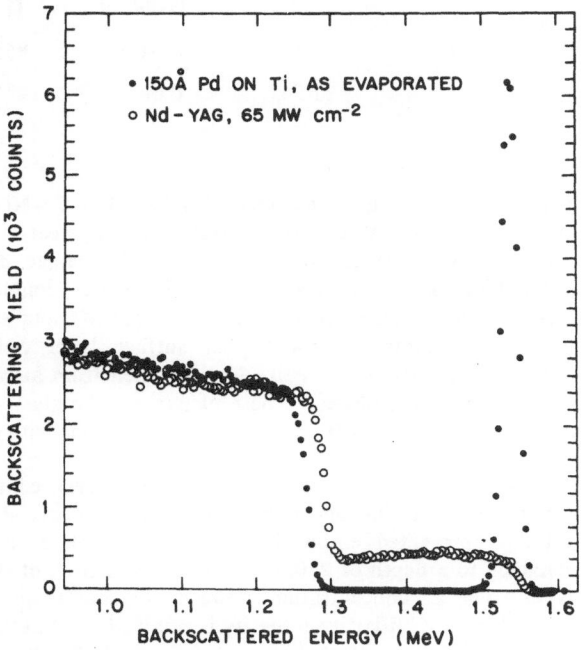

Figure 11   Rutherford backscattering spectra for 150Å of Pd on polycrystalline Ti before (solid circles) and after (open circles) Q-switched Nd-YAG laser alloying. The resulting surface alloy has a peak Pd concentration of 4 at.%. Alloying extends to at least 3000Å.

Pd-Ti

It is well known that the mixed potential produced by noble metal alloying with Ti drastically lowers the corrosion rate of Ti in reducing acid environments. Bulk Ti-Pd alloys are commonly used in the chemical industry. Surface alloying methods can lower corrosion rates at considerably reduced precious metal consumption as has been recently demonstrated for implantation (Hubler and McCafferty, 1980) and laser alloying (Draper *et al.*, 1981c).

Figure 11 and Table II summarize surface alloying and corrosion tests on the Pd-Ti system. Following Q-switched Nd-YAG irradiation of 150Å Pd on Ti a 4 at.% alloy extending several thousand Å is produced. The corrosion rates in boiling hydrochloric acid and the evolution of the surface enriched Pd distribution have been studied. The corrosion rate is reduced significantly and the selective dissolution of Ti results in Pd retention at a surface concentration determined by the anode/cathode surface area factors.

TABLE II

Corrosion Rate and Penetration Depths
for Ti and the Pd-Ti Surface Alloy in Boiling HCl Solutions

Test	Molarity (M)	Total Time (hr)	Corr Rate (mg/cm²/hr)		Penetration ($\times 10^{-3}$ inch/year)	
			Ti	Ti-Pd	Ti	Ti-Pd
C1	1.1	2	1.12	0.04	855	28
C2	2.0	3	1.94	0.06	1485	42

Cu-Zr and Zr-Ni

Both Cu and Ni are well known glass formers with Zr. The Zr-Ni phase diagram in Figure 4 is typical for both systems. Within concentration ranges near eutectic points glass formation is often observed. Cu surface concentrations, in the range 30-60%, have been produced by melting Cu films on Zr substrates with a frequency doubled Nd-YAG laser (Draper *et al.*, 1982a). Transmission electron microscopy (TEM) and selected area x-ray diffraction (XRD) on thinned samples show a near surface layer rich with interesting microstructures. The diffraction pattern in Figure 12 was taken from an area of the 32 at.% sample demonstrating three identified phases. These phases are the glass, a metastable ω-Zr phase, and a hexagonal Zr phase with a c/a ratio the same as Zr but with a 6% larger lattice constant.

The formation of a surface glass in the Zr-Ni case has also been confirmed by glancing angle XRD. Rutherford backscattering and channeling measurements show quite different behavior from the Ni data presented earlier in this article. The channeled and random spectra (Figure 13) coincide to a depth of 2500Å giving an indication of the thickness of the amorphous surface layer. The Zr concentration, however, extends to much greater depths. The Zr concentration at which the Ni lattice integrity is retained is 33 at.% which agrees well with limiting Zr concentrations for producing amorphous Zr-Ni by conventional melt quenching reported by Dong *et al.* (1981).

Cr-Cu

Copper and its alloys are utilized in a wide range of applications. Such is the diversity that they are probably second to only the ferrous based alloys. The passive and impervious nature of chromium oxide plays a crucial role in the application of many ferrous alloys. However, as the Cr-Cu phase diagram in Figure 14 shows there is a region of liquid immiscibility, a two phase regime below the liquidus that spans nearly the entire diagram and a sharply decreasing solid solubility for Cr in Cu with decreasing temperature. It is therefore not possible by conventional bulk metallurgical manufacturing methods to produce single

Figure 12 Selected area electron diffraction pattern from a region of laser surface alloyed Cu-Zr. In this area both glassy and polycrystalline phases exist. Nearer the surface only the glassy phase is found.

phase alloys of Cr in Cu at concentrations greater than about 1 at.%. This would appear to be an attractive binary system in which to prepare surface alloys and preliminary (Draper *et al.*, 1982b) results are encouraging.

Figure 15 presents Rutherford backscattering and channeling on single crystal Cu with a multilayer Cr + Cu film structure which has been laser alloyed with Q-switched frequency doubled Nd-YAG laser pulses. The top spectra show the as-deposited Cr-Cu sandwich (see inset) and the underlying Cu substrate. The backscattering mass resolution and detector resolution are such that the multilayer structure is not resolved. The lower set of spectra show striking changes following laser alloying. The Cr diffuses over 2400Å with a peak concentration of 18 at.% at 700Å and an average concentration of 8 at.%. The surface layer is epitaxial. It would appear that most of the Cr is substitutional. However, with the large amount of dechanneling in this strained layer it is difficult to quantify the degree of substitutionality.

In order to characterize the microstructural nature, the laser mixed multilayer structures were examined by TEM. For the TEM observations the samples were raster scanned with incomplete overlap so that individual melt spots were produced. In the bright field transmission electron micrograph, Figure 16, the beam direction is near Cu (110). The central bright area is a circular laser-melted zone approximately 8 μm in diameter. On the periphery of this is a narrow darker heat-affected zone surrounded by virgin material. The inset, a selected area diffraction pattern from the virgin material shows expected (111) Cu substrate spots with evidence for texturing in the polycrystalline Cu deposited layers with grains only a few hundred Å in size. As one enters the heat affected zone the grain size increases, reaching thousands of Å, with signs of greater elongation radially, as might be expected from heat flow considerations. The laser melted zone itself contains few discernible grains at this magnification although the diffraction pattern reveals a broad (i.e., fine grain) polycrystalline ring, very close to (111) Cu. Grains ~100Å can be resolved at higher magnifications. However, it is clear that most of the material in this zone is single crystal of

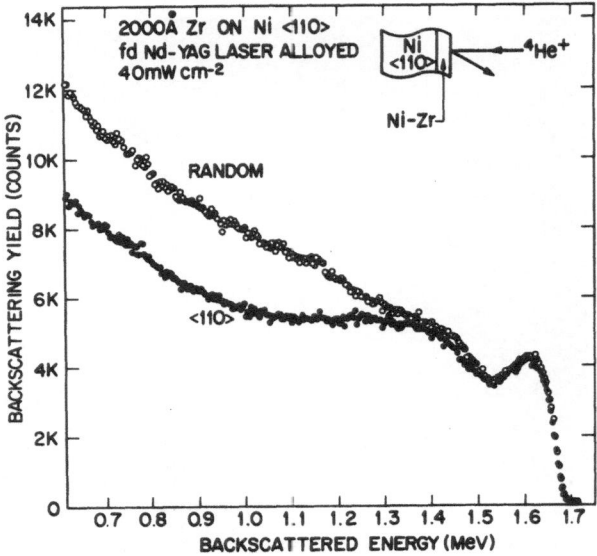

Figure 13   Rutherford backscattering and channeling spectra for 2000Å Zr on <110> Ni surface alloyed with the
Q-switched frequency doubled Nd-YAG laser. The Zr concentration is peaked near the surface with a
sharply falling concentration gradient extending about 8000Å. There is no channeling in the Zr. The
Ni yields indicate disorder for 2500Å where the channeled spectrum breaks from the random indicating
that below the disordered layer the Ni substrate remained single crystal. The estimated Zr concentration
at this depth is 33 at.%.

(110) orientation without any second phase Cr precipitates. The channeling and TEM data,
therefore, indicate that a solid solution of Cr-Cu has been formed.

Preliminary electrochemical polarization and sulphidation studies on polycrystalline Cu
alloy substrates laser surface alloyed with Cr and Cr + Ni have been carried out. In the
electrochemical tests, samples (5-10 at.% Cr) were subjected to anodic and cathodic potential
scans in a 1M H_2SO_4 solution. The polarization scans indicate that a thin chromium oxide
film is present on the surface which offers protection to the alloy against anodic dissolution.
However, this protection is removed by cathodic polarization, which reduces the protective
oxide. In the sulphidation tests, samples were exposed to moist, purified air containing a
controlled concentration of hydrogen sulphide. Although some copper sulphide film growth
occurred on the laser alloyed samples the samples could qualitatively be characterized as inert
relative to the typical behavior of conventional copper alloys.

V. METAL SILICIDES

One obvious area of application of surface alloying is the fabrication of metal contacts on
Si to form either Ohmic or Schottky electrical contacts. The first measurements of silicide
formation by laser alloying are due to Poate *et al.* (1978) who alloyed Pt, Pd and Ni films to
Si using Q-switched Nd-YAG irradiation. The surfaces were melted to produce laterally very

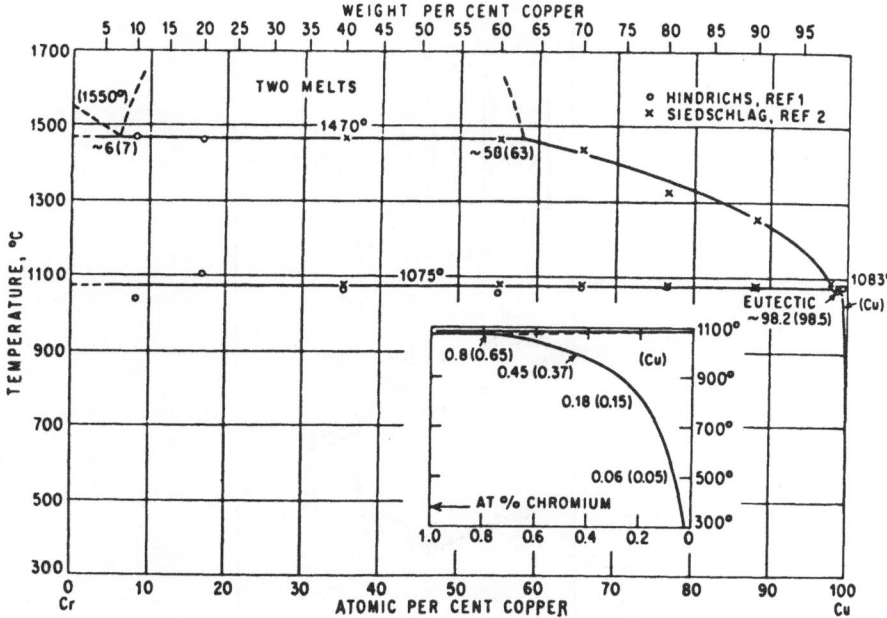

Figure 14 The Cr-Cu phase diagram. Note the miscibility gap in the liquid between 6 and 58% Cu and the very limited solid solubility for Cr in Cu.

uniform alloy layers whose average composition could be changed over a wide range by varying film thickness and laser power. The reacted layer, however, consisted not of a single phase but of cellular microstructures with typical diameters of 1000Å. These cells result from constitutional supercooling as discussed in Chapter 4 and are due to the exceptionally small solid state solubilities of metals in Si and low segregations coefficients. It is possible (Sigmon, 1981) to form single phase silicides by heating in the solid phase with CW lasers.

Tu *et al.* (1981) have recently proposed that the difference in reflectivity of metal films and their silicides could be used for archival optical storage. They have measured reflectivity differences as large as 53% between Pd and Pd-Si. Dark silicides dots could, therefore, be written onto a bright sea of metal film on Si by use of laser alloying. The success of this application will depend on the thermal stability of the metal films on Si over many years of storage.

VI. CONCLUSIONS

The laser alloying of deposited films on metals and semiconductors is a powerful adjunct to the materials science of surface modification. If the elements are miscible in the liquid phase, novel phases can be produced by rapid solidification. Surface alloys can be produced of greater thickness and concentration than those formed by ion beams. In some areas therefore, laser alloying offers greater opportunities for successful application.

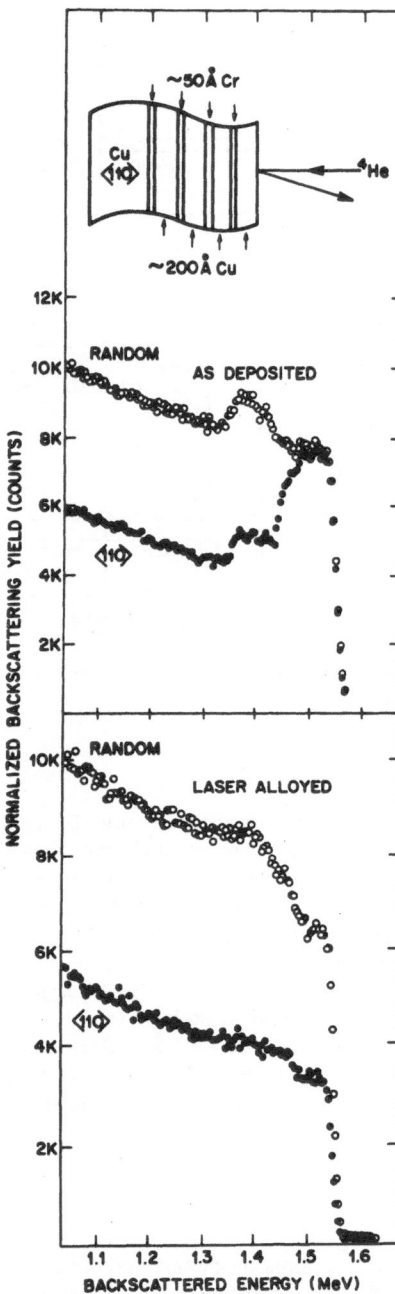

Figure 15   Rutherford backscattering and channeling spectra for a Cu-Cr multilayer on <110> Cu (see inset
schematic) laser alloyed with the Q-switched frequency doubled Nd-YAG laser. The upper spectra
correspond to the as-deposited multilayer. The lower spectra correspond to the laser mixed area which
has regrown epitaxially from the <110> Cu substrate. The metastable single phase alloy has a peak Cr
concentration of ~18 at.%.

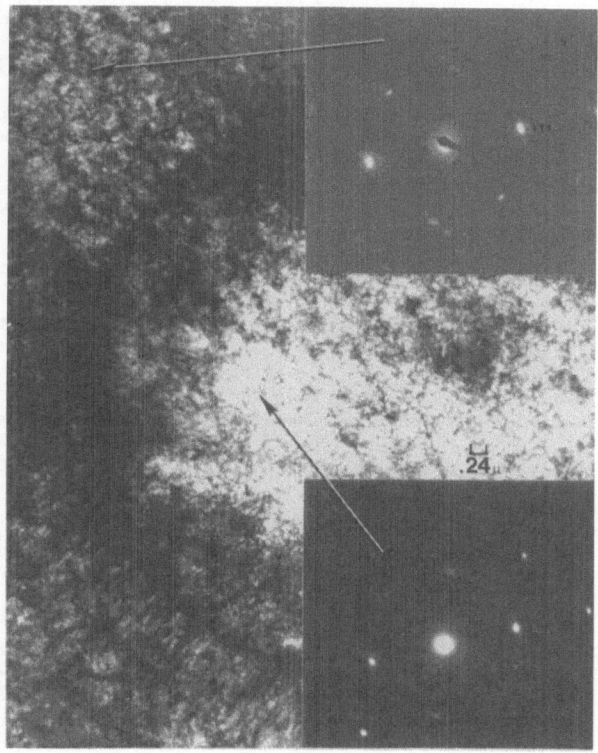

Figure 16   Bright field transmission electron micrograph (same sample as Fig. 15) with selected area diffraction patterns for a Cu-Cr multilayer on <110> Cu (see insert in Fig. 15) laser alloyed with the Q-switched frequency doubled Nd-YAG laser. The upper SAED is from an as-deposited region. The lower SAED is from a laser mixed area which has regrown epitaxially from the <110> Cu substrate.

REFERENCES

Belotskii, A. V., Kovalenko, V. S., Volgin, V. I. and Pshenichnyi, V. I. (1977), Fiz. Khim. Obrab. Mater. *3*, 24.

Bergmann, H. W. and Mordike, B. L. (1980), Z. Metallkde. *71*, 658.

Bergmann, H. W. and Mordike, B. L. (1981), J. Mater. Sci. *16*, 863.

Borodina, G. G. *et al.* (1981), Dokl. Akad. Nauk. *USSR* 259, 826.

Buene, L., Draper, C. W., Jacobson, D. C. and Poate, J. M. (1980a), unpublished.

Buene, L., Poate, J. M., Jacobson, D. C., Draper, C. W. and Hirvonen, J. K. (1980b), Appl. Phys. Lett. *37*, 385.

Dong, Y. D., Gregan, G. and Scott, M. G. (1981), J. Non-Cryst. Solids *43*, 403.

Draper, C. W. (1981a), NBS Spec. Pub. *620*, 210.

Draper, C. W. (1981b), Appl. Opt. *20*, 3093.

Draper, C. W., Jacobson, D. C. and Poate, J. M. (1981a), unpublished.

Draper, C. W., Meyer, L. S., Buene, L., Jacobson, D. C. and Poate, J. M. (1981b), Appl. Surf. Sci. *7*, 276.

Draper, C. W., Meyer, L. S., Jacobson, D. C., Buene, L. and Poate, J. M. (1981c), Thin Solid Films, *75*, 237.

Draper, C. W., Den Broeder, F. J. A., Jacobson, D. C., Kaufmann, E. N., Vandenberg, J. M. and McDonald, M. L. (1982a), in "Laser and Electron Beam Interactions with Solids," (B. R. Appleton and G. K. Cellar, eds.), North Holland, New York.

Draper, C. W., Jacobson, D. C., Gibson, J. M., Poate, J. M., Vandenberg, J. M. and Cullis, A. G. (1982b), in "Laser and Electron Beam Interactions with Solids," (B. R. Appleton and G. K. Cellar, eds.), North Holland, New York.

Ferris, S. D., Leamy, H. J. and Poate, J. M. (1979), Editors "Laser-Solid Interactions and Laser Processing-1978," AIP Conf. Proceedings 50.

Gibbons, J. F., Hess, L. D. and Sigmon, T. W. (1981), Editors "Laser and Electron-Beam Solid Introduction and Materials Processing," MRS Symposia Proceedings Vol. 1. North Holland, New York.

Hsu, S. C., Chakravorty, S. and Mehrabian, R. (1978) Met. Trans. *9B*, 221.

Hubler, G. K. and McCafferty, E. (1981), Corros. Sci. 20, 103.

Lumsden, J. B. and Gnanamuthu, D. S. (1982), in "Corrosion of Metals Processed by Directed Energy Beams," (C. R. Clayton and C. M. Preece, eds.), TMS-AIME, Warrendale, PA.

Moore, P. G. and McCafferty, E. (1981), J. Electrochem. Soc. *128*, 1391.

Poate, J. M. and Cullis, A. G. (1980), in "Treatise on Materials Science and Technology," (J. K. Hirvonen, ed.), p. 85, Academic Press, New York.

Poate, J. M., Leamy, H. J., Sheng, T. T. and Celler, G. K. (1978), Appl. Phys. Lett. *33*, 918.

Sigmon, T. (1981), p. 511, in Gibbons *et al.* (1981).

Swager, W. L. (1980), Battelle Today, *18*, 3.

Tu, K. N., Ahn, K. Y. and Herd, S. R. (1981), Appl. Phys. Lett. *39*, 927.

White, C. W. and Peercy, P. S. (1980), Editors "Laser and Electron Beam Processing of Materials Academic Press, New York.

TREVI INSTITUTE:
PARTICIPANTS AND CONTRIBUTORS

H. H. Andersen
Institute of Physics
University of Aarhus
8000 Aarhus C
DENMARK

*†V. Ashworth
Corrosion and Protection Center
UMIST.
P.O. Box 88
Manchester M60 IQD
ENGLAND

J. Bøttiger
Institute of Physics
University of Aarhus
8000 Aarhus C
DENMARK

P. Baeri
Istituto di Struttura della Materia
Università di Catania
Corso Italia, 57
I 95129 Catania
Sicily
ITALY

J. E. E. Baglin
IBM Thomas J. Watson
   Research Center
Yorktown Heights, New York 10598
U.S.A.

H. Bernas
Institut de Physique Nucleaire
BP-1
91406-ORSAY
FRANCE

*W. Buckel
Kernforschungsanlage, Julich
D-5170 Julich
WEST GERMANY

L. Buene
Veritas
Veritas Veien 1
P.O. Box 300
N-1322 Hovik
NORWAY

S. U. Campisano
Istituto di Struttura della Materia
Universita di Catania
Corso Italia, 57
I 95129 Catania
Sicily
ITALY

C. Carbone
C.N.R.
Gruppo Nazionale di Struttura
   della Materia
Viale dell'Universita 11
Roma
ITALY

C. R. Clayton
Department of Materials Science
   & Engineering
SUNY Stony Brook
Stony Brook, New York 11794
U.S.A.

A. G. Cullis
R.S.R.E.
Malvern, Worcestershire
ENGLAND

---

*Contributor only
†Committee Member

J. A. Davies
Chalk River Nuclear Laboratories
Chalk River, Ontario K0J1J0
CANADA

*†G. Dearnaley
AERE Harwell
Didcot, Oxfordshire, OXII ORA
ENGLAND

G. Della Mea
Istituto di Fisica "G. Galilei"
Università di Padova
Via Marzolo, 8
I 40100, Padova
ITALY

L. F. Donà Dalle Rose
Istituto di Fisica
Università di Padova
Padova
ITALY

C. W. Draper
Western Electric Engineering
    Research Center
Princeton, New Jersey
U.S.A.

G. Falcone
Istituto di Fisica
Università di Calabria
Cosenza
ITALY

†G. Foti
Istituto di Struttura della Materia
Università di Catania
Corso Italia, 57
I 95129 Catania
Sicily
ITALY

R. Galloni
Laboratorio Lamel - C.N.R.
Via de Castagnoli
I 40126
Bologna
ITALY

*J. F. Gibbons
Stanford Electronics Laboratories
Stanford University
Stanford, California 94305
U.S.A.

*P. Heilmann
Ohio State University
Columbus, Ohio
U.S.A.

*H. Herman
State University of New York
Stony Brook
New York
U.S.A.

†J. K. Hirvonen
Naval Research Laboratory
Washington, D.C. 20375
U.S.A.

W. O. Hofer
Kernforschungsanlage, Julich
D-5170 Julich
GERMANY

*T. Hussain
Kernforshungszentrum, Karlsruhe IAK
P.O. Box 3640
7500 Karlsruhe
GERMANY

K. A. Jackson
Bell Laboratories
600 Mountain Avenue
Murray Hill, New Jersey 07974
U.S.A.

*D. C. Jacobson
Bell Laboratories
600 Mountain Avenue
Murray Hill, New Jersey 07974
U.S.A.

*J. A. Knapp
Sandia National Laboratories
Albuquerque, New Mexico 87185
U.S.A.

*Contributor only
†Committee Member

*N. Q. Lam
Materials Science Division
Argonne National Laboratory
Argonne, Illinois
U.S.A.

S. S. Lau
Department of Electrical Engineering
University of California, San Diego
La Jolla, California 92093
U.S.A.

G. Linker
Kernforschungszentrum, Karlsruhe IAK
P.O. Box 3640
7500 Karlsruhe
GERMANY

U. Littmark
Kernforschungsanlage, Julich
D-5170 Julich
GERMANY

A. D. Marwick
AERE Harwell
Didcot, Oxfordshire, OX11 ORA
ENGLAND

†J. W. Mayer
Materials Science and Engineering
Bard Hall
Cornell University
Ithaca, New York 14853
U.S.A.

†P. Mazzoldi
Unita GNSM - CNR
Istituto di Fisica
Università di Padova
Padova
ITALY

†O. Meyer
Kernforschungszentrum, Karlsruhe IAK
P.O. Box 3640
7500 Karlsruhe
GERMANY

S. Myers
Sandia Laboratories
Albuquerque, New Mexico 87185
U.S.A.

G. Ottaviani
Istituto di Fisica
Università di Modena
Via Vivaldi, 70
I 41100 - Modena
ITALY

*P. S. Peercy
Sandia National Laboratories
Albuquerque, New Mexico 87185
U.S.A.

H. W. Pickering
Department of Materials Science
Penn State
University Park, Pennsylvania 16802
U.S.A.

S. T. Picraux
Sandia Laboratories
Albuquerque, New Mexico 87185
U.S.A.

†J. M. Poate
Bell Laboratories
600 Mountain Avenue
Murray Hill, New Jersey 07974
U.S.A.

*L. E. Rehn
Materials Science Division
Argonne National Laboratory
Argonne, Illinois 60439
U.S.A.

D. A. Rigney
Ohio State University
Columbus, Ohio
U.S.A.

†E. Rimini
Istituto di Struttura della Materia
Università di Catania
Corso Italia, 57
I 95129 Catania
Sicily
ITALY

*Contributor only
†Committee Member

F. Saris
FOM Institut
Postbus 41883
1009 DB
Amsterdam,
NETHERLANDS

P. Siffert
Centre de Recherches Nucleaires
Groupe de Physique et. Applications
des Semiconductors (PHASE)
67037 Strasbourg Cedex
FRANCE

*D. K. Sood
Bhabha Atomic Research Center
Nuclear Physics Division
Trombay, Bombay 400085
INDIA

B. Stritzker
Kernforschungsanlage, Julich
D-5170 Julich
GERMANY

A. Trovato
Istituto di Struttura della Materia
Universita di Catania
Corso Italia, 57
I 95129 Catania
Sicily
ITALY

K. N. Tu
IBM Thomas J. Watson
   Research Center
Yorktown Heights, New York 10598
U.S.A.

D. Turnbull
Division of Applied Sciences
Pierce Hall
Harvard University
Cambridge, Massachusetts 02138
U.S.A.

M. von Allmen
Institut Fur Angewandte Physik
Universitat Bern
Sidlerstrasse 5
CH 3000 Bern
SWITZERLAND

*W. R. Wampler
Sandia National Laboratories
Albuquerque, New Mexico 87185
U.S.A.

C. W. White
Solid State Division
Oak Ridge National Laboratory
Oak Ridge, Tennessee 37830
U.S.A.

H. Wiedersich
Materials Science Division
Argonne National Laboratories
Argonne, Illinois 60439
U.S.A.

J. S. Williams
Communications and Electrical
   Engineering Department
Royal Melbourne Institute
   of Technology
Melbourne, Victoria
AUSTRALIA

G. K. Wolf
Physikalisch-Chemisches Institut
IM Neuenheimer Feld 500
University of Heidelberg
Heidelberg
GERMANY

*D. M. Zehner
Solid State Division
Oak Ridge National Laboratory
Oak Ridge, Tennessee 37830
U.S.A.

*P. Ziemann
Kernforschungsanlage, Julich
D-5170 Julich
WEST GERMANY

---

*Contributor only
†Committee Member

INDEX

Z